Helical Wormlike Chains

Springer

Berlin
Heidelberg
New York
Barcelona
Budapest
Hong Kong
London
Milan
Paris
Santa Clara
Singapur
Tokyo

Hiromi Yamakawa

Helical Wormlike Chains
in Polymer Solutions

With 24 Tables and 149 Figures

 Springer

Professor Emeritus Hiromi Yamakawa

Department of Polymer Chemistry
Kyoto University
Kyoto 606-8501, Japan

ISBN 3-540-62960-2 Springer-Verlag Berlin Heidelberg New York

DieDeutsche Bibliothek - CIP-Einheitsaufnahme

Yamakawa, Hiromi:
Helical wormlike chains in polymer solutions / Hiromi Yamakawa. - Berlin ; Heidelberg ; New York ; Barcelona ; Budapest ; Hong Kong ; London ; Milan ; Paris ; Santa Clara ; Singapur ; Tokyo : Springer, 1997
    ISBN 3-540-62960-2

CIP data applied for

Typesetting: Camera-ready by author
Cover: Teichmann, Heidelberg

SPIN:10497194   02/3020 - 5 4 3 2 1 0 - Printed on acid -free paper

Preface

This book is intended to give a comprehensive and systematic description of the statistical-mechanical, transport, and dynamic theories of dilute-solution properties of both flexible and semiflexible polymers, including oligomers. This description was developed on the basis of the "helical wormlike chain" model along with an analysis of extensive experimental data. Chapter 2 and the fundamental parts of Chapters 3, 4, 6, and 9 are based on the author's lecture notes for courses in polymer statistical mechanics given at the Graduate School of Kyoto University from 1978 through 1994. Much of the material in the book arises from his research reported since the time of publication (1971) of his earlier book, *Modern Theory of Polymer Solutions.*

The accomplishment of this research was made possible by the collaboration of his students, postdoctoral students, and other collaborators, especially F. Abe, Y. Einaga, M. Fujii, W. Gobush, T. Konishi, K. Nagasaka, M. Osa, N. Sawatari, J. Shimada, W. H. Stockmayer, Y. Takaeda, and T. Yoshizaki. Professor W. H. Stockmayer gave the author an opportunity to stay in Hanover in 1971/2, and indeed the research in this book was started at that time. Professor T. Norisuye, who is not a collaborator of the author but his colleague, kindly provided him with the unpublished material on the third virial coefficient dealt with in Chapter 8.

The author is indebted to Dr. J. Shimada for his careful checking of the mathematical equations in the manuscript. Finally, it is a pleasure to acknowledge the assistance of Prof. T. Yoshizaki and Mr. M. Osa who prepared some of the tables and figures and also translated the manuscript into the compuscript with Miss M. Fukui.

Kyoto
January 1997

Hiromi Yamakawa

Contents

1 Introduction

1.1 Historical Survey

A first stage in the study of *polymer solution science* ended with the worldwide acceptance of the concept of the excluded-volume effect in the mid 1950s shortly after the publication of the celebrated book by Flory [1] in 1953. In the next stage, activity was centered mainly in the study of dilute solution behavior of flexible polymers within the Flory framework which consists of that concept for the Gaussian chain and the universality of its Θ state without that effect. The theoretical developments were then made by an application of orthodox but rather classical techniques in statistical mechanics for many-body problems with a more rigorous consideration of chain connectivity, thus all leading except for a few cases to the so-called two-parameter (TP) theory, which predicts that all dilute solution properties may be expressed in terms of the unperturbed (Θ) dimension of the chain and its total effective excluded volume. The results derived until the late 1960s are summarized in Yamakawa's 1971 book [2] along with a comparison with experimental data. In the meantime, on the other hand, an experimental determination of the (asymptotic) unperturbed chain dimension for a wide variety of long flexible polymers [3] brought about great advances in its theoretical evaluation for arbitrary chain length on the basis of the rotational isomeric state (RIS) model [4], and all related properties are sophisticatedly treated in Flory's second (1969) book [5].

Subsequently, the advances in the field have been diversified in many directions. The foremost of these is a new powerful theoretical approach to the excluded-volume problems by an application of scaling concepts and renormalization group theory, which began in the early 1970s when the analogy between many-body problems in the Gaussian chain and magnetic systems was discovered [6, 7]. These techniques enable us to derive asymptotic forms for various molecular properties as functions of chain length (or molecular weight) for long enough chains. The basic scaling ideas and their applications to polymer problems are plainly explained by de Gennes [8] in his renowned third book, while the details of the methods and results of the polymer renormalization group theory are described in the review article by Oono [9] and also in the books by Freed [10] and by des Cloizeaux and Jannink [11].

At about the same time there occurred new developments in the dynamics of polymer constrained systems in two directions. One concerns dilute solutions, and the other concentrated solutions and melts. In particular, there have been significant advances in the latter. Although the single-chain dynamics was first formulated by Kirkwood [12] in 1949 for realistic chains with rigid constraints on bond lengths and bond angles, dynamical properties related to global chain motions in dilute solution have long been discussed on the basis of the Gaussian (spring-bead) chain [2,13–16]. However, the study of constrained-chain dynamics was initiated by Fixman and Kovac [17] in 1974 in order to treat local properties, and much progress in actual calculations has been made possible by slight coarse-graining of the conventional bond chain [18,19] (see below).

In condensed or many-chain systems, on the other hand, intermolecular constraints, that is, entanglement effects play an important role, and the chain motion in such an environment is very difficult to treat by considering intermolecular forces of the ordinary dispersion type. Indeed, this had been for long one of the unsolved problems in polymer science until 1971 when a breakthrough was brought about by de Gennes [8, 20], who introduced the concept of the reptation in a tube, the concept of the tube itself being originally due to Edwards [21]. In their book, Doi and Edwards [22] summarize comprehensively the successful applications of the tube model to viscoelastic properties of concentrated solutions and melts of long flexible polymers.

Necessarily, if the chain length is decreased, the (static) chain stiffness becomes an important factor even for ordinary flexible chain polymers as well as for typical stiff or semiflexible macromolecules such as DNA and α-helical polypeptides. The stiffness arises from the structural constraints mentioned above and hindrances to internal rotations in the chain. The RIS model or its equivalents on the atomic level must then be the best to consider this effect, and therefore to mimic the equilibrium conformational behavior of real chains of arbitrary length. However, for many equilibrium and steady-state transport problems on such stiff chains, the structural details are not amenable to mathematical treatments, and moreover, are often unnecessary to consider. Some coarse-graining may then be introduced to replace these discrete chains by continuous models, although the discreteness must be, to some extent, retained in the study of the dynamics, especially of local chain motions, as mentioned below.

The foremost of these continuous models is the wormlike chain proposed by Kratky and Porod (KP) [23] in 1949. Since the mid-1960s there has been renewal of activity in studying this model and some new aspects have evolved. The theoretical developments thus made for the KP chain and its numerous modifications are reviewed by Freed [24] and critically by Yamakawa [2, 18, 25, 26].

One of these modifications is the helical wormlike (HW) chain [18, 19, 27, 28] proposed in 1976, which is the subject of the present book. It is a generalization of the KP chain and includes the latter as a special case. In fact, in the early 1970s, a comparison of the KP chain with the RIS model proved that the application of the former to flexible chains is limited to only very symmetric chains such as polymethylene. This was the motivation of the generalization. Now the HW chain may describe equilibrium conformational and steady-state transport properties of all kinds of real chains, both flexible and stiff, on the bond length or somewhat longer scales, thus bridging a gap between them and the KP chain. When the excluded-volume effect and steady-state transport properties are treated, beads are arrayed touching one after another on the chain contour (touched-bead model) or a cylinder whose axis coincides with the contour is considered (cylinder model), as the case may be [19]. For study of the dynamics, however, the discreteness must be recovered to give a kind of coarse-grained bond chain, as mentioned above. This was done by modeling the real chain by a (discrete) chain of rigid bodies (motional units), each corresponding to the monomer unit and each center being located on the (continuous) HW chain contour. This is the dynamic HW model [18, 19, 29].

Also on the experimental side, since the mid-1970s there have been obtained many important and exciting results in various branches of the field, including those mentioned above. Some are presented in the book edited by Nagasawa [30] and also reviewed in the most recent book by Fujita [31]. For convenience, we confine ourselves here to dilute solution problems, in particular a few noteworthy topics. Most important is the fact that accurate measurements demonstrated that the TP theory for perturbed flexible polymers is not always valid even for fairly large molecular weights, especially in good solvents [31]. This must be regarded as arising from effects of chain stiffness. In this connection, it is important to note that the renormalization group approach still leads essentially to the TP theory. Further, since the mid-1980s precise measurements have been extended to the oligomer region [32–35]. These have been made possible because well-characterized samples, including oligomers, have become available owing to the progress in polymer synthesis and characterization techniques such as ionic polymerization, GPC, and NMR and also because new experimental tools such as neutron and dynamic light scattering methods have been added to the classical ones such as static light and small-angle X-ray scattering.

As a result, on the one hand it has been confirmed that strictly, the TP theory is valid only in the asymptotic limit of large molecular weight [35], and on the other hand it has been shown that for (unperturbed) flexible chains neither the characteristic ratio [5] nor the Kuhn segment length is in general a measure of chain stiffness [34]. All of these and other recent findings indicate that the Flory framework, including the RIS description of the unperturbed chain conformation, breaks down. They may probably be due to chain stiffness and may therefore be explained by adopting the HW model. It must also be mentioned that since the mid-1970s there have been carried out extensive accurate measurements of equilibrium conformational and transport properties of typical stiff chains such as DNA, poly(n-alkyl isocyanate)s, and polysaccharides to determine their stiffness mainly on the basis of the KP model [31, 36].

Now the renormalization (coarse-graining and scaling) process works for long flexible (Gaussian) chains and leads to the universality as represented by the asymptotic forms (exponent and prefactor) for their physical properties. The theory of this type in principle fails to take into account possible effects of chain stiffness. As is well established for flexible polymers, the behavior of the unperturbed chain dimension already reaches its Gaussian limit at molecular weights smaller than about 10^5. It is also true that the exponent laws for the chain dimension, intrinsic viscosity, and other properties in Θ and good solvents hold over a wide range of molecular weight for typical flexible polymers such as polystyrene and polyisobutylene, though no longer for many cases at the present time. These facts have misled many polymer scientists

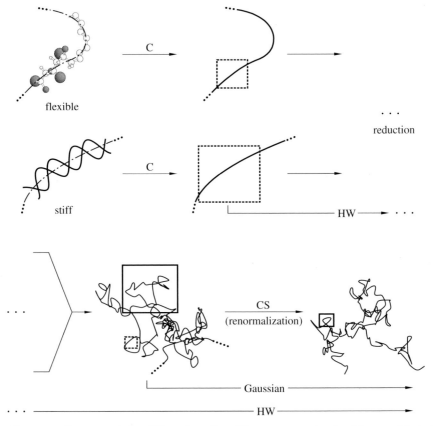

Fig. 1.1. Coarse-graining (C) and scaling (S) processes of a flexible or stiff chain and the ranges of application of the HW and Gaussian chain models

to the view that universality is the very essential and fundamental feature of the polymer behavior. However, the recent experimental findings mentioned above show that the effect of chain stiffness on the excluded-volume effect remains rather appreciable even for molecular weights larger than 10^5, up to about 10^6, so that the TP theory, including the renormalization group theory, breaks down even for molecular weights ordinarily of interest. A large part of the range in which the Gaussian chain theory breaks down may be covered much more easily by the HW chain than by any other model on the atomic level, for instance, the RIS model. The situation is schematically depicted in Fig. 1.1.

Finally, brief mention should be made of some other topics on semidilute and concentrated solutions. Recently, there have been reported some experimental results which indicate that the osmotic universality breaks down in semidilute solutions [37]. This may also probably be due to the effect of chain stiffness. Another branch of the field in which many investigations have been performed is liquid crystals, where, for instance, the semiflexibility of rodlike polymers, including polyelectrolytes, has been considered on the basis of the KP model [38]. However, this may be part of the newly growing, interesting, and vast field called complex fluids [39–41], which covers ordering and pattern formation, phase separation, gelation, and so forth in both systems of simple and polymer molecules.

1.2 Scope

This book is intended to formulate the HW chain model, including the KP chain as a special case, to treat almost all static, transport, and dynamical properties of both flexible and stiff polymers in dilute solution on the small to large length and time scales. A comprehensive description of the statistical-mechanical, hydrodynamic, and dynamic theories developed for them on the basis of this model is given along with a comparison with other models for some cases. The theoretical methods and derivation are described as simply as possible but without loss of the lowest rigor. There are also given analyses of recent experimental data by use of these theories for flexible polymers over a wide range of molecular weight, including the oligomer region, and for stiff polymers, including biological macromolecules such as DNA. In particular, one of the purposes of the book is to show that a new theory of the excluded-volume effects for the HW chain in dilute solution may give an explanation of recent extensive experimental results which indicate that the TP theory breaks down. The book contains a reasonable number of theoretical equations, tables, and figures, enough to provide an understanding of the basic theories and to facilitate their applications to experimental data for polymer molecular characterization. For the latter purpose, computer-aided forms are also given for some of the theoretical expressions, along with necessary numerical tables (in the text and Appendixes I–V). Use of familiar terminology which already appeared in the present author's earlier book, *Modern Theory of Polymer Solutions* (MTPS) [2], is made without explanation except for those cases in which redefinition or reconsideration is needed.

In Chap. 2, which is also an introduction to the following two chapters, a brief survey is made of several fundamental discrete and continuous models for polymer chains in a form convenient for later developments. Chapters 3 and 4 deal rather in detail with the statistical mechanics of the KP and HW chains, respectively. The Fokker–Planck diffusion equations for various distribution functions are derived by analogy with certain quantum particles with the use of path integrals. An operational method for effectively computing the moments is presented. In particular, in Chap. 4 various approximations to the distribution functions for the HW chain are given for practical use and adaptation to real chains is discussed in detail. In Chaps. 5 and 6 equilibrium conformational and steady-state transport properties of unperturbed chains without excluded volume are treated, respectively. The former includes the mean-square radius of gyration, scattering function, mean-square optical anisotropy, and so forth, and the later includes the intrinsic viscosity and translational friction and diffusion coefficients. In Chap. 7, which may digress from the subject of the book, there are some interesting applications of the model to circular DNA, in particular to the statistical and transport behavior of its topoisomers. Chapter 8 presents treatments of excluded-volume effects, that is, various radius expansion factors and the second and third virial coefficients within the new framework mentioned above.

The final two chapters are devoted to dynamics, where discreteness is introduced in the chain to give the dynamic HW model. Chapter 9 begins with a general discussion of the dynamics of constrained chains. Then the diffusion equations describing the time evolution of the distribution functions for this model are derived in the classical diffusion (Smoluchowski) limit, and relevant eigenvalue problems and time-correlation functions are formulated. A coarse-grained dynamic HW model is also presented to treat global chain motions of somewhat shorter wavelengths than those treated by the spring-bead model. In Chap. 10 there are treated various dynamical properties of unperturbed flexible and semiflexible polymers. They include local properties such as dielectric and nuclear magnetic relaxation, fluorescence depolarization, and dynamic depolarized light scattering, and also global ones such as the first cumulant of the dynamic structure factor, and so forth.

References

1. P. J. Flory: *Principles of Polymer Chemistry* (Cornell Univ. Press, Ithaca, N.Y., 1953).
2. H. Yamakawa: *Modern Theory of Polymer Solutions* (Harper & Row, New York, 1971).

3. M. Kurata and W. H. Stockmayer: Adv. Polym. Sci. **3**, 196 (1963).
4. M. V. Volkenstein: *Configurational Statistics of Polymeric Chains* (Interscience, New York, 1963).
5. P. J. Flory: *Statistical Mechanics of Chain Molecules* (Interscience, New York, 1969).
6. P.-G. de Gennes: Phys. Lett. **38A**, 339 (1972).
7. J. des Cloizeaux: J. Phys. (Paris) **36**, 281 (1975).
8. P.-G. de Gennes: *Scaling Concepts in Polymer Physics* (Cornell Univ. Press, Ithaca, N.Y., 1979).
9. Y. Oono: Adv. Chem. Phys. **61**, 301 (1985).
10. K. F. Freed: *Renormalization Group Theory of Macromolecules* (John Wiley & Sons, New York, 1987).
11. J. des Cloizeaux and G. Jannink: *Polymers in Solution. Their Modelling and Structure* (Clarendon Press, Oxford, 1990).
12. J. G. Kirkwood: Rec. Trav. Chim. **68**, 649 (1949); J. Polym. Sci. **12**, 1 (1954).
13. P. E. Rouse, Jr.: J. Chem. Phys. **21**, 1272 (1953).
14. B. H. Zimm: J. Chem. Phys. **24**, 269 (1956).
15. M. Fixman and W. H. Stockmayer: Ann. Rev. Phys. Chem. **21**, 407 (1970).
16. W. H. Stockmayer: In *Molecular Fluids—Fluides Moléculaires*, R. Balian and G. Weill, eds. (Gordon & Breach, New York, 1976), p. 107.
17. M. Fixman and J. Kovac: J. Chem. Phys. **61**, 4939 (1974).
18. H. Yamakawa: Ann. Rev. Phys. Chem. **35**, 23 (1984).
19. H. Yamakawa: In *Molecular Conformation and Dynamics of Macromolecules in Condensed Systems*, M. Nagasawa, ed. (Elsevier, Amsterdam, 1988), p. 21.
20. P.-G. de Gennes: J. Chem. Phys. **55**, 572 (1971).
21. S. F. Edwards: Proc. Phys. Soc. **92**, 9 (1967); In *Molecular Fluids—Fluides Moléculaires*, R. Balian and G. Weill, eds. (Gordon & Breach, New York, 1976), p. 151.
22. M. Doi and S. F. Edwards: *The Theory of Polymer Dynamics* (Clarendon Press, Oxford, 1986).
23. O. Kratky and G. Porod: Rec. Trav. Chim. **68**, 1106 (1949).
24. K. F. Freed: Adv. Chem. Phys. **22**, 1 (1972).
25. H. Yamakawa: Ann. Rev. Phys. Chem. **25**, 179 (1974).
26. H. Yamakawa: Pure Appl. Chem. **46**, 135 (1976).
27. H. Yamakawa and M. Fujii: J. Chem. Phys. **64**, 5222 (1976).
28. H. Yamakawa: Macromolecules **10**, 692 (1977).
29. H. Yamakawa and T. Yoshizaki: J. Chem. Phys. **75**, 1016 (1981).
30. M. Nagasawa, ed.: *Molecular Conformation and Dynamics of Macromolecules in Condensed Systems*, (Elsevier, Amsterdam, 1988).
31. H. Fujita: *Polymer Solutions* (Elsevier, Amsterdam, 1990).
32. K. Huber, W. Burchard, and A. Z. Akcasu: Macromolecules **18**, 2743 (1985).
33. K. Huber and W. H. Stockmayer: Macromolecules **20**, 1400 (1987).
34. T. Konishi, T. Yoshizaki, J. Shimada, and H. Yamakawa: Macromolecules **22**, 1921 (1989); and succeeding papers.
35. F. Abe, Y. Einaga, T. Yoshizaki, and H. Yamakawa: Macromolecules **26**, 1884 (1993); and succeeding papers.
36. T. Norisuye: Prog. Polym. Sci. **18**, 543 (1993).
37. G. Merkle, W. Burchard, P. Lutz, K. F. Freed, and J. Gao: Macromolecules **26**, 2736 (1993).
38. T. Odijk: Macromolecules **19**, 2313 (1986).
39. S. A. Safran and N. A. Clark, eds.: *Physics of Complex and Supermolecular Fluids* (John Wiley & Sons, New York, 1987).

40. F. Tanaka, M. Doi, and T. Ohta, eds.: *Space–Time Organization in Macro-molecular Fluids* (Springer, Berlin, 1989).
41. A. Onuki and K. Kawasaki, eds.: *Dynamics and Patterns in Complex Fluids* (Springer, Berlin, 1990).

2 Models for Polymer Chains

In this chapter a brief description is given of several fundamental models for polymer chains, both discrete and continuous, the latter being obtained as a continuous limit of the former under certain conditions. The *unperturbed* chains without excluded volume are considered throughout the chapter but all basic equations are valid for both unperturbed and *perturbed* chains unless otherwise noted. Thus the symbol $\langle\ \rangle$ is used without the subscript 0 to denote a conformational average even in the unperturbed state, for simplicity.

2.1 Discrete Models

2.1.1 Average Chain Dimensions

Consider a single chain composed of $n + 1$ identical main chain atoms, say carbon atoms, which are joined successively by single bonds and which are numbered $0, 1, 2, \cdots, n$ from one end to the other. Let $\mathbf{r}_i$ be the vector position of the ith carbon atom $(i = 0, 1, \cdots, n)$ in the instantaneous configuration of the chain, as depicted in Fig. 2.1. The vector $\mathbf{l}_i$ defined by

$$\mathbf{l}_i = \mathbf{r}_i - \mathbf{r}_{i-1} \qquad (i = 1, 2, \cdots, n) \tag{2.1}$$

is called the ith bond vector, whose magnitude l_i is the bond length. The

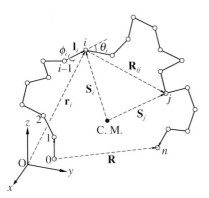

Fig. 2.1. Instantaneous configurations of a discrete chain and its various configurational quantities

angle θ_i ($i = 1, 2, \cdots, n-1$) between the vectors $\mathbf{l}_i$ and $\mathbf{l}_{i+1}$ is the supplement of the ith bond angle. The angle between the two planes containing $\mathbf{l}_{i-1}$ and $\mathbf{l}_i$, and $\mathbf{l}_i$ and $\mathbf{l}_{i+1}$, respectively, defines the internal rotation angle ϕ_i ($i = 2, \cdots, n-1$) about the ith bond, which is chosen to be zero when $\mathbf{l}_{i-1}$ and $\mathbf{l}_{i+1}$ are situated in the *trans* position with respect to each other. For the present purpose, both l_i and θ_i may be fixed at constant values (except for hypothetical cases), ignoring atomic vibrations. This is the *discrete model*. In what follows, we assume that $l_i = l$ and $\theta_i = \theta$ for all i, for simplicity.

Now the end-to-end vector $\mathbf{R} = \mathbf{r}_n - \mathbf{r}_0$ of the chain is the resultant of n bond vectors, that is,

$$\mathbf{R} = \sum_{i=1}^{n} \mathbf{l}_i \,, \tag{2.2}$$

so that the *mean-square end-to-end distance* $\langle R^2 \rangle$ as a measure of the average chain dimension is given by

$$\langle R^2 \rangle = \sum_{i=1}^{n} \sum_{j=1}^{n} \langle \mathbf{l}_i \cdot \mathbf{l}_j \rangle$$
$$= nl^2 + 2 \sum \sum_{1 \leq i < j \leq n} \langle \mathbf{l}_i \cdot \mathbf{l}_j \rangle \,. \tag{2.3}$$

Further, if $\mathbf{S}_i$ is the vector distance of the ith carbon atom from the center of mass (C.M.) of the chain, the *radius of gyration* S is defined by

$$S^2 = \frac{1}{n+1} \sum_{i=0}^{n} S_i{}^2 \,. \tag{2.4}$$

Assume that an equal mass is centered at each carbon atom, and by definition, we have

$$\sum_{i=0}^{n} \mathbf{S}_i = \mathbf{0} \,. \tag{2.5}$$

If $\mathbf{R}_{ij} = \mathbf{S}_j - \mathbf{S}_i$ is the vector distance between the ith and jth carbon atoms, then we obtain the well-known formula for the *mean-square radius of gyration* $\langle S^2 \rangle$, which is another measure of the average chain dimension,

$$\langle S^2 \rangle = \frac{1}{(n+1)^2} \sum \sum_{0 \leq i < j \leq n} \langle R_{ij}^2 \rangle \,. \tag{2.6}$$

The simple derivation of Eq. (2.6) is given in the earlier book (MTPS) [1].

Note that Eqs. (2.3) and (2.6) are valid for both unperturbed and perturbed chains.

2.1.2 Random-Flight Chains – The Gaussian Chain

The simplest hypothetical discrete model is obtained by setting $\langle \mathbf{l}_i \cdot \mathbf{l}_j \rangle = 0$ for $i \neq j$ in Eqs. (2.3); thus it has no correlations between any two bonds even with the uniform distribution of θ in its possible range from 0 to π, so that

$$\langle R^2 \rangle = nl^2 . \tag{2.7}$$

This is the *random-flight chain* or the *freely jointed chain*; it is also called the *random-coil model* [2]. For this chain $\langle R^2 \rangle$ is proportional to n.

Now suppose that the initial (0th) carbon atom is fixed at the origin of a Cartesian coordinate system and let $P(\mathbf{R})d\mathbf{R}$ then be the probability of finding the terminal (nth) carbon atom in the volume element $d\mathbf{R} = dxdydz$ at $\mathbf{R}(x, y, z)$. In the asymptotic limit of large n the distribution function $P(\mathbf{R})$ of $\mathbf{R}$ for this chain is found to be [1, 3, 4]

$$P(\mathbf{R}) = \left(\frac{3}{2\pi nl^2} \right)^{3/2} \exp\left(-\frac{3R^2}{2nl^2} \right) . \tag{2.8}$$

That is, in this limit the random-flight chain becomes the *Gaussian chain*. As is readily seen from Eq. (2.8), the latter chain has the same second moment $\langle R^2 \rangle = nl^2$ as the former for arbitrary n. For the Gaussian chain we also have the well-known relation [1, 5]

$$\langle S^2 \rangle = \frac{1}{6} \langle R^2 \rangle . \tag{2.9}$$

2.1.3 Freely Rotating Chains

Next we consider a model in which both l and θ are fixed but in which the distribution of ϕ_i is uniform in its range from $-\pi$ to π. This model is called the *freely rotating chain*. In this case it is easy to show that $\langle \mathbf{l}_i \cdot \mathbf{l}_{i+1} \rangle = l^2 \cos \theta$ ($i = 1, \cdots, n - 1$) and in general $\langle \mathbf{l}_i \cdot \mathbf{l}_j \rangle = l^2 \cos^{j-i} \theta$ ($i < j$). On performing the summations in the second line of Eqs. (2.3) after substitution of this result, we obtain [6–8]

$$\langle R^2 \rangle = nl^2 \frac{1 + \cos \theta}{1 - \cos \theta} - 2l^2 \cos \theta \frac{1 - \cos^n \theta}{(1 - \cos \theta)^2} . \tag{2.10}$$

For this case, if $0 < \theta < \pi/2$, $\langle R^2 \rangle / n$ increases monotonically with increasing n and approaches the constant $l^2(1 + \cos \theta)/(1 - \cos \theta)$; that is [2],

$$\langle R^2 \rangle = nl^2 \frac{1 + \cos \theta}{1 - \cos \theta} \qquad \text{for } n \gg 1 . \tag{2.11}$$

Similarly, we can calculate the average $\langle \mathbf{R} \cdot \mathbf{u}_0 \rangle$ with $\mathbf{u}_0 = \mathbf{l}_1/l$ being the unit vector in the direction of the first bond $\mathbf{l}_1$ as follows,

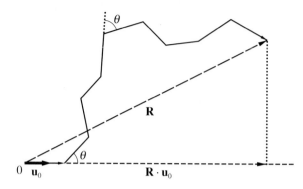

Fig. 2.2. Persistence of the component of l_i in the direction $\mathbf{u}_0 = \mathbf{l}_1/l$ of the first bond in a freely rotating chain

$$\langle \mathbf{R} \cdot \mathbf{u}_0 \rangle = l^{-1} \sum_{i=1}^{n} \langle \mathbf{l}_1 \cdot \mathbf{l}_i \rangle = l \frac{1 - \cos^n \theta}{1 - \cos \theta} . \tag{2.12}$$

The scalar product $\mathbf{R} \cdot \mathbf{u}_0$ is the projection of $\mathbf{R}$ in the direction of $\mathbf{l}_1$, as depicted in Fig. 2.2. If we assume again that $0 < \theta < \pi/2$, we have

$$\lim_{n \to \infty} \langle \mathbf{R} \cdot \mathbf{u}_0 \rangle = \frac{l}{1 - \cos \theta} . \tag{2.13}$$

The quantity on the left-hand side of Eq. (2.13) is called the *persistence length* [9, 10], and we denote it by q. The origin of this term is that the situation is similar to that encountered in the kinetic theory of gases, in which the component of the velocity $\mathbf{u}$ of a particle in its initial direction $\mathbf{u}_0$ persists after collisions. We have $q > l$ for the freely rotating chain and $q = l$ for the random-flight chain. Thus q is often used as a measure of *chain stiffness* but this is not always correct [11, 12] (see also below).

2.1.4 Chains with Coupled Rotations – The Rotational Isomeric State Model

In the real chain with fixed bond lengths and bond angles there are hindrances to internal rotations and ϕ_i is not uniformly distributed. In this case the chain has a potential energy $E(\{\phi_{n-2}\})$ as a function of all $(n-2)$ internal rotation angles $\{\phi_{n-2}\} = \phi_2, \cdots, \phi_{n-1}$, so that the average $\langle \mathbf{l}_i \cdot \mathbf{l}_j \rangle$ in Eqs. (2.3) must be calculated statistical-mechanically with the Boltzmann factor or the chain conformational partition function Z,

$$Z = \int \exp[-E(\{\phi_{n-2}\})/k_B T] d\{\phi_{n-2}\} , \tag{2.14}$$

where k_B is the Boltzmann constant, T is the absolute temperature, and $d\{\phi_{n-2}\} = d\phi_2 \cdots d\phi_{n-1}$.

For illustration, we consider the simplest of such *chains with coupled rotations*, that is, the one only with the rotational potential E_1 about each

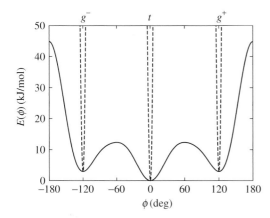

Fig. 2.3. Internal-rotational potential of *n*-butane. The *dashed lines* represent the RIS approximation

bond and the correlation E_2 between rotations about two successive bonds. In fact, higher-order neighbor interactions of this kind may be neglected in many cases. The total rotational potential E may then be written in the form

$$E(\{\phi_{n-2}\}) = \sum_{i=2}^{n-1} E_i(\phi_{i-1}, \phi_i), \qquad (2.15)$$

where

$$E_i(\phi_{i-1}, \phi_i) = E_{1i}(\phi_i) + E_{2i}(\phi_{i-1}, \phi_i) \qquad (2.16)$$

with $E_{22} = 0$. In particular, a hypothetical chain with $E_i = E_{1i}$ is called the *chain with independent rotations*. The potential E_{1i} may be regarded as close to the potential $E(\phi)$ about the central C–C bond in *n*-butane, which has three minima corresponding to the three stable conformations or rotational isomers: *trans* (t) ($\phi \simeq 0°$), *gauche*$^+$ (g^+) ($\phi \simeq 120°$), and *gauche*$^-$ (g^-) ($\phi \simeq -120°$), as illustratively shown in Fig. 2.3. On the other hand, E_{2i} becomes very large for the conformation (ϕ_{i-1}, ϕ_i) $=(g^+, g^-)$ or (g^-, g^+). This is the so-called *pentane effect* [13, 14].

With such potentials, however, the mathematical treatments become very difficult. It is therefore convenient to introduce an assumption such that ϕ_i takes only the three or more discrete values corresponding to the potential minima, t, g^+, g^-, and so on, thus approximating E_{1i} by a finite set of discrete levels or sharp square-well potentials at those ϕ_i, as shown in the dashed lines in Fig. 2.3 [15, 16]. This is the *rotational isomeric state* (RIS) *model*. Equation (2.14) with Eq. (2.15) may then be reduced to

$$Z = \sum_{\{\phi_{n-2}\}} \prod_{i=2}^{n-1} u_{\mu\nu,i}, \qquad (2.17)$$

where

$$u_{\mu\nu,i} = \exp[-E_i(\mu, \nu)/k_B T] \qquad (2.18)$$

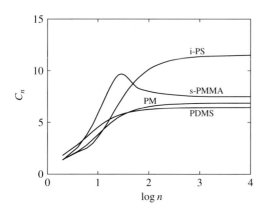

Fig. 2.4. Characteristic ratio C_n for typical flexible polymers

with μ, $\nu = t$, g^+, g^-, $\cdots$. Thus the problem becomes equivalent to that of the one-dimensional Ising model [17–21].

The results for $\langle R^2 \rangle$ are often given for the *characteristic ratio* C_n defined by [21]

$$C_n = \langle R^2 \rangle / nl^2 . \tag{2.19}$$

In Fig. 2.4 values of C_n so calculated are plotted against $\log n$ for polymethylene (PM), poly(dimethylsiloxane) (PDMS), isotactic polystyrene (i-PS), and syndiotactic poly(methyl methacrylate) (s-PMMA), for illustration.

2.2 Continuous Models

For a discrete chain of n bonds, each of length l, we define the total *contour length* L and the contour distance s ($0 \leq s \leq L$) of the ith carbon atom from the initial (0th) one along the chain by the equations

$$L = nl , \quad s = li , \tag{2.20}$$

respectively. We let $n \to \infty$ and $l \to 0$ at constant L (and $i \to \infty$ at constant s) in such a way that the discrete chain contour becomes a continuous and differentiable space curve. This is the *continuous model* which is mainly considered in this book. It is specified by certain additional conditions imposed in the limiting process [9, 11, 12, 22].

In any case we can define the unit vector $\mathbf{u}(s)$ tangent to the curve at the contour point s, that is,

$$\mathbf{u}(s) = \frac{d\mathbf{r}(s)}{ds} \tag{2.21}$$

with $\mathbf{r}_i = \mathbf{r}(s)$ being the radius vector, as depicted in Fig. 2.5. Equation (2.21) is the continuous limit of $\mathbf{l}_i/l$ with $\mathbf{l}_i$ being given by Eq. (2.1). The end-to-end vector $\mathbf{R}$ may then be expressed in terms of $\mathbf{u}$ as

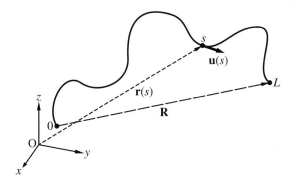

Fig. 2.5. Instantaneous configurations of a continuous chain and its configurational quantities

$$\mathbf{R} = \mathbf{r}(L) - \mathbf{r}(0) = \int_0^L \mathbf{u}(s)ds \qquad (2.22)$$

as the continuous limit of Eq. (2.2), so that we have

$$\langle R^2 \rangle = \int_0^L \int_0^L \langle \mathbf{u}(s_1) \cdot \mathbf{u}(s_2) \rangle ds_1 ds_2 \qquad (2.23)$$

as the continuous limit of Eqs. (2.3). If $\mathbf{R}(s_1, s_2)$ is the vector distance between the contour points s_1 and s_2 ($s_1 \leq s_2$), Eq. (2.6) becomes

$$\langle S^2 \rangle = \frac{1}{L^2} \int_0^L ds_1 \int_{s_1}^L ds_2 \langle R^2(s_1, s_2) \rangle . \qquad (2.24)$$

For the unperturbed chain there holds the relation $\langle R^2(s_1, s_2) \rangle = \langle R^2(s) \rangle$ with $s = s_2 - s_1$ [1], and therefore Eq. (2.24) reduces to

$$\langle S^2 \rangle = \frac{1}{L^2} \int_0^L (L - s)\langle R^2(s) \rangle ds. \qquad (2.25)$$

Necessarily, we have, for the continuous chain above,

$$\mathbf{u}^2(s) = 1 \quad \text{for } 0 \leq s \leq L. \qquad (2.26)$$

Indeed, this relation is satisfied by the *Kratky–Porod* (KP) *wormlike chain* [9] and also the *helical wormlike* (HW) *chain* [11, 12] treated in this book. In anticipation of the results we here note only that the former is obtained as a continuous limit of the freely rotating chain, while the latter is obtained from a discrete chain with coupled rotations by some coarse-graining, followed by the continuous-limiting process. However, in order to make the model more tractable, the constraint of Eq. (2.26) is often relaxed, thereby leading to various modifications [11, 23, 24] of the former. Further, we note that for the continuous chain the *Kuhn segment length* A_K and the *persistence length* q are defined by

$$A_K = \lim_{L \to \infty} \left(\langle R^2 \rangle / L \right), \tag{2.27}$$

$$q = \lim_{L \to \infty} \langle \mathbf{R} \cdot \mathbf{u}_0 \rangle \tag{2.28}$$

with $\mathbf{u}_0 = \mathbf{u}(0)$, which in general satisfy the relation

$$A_K = 2q, \tag{2.29}$$

but that neither A_K nor q is a measure of chain stiffness except for the KP chain [11, 12].

Finally, we consider a *continuous* Gaussian chain and give some fundamentals related to it. Its $\langle R^2 \rangle$ must be given by

$$\langle R^2 \rangle = lL \,. \tag{2.30}$$

This is rather the defining equation for the "bond length" l for the chain of total contour length L. It is evident that the above-mentioned continuous limit of the random-flight chain cannot be taken to obtain this continuous chain when $\langle R^2 \rangle$ and L are given. Its rigorous treatment requires an elaborate maneuver [23,25]. Thus we follow instead a simple fashion to regard L merely as a continuous variable for very large n in the discrete random-flight chain, as usually done [1,4,26]. This suffices for the present purpose (see also Appendix 3.A). It is then convenient to rewrite Eq. (2.8) as

$$P(\mathbf{R}; L) = \left(\frac{3}{2\pi lL} \right)^{3/2} \exp\left(-\frac{3R^2}{2lL} \right). \tag{2.31}$$

This $P(\mathbf{R}; L)$ is the solution of the differential equation

$$\left(\frac{\partial}{\partial L} - \frac{l}{6} \nabla_R^2 \right) P(\mathbf{R}; L) = 0 \tag{2.32}$$

subject to the boundary condition

$$P(\mathbf{R}; 0) = \delta(\mathbf{R}), \tag{2.33}$$

where ∇_R^2 is the Laplacian operator with respect to $\mathbf{R}$ and $\delta(\mathbf{R})$ is a three-dimensional Dirac delta function.

Now we introduce the *Green function* $G(\mathbf{R}; L)$ defined by

$$G(\mathbf{R}; L) = P(\mathbf{R}; L) \qquad \text{for } L > 0$$
$$= 0 \qquad\qquad \text{for } L < 0, \tag{2.34}$$

where $P(\mathbf{R}; L)$ is given by Eq. (2.31). Equation (2.32) with Eq. (2.33) may then be reduced to

$$\left(\frac{\partial}{\partial L} - \frac{l}{6} \nabla_R^2 \right) G(\mathbf{R}; L) = \delta(L)\delta(\mathbf{R}). \tag{2.35}$$

If we integrate both sides of Eq. (2.35) over L from $-\epsilon$ to ϵ with ϵ being positive and small, then the left-hand side becomes

$$\int_{-\epsilon}^{\epsilon} \frac{\partial G(\mathbf{R}; L)}{\partial L} dL = G(\mathbf{R}; \epsilon), \qquad (2.36)$$

and the right-hand side becomes $\delta(\mathbf{R})$, so that we have

$$\lim_{L \to +0} G(\mathbf{R}; L) = \delta(\mathbf{R}). \qquad (2.37)$$

Thus the solution of Eq. (2.35) is indeed that of Eq. (2.32) subject to the boundary condition of Eq. (2.33). If L is regarded as "time," Eq. (2.32) or Eq. (2.35) is just the diffusion equation associated with the random process (position) $\mathbf{r}(s)$ of a Brownian particle with diffusion coefficient $l/6$ at a long time.

Note that for this chain the *bond correlation function* $\langle \mathbf{u}(s_1) \cdot \mathbf{u}(s_2) \rangle$ is *formally* given by

$$\langle \mathbf{u}(s_1) \cdot \mathbf{u}(s_2) \rangle = l\delta(s_1 - s_2), \qquad (2.38)$$

because substitution of this equation into Eq. (2.32) recovers Eq. (2.30).

References

1. H. Yamakawa: *Modern Theory of Polymer Solutions* (Harper & Row, New York, 1971).
2. W. Kuhn: Kolloid Z. **68**, 2 (1934).
3. Lord Rayleigh: Phil. Mag. **37**, 321 (1919).
4. S. Chandrasekhar: Rev. Mod. Phys. **15**, 1 (1943).
5. P. Debye: J. Chem. Phys. **14**, 636 (1946).
6. H. Eyring: Phys. Rev. **39**, 746 (1932).
7. R. M. Fuoss and J. G. Kirkwood: J. Am. Chem. Soc. **63**, 385 (1941).
8. F. T. Wall: J. Chem. Phys. **11**, 67 (1943).
9. O. Kratky and G. Porod: Rec. Trav. Chim. **68**, 1106 (1949).
10. See also S. E. Bresler and Ya. I. Frenkel: Acta Phys.-Chim. USSR **11**, 485 (1939).
11. H. Yamakawa: Ann. Rev. Phys. Chem. **35**, 23 (1984).
12. H. Yamakawa: In *Molecular Conformation and Dynamics of Macromolecules in Condensed Systems*, M. Nagasawa, ed. (Elsevier, Amsterdam, 1988), p. 21.
13. K. S. Pitzer: J. Chem. Phys. **8**, 711 (1940).
14. W. J. Taylor: J. Chem. Phys. **16**, 257 (1948).
15. M. V. Volkenstein: Dokl. Akad. Nauk SSSR **78**, 879 (1951); *Configurational Statistics of Polymeric Chains* (Interscience, New York, 1963).
16. See also R. Kubo: J. Phys. Soc. Japan **4**, 319 (1949).
17. S. Lifson: J. Chem. Phys. **30**, 964 (1959).
18. K. Nagai: J. Chem. Phys. **31**, 1169 (1959); **37**, 490 (1962).
19. T. M. Birshtein and O. B. Ptitsyn: Zh. Tekhn. Fiz. **29**, 1048 (1959); T. M. Birshtein, Vysokomolekul. Soedin. **1**, 798 (1959); T. M. Birshtein and E. A. Sokolova, Vysokomolekul. Soedin. **1**, 1086 (1959).
20. C. A. J. Hoeve: J. Chem. Phys. **32**, 888 (1960).

21. P. J. Flory: *Statistical Mechanics of Chain Molecules* (Interscience, New York, 1969).
22. A. Miyake and Y.Hoshino: J. Phys. Soc. Japan **46**, 1324 (1979).
23. K. F. Freed: Adv. Chem. Phys. **22**, 1 (1972).
24. H. Yamakawa: Ann. Rev. Phys. Chem. **25**, 179 (1974); Pure Appl. Chem. **46**, 135 (1976).
25. K. F. Freed: *Renormalization Group Theory of Macromolecules* (John Wiley & Sons, New York, 1987).
26. M. Doi and S. F. Edwards: *The Theory of Polymer Dynamics* (Clarendon Press, Oxford, 1986).

3 Chain Statistics – Wormlike Chains

This chapter presents the foundation of the statistical mechanics of the KP wormlike chain. In particular, there is a detailed description of the formulation of the model and the theoretical methods which can also be applied to the statistical mechanics of the HW chain developed in the next chapter. Its static and transport properties are treated in later chapters as special cases of those of the HW chain. The unperturbed chain without excluded volume is considered throughout the chapter.

3.1 Definition of the Model

Consider the freely rotating chain composed of n bonds with bond length l and bond angle $\pi - \theta$. Its persistence length q, which we set equal to $(2\lambda)^{-1}$, is given by Eq. (2.13),

$$q \equiv \frac{1}{2\lambda} = \frac{l}{1 - \cos\theta}, \tag{3.1}$$

so that

$$\cos\theta = 1 - 2\lambda l = 1 - \frac{2\lambda L}{n}. \tag{3.2}$$

The KP chain is defined as a limiting continuous chain formed from this discrete chain by letting $n \to \infty$, $l \to 0$, and $\theta \to 0$ under the restriction that $L = nl$ and λ remain constant [1].

Now, for the freely rotating chain a function of n, l, and θ may thus be considered that of n, L, and λ, and therefore any dimensional quantity for the KP chain may be obtained by taking the limit of $n \to \infty$ at constant L and λ in that quantity as a function of n, L, and λ for the former. Thus, if we note that

$$\lim_{\substack{n \to \infty \\ \theta \to 0}} \cos^n \theta = \lim_{n \to \infty} \left(1 - \frac{2\lambda L}{n}\right)^n = e^{-2\lambda L}, \tag{3.3}$$

then for the KP chain we have, from Eqs. (2.12) and (2.10) [2],

$$\langle \mathbf{R} \cdot \mathbf{u}_0 \rangle = \frac{1}{2\lambda}(1 - e^{-2\lambda L}), \tag{3.4}$$

$$\langle R^2 \rangle = \frac{L}{\lambda} - \frac{1}{2\lambda^2}(1 - e^{-2\lambda L}) . \tag{3.5}$$

We note that since we have $d\langle R^2 \rangle = 2\langle \mathbf{R} \cdot \mathbf{u} \rangle dL$ from Eq. (2.22), Eq. (3.5) may also be obtained by integration of $2\langle \mathbf{R} \cdot \mathbf{u_0} \rangle$ over L [1]. Substituting Eq. (3.5) with $L = s$ into Eq. (2.25) and performing the integration, we also obtain [3]

$$\langle S^2 \rangle = \frac{L}{6\lambda} - \frac{1}{4\lambda^2} + \frac{1}{4\lambda^3 L} - \frac{1}{8\lambda^4 L^2}(1 - e^{-2\lambda L}) . \tag{3.6}$$

In the limits of $\lambda L \to 0$ (rigid rod) and of $\lambda L \to \infty$ (random coil), we have, from Eqs. (3.5) and (3.6),

$$\langle R^2 \rangle = 12\langle S^2 \rangle = L^2 \quad \text{for } \lambda L \to 0 , \tag{3.7}$$

$$\langle R^2 \rangle = 6\langle S^2 \rangle = \frac{L}{\lambda} \quad \text{for } \lambda L \to \infty . \tag{3.8}$$

As shown in Fig. 3.1, the dimensionless ratios $\lambda\langle R^2 \rangle/L$ and $6\lambda\langle S^2 \rangle/L$ increase monotonically from 0 to 1 as λL is increased from 0 to ∞, and thus the KP model is an interpolation from the two extremes, rigid-rod limit and random-coil limit. It is therefore a good model for most typical stiff polymers, and also mimics those flexible polymers for which the characteristic ratio C_n defined by Eq. (2.19) increases monotonically with increasing n and levels off to its asymptotic value C_∞, since $\lambda\langle R^2 \rangle/L$ corresponds to C_n/C_∞ if L is properly converted to n. As easily recognized, however, even when the behavior of the chain contour or of C_n can be explained by this model, it is impossible to assign, for instance, local dipole moments and polarizabilities to it unless they are parallel to and cylindrically symmetric about the chain contour, respectively.

For the KP chain the Kuhn segment length A_K defined by Eq. (2.27) and the persistence length q defined by Eq. (2.28), which is naturally the same as that of the original freely rotating chain, are obtained from Eqs. (3.4) and (3.5) as

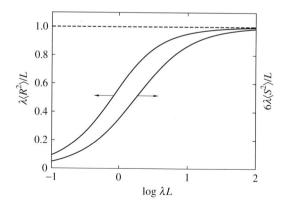

Fig. 3.1. $\lambda\langle R^2 \rangle/L$ and $6\lambda\langle S^2 \rangle/L$ plotted against $\log \lambda L$ for the KP chain

$$A_K = 2q = \lambda^{-1}. \qquad (3.9)$$

In the next section λ is defined from a different point of view, and then the parameter λ^{-1} (having the dimension of length) in general proves to be a measure of (static) chain stiffness and is referred to as the *stiffness parameter*; the stiffer the chain, the larger the parameter λ^{-1}. Thus, for the KP chain both A_K and q are just measures of chain stiffness. Of course, it is easy to understand from the above discussion that for this model λ^{-1} represents the chain stiffness.

3.2 Diffusion Equations

3.2.1 Green Functions

We can define a conditional distribution function $P(\mathbf{R}, \mathbf{u}, \mathbf{u}_0; L)/P(\mathbf{u}_0)$ of the radius vector $\mathbf{r}(L) = \mathbf{R}$ and the unit tangent vector $\mathbf{u}(L) = \mathbf{u}$ at the terminal end of the KP chain of contour length L when $\mathbf{r}(0) = \mathbf{0}$ and $\mathbf{u}(0) = \mathbf{u}_0$ at the initial end. This is the Green function $G(\mathbf{R}, \mathbf{u}, L \,|\, \mathbf{0}, \mathbf{u}_0, 0)$, which we simply denote by $G(\mathbf{R}, \mathbf{u} \,|\, \mathbf{u}_0; L)$ and which is normalized as

$$\int G(\mathbf{R}, \mathbf{u} \,|\, \mathbf{u}_0; L)d\mathbf{R}d\mathbf{u} = 1, \qquad (3.10)$$

where $d\mathbf{u} = \sin\theta d\theta d\phi$ with $\mathbf{u} = (1, \theta, \phi)$ in spherical polar coordinates. The *characteristic function* $I(\mathbf{k}, \mathbf{u} \,|\, \mathbf{u}_0; L)$, that is, the Fourier transform of G with respect to $\mathbf{R}$ is defined by

$$I(\mathbf{k}, \mathbf{u} \,|\, \mathbf{u}_0; L) = \int G(\mathbf{R}, \mathbf{u} \,|\, \mathbf{u}_0; L)\exp(i\mathbf{k} \cdot \mathbf{R})d\mathbf{R} \qquad (3.11)$$

with i the imaginary unit, so that

$$G(\mathbf{u} \,|\, \mathbf{u}_0; L) = \int G(\mathbf{R}, \mathbf{u} \,|\, \mathbf{u}_0; L)d\mathbf{R} = I(\mathbf{0}, \mathbf{u} \,|\, \mathbf{u}_0; L). \qquad (3.12)$$

Similarly, the distribution functions $G(\mathbf{R} \,|\, \mathbf{u}_0; L)$, $G(\mathbf{R}, \mathbf{u}; L)$, and $G(\mathbf{R}; L)$ may be obtained as

$$G(\mathbf{R} \,|\, \mathbf{u}_0; L) = \int G(\mathbf{R}, \mathbf{u} \,|\, \mathbf{u}_0; L)d\mathbf{u}, \qquad (3.13)$$

$$G(\mathbf{R}, \mathbf{u}; L) = (4\pi)^{-1} \int G(\mathbf{R}, \mathbf{u} \,|\, \mathbf{u}_0; L)d\mathbf{u}_0, \qquad (3.14)$$

$$G(\mathbf{R}; L) = (4\pi)^{-1} \int G(\mathbf{R}, \mathbf{u} \,|\, \mathbf{u}_0; L)d\mathbf{u}d\mathbf{u}_0. \qquad (3.15)$$

Note that $G(\mathbf{R}; L)$ may also be obtained by averaging $G(\mathbf{R} \,|\, \mathbf{u}_0; L)$ over the orientation of $\mathbf{R}$ since the former is a function only of $|\mathbf{R}| = R$. The corresponding characteristic functions $I(\mathbf{k} \,|\, \mathbf{u}_0; L)$, $I(\mathbf{k}, \mathbf{u}; L)$, and $I(\mathbf{k}; L)$ may readily be written down. The distribution function dependent on $\mathbf{R}$ may be obtained by Fourier inversion of its characteristic function; for example,

$$G(\mathbf{R}, \mathbf{u} \,|\, \mathbf{u}_0; L) = (2\pi)^{-3} \int I(\mathbf{k}, \mathbf{u} \,|\, \mathbf{u}_0; L) \exp(-i\mathbf{k} \cdot \mathbf{R}) d\mathbf{k} \,. \quad (3.16)$$

The Green functions $G(\mathbf{R}, \mathbf{u} \,|\, \mathbf{u}_0; L)$ and $G(\mathbf{u} \,|\, \mathbf{u}_0; L)$ satisfy the *Fokker–Planck equations*, which are of the diffusion type.

3.2.2 Fokker–Planck Equations

As in the case of the Gaussian chain considered in Sect. 2.2, $\mathbf{r}(s)$ and also $\mathbf{u}(s)$ may be regarded as Markov random processes on the proper "time" scale of s (or L). We disregard temporarily the boundary conditions simply to put $G(\mathbf{R}, \mathbf{u} \,|\, \mathbf{u}_0; L) = P(\mathbf{R}, \mathbf{u}; L)$. Then it satisfies the *Markov integral equation* [4,5]

$$P(\mathbf{R}, \mathbf{u}; L+l) = \int P(\mathbf{R} - \Delta\mathbf{R}, \mathbf{u} - \Delta\mathbf{u}; L)$$
$$\times \Psi(\Delta\mathbf{R}, \Delta\mathbf{u} \,|\, \mathbf{R} - \Delta\mathbf{R}, \mathbf{u} - \Delta\mathbf{u}; l) d(\Delta\mathbf{R}) d(\Delta\mathbf{u}) \,, \quad (3.17)$$

where $\Psi(\Delta\mathbf{R}, \Delta\mathbf{u} \,|\, \mathbf{R}, \mathbf{u}; l)$ is the *transition probability* of $\mathbf{R}$ and $\mathbf{u}$ from $(\mathbf{R}, \mathbf{u})$ to $(\mathbf{R} + \Delta\mathbf{R}, \mathbf{u} + \Delta\mathbf{u})$ (by $\Delta\mathbf{R}$ and $\Delta\mathbf{u}$) in "time" l. For sufficiently small l, we have $\Delta\mathbf{R} = l\Delta\mathbf{u}$, so that Ψ may be written in the form

$$\Psi = \delta(\Delta\mathbf{R} - l\Delta\mathbf{u})\psi(\Delta\mathbf{u} \,|\, \mathbf{u} - \Delta\mathbf{u}; l) \,, \quad (3.18)$$

where ψ is the transition probability only of $\mathbf{u}$. Then Eq. (3.17) reduces to

$$P(\mathbf{R}, \mathbf{u}; L+l) = \int P(\mathbf{R} - l\mathbf{u}, \mathbf{u} - \Delta\mathbf{u}; L)\psi(\Delta\mathbf{u} \,|\, \mathbf{u} - \Delta\mathbf{u}; l) d(\Delta\mathbf{u}) \,. \quad (3.19)$$

If we expand P and ψ on both sides of Eq. (3.19) in Taylor series following a standard procedure [4,5], we obtain the Fokker–Planck equation

$$\frac{\partial P}{\partial L} + \mathbf{u} \cdot \nabla_R P = -\nabla_u \cdot (\mathbf{a}^{(1)} P) + \frac{1}{2}\nabla_u \nabla_u : (\mathbf{a}^{(2)} P) \quad (3.20)$$

with

$$\mathbf{a}^{(1)} = \lim_{l \to 0} l^{-1}\langle \Delta\mathbf{u} \rangle_1 \,, \quad (3.21)$$

$$\mathbf{a}^{(2)} = \lim_{l \to 0} l^{-1}\langle (\Delta\mathbf{u})(\Delta\mathbf{u}) \rangle_1 \,, \quad (3.22)$$

where ∇_R and ∇_u are the gradient operators with respect to $\mathbf{R}$ and $\mathbf{u}$, respectively, and $\langle\ \rangle_1$ denotes an average over $\psi(\Delta\mathbf{u} \,|\, \mathbf{u}; l)$. Note that $\mathbf{a}^{(1)}$ and $\mathbf{a}^{(2)}$ are vector and tensor moments, respectively.

Now, if we put $\mathbf{l}_i/l = \mathbf{u}_i$ (unit bond vector) and let $\theta \to 0$ in the freely rotating chain, then $\mathbf{u}_i \cdot \Delta\mathbf{u}_i$ with $\Delta\mathbf{u}_i = \mathbf{u}_{i+1} - \mathbf{u}_i$ becomes $\mathbf{u} \cdot \Delta\mathbf{u}$, which

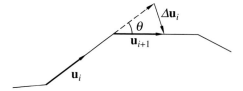

Fig. 3.2. Unit bond (tangent) vector $\mathbf{u}_i$ and its change $\Delta\mathbf{u}_i$ in (the continuous limit of) the freely rotating chain

vanishes; that is, $\Delta\mathbf{u}$ is perpendicular to $\mathbf{u}$, as depicted in Fig. 3.2. Since ψ is symmetric about $\mathbf{u}$, we then have $\langle\Delta\mathbf{u}\rangle_1 = \mathbf{0}$, so that

$$\mathbf{a}^{(1)} = \mathbf{0}\,. \tag{3.23}$$

If we expand $\cos\theta$ around $\theta = 0$, we obtain, from the first of Eqs. (3.2),

$$\theta^2 \rightarrow 4\lambda l\,. \tag{3.24}$$

As seen from Fig. 3.2, on the other hand, we have

$$|\Delta\mathbf{u}_i| \rightarrow \theta\,. \tag{3.25}$$

From Eqs. (3.24) and (3.25), we have

$$\lim_{l\to 0} l^{-1}\langle(\Delta\mathbf{u})^2\rangle_1 = \lim_{\substack{l\to 0 \\ \theta\to 0}} l^{-1}\theta^2 = 4\lambda\,. \tag{3.26}$$

Further, in a Cartesian coordinate system (x', y', z') with the z' axis in the direction of $\mathbf{u}$, we have

$$\langle(\Delta\mathbf{u}')(\Delta\mathbf{u}')\rangle_1 = \frac{1}{2}\langle(\Delta\mathbf{u})^2\rangle_1 \begin{pmatrix} 1 & 0 & 0 \\ 0 & 1 & 0 \\ 0 & 0 & 0 \end{pmatrix}\,. \tag{3.27}$$

Since $\nabla_u\nabla_u : (\mathbf{a}^{(2)}P)$ is invariant to rotation of the coordinate system, we obtain, from Eqs. (3.22), (3.26), and (3.27),

$$\frac{1}{2}\nabla_u\nabla_u : (\mathbf{a}^{(2)}P) = \lambda\nabla_{u'}^2 P = \lambda\nabla_u^2 P\,. \tag{3.28}$$

Substitution of Eqs. (3.23) and (3.28) into Eq. (3.20) leads to

$$\frac{\partial P}{\partial L} = \lambda\nabla_u^2 P - \mathbf{u}\cdot\nabla_R P\,, \tag{3.29}$$

where the Laplacian operator ∇_u^2 in spherical polar coordinates is given by

$$\nabla_u^2 = \frac{1}{\sin\theta}\frac{\partial}{\partial\theta}\sin\theta\frac{\partial}{\partial\theta} + \frac{1}{\sin^2\theta}\frac{\partial^2}{\partial\phi^2}\,, \tag{3.30}$$

since we have the constraint of Eq. (2.26), $\mathbf{u}^2 = 1$. Equation (3.29) is the differential equation first derived by Hermans and Ullman [6]. A differential equation for $P(\mathbf{R}; L)$, or $G(\mathbf{R} \mid \mathbf{u}_0; L)$, was derived by Daniels [7] before them, but we do not treat it here.

Thus the Green function $G(\mathbf{R}, \mathbf{u} \mid \mathbf{u}_0; L)$ of Eq. (3.29) satisfies the differential equation

$$\left(\frac{\partial}{\partial L} - \lambda \nabla_u^2 + \mathbf{u} \cdot \nabla_R \right) G(\mathbf{R}, \mathbf{u} \mid \mathbf{u}_0; L) = \delta(L)\delta(\mathbf{R})\delta(\mathbf{u} - \mathbf{u}_0). \quad (3.31)$$

Taking the Fourier transform of both sides of Eq. (3.31), we obtain

$$\left(\frac{\partial}{\partial L} - \lambda \nabla_u^2 - i\mathbf{k} \cdot \mathbf{u} \right) I(\mathbf{k}, \mathbf{u} \mid \mathbf{u}_0; L) = \delta(L)\delta(\mathbf{u} - \mathbf{u}_0), \quad (3.32)$$

and therefore also

$$\left(\frac{\partial}{\partial L} - \lambda \nabla_u^2 \right) G(\mathbf{u} \mid \mathbf{u}_0; L) = \delta(L)\delta(\mathbf{u} - \mathbf{u}_0). \quad (3.33)$$

These are the basic starting equations in the statistical mechanics of the KP chain.

3.2.3 Path Integrals and Formal Solutions

It is seen that Eq. (3.32) is just the "Schrödinger" equation (in units of $\hbar = h/2\pi$ with h the Planck constant) for the "quantum-mechanical amplitude" or "kernel" I for a rigid electric dipole $\mathbf{u}$ in an electric field $\mathbf{k}$, and Eq. (3.33) is that for I for a free rigid dipole with $\mathbf{k} = \mathbf{0}$. The kernel (wave function) may be written in the Feynman *path integral* form [8, 9], a short sketch of which is given in Appendix 3.A. Thus, in the present case, the characteristic function $I(\mathbf{k}, \mathbf{u} \mid \mathbf{u}_0; L)$ may be expressed in terms of the path integral over all possible paths (configurations) $\mathbf{u}(s)$ from $\mathbf{u}(0) = \mathbf{u}_0$ to $\mathbf{u}(L) = \mathbf{u}$ as follows,

$$I(\mathbf{k}, \mathbf{u} \mid \mathbf{u}_0; L) = \int_{\mathbf{u}(0)=\mathbf{u}_0}^{\mathbf{u}(L)=\mathbf{u}} \exp\left(i \int_0^L \mathcal{L} ds \right) \mathcal{D}[\mathbf{u}(s)] \quad (3.34)$$

(in units of $\hbar$) with $\mathcal{L}$ being the "Lagrangian" given by

$$\mathcal{L} = iU/k_B T + \mathbf{k} \cdot \mathbf{u}, \quad (3.35)$$

where

$$U = \frac{1}{2}\alpha [\dot{\mathbf{u}}(s)]^2 \quad (3.36)$$

with

$$\alpha = k_B T/2\lambda. \quad (3.37)$$

In Eq. (3.36), the over dot denotes the derivative with respect to s. The first term on the right-hand side of Eq. (3.35) is the "kinetic energy" of the "particle" (rigid dipole), and the second term is the negative of its "potential energy." Such an analogy with the formalism in quantum mechanics was first used by Saito et al. [10].

Now Eq. (3.34) may be given a statistical-mechanical interpretation. The U given by Eq. (3.36) is just the bending energy, per unit length, of an elastic wire with bending force constant α [11], and therefore the total potential energy E of the KP chain as the wire is given by

$$E = \int_0^L U ds = \frac{\alpha}{2} \int_0^L \dot{\mathbf{u}}^2 ds . \tag{3.38}$$

The Green function $G(\mathbf{R}, \mathbf{u} \,|\, \mathbf{u}_0; L)$ may be expressed as the sum of the Boltzmann factor $\exp(-E/k_{\mathrm{B}}T)$ over all possible configurations, and hence as its path integral over $\mathbf{u}(s)$ from $\mathbf{u}(0) = \mathbf{u}_0$ to $\mathbf{u}(L) = \mathbf{u}$ subject to the condition of Eq. (2.22), that is,

$$\int_0^L \mathbf{u} ds = \mathbf{R} . \tag{3.39}$$

Thus it may be written in the form

$$G(\mathbf{R}, \mathbf{u} \,|\, \mathbf{u}_0; L) = \int_{\mathbf{u}(0)=\mathbf{u}_0}^{\mathbf{u}(L)=\mathbf{u}} \delta\left(\mathbf{R} - \int_0^L \mathbf{u} ds \right) \exp\left(-\frac{1}{k_{\mathrm{B}}T} \int_0^L U ds \right) \mathcal{D}\big[\mathbf{u}(s) \big] . \tag{3.40}$$

Taking the Fourier transform of both sides of Eq. (3.40), we obtain Eq. (3.34). The treatment of the KP chain as the elastic wire with bending energy was first made by Bresler and Frenkel [12], Landau and Lifshitz [13], Harris and Hearst [14], and also Saito et al. [10]. The parameter λ^{-1} is now proportional to the bending force constant relative to the thermal energy, and Eq. (3.37) is its general definition for the continuous models treated in this book.

Next we obtain the formal solution of Eq. (3.32) in a series form [15, 16]. Treating the potential energy part $-\mathbf{k} \cdot \mathbf{u}$ of the Lagrangian $\mathcal{L}$ of the "dipole" as a perturbation, we can readily obtain, from Eq. (3.34), an integral equation for the kernel [8] as in collision theory [17]. The result is

$$I(\mathbf{k}, \mathbf{u} \,|\, \mathbf{u}_0; L) = G(\mathbf{u} \,|\, \mathbf{u}_0; L) + i\mathbf{k} \cdot \int_0^L \int \mathbf{u}_1 G(\mathbf{u} \,|\, \mathbf{u}_1; L - s_1)$$
$$\times I(\mathbf{k}, \mathbf{u}_1 \,|\, \mathbf{u}_0; s_1) ds_1 d\mathbf{u}_1 . \tag{3.41}$$

Thus $G(\mathbf{u} \,|\, \mathbf{u}_0; L)$ is the "free-particle" Green function. (In the present case, it is the free rigid dipole or dumbbell rotor.) An integral equation for $I(\mathbf{k}, \mathbf{u}; L)$ may also be obtained by integrating both sides of Eq. (3.41) over $\mathbf{u}_0$ and dividing them by 4π, but the result is omitted. The formal solution of Eq. (3.41), which is equivalent to that of Eq. (3.32), may be obtained by iteration if the free-particle Green function is known. In what follows, all lengths are measured in units of λ^{-1} unless otherwise noted, for simplicity, so that, for instance, λL is replaced by (reduced) L. The solution of Eq. (3.33) for the free-particle Green function is well known and is given by

$$G(\mathbf{u}\,|\,\mathbf{u}_0; L) = \sum_{l=0}^{\infty} \exp\left[-l(l+1)L\right] \sum_{m=-l}^{l} Y_l^m(\theta, \phi)Y_l^{-m}(\theta_0, \phi_0), \quad (3.42)$$

where Y_l^m is the normalized spherical harmonics and $\mathbf{u}_0 = (1, \theta_0, \phi_0)$ in spherical polar coordinates. This solution and the definition of Y_l^m adopted in this book are given in Appendix 3.B. Note that the known part of the integral equation for $I(\mathbf{k}, \mathbf{u}; L)$ is then $(4\pi)^{-1/2}Y_0^0(\theta, \phi)$.

If we choose $\mathbf{u}_0$ to be in the direction of the z axis of a Cartesian coordinate system ($\mathbf{u}_0 = \mathbf{e}_z$), both $I(\mathbf{k}, \mathbf{u}\,|\,\mathbf{u}_0; L)$ and $I(\mathbf{k}, \mathbf{u}; L)$, which we simply denote by $I(L)$, may be expanded in terms of $Y_l^m(\theta, \phi)$,

$$I(L) = \sum_{l=0}^{\infty} \sum_{m=-l}^{l} K_l^m(L)Y_l^m(\theta, \phi), \quad (3.43)$$

where $K_l^m(L)$ stands for $K_l^m(\mathbf{k}\,|\,\mathbf{u}_0; L)$ or $K_l^m(\mathbf{k}; L)$, as the case may be. Further, if $\mathbf{e}_k = \mathbf{k}/k$ is the unit vector in the direction of $\mathbf{k}$ with $\mathbf{e}_k = (1, \chi, \omega)$ in spherical polar coordinates, we have, from Eq. (3.B.15),

$$\mathbf{e}_k \cdot \mathbf{u}_1 = \frac{4\pi}{3} \sum_{m=-1}^{1} Y_1^m(\chi, \omega)Y_1^{-m}(\theta_1, \phi_1). \quad (3.44)$$

Substitution of Eqs. (3.43) and (3.44) into Eq. (3.41) [and the corresponding equation for $I(\mathbf{k}, \mathbf{u}; L)$] and integration over $\mathbf{u}_1$ leads to the integral equations for $K_l^m(L)$,

$$K_l^m = f_l^m + i\bar{k}f_l * \mathcal{L}K_l^m, \quad (3.45)$$

where the asterisk indicates a convolution integration,

$$f * g = \int_0^L f(L-s)g(s)ds, \quad (3.46)$$

and

$$\bar{k} = \left(\frac{2\pi}{3}\right)^{1/2} k, \quad (3.47)$$

$$f_l = \exp\left[-l(l+1)L\right], \quad (3.48)$$

$$f_l^m = \left(\frac{2l+1}{4\pi}\right)^{1/2} \delta_{m0} f_l \qquad \text{for } K_l^m = K_l^m(\mathbf{k}\,|\,\mathbf{u}_0; L)$$

$$= (4\pi)^{-1/2}\delta_{l0}\delta_{m0} f_0 \qquad \text{for } K_l^m = K_l^m(\mathbf{k}; L) \quad (3.49)$$

with δ_{lm} being the Kronecker delta. In Eq. (3.45), $\mathcal{L}$ is an operator (not the Lagrangian) defined by

$$\mathcal{L} = 2^{1/2}Y_1^0(a_1^0 + a_{-1}^0) + Y_1^1(a_1^1 - a_{-1}^1) + Y_1^{-1}(a_1^{-1} - a_{-1}^{-1}), \quad (3.50)$$

where the arguments of Y_l^m are χ and ω; and a_μ^ν ($\mu = \pm 1$; $\nu = 0, \pm 1$) are *creation* and *annihilation operators* which operate on f_l as

$$a_\mu^\nu f_l = f_{l+\mu} , \tag{3.51}$$

and on K_l^m and f_l^m as

$$a_\mu^0 f_l^m = A_{l+(1/2)(\mu-1)}^{|m|} f_{l+\mu}^m ,$$
$$a_\mu^\nu f_l^m = [2h(\nu m) - 1] E_{l+(1/2)(\mu-1)}^{-\mu\nu[m-(\nu/2)(\mu-1)]} f_{l+\mu}^{m+\nu} \qquad (\nu \neq 0) \tag{3.52}$$

with h being a unit step function such that $h(x) = 1$ for $x \geq 0$ and $h(x) = 0$ for $x < 0$, and with

$$A_l^m = \left[\frac{(l+m+1)(l-m+1)}{(2l+1)(2l+3)}\right]^{1/2} ,$$
$$E_l^m = \left[\frac{(l-m+1)(l-m+2)}{(2l+1)(2l+3)}\right]^{1/2} . \tag{3.53}$$

The solution for K_l^m may then be expressed as

$$K_l^m = \sum_{n=0}^{\infty} (i\bar{k})^n (f_l * \mathcal{L})^n f_l^m . \tag{3.54}$$

In particular, integration of Eq. (3.43) with Eq. (3.54) over $\mathbf{u}$ leads to

$$I(\mathbf{k} \,|\, \mathbf{u}_0; L) = (4\pi)^{1/2} \sum_{n=0}^{\infty} (i\bar{k})^n (f_0 * \mathcal{L})^n f_0^0 , \tag{3.55}$$

$$I(\mathbf{k}; L) = (4\pi)^{1/2} \sum_{m=0}^{\infty} (-1)^m \bar{k}^{2m} (f_0 * \mathcal{L})^{2m} f_0^0 , \tag{3.56}$$

where f_0^0 in Eqs. (3.55) and (3.56) are given by the first and second lines of Eqs. (3.49), respectively. Equations (3.43), (3.55), and (3.56) are the desired results for I. By Fourier inversion of them, the corresponding distribution functions G may in principle be obtained. As seen later, however, the results are very complicated. This arises from the constraint of $\mathbf{u}^2 = 1$, and therefore various modifications of the KP chain were presented by relaxing this constraint. They are briefly discussed in Appendix 3.C. Naturally, however, they do not give all of the exact moments $\langle R^2 \rangle$ and $\langle S^2 \rangle$ and also those derived in the next section for the KP chain, and thus we do not pursue them further in later chapters.

3.3 Moments

By the continuous limiting process from the freely rotating chain, the moment $\langle R^4 \rangle$ for the KP chain may be evaluated as well as $\langle \mathbf{R} \cdot \mathbf{u}_0 \rangle$ and $\langle R^2 \rangle$ [2]. However, the evaluation of higher moments becomes extremely difficult. Thus, from Eq. (3.29), Hermans and Ullman [6] derived a recurrence formula for $\langle (\mathbf{R} \cdot \mathbf{u}_0)^{k-m} R^{2m} \rangle$, from which the first three of $\langle R^{2m} \rangle$ were readily obtained [6, 18]. With this formula, however, the analytical evaluation still becomes extremely laborious as m is increased, and therefore Nagai [19] evaluated from it numerically $\langle R^{2m} \rangle$ as a function of L for $m \leq 20$ by the use of a computer. On the other hand, from Eqs. (3.55) and (3.56), we can derive formal expressions for the moments [15, 16], from which they can be operationally and hence more efficiently evaluated by the use of a computer. This *operational method* may, of course, also be used for the evaluation of the distribution functions from Eqs. (3.55) and (3.56) or their alternatives, as done in the next section.

3.3.1 $\langle (\mathbf{R} \cdot \mathbf{u}_0)^n \rangle$

The characteristic function $I(\mathbf{k} \mid \mathbf{u}_0; L)$ may be expanded in terms of the moments $\langle (\mathbf{R} \cdot \mathbf{e}_k)^n \rangle$ as in Eq. (4.13) of MTPS [20]; that is,

$$I(\mathbf{k} \mid \mathbf{u}_0; L) = \sum_{n=0}^{\infty} \frac{1}{n!} \langle (\mathbf{R} \cdot \mathbf{e}_k)^n \rangle (ik)^n , \tag{3.57}$$

where

$$\langle (\mathbf{R} \cdot \mathbf{e}_k)^n \rangle = \int (\mathbf{R} \cdot \mathbf{e}_k)^n G(\mathbf{R} \mid \mathbf{u}_0; L) d\mathbf{R} . \tag{3.58}$$

Thus, in order to obtain the equivalent expansion from Eq. (3.55), we expand $(f_0 * \mathcal{L})^n f_0$ to have

$$I(\mathbf{k} \mid \mathbf{u}_0; L) = \sum_{n=0}^{\infty} (ik)^n \sum_{q \leq n} (2q+1)^{1/2} \sum_{\substack{\text{paths} \\ (0 \to q)}} \sum_{\nu} (-1)^x$$

$$\times 2^{-(n-n^0)} C_\mu^\nu \, \Gamma_{0 \cdots q}(L) \cos^{n^0} \chi \, \sin^{(n-n^0)} \chi , \tag{3.59}$$

where

$$C_\mu^\nu = (f_q^0)^{-1} a_{\mu_n}^{\nu_n} a_{\mu_{n-1}}^{\nu_{n-1}} \cdots a_{\mu_2}^{\nu_2} a_{\mu_1}^{\nu_1} f_0^0 , \tag{3.60}$$

$$\Gamma_{0 \cdots q}(L) = f_0 * f_1 * f_{l_2} * \cdots * f_{l_{n-1}} * f_q , \tag{3.61}$$

$$l_j = \sum_{i=1}^{j} \mu_i \geq 0 \qquad (l_0 = 0, \ l_1 = \mu_1 = 1, \ l_n = q) , \tag{3.62}$$

$$\sum_{i=1}^{n} \nu_i = 0 . \tag{3.63}$$

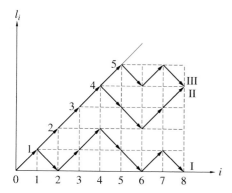

Fig. 3.3. Stone-fence diagram. One $(0 \to 0)$ path I (010121010) and two $(0 \to 4)$ paths II (012343234) and III (012345454) for $n = 8$ are depicted as examples, I being also a $(0 \to 0)$ one for $m = 4$

In Eq. (3.59), n^0 and x are the numbers of a_μ^0 and $a_{-1}^\nu (\nu \neq 0)$ in C_μ^ν, respectively, the third sum is taken over all possible paths $(01l_2 \cdots l_{n-1}q)$ from 0 to q, and the fourth sum is taken over $\nu_1, \cdots, \nu_n$ compatible with Eq. (3.63). If n is even, n^0 is even and $q = 0, 2, 4, \cdots, n$; and if n is odd, n^0 is odd and $q = 1, 3, 5, \cdots, n$. Note that Eq. (3.63) must hold because of the first of Eqs. (3.49) and that C_μ^ν is a constant independent of L. Note also that if one of the paths is given, the corresponding values of $\mu_1, \cdots, \mu_n$ are determined uniquely since $\mu_j = l_j - l_{j-1}$, so that Eq. (3.59) does not involve the sum over μ. The paths $(01l_2 \cdots l_{n-1}q)$ may be conveniently represented by *stone-fence diagrams* in an (i, l_i)-plane, as shown in Fig. 3.3, where one $(0 \to 0)$ path I (010121010) and two $(0 \to 4)$ paths II (012343234) and III (012345454) for $n = 8$ are depicted as examples. The diagram is equivalent to the two-dimensional representation of one-dimensional random walks with a reflecting barrier at the origin [4].

Now, if we choose $\mathbf{k}$ to be in the direction of $\mathbf{u}_0$, that is, $\chi = 0$, then we obtain the expansion of Eq. (3.57) for $I(\mathbf{k}|\mathbf{u}_0; L)$ in terms of $\langle (\mathbf{R} \cdot \mathbf{u}_0)^n \rangle$ with

$$\langle (\mathbf{R} \cdot \mathbf{u}_0)^n \rangle = n! \sum_{q \leq n} (2q+1)^{1/2} \sum_{\substack{\text{paths} \\ (0 \to q)}} C_\mu^0 \, \Gamma_{0 \cdots q}(L), \qquad (3.64)$$

where only the paths from 0 to q with $\nu_i = 0$ for all i make contribution since $n = n^0$. We note that if $p_i(L)$ are the residues of the function $Q(p)$,

$$Q(p) = \frac{e^{Lp}}{p \displaystyle\prod_{j=1}^{n} [p + l_j(l_j + 1)]}, \qquad (3.65)$$

then $\Gamma_{0 \cdots q}(L)$ may be expressed as

$$\Gamma_{0 \cdots q}(L) = \sum_i p_i(L). \qquad (3.66)$$

Equation (3.64) with Eq. (3.66) is the desired formal expression for $\langle (\mathbf{R} \cdot \mathbf{u}_0)^n \rangle$.

3.3.2 $\langle R^{2m} \rangle$

The characteristic function $I(\mathbf{k}; L)$ may be expanded in terms of the moments $\langle R^{2m} \rangle$ as in Eq. (4.17) of MTPS; that is,

$$I(\mathbf{k}; L) = \sum_{m=0}^{\infty} \frac{(-1)^m}{(2m+1)!} \langle R^{2m} \rangle k^{2m} , \qquad (3.67)$$

where

$$\langle R^{2m} \rangle = \int R^{2m} G(\mathbf{R}; L) d\mathbf{R} . \qquad (3.68)$$

By expanding $(f_0 * \mathcal{L})^{2m} f_0^0$ as in Eq. (3.59), Eq. (3.56) may be reduced to Eq. (3.67) with

$$\langle R^{2m} \rangle = (2m+1)! \sum_{\substack{\text{paths} \\ (0 \to 0)}} C_\mu^0 \, \Gamma_{0 \cdots 0}(L) , \qquad (3.69)$$

where only the paths from 0 to 0 with $\nu_i = 0$ for all i, the number of which is equal to $(2m)!/m!(m+1)!$, make contribution because of the second of Eqs. (3.49), and C_μ^0 is explicitly given by

$$C_\mu^0 = \prod_{j=1}^{2m} A_{l_j - 1 + (1/2)(\mu_j - 1)}^0 \qquad (3.70)$$

with $\mu_1 = 1$ and $\mu_{2m} = -1$. In Fig. 3.3, the path I is a $(0 \to 0)$ one for $m = 4$. If x_j is the number of the factors with $l_i = j$ in the denominator of Eq. (3.65), the formula for residues gives

$$\Gamma_{0 \cdots 0}(L) = \sum_{\substack{j=0 \\ x_j \neq 0}}^{m} \frac{1}{(x_j - 1)!} \left\{ \frac{d^{x_j - 1}}{dp^{x_j - 1}} \frac{\left[p + j(j+1) \right]^{x_j} e^{Lp}}{\prod_{i=0}^{m} \left[p + i(i+1) \right]^{x_j}} \right\}_{p = -j(j+1)} . \qquad (3.71)$$

Thus $\langle R^{2m} \rangle$ may be finally written in the form

$$\langle R^{2m} \rangle = \sum_{j=0}^{m} \sum_{i=j}^{m} A_{ij}^{(m)} L^{i-j} \exp\left[-j(j+1)L \right] , \qquad (3.72)$$

where $A_{ij}^{(m)}$ are numerical coefficients independent of L and may be expressed in terms of C_μ^0 and x_j, the result being omitted. For $m = 1 - 3$, Eq. (3.72) gives the results mentioned above [1, 2, 6, 18].

The coefficients $A_{ij}^{(m)}$ for $m \leq 11$ may be evaluated efficiently by the use of a computer; it consists of generating all possible $(0 \to 0)$ paths and counting the number x_j [21]. Their values thus obtained for $m \leq 5$ are given as fractional numbers in Appendix I. Nagai's results for $m \leq 6$ [19] are in good agreement with the exact values from Eq. (3.72) (to order 10^{-5}) for all values of L, but those for higher m are much less accurate, especially at small L.

3.4 Distribution Functions

3.4.1 Asymptotic Behavior – Daniels-Type Distributions

The correction to the Gaussian distribution G for large L may be expanded in inverse powers of L to give the so-called *Daniels-type distribution function*. Daniels [7] first solved the differential equation for $G(\mathbf{R} \,|\, \mathbf{u}_0; L)$ to derive its asymptotic expansion of this kind to terms of $\mathcal{O}(L^{-3/2})$. Many years later, Gobush and co-workers [22] attempted to solve the differential equation for the Laplace transform $\tilde{I}(\mathbf{k}, \mathbf{u} \,|\, \mathbf{u}_0; p)$ of $I(\mathbf{k}, \mathbf{u} \,|\, \mathbf{u}_0; L)$ with respect to L by an application of operational techniques developed by Prigogine and co-workers [23, 24] in attacking the Liouville equation, and obtained the expansion of $G(\mathbf{R}, \mathbf{u} \,|\, \mathbf{u}_0; L)$ to terms of $\mathcal{O}(L^{-2})$.

We begin by reformulating the Gobush expansion along the line of the preceding sections. This is convenient for a comparison with the expansion of Eq. (3.43) with Eq. (3.54) and also facilitates computer calculations of the expansion coefficients of the distribution function. We again choose $\mathbf{u}_0$ to be in the direction of the z axis of a Cartesian coordinate system ($\mathbf{u}_0 = \mathbf{e}_z$) and express $\mathbf{u}$ and $\mathbf{k}$ as $\mathbf{u} = (1, \theta, \phi)$ and $\mathbf{k} = (k, \chi, \phi - \psi)$, respectively, in spherical polar coordinates. (All lengths are still measured in units of λ^{-1}.) We define an operator $\mathcal{L}$ by

$$\mathcal{L} = \mathcal{L}_0 + \delta\mathcal{L}, \tag{3.73}$$

where $\mathcal{L}_0 = \nabla_u^2$ and $\delta\mathcal{L}$ is given by

$$\delta\mathcal{L} = i\mathbf{k} \cdot \mathbf{u} \equiv ik\delta\bar{\mathcal{L}}, \tag{3.74}$$

so that the left-hand side of Eq. (3.32) becomes $(\partial/\partial L - \mathcal{L})I$. The solution for the Laplace transform $\tilde{I}(p)$,

$$\tilde{I}(\mathbf{k}, \mathbf{u} \,|\, \mathbf{u}_0; p) = \int_0^\infty I(\mathbf{k}, \mathbf{u} \,|\, \mathbf{u}_0; L) \exp(-pL) dL, \tag{3.75}$$

may then be expanded as follows [22],

$$\tilde{I}(p) = \sum_{n=0}^\infty \sum_{l=0}^\infty (ik)^n \left(\frac{2l+1}{4\pi} \right)^{1/2} Q(\delta\bar{\mathcal{L}}Q)^n Y_l^0(\theta, \phi), \tag{3.76}$$

where

$$Q = -(\mathcal{L}_0 - p)^{-1}. \tag{3.77}$$

In the present notation, the operator $\delta\bar{\mathcal{L}}$ may be written as

$$\delta\bar{\mathcal{L}} = (\cos\chi)(a_1^0 + a_{-1}^0) + \frac{1}{2}(\sin\chi)\left[(a_1^1 - a_{-1}^1) + (a_1^{-1} - a_{-1}^{-1})\right]. \tag{3.78}$$

a_μ^ν operate on f_l and g_l^m (instead of f_l^m) in the same way as in Eqs. (3.51) and (3.52), where g_l^m is defined by

$$g_l^m = f_l Y_l^m(\theta, \psi). \tag{3.79}$$

Then Laplace inversion of Eq. (3.76) leads to

$$I(\mathbf{k}, \mathbf{u} \,|\, \mathbf{u}_0; L) = \sum_{n=0}^{\infty} \sum_{l=0}^{\infty} (ik)^n \left(\frac{2l+1}{4\pi}\right)^{1/2} (f_l * \delta\bar{\mathcal{L}})^n g_l^0. \tag{3.80}$$

The structure of the cascade of the successive a_μ^ν operations involved in Eq. (3.80) seems different from that in Eq. (3.43) with Eq. (3.54), but both are equivalent. This may be observed more explicitly by considering $I(\mathbf{k} \,|\, \mathbf{u}_0; L)$. From Eq. (3.80), we have

$$I(\mathbf{k} \,|\, \mathbf{u}_0; L) = \sum_{n=0}^{\infty} \sum_{l=0}^{\infty} (ik)^n (2l+1)^{1/2} \int g_0^0 (f_l * \delta\bar{\mathcal{L}})^n g_l^0 d\mathbf{u}. \tag{3.81}$$

It can be seen that if the integrand of Eq. (3.81) is expanded, there is one-to-one correspondence between the terms in Eqs. (3.59) and (3.81); each path of the stone-fence diagram in the latter is just the reversal of the corresponding $(0 \to l)$ path in the former.

Similarly, Eq. (3.76) gives

$$I(\mathbf{k}; L) = \sum_{m=0}^{\infty} (-1)^m k^{2m} (g_0^0)^{-1} (f_0 * \delta\bar{\mathcal{L}})^{2m} g_0^0, \tag{3.82}$$

where only the paths from 0 to 0 with $\nu_i = 0$ for all i make contribution as in Eq. (3.67) with Eq. (3.69) if $(f_0 * \delta\bar{\mathcal{L}})^{2m} g_0^0$ is expanded. Of course, Eqs. (3.67) [or (3.56)] and (3.82) are equivalent.

Thus, by Fourier inversion of any of these I, we can obtain the corresponding Daniels-type distribution function G. However, it is convenient to obtain $G(\mathbf{R}, \mathbf{u} \,|\, \mathbf{u}_0; L)$ by inversion of Eq. (3.80), from which the other G may readily be derived. If we express $\mathbf{R}$ and $\mathbf{u}$ as $\mathbf{R} = (R, \Theta, \Phi)$ and $\mathbf{u} = (1, \theta, \varphi + \Phi)$, respectively, in spherical polar coordinates (with $\mathbf{u}_0 = \mathbf{e}_z$), the result may be written in the form [22]

$$G(\mathbf{R}, \mathbf{u} \,|\, \mathbf{u}_0; L) = (4\pi)^{-1/2} \left(\frac{3}{2\pi L}\right)^{3/2} \exp\left(-\frac{3R^2}{2L}\right)$$

$$\times \sum_{l=0}^{\infty} \sum_{m=-l}^{l} Y_l^m(\theta, \varphi) F_l^{|m|}(R, \Theta). \tag{3.83}$$

An approximation such that terms of $F_l^{|m|}$ are retained to $\mathcal{O}(L^{-s})$ is referred to as the *sth Daniels approximation* with s being a positive integer. [Note that R^2 is of $\mathcal{O}(L)$.] The second Daniels approximation to F_0^0 is given by

$$F_0^0(R, \Theta) = F(R) + \left(\frac{3R}{2L} - \frac{25R}{16L^2} + \frac{153R^3}{40L^3} - \frac{99R^5}{80L^4} \right) P_1(\cos \Theta)$$

$$+ \left(\frac{R^2}{2L^2} - \frac{67R^2}{60L^3} + \frac{961R^4}{560L^4} - \frac{33R^6}{80L^5} \right) P_2(\cos \Theta)$$

$$+ \frac{3R^3}{40L^3} P_3(\cos \Theta) + \frac{9R^4}{1400L^4} P_4(\cos \Theta) + \cdots , \qquad (3.84)$$

where P_l is the Legendre polynomial (see Appendix 3.B) and $F(R)$ is given by

$$F(R) = 1 - \frac{5}{8L} + \frac{2R^2}{L^2} - \frac{33R^4}{40L^3} - \frac{79}{640L^2} - \frac{329R^2}{240L^3} + \frac{6799R^4}{1600L^4}$$

$$- \frac{3441R^6}{1400L^5} + \frac{1089R^8}{3200L^6} . \qquad (3.85)$$

Note that F_0^0 is the function given by Daniels [7] (with oversight of the P_3 term) to $\mathcal{O}(L^{-3/2})$.

Although the other $F_l^{|m|}$ have also been obtained in the second Daniels approximation [22], we give them in the first approximation, for simplicity:

$$F_1^0(R, \Theta) = \frac{\sqrt{3}}{2} \left[-\frac{1}{2L} + \frac{R^2}{2L^2} - \frac{R}{L} P_1(\cos \Theta) + \frac{R^2}{L^2} P_2(\cos \Theta) + \cdots \right],$$

$$F_1^1(R, \Theta) = \frac{\sqrt{6}}{4} R \sin \Theta \left[\frac{1}{L} + \frac{3R}{2L^2} P_1(\cos \Theta) + \cdots \right],$$

$$F_2^0(R, \Theta) = \frac{\sqrt{5}}{10} \frac{R^2}{L^2} P_2(\cos \Theta) + \cdots , \qquad (3.86)$$

$$F_2^1(R, \Theta) = \frac{\sqrt{30}}{20} \frac{R^2}{L^2} \sin \Theta P_1(\cos \Theta) + \cdots ,$$

$$F_2^2(R, \Theta) = \frac{\sqrt{30}}{60} \frac{R^2}{L^2} [1 - P_2(\cos \Theta) + \cdots] .$$

(We note that in the original paper [22] the exponent 2 of L in the second term of F_1^0 is missing and the coefficient $298/105$ in F_2^2 should be replaced by $19/15$.)

From Eq. (3.83), we obtain for the other distribution functions

$$G(\mathbf{R} \,|\, \mathbf{u}_0; L) = \left(\frac{3}{2\pi L} \right)^{3/2} \exp\left(-\frac{3R^2}{2L} \right) F_0^0(R, \Theta), \qquad (3.87)$$

$$G(\mathbf{R}; L) = \left(\frac{3}{2\pi L} \right)^{3/2} \exp\left(-\frac{3R^2}{2L} \right) F(R). \qquad (3.88)$$

In the sth Daniels approximation, $F(R)$ may in general be written in the form

$$F(R) = 1 + \sum_{i=1}^{s} \sum_{j=0}^{2i} \frac{C_{ji}}{L^i} \left(\frac{R^2}{L} \right)^j , \qquad (3.89)$$

where C_{ji} are numerical coefficients independent of R and L. These coefficients have been evaluated for $s \leq 10$ by the use of a computer, generating necessary paths [25], but the results are not reproduced. However, note that C_{ji} for $j \leq 4$ and $i \leq 2$ have already appeared in Eq. (3.85).

We readily obtain the *ring-closure probability* $G(\mathbf{0}; L)$ from Eq. (3.88) and can also evaluate the moment $\langle R^{-1} \rangle$ (mean reciprocal of the end-to-end distance) by the use of the same equation, both in the sth Daniels approximation, as follows,

$$G(\mathbf{0}; L) = \left(\frac{3}{2\pi L} \right)^{3/2} F(0)$$

$$= \left(\frac{3}{2\pi L} \right)^{3/2} \left(1 + \sum_{i=1}^{s} \frac{C_{0i}}{L^i} \right) , \qquad (3.90)$$

$$\langle R^{-1} \rangle = \left(\frac{6}{\pi L} \right)^{1/2} \left[1 + \sum_{i=1}^{s} \sum_{j=0}^{2i} j! \left(\frac{2}{3} \right)^j \frac{C_{ji}}{L^i} \right] . \qquad (3.91)$$

These quantities also serve to examine the convergence of the Daniels approximation.

Now, in general, the correction to the Gaussian distribution $G(\mathbf{R}; L)$ for $L \to \infty$ may also be expanded in terms of the moments $\langle R^{2m} \rangle$ [26], or of Hermite polynomials [27,28]. This gives the so-called *moment-based distribution function* of $\mathbf{R}$. Its sth approximation involves the moments $\langle R^{2m} \rangle$ with $m \leq s$, and its convergence may also be examined by the use of $G(\mathbf{0}; L)$ and $\langle R^{-1} \rangle$ obtained from this $G(\mathbf{R}; L)$ with the moments given by Eq. (3.72), although the explicit expressions for them are not reproduced. Figures 3.4 and 3.5 show plots of $G(\mathbf{0}; L)$ against L in the sth Daniels approximations

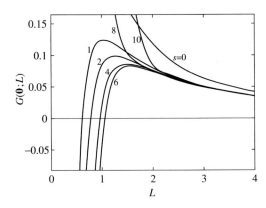

Fig. 3.4. Ring-closure probability $G(\mathbf{0}; L)$ plotted against (reduced) L for the KP chain in the sth Daniels approximations

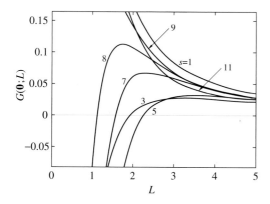

Fig. 3.5. Ring-closure probability $G(\mathbf{0}; L)$ plotted against
(reduced) L for the KP chain
in the sth Hermite polynomial
approximations

with $s \leq 10$ and in the sth Hermite polynomial approximations with $s \leq 11$,
respectively. Figure 3.6 shows plots of $L\langle R^{-1}\rangle$ against the degree s of approximation in the Daniels and Hermite polynomial approximations for the
indicated values of L. It is seen that the Daniels approximation is convergent
for $L \gtrsim 3$, while the convergence of the Hermite polynomial approximation is
much worse, it being convergent only for $L \gtrsim 10$. However, this is not always
the case with the HW chain, and the moment-based distribution functions
are considered in detail in the next chapter. Of course, both approximations
are divergent near the rod limit of $L \to 0$, and this region must be treated in
a different way.

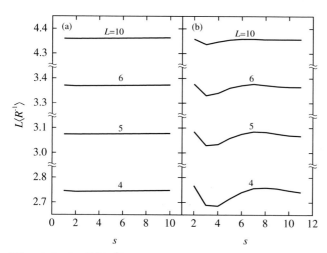

Fig. 3.6a,b. $L\langle R^{-1}\rangle$ plotted against the degree s of approximation for the KP
chain: **a** in the Daniels; **b** and Hermite polynomial approximations for the indicated
values of (reduced) L

3.4.2 Near the Rod Limit

A possible method of obtaining the distribution functions valid near the rod limit of $L \to 0$ may be the use of the WKB approximation, in which the Schrödinger Eq. (3.32) is solved in the "classical" limit of $k \to \infty$ (corresponding to the limit of $\hbar \to 0$). If we let k approach infinity and suppress the Laplacian in Eq. (3.32), it gives the distribution function for the rod that is a delta function. If we adopt the path integral approach, this limit is obtained from the "classical" path $\bar{u}(s)$ which satisfies the extremum condition

$$\delta \int_0^L \mathcal{L} ds = 0 \,, \tag{3.92}$$

where $\mathcal{L}$ is the Lagrangian given by Eq. (3.35) with Eq. (3.36) (see Appendix 3.A). Then the WKB approximation consists of taking into account the deviation of the "potential energy" part $-\mathbf{k} \cdot \mathbf{u}$ of $\mathcal{L}$ from its classical value to second order [8]. Thus Eq. (3.34) is reduced to the form

$$I(\mathbf{k}, \mathbf{u} \,|\, \mathbf{u}_0; L) = f(\mathbf{k}, L) \exp\left(i \int_0^L \bar{\mathcal{L}} ds \right) , \tag{3.93}$$

where $\bar{\mathcal{L}}$ is the classical value of $\mathcal{L}$ with $\mathbf{u} = \bar{\mathbf{u}}$ and $f(\mathbf{k}, L)$ is the normalization factor.

The result thus found by Fourier inversion of Eq. (3.93) is [29]

$$G(\mathbf{R}, \mathbf{u} \,|\, \mathbf{u}_0; L) = \left(\frac{3}{4\pi^2 L^4} \right) \left(1 + \frac{2}{3} L \right) \left(\frac{45}{4\pi L^3 \theta^2} \right)^{1/2}$$

$$\times \exp\left[-\frac{\theta^2}{4L} - \frac{3}{L^3} \left(x - \frac{1}{2} L\theta \cos\phi \right)^2 - \frac{3}{L^3} \left(y - \frac{1}{2} L\theta \sin\phi \right)^2 \right.$$

$$\left. - \frac{45}{4L^3\theta^2} \left(z - L + \frac{1}{6} L\theta^2 \right)^2 \right] \left[1 - \frac{15}{2L\theta^2} \left(z - L + \frac{1}{6} L\theta^2 \right) + \cdots \right] \tag{3.94}$$

(in units of λ^{-1}), where $\mathbf{R} = (x, y, z)$ in Cartesian coordinates and $\mathbf{u} = (1, \theta, \phi)$ in spherical polar coordinates with $\mathbf{u}_0 = \mathbf{e}_z$. It is impossible to integrate this G over $\mathbf{u}$ and $\mathbf{u}_0$ to obtain analytical expressions for $G(\mathbf{R} \,|\, \mathbf{u}_0; L)$ and $G(\mathbf{R}; L)$. However, it is seen that Eq. (3.94) gives

$$\lim_{L \to 0} G(\mathbf{R} \,|\, \mathbf{u}_0; L) = \delta(x)\delta(y)\delta(z - L) \,. \tag{3.95}$$

This is the distribution function of $\mathbf{R}$ for the rigid rod of length L oriented in the direction of the z axis.

The distribution function given by Eq. (3.94) yields the correct first-order corrections to the rod limits of all the moments,

$$\left\langle (\mathbf{R} \cdot \mathbf{u}_0)^n \right\rangle = \langle z^n \rangle = L^n \left[1 - nL + \mathcal{O}(L^2) \right],$$

$$\langle R^{2m} \rangle = L^{2m} \left[1 - \frac{2}{3} mL + \mathcal{O}(L^2) \right], \tag{3.96}$$

$$\langle R^{-1} \rangle = \frac{1}{L} \left[1 + \frac{1}{3} L + \mathcal{O}(L^2) \right].$$

Note that the third of Eqs. (3.96) was first obtained by Hearst and Stockmayer [30] by a different method.

If we confine ourselves to $G(\mathbf{R}; L)$, higher-order approximations, although formal, may be easily obtained, as done by Norisuye and co-workers [31]. Substitution of the expansion of $\langle R^{2m} \rangle$ in powers of L into Eq. (3.67) and summation leads to

$$I(\mathbf{k}; L) = j_0(z) + \frac{z}{3} j_1(z) L + \frac{z}{90} \left[6j_1(z) - 7zj_0(z) \right] L^2$$

$$+ \frac{z}{1890} \left[(24 - 31z^2)j_1(z) + 34zj_0(z) \right] L^3$$

$$+ \frac{z}{37800} \left[(212z^2 + 120)j_1(z) + z(127z^2 - 320)j_0(z) \right] L^4$$

$$+ \frac{z}{3742200} \left[(2555z^4 - 8136z^2 + 4320)j_1(z) \right.$$

$$\left. - 4z(3053z^2 - 1620)j_0(z) \right] L^5 + \cdots \tag{3.97}$$

with

$$z = Lk, \tag{3.98}$$

where $j_l(z)$ is the spherical Bessel function of the first kind. By Fourier inversion of Eq. (3.97), we obtain

$$G(\mathbf{R}; L) = \frac{h_0(L)}{4\pi L^2} \delta(R - L) + \frac{1}{4\pi R} \sum_{n=1}^{\infty} h_n(L) \delta^{(n)}(R - L), \tag{3.99}$$

where

$$h_0 = 1 + \frac{L}{3} + \frac{L^2}{15} + \frac{4L^3}{315} + \frac{L^4}{315} + \frac{4L^5}{3465} + \cdots,$$

$$h_1 = h_0 - 1,$$

$$h_2 = \frac{7L^3}{90} \left(1 - \frac{L}{49} + \frac{9L^2}{245} + \frac{46L^3}{8085} + \cdots \right),$$

$$h_3 = \frac{31L^5}{1890} \left(1 - \frac{53L}{155} + \frac{226L^2}{1705} + \cdots \right), \tag{3.100}$$

$$h_4 = \frac{127L^7}{37800} \left(1 - \frac{1073L}{1397} + \cdots \right),$$

$$h_5 = \frac{73L^9}{106920} + \cdots .$$

Note that the nth derivative of the delta function, $\delta^{(n)}(x) = d^n\delta(x)/dx^n$, is defined by

$$\int_{-\infty}^{\infty} f(x)\delta^{(n)}(x)dx = (-1)^n f^{(n)}(0)\,, \qquad (3.101)$$

where $f(x)$ is a function whose nth derivative is continuous.

In the limit of $L \to 0$, Eq. (3.99) reduces to

$$\lim_{L\to 0} G(\mathbf{R}; L) = \frac{1}{4\pi L^2}\delta(R - L)\,. \qquad (3.102)$$

This is the distribution function of $\mathbf{R}$ for the rigid rod of length L without orientation. By the use of Eq. (3.99), $\langle R^{-1}\rangle$ is evaluated to be

$$\langle R^{-1}\rangle = \frac{h_0(L)}{L}\,. \qquad (3.103)$$

It is interesting to note that Eq. (3.103) is formally obtained from the expansion of $\langle R^{2m}\rangle$ in powers of L by putting $m = -1/2$.

Now it is evident that neither Eq. (3.94) nor Eq. (3.99) can yet give the correct ring-closure probability $G(\mathbf{0}; L)$, although the rod limits of the moments and the corrections to them have been correctly evaluated. The evaluation of $G(\mathbf{0}; L)$ still requires a different approach, which is considered in relation to the problems of circular DNA in Chap. 7. Further, in the next chapter there are presented a simple and more powerful method for evaluating the moments and characteristic function of the distribution function $G(\mathbf{R}; L)$ [or generally $G(\mathbf{R}, \Omega \,|\, \Omega_0; L)$] near the rod limit for any model and also a method of interpolation from this limit and the Daniels approximation. We also note that we can construct approximately distribution functions for KP wormlike rings since we have derived the two expansions of those for linear chains from the random-coil and rod limits. The results are given in Appendix 3.D.

Appendix 3.A Path Integrals

In this appendix we give a short sketch of the path integral formalism following Feynman and Hibbs [8]. Consider a particle of mass m in a potential $V(\mathbf{R}, t)$ as a function of its position $\mathbf{R}$ and time t. The Green function $G(\mathbf{R}_2, t_2 \,|\, \mathbf{R}_1, t_1)$ of the Schrödinger equation satisfies the differential equation

$$\left(\frac{\partial}{\partial t_2} - \frac{i\hbar}{2m}\nabla_2^2 + \frac{i}{\hbar}V\right)G(\mathbf{R}_2, t_2 \,|\, \mathbf{R}_1, t_1) = \delta(t_2 - t_1)\delta(\mathbf{R}_2 - \mathbf{R}_1)\,, \quad (3.A.1)$$

where ∇_2^2 is the Laplacian operator with respect to $\mathbf{R}_2$ and $V = V(\mathbf{R}_2, t_2)$. The function G is the quantum-mechanical amplitude or kernel, and $|G|^2$ is the probability density that the particle which was at $\mathbf{R}_1$ at time t_1 arrives at $\mathbf{R}_2$ at time t_2.

For simplicity, we consider the Green function $G(x_b, t_b \,|\, x_a, t_a) \equiv G(b\,|\,a)$ for the quantal motion from a to b (instead of from 1 to 2) in one dimension. The coordinate $x(t)$ of the particle is a function of t with the boundary conditions

$$x(t_a) = x_a, \qquad x(t_b) = x_b. \tag{3.A.2}$$

The function $x(t)$ may be represented by a curve in an (x, t)-plane, and it is called the *path* $x(t)$ from a to b.

In classical mechanics, the classical path $\bar{x}(t)$ is only possible path and is determined by the minimization of the action integral S (the principle of least action),

$$\delta S = 0 \tag{3.A.3}$$

subject to $\delta x(t_a) = \delta x(t_b) = 0$, where

$$S = \int_{t_a}^{t_b} \mathcal{L}(\dot{x}, x, t)dt \tag{3.A.4}$$

with $\mathcal{L}$ the Lagrangian,

$$\mathcal{L} = \frac{1}{2}m\dot{x}^2 - V(x, t), \tag{3.A.5}$$

the over dot indicating the derivative with respect to t as usual. Equation (3.A.3) with Eqs. (3.A.4) and (3.A.5) (the variational principle) gives the Lagrange equation of motion (Euler's equation in mathematics), and $\bar{x}(t)$ is its solution with the boundary conditions given by Eqs. (3.A.2).

In quantal motions, various paths other than $\bar{x}(t)$ may be realized. Then we postulate that $G(b\,|\,a)$ is given by

$$G(b\,|\,a) = \text{const.} \sum_{\substack{\text{all paths} \\ (a \to b)}} \exp\left\{ \frac{i}{\hbar} S[x(t)] \right\}, \tag{3.A.6}$$

where $S[x(t)]$ indicates that S is a functional of $x(t)$. The sum in Eq. (3.A.6) may be reduced to the path integral (functional integral) form. Divide the interval $[t_a, t_b]$ into N intervals of width ϵ with $x(t_j) = x_j$ ($j = 0, 1, \cdots, N$; $t_0 = t_a$, $t_N = t_b$, $x_0 = x_a$, $x_N = x_b$), as shown in Fig. 3.A.1. Equation (3.A.6) may then be rewritten as

$$G(b\,|\,a) = \lim_{\epsilon \to 0} C^{-N} \int \exp\left[\frac{i}{\hbar} S(b\,|\,a) \right] \prod_{j=1}^{N-1} dx_j, \tag{3.A.7}$$

where C^{-N} is the normalization constant and $S(b\,|\,a) = S[x(t)]$. As shown later, the limit in Eq. (3.A.7) exists if C is chosen to be

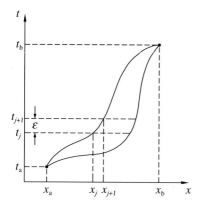

Fig. 3.A.1. Paths $x(t)$ from a to b in the one-dimensional case

$$C = \left(\frac{2\pi i \hbar \epsilon}{m}\right)^{1/2}. \tag{3.A.8}$$

For convenience, we write Eq. (3.A.7) as

$$G(b\,|\,a) = \int_{x(t_a)=x_a}^{x(t_b)=x_b} \exp\left[\frac{i}{\hbar}S(b\,|\,a)\right]\mathcal{D}[x(t)]. \tag{3.A.9}$$

This is the path integral representation of G.

Now the problem is to show that the G given by Eq. (3.A.9) satisfies the Schrödinger equation. From Eq. (3.A.4), we have

$$S(b\,|\,a) = S(b\,|\,c) + S(c\,|\,a), \tag{3.A.10}$$

so that $G(b\,|\,a)$ may be expressed as a convolution integral,

$$G(b\,|\,a) = \int_{-\infty}^{\infty} G(b\,|\,c)G(c\,|\,a)dx_c. \tag{3.A.11}$$

Continuing this process, we arrive at the expression

$$G(b\,|\,a) = \int G(b\,|\,N-1)G(N-1\,|\,N-2)\cdots G(1\,|\,a)\prod_{j=1}^{N-1} dx_j. \tag{3.A.12}$$

By a comparison of Eq. (3.A.7) with Eq. (3.A.12), we have

$$G(j+1\,|\,j) = C^{-1}\exp\left[\frac{i\epsilon}{\hbar}\mathcal{L}\left(\frac{\Delta x_j}{\epsilon}, x_j, t_j\right)\right] \tag{3.A.13}$$

with $\Delta x_j = x_{j+1} - x_j$. Since G is the wave function, we put $G(x,t\,|\,x-\Delta x, t-\epsilon) = \psi(x,t)$ to have, from Eq. (3.A.11),

$$\psi(x, t+\epsilon) = \int_{-\infty}^{\infty} \psi(x-\Delta x, t)G(x\,|\,x-\Delta x; \epsilon)d(\Delta x), \tag{3.A.14}$$

where G is given, from Eq. (3.A.13) with Eq. (3.A.5), by

$$G(x \mid x - \Delta x; \epsilon) = C^{-1} \exp\left[\frac{im(\Delta x)^2}{2\hbar\epsilon}\right]\left[1 - \frac{i\epsilon}{\hbar}V(x,t)\right]. \quad (3.A.15)$$

Equation (3.A.14) is of the same form as the Markov integral equation, and therefore a differential equation satisfied by ψ may be derived from it in a manner similar to that used in the derivation of the Fokker–Planck equation. If we note that

$$\int_{-\infty}^{\infty} x^{2m} e^{-ax^2} dx = \frac{(2m-1)!!}{2^m}\left(\frac{\pi}{a^{2m+1}}\right)^{1/2} \quad \text{for Re } a \geq 0, \quad (3.A.16)$$

where Re indicates the real part, then we obtain, from Eq. (3.A.14) with Eqs. (3.A.15) and (3.A.8),

$$i\hbar\frac{\partial\psi}{\partial t} = -\frac{\hbar^2}{2m}\frac{\partial^2\psi}{\partial x^2} + V(x,t)\psi. \quad (3.A.17)$$

This is just the Schrödinger equation for the system under consideration.

Finally, we show two applications of this formalism to polymer chains. We first consider the Gaussian chain of total contour length L whose bond probability τ_i is given by the Gaussian function, Eq. (5.35) of MTPS [20]. The distribution function $P(\{\mathbf{r}_{n+1}\})$ for the entire chain may then be given by

$$P(\{\mathbf{r}_{n+1}\}) = \prod_{i=1}^{n} \tau_i(\mathbf{r}_i - \mathbf{r}_{i-1})$$
$$= \exp(-E/k_\mathrm{B}T), \quad (3.A.18)$$

where E is the total potential (configurational) energy and is given by

$$E = \frac{3k_\mathrm{B}T}{2}\sum_{i=1}^{n}\left(\frac{\mathbf{r}_i - \mathbf{r}_{i-1}}{l}\right)^2, \quad (3.A.19)$$

where a constant term has been omitted and l is the root-mean-square bond length. In the continuous limit, Eq. (3.A.19) may be written in the form

$$E = \frac{3k_\mathrm{B}T}{2l}\int_0^L \dot{\mathbf{r}}^2 ds. \quad (3.A.20)$$

The Green function $G(\mathbf{R}; L)$ of the end-to-end distance $\mathbf{R}$ is the sum of the Boltzmann factor $\exp(-E/k_\mathrm{B}T)$ over all possible configurations or paths $\mathbf{r}(s)$ subject to $\mathbf{r}(0) = \mathbf{0}$ and $\mathbf{r}(L) = \mathbf{R}$, so that it may be written in the path integral form

$$G(\mathbf{R}; L) = \int_{\mathbf{r}(0)=\mathbf{0}}^{\mathbf{r}(L)=\mathbf{R}} \exp\left(-\frac{3}{2l}\int_0^L \dot{\mathbf{r}}^2 ds\right)\mathcal{D}[\mathbf{r}(s)]. \quad (3.A.21)$$

Thus, if we put formally $im/2\hbar \to -3/2l$ and $V \to 0$ (regarding t as L), the Schrödinger Eq. (3.A.1) becomes Eq. (2.35).

The second example is the KP chain. In this case, suppose that the path integral representation of the characteristic function $I(\mathbf{k}, \mathbf{u} \,|\, \mathbf{u}_0; L)$ is given by Eq. (3.34) with Eq. (3.35). Then, if we put $\hbar \to 1$, $i/2m \to \lambda$, $V \to -\mathbf{k} \cdot \mathbf{u}$, and $\mathbf{R}_2 \to \mathbf{u}$ (with $|\mathbf{R}_2| = 1$), Eq. (3.A.1) becomes Eq. (3.32).

Appendix 3.B Spherical Harmonics and the Free-Particle Green Function

Throughout this book it is sufficient to choose the spherical harmonics $Y_l^m(\theta, \phi)$ $(l = 0, 1, 2, \cdots; m = -l, -l+1, \cdots, l)$ to be

$$Y_l^m(\theta, \phi) = \left[\frac{2l+1}{4\pi} \frac{(l-|m|)!}{(l+|m|)!}\right]^{1/2} P_l^{|m|}(\cos\theta)e^{im\phi} \qquad (3.B.1)$$

without the phase factor, where $P_l^m(x)$ is the associated Legendre function,

$$P_l^m(x) = (1-x^2)^{m/2}\frac{d^m}{dx^m}P_l(x) \qquad (|x| \leq 1) \qquad (3.B.2)$$

with $P_l(x)$ being the Legendre polynomial,

$$P_l(x) = \frac{1}{2^l l!}\frac{d^l}{dx^l}(x^2-1)^l \qquad (|x| \leq 1). \qquad (3.B.3)$$

We therefore have the complex conjugation

$$Y_l^{m*} = Y_l^{-m}, \qquad (3.B.4)$$

and also the orthonormality and closure relations

$$\int Y_l^{m*}Y_{l'}^{m'}\,d\mathbf{u} = \int_0^{2\pi} d\phi \int_0^\pi \sin\theta d\theta\, Y_l^{m*}(\theta, \phi)Y_{l'}^{m'}(\theta, \phi)$$

$$= \delta_{ll'}\delta_{mm'}, \qquad (3.B.5)$$

$$\sum_{l=0}^\infty \sum_{m=-l}^l Y_l^{m*}(\theta, \phi)Y_l^m(\theta', \phi') = \frac{1}{\sin\theta}\delta(\theta-\theta')\delta(\phi-\phi') = \delta(\mathbf{u}-\mathbf{u}'), \qquad (3.B.6)$$

where δ_{lm} and δ are the Kronecker delta and a Dirac delta function, respectively. We note that P_l^m and P_l have the orthonormality properties,

$$\int_{-1}^1 P_l^m(x)P_{l'}^m(x)dx = \frac{2}{2l+1}\frac{(l+m)!}{(l-m)!}\delta_{ll'}, \qquad (3.B.7)$$

$$\int_{-1}^1 P_l(x)P_{l'}(x)dx = \frac{2}{2l+1}\delta_{ll'}. \qquad (3.B.8)$$

Now we find the solution of Eq. (3.33). We expand the free-particle Green function $G(\mathbf{u} \,|\, \mathbf{u}_0; L)$ in terms of Y_l^m as

$$G(\mathbf{u} \,|\, \mathbf{u}_0; L) = \sum_{l=0}^{\infty} \sum_{m=-l}^{l} C_l^m(\mathbf{u}_0; L) Y_l^m(\theta, \phi). \qquad (3.B.9)$$

The spherical harmonics are the eigenfunctions of the Laplacian operator ∇_u^2 [32], that is

$$\nabla_u^2 Y_l^m = -l(l+1) Y_l^m. \qquad (3.B.10)$$

From Eqs. (3.33) (for $L > 0$), (3.B.9), and (3.B.10), we obtain the solution for C_l^m,

$$C_l^m(\mathbf{u}_0; L) = A_l^m(\mathbf{u}_0) \exp\left[-\lambda l(l+1)L\right]. \qquad (3.B.11)$$

From Eqs. (3.33) and (3.B.6), we have the boundary condition

$$G(\mathbf{u} \,|\, \mathbf{u}_0; 0) = \delta(\mathbf{u} - \mathbf{u}_0) = \sum_{l=0}^{\infty} \sum_{m=-l}^{l} Y_l^{m*}(\theta_0, \phi_0) Y_l^m(\theta, \phi), \qquad (3.B.12)$$

so that, from Eqs. (3.B.9) and (3.B.12),

$$C_l^m(\mathbf{u}_0; 0) = A_l^m(\mathbf{u}_0) = Y_l^{m*}(\theta_0, \phi_0). \qquad (3.B.13)$$

Substitution of Eq. (3.B.11) with Eq. (3.B.13) into Eq. (3.B.9) leads to Eq. (3.42) (in units of λ^{-1}).

Finally, we note that the bond correlation function $\langle \mathbf{u}(s_1) \cdot \mathbf{u}(s_2) \rangle$ $(s_1 < s_2)$ may be evaluated by the use of Eq. (3.42) as follows,

$$\begin{aligned}
\langle \mathbf{u}(s_1) \cdot \mathbf{u}(s_2) \rangle &= \langle \mathbf{u}_0(0) \cdot \mathbf{u}(s_2 - s_1) \rangle \\
&= \int \cos\theta \, G(\mathbf{u} \,|\, \mathbf{u}_0; s_2 - s_1) d\mathbf{u} \\
&= \exp\left[-2(s_2 - s_1)\right],
\end{aligned} \qquad (3.B.14)$$

where $\mathbf{u}_0$ has been chosen to be in the direction of the z axis of a Cartesian coordinate system. Substitution of the third line of Eqs. (3.B.14) into Eq. (2.23) and integration leads to Eq. (3.5).

We also note that if α is the angle between the unit vectors $\mathbf{u}_1$ and $\mathbf{u}_2$, that is, $\mathbf{u}_1 \cdot \mathbf{u}_2 = \cos\alpha = P_1(\cos\alpha)$ with $\mathbf{u}_i = (1, \theta_i, \phi_i)$, then there hold the relations

$$P_l(\cos\alpha) = \frac{4\pi}{2l+1} \sum_{m=-l}^{l} Y_l^{m*}(\theta_1, \phi_1) Y_l^m(\theta_2, \phi_2), \qquad (3.B.15)$$

$$\exp(i\mathbf{r}_1 \cdot \mathbf{r}_2) = \sum_{l=0}^{\infty} (2l+1) i^l j_l(r_1 r_2) P_l(\cos\alpha), \qquad (3.B.16)$$

where $\mathbf{r}_i = r_i \mathbf{u}_i$ and j_l is the spherical Bessel function of the first kind.

Appendix 3.C Modified Wormlike Chains

Since it is impossible to find the exact solution of Eq. (3.31) or (3.32) in a closed form, various attempts have been made to relax the constraint of Eq. (2.26), $\mathbf{u}^2 = 1$. In this appendix we briefly discuss these modified worm-like chains [16]. The unnormalized and unconditional characteristic function $I(\mathbf{k}, \mathbf{u}, \mathbf{u}_0; L)$ for them may be written in the path integral form of Eq. (3.34) with the Lagrangian,

$$\mathcal{L} = \frac{i}{k_B T}\left(\frac{1}{2}\alpha\dot{\mathbf{u}}^2 + U'\right) + \mathbf{k}\cdot\mathbf{u}, \tag{3.C.1}$$

where U' is an additional true potential energy of the chain associated with the relaxation of the constraint, so that $-(iU'/k_B T + \mathbf{k}\cdot\mathbf{u})$ is the "potential energy" of the "particle." Then the Schrödinger equation is of the form

$$\left(\frac{\partial}{\partial L} - \frac{1}{2\alpha}\nabla_u^2 + \frac{1}{k_B T}V - i\mathbf{k}\cdot\mathbf{u}\right)I(\mathbf{k}, \mathbf{u}, \mathbf{u}_0; L) = \delta(L)\delta(\mathbf{u} - \mathbf{u}_0), \tag{3.C.2}$$

where V is determined from U'. Note that Eqs. (3.C.1) and (3.C.2) (and hence also ∇_u^2) are no longer subject to the condition $\mathbf{u}^2 = 1$.

Harris and Hearst (HH) [14] permitted Rouse-type stretching [20, 33] as well as bending of the chain, so that their U' and V are given by

$$U'^{(\mathrm{HH})} = V^{(\mathrm{HH})} = \frac{1}{2}\beta\mathbf{u}^2 \tag{3.C.3}$$

with β the stretching force constant. For this model, α is equated to $3k_B T/4\lambda$, and β is determined as a function of L and λ; and L should be regarded as the contour length in the unstretched state. Equation (3.C.2) with $V = V^{(\mathrm{HH})}$ was first derived by Freed [9], and therefore his model is the same as that of Harris and Hearst except for the equations determining α and β. This model becomes invalid for high stiffness; near the rod limit, it cannot give correctly the KP wormlike moments other than $\langle R^2 \rangle$. In particular, the contour length increases indefinitely if an external force is applied and increased. Thus Noda and Hearst [34] attempted to remedy this defect by forcing β to depend on, for instance, rate of shear.

Fixman and Kovac (FK) [35] considered a more general modification by introducing an external potential $-\mathbf{R}\cdot\mathbf{f}$ acting on the end-to-end vector $\mathbf{R}$, so that

$$U'^{(\mathrm{FK})} = V^{(\mathrm{FK})} = \frac{1}{2}\beta\mathbf{u}^2 - \mathbf{f}\cdot\mathbf{u}. \tag{3.C.4}$$

For this model, α is still equated to $3k_B T/4\lambda$, and β is determined as a function of L, λ, and also the force $\mathbf{f}$ so that L and $\langle \mathbf{R} \rangle$ do not increase indefinitely with f. When $f = 0$, this model reduces to the HH model. Now Eq. (3.C.2) with $V = V^{(\mathrm{FK})}$ is just the Schrödinger equation for a harmonic oscillator in an external force field $\mathbf{k} - i\mathbf{f}$, and its solution is well known

[8]. Thus we readily have for the normalized but unconditional characteristic function $I(\mathbf{k}, \mathbf{u}, \mathbf{u}_0; L)$ for finite $\mathbf{f}$

$$I(\mathbf{k}, \mathbf{u}, \mathbf{u}_0; \mathbf{f}, L) = P(\mathbf{u}, \mathbf{u}_0; \mathbf{f}, L) \exp\left\{-\frac{Lk^2}{2\beta}\left(1 - \frac{1}{a}\tanh a\right)\right.$$
$$\left. + i\mathbf{k} \cdot \left[\frac{L}{\beta}\left(1 - \frac{1}{a}\tanh a\right)\mathbf{f} + \frac{L}{2a}(\tanh a)(\mathbf{u} + \mathbf{u}_0)\right]\right\}, \quad (3.C.5)$$

where

$$P(\mathbf{u}, \mathbf{u}_0; \mathbf{f}, L) = \left(\frac{b}{\pi}\right)^3 \exp\left\{-\frac{b}{\sinh 2a}\left[(\cosh 2a)(\mathbf{u}^2 + \mathbf{u}_0{}^2) - 2\mathbf{u} \cdot \mathbf{u}_0\right]\right.$$
$$\left. + \frac{L}{2a}(\tanh a)\mathbf{f} \cdot (\mathbf{u} + \mathbf{u}_0) - \frac{L}{2a\beta}(\tanh a)\mathbf{f}^2\right\}, \quad (3.C.6)$$

$$a = \frac{L}{2}\left(\frac{\beta}{\alpha}\right)^{1/2},$$
$$b = \frac{1}{2}(\alpha\beta)^{1/2}. \quad (3.C.7)$$

When $\mathbf{f} = 0$, the I given by Eq. (3.C.5) is identical with that of Freed [9] except for the normalization constant. By Fourier inversion of Eq. (3.C.5), we obtain for the normalized trivariate distribution function $P(\mathbf{R}, \mathbf{u}, \mathbf{u}_0; \mathbf{f}, L)$

$$P(\mathbf{R}, \mathbf{u}, \mathbf{u}_0; \mathbf{f}, L) = P(\mathbf{u}, \mathbf{u}_0; \mathbf{f}, L)\left[\frac{\beta}{2\pi L(1 - a^{-1}\tanh a)}\right]^{3/2}$$
$$\times \exp\left\{-\frac{\beta}{2L(1 - a^{-1}\tanh a)}\left[\mathbf{R} - \frac{L}{2a}(\tanh a)(\mathbf{u} + \mathbf{u}_0)\right.\right.$$
$$\left.\left. - \frac{L}{\beta}\left(1 - \frac{1}{a}\tanh a\right)\mathbf{f}\right]^2\right\}. \quad (3.C.8)$$

The distribution function $P(\mathbf{R}, \mathbf{u}, \mathbf{u}_0; L)$ given by Eq. (3.C.8) with $\mathbf{f} = 0$ may also be obtained from the formulation of Harris and Hearst. If the radius vector $\mathbf{r}(s)$ is expanded in terms of the eigenfunctions ψ_i for the equation of motion and if ϵ_i are the expansion coefficients, then the instantaneous distribution function for the entire free HH chain may be expressed as a product of Gaussian distributions of ϵ_i [14]. From this, we can therefore derive the trivariate Gaussian distribution $P(\mathbf{R}, \mathbf{u}, \mathbf{u}_0; L)$ by the use of the Wang–Uhlenbeck theorem [20]. Thus it is explicitly recognized that the Freed model is exactly equivalent to the HH model. However, their expressions for the moments, for example, $\langle R^2 \rangle$ as functions of α and β are different from each other. This arises from the fact that Freed regarded erroneously the above $P(\mathbf{R}, \mathbf{u}, \mathbf{u}_0; L)$ as the conditional distribution $P(\mathbf{R}, \mathbf{u} \mid \mathbf{u}_0; L)$ and evaluated averages with $P(\mathbf{R}, \mathbf{u}, \mathbf{u}_0; L)P(\mathbf{u}_0; L)$.

Tagami (T) [36] assumed $G(\mathbf{R}, \mathbf{u} \,|\, \mathbf{u}_0; L)$ to be the same as the Green function for a free Brownian particle with $\mathbf{R}$ the position and $\mathbf{u}$ the velocity. Then the Fokker–Planck equation satisfied by this G and also its closed-form solution are well known [4]. The Lagrangian of this system was already given by Saito and co-workers [37, 38], and we have

$$U'^{(\mathrm{T})} = \frac{1}{2}\beta \mathbf{u}^2 + (\alpha\beta)^{1/2}\mathbf{u} \cdot \left(\frac{d\mathbf{u}}{ds}\right),$$

$$V^{(\mathrm{T})} = -\left(\frac{\beta}{\alpha}\right)^{1/2} \nabla_u \cdot \mathbf{u}$$

(3.C.9)

with $\alpha = 3k_{\mathrm{B}}T/4\lambda$ and $\beta = \lambda k_{\mathrm{B}}T$. It is seen that the stretching energy is still of the Rouse type but that there is coupling between bending and stretching. However, it is not clear what physical property of the real chain this coupling reflects. We also note that this model gives the correct rod limits of the moments, but not the correct first-order corrections to these limits (see Sect. 3.3.2).

Further, we consider three other models, which are somewhat different from the above modified wormlike chains in nature. In all of these, the minimum of the stretching energy is located at $\mathbf{u} = \mathbf{0}$. However, Saito, Takahashi, and Yunoki (STY) [10] instead introduced the stretching energy whose minimum is at $|\mathbf{u}| = 1$, so that

$$U'^{(\mathrm{STY})} = V^{(\mathrm{STY})} = \frac{1}{2}\beta(|\mathbf{u}| - 1)^2.$$

(3.C.10)

Although the determination of α and β is not yet explicit, we must have $\alpha = k_{\mathrm{B}}T/2\lambda$ for $\beta \to \infty$ since this chain reduces to the KP chain in this limit. Its mathematical treatment is not necessarily easier than that of the latter. Soda [39] also considered the potential given by Eq. (3.C.10) but imposed the constraint on the bond angle (supplement) θ instead of on $\cos\theta$ as done in all other models. The differential equation for the distribution function that results is nonlinear and is much less tractable than Eq. (3.31).

Finally, we discuss the model of Winkler, Reineker, and Harnau (WRH) [40], who introduced an *end effect* into the HH chain in such a way that

$$U'^{(\mathrm{WRH})} = U'^{(\mathrm{HH})} + U_0'$$

(3.C.11)

with

$$U_0' = \beta_0\big[\delta(s) + \delta(L - s)\big]\mathbf{u}^2,$$

(3.C.12)

where β_0 is another stretching force constant. We then have

$$V^{(\mathrm{WRH})} = U'^{(\mathrm{HH})} = V^{(\mathrm{HH})},$$

(3.C.13)

and $\alpha = 3k_{\mathrm{B}}T/4\lambda$, $\beta = 3\lambda k_{\mathrm{B}}T$, and $\beta_0 = 3k_{\mathrm{B}}T/2$, so that the differential equation for I is exactly the same as that of Harris and Hearst, that is, Eq. (3.C.2) with $V = V^{(\mathrm{HH})}$. Thus a trick for the end effect U_0', which has no physical meaning, must be made in the HH distribution function, thereby leading accidentally to the exact KP wormlike moments $\langle R^2 \rangle$ and $\langle S^2 \rangle$. It is pertinent to note here that their wrong statement [40] concerning the above analysis [16] of the Freed model arises from their misunderstanding of it.

Appendix 3.D Wormlike Rings

In this appendix we derive approximately distribution functions and moments for KP wormlike rings [41]. Consider two contour points 1 and 2 separated by the contour distance s on the KP ring of total contour length L. We define a conditional distribution function $P(\mathbf{R}, \mathbf{u} \,|\, \mathbf{u}_0; s, L)$ of both the vector distance $\mathbf{R}$ between the points 1 and 2 and the unit tangent $\mathbf{u}$ at the point 2 with the unit tangent vector $\mathbf{u}_0$ at the point 1 fixed. Note that this P is not the Green function. (All lengths are measured in units of λ^{-1}.) As seen from Fig. 3.D.1, P may be expressed in the form

$$P(\mathbf{R}, \mathbf{u} \,|\, \mathbf{u}_0; s, L) = CG(\mathbf{R}, \mathbf{u} \,|\, \mathbf{u}_0; s)G(\mathbf{R}, -\mathbf{u} \,|\, -\mathbf{u}_0; L - s)\,, \quad (3.D.1)$$

where $C = G^{-1}(0, \mathbf{u}_0 \,|\, \mathbf{u}_0; L)$ is the normalization constant and G are the Green functions for the linear chain. We first consider two limiting cases: (1) $s \gg 1$ and $L - s \gg 1$ and (2) $s \ll 1$ and $L - s \gg 1$.

In the first case, we use the first Daniels approximations to the two G in Eq. (3.D.1). Integrating P over $\mathbf{u}$ and $\mathbf{u}_0$, we then obtain for the distribution function $P(\mathbf{R}; s, L)$ of $\mathbf{R}$

$$
\begin{aligned}
P(\mathbf{R}; s, L) = {}& \left[\frac{3L}{2\pi s(L - s)} \right]^{3/2} \left(1 - \frac{11}{8L} \right)^{-1} \exp\left[-\frac{3LR^2}{2s(L - s)} \right] \\
& \times \left[1 - \frac{5}{8s} - \frac{5}{8(L - s)} + \frac{2R^2}{s^2} + \frac{2R^2}{(L - s)^2} \right. \\
& \left. - \frac{3R^2}{2s(L - s)} - \frac{33R^4}{40s^3} - \frac{33R^4}{40(L - s)^3} + \cdots \right]. \quad (3.D.2)
\end{aligned}
$$

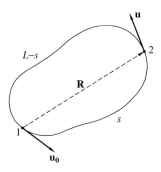

Fig. 3.D.1. The distance $\mathbf{R}(s)$ between the contour points 1 and 2 on a KP wormlike ring and the unit tangent vectors $\mathbf{u}_0$ and $\mathbf{u}$ there

In the second case, we use the WKB approximation to $G(\mathbf{R}, \mathbf{u} \,|\, \mathbf{u}_0; s)$ and the first Daniels approximation to the other G. We choose $\mathbf{u}_0$ to be in the direction of the z axis of a Cartesian coordinate system and express $\mathbf{R}$ and $\mathbf{u}$ as $\mathbf{R} = (x, y, z)$ and $\mathbf{u} = (1, \theta, \phi)$ in Cartesian and spherical polar coordinates, respectively. Since $\theta^2 = \mathcal{O}(s)$, we expand $\cos\theta$ and $\sin\theta$ in the first Daniels $G(\mathbf{R}, -\mathbf{u} \,|\, -\mathbf{u}_0; L-s)$ in powers of θ and retain terms to $\mathcal{O}(\theta^2)$. Then $P(\mathbf{R}, \mathbf{u} \,|\, \mathbf{u}_0; s, L)$ may be expressed as

$$
P(\mathbf{R}, \mathbf{u} \,|\, \mathbf{u}_0; s, L) = C'(s, L) \left(\frac{3}{\pi s^3}\right) \left(\frac{45}{4\pi s^3 \theta^2}\right)^{1/2}
$$

$$
\times \exp\left[-\frac{\theta^2}{4s} - \frac{3}{s^3}\left(x - \frac{1}{2}s\theta\cos\phi\right)^2 \right.
$$

$$
\left. - \frac{3}{s^3}\left(y - \frac{1}{2}s\theta\sin\phi\right)^2 - \frac{45}{4s^3\theta^2}\left(z - s + \frac{1}{6}s\theta^2\right)^2 \right]
$$

$$
\times \left[1 - \frac{15}{2s\theta^2}\left(z - s + \frac{1}{6}s\theta^2\right) + \cdots\right]
$$

$$
\times \left[1 - \frac{11}{8(L-s)} - \frac{2z}{L-s} + \frac{3\theta^2}{8(L-s)} + \cdots\right] \tag{3.D.3}
$$

with $C'(s, L)$ the normalization constant. For this case, it is impossible to integrate P over $\mathbf{u}$ and $\mathbf{u}_0$ to obtain an analytical expression for $P(\mathbf{R}; s, L)$. We note that the distribution function $P(\mathbf{R}, \mathbf{u} \,|\, \mathbf{u}_0; s, L)$ for $s \gg 1$ and $L-s \ll 1$ may be obtained from Eq. (3.D.3) by exchanging s for $L-s$.

By the use of Eqs. (3.D.2) and (3.D.3), we obtain for the mean-square distance $\langle R^2(s) \rangle$ between the two points on the ring in the two limiting cases

$$
\langle R^2(s) \rangle = \frac{s(L-s)}{L} - \frac{1}{2} + \frac{11s(L-s)}{6L^2} \quad \text{for } s \gg 1 \text{ and } L-s \gg 1
$$

$$
= s^2\left(1 - \frac{2}{3}s + \cdots\right) \quad \text{for } s \ll 1 \text{ and } L-s \gg 1. \tag{3.D.4}
$$

From a comparison of the second line of Eqs. (3.D.4) with the second of Eqs. (3.96), it is seen that the first-order correction to the rigid-ring limit of $\langle R^2(s) \rangle$ is the same as that to the rigid-rod limit.

Now we join the two $\langle R^2(s) \rangle$ given by Eqs. (3.D.4) to complete an approximate expression for $\langle R^2(s) \rangle$ following the procedure of Hearst and Stockmayer [30]. That is

$$
\langle R^2(s) \rangle = \frac{s(L-s)}{L} - \frac{1}{2} + \frac{11s(L-s)}{6L^2} \quad \text{for } \alpha < s \le \frac{L}{2}
$$

$$
= s^2\left(1 - \frac{2}{3}s + k_2 s^2 + k_3 s^3\right) \quad \text{for } 0 \le s \le \alpha, \tag{3.D.5}
$$

where α, k_2, and k_3 are determined as functions of L in such a way that the two $\langle R^2(s) \rangle$ given by Eqs. (3.D.5) have the same value and the same first and second derivatives at $s = \alpha$. The results are

$$\alpha = 1.81892 - \frac{6.53529}{L} + \frac{13.6768}{L^2} - \frac{10.1456}{L^3}, \tag{3.D.6}$$

and

$$k_2 = \frac{1}{\alpha}\left[\frac{4}{3} - \frac{3}{\alpha} - \frac{5}{2\alpha^3} + \frac{1}{\alpha^2}\left(1 + \frac{11}{6L}\right)\left(4 - \frac{3\alpha}{L}\right) \right],$$

$$k_3 = \frac{1}{\alpha^2}\left[-\frac{1}{2} + \frac{5}{4\alpha} + \frac{3}{4\alpha^3} - \frac{1}{4\alpha^2}\left(1 + \frac{11}{6L}\right)\left(6 - \frac{5\alpha}{L}\right) \right]. \tag{3.D.7}$$

Substitution of Eqs. (3.D.5) into Eq. (2.25), which is valid also for a ring, and integration leads to

$$\langle S^2 \rangle = \left(1 + \frac{11}{6L}\right)\left(\frac{L}{12} - \frac{\alpha^2}{2L} + \frac{\alpha^3}{3L^2}\right) - \frac{1}{4} + \frac{\alpha}{2L}$$

$$+ \frac{\alpha^3}{L}\left(\frac{1}{3} - \frac{\alpha}{6} + \frac{k_2}{5}\alpha^2 + \frac{k_3}{6}\alpha^3 \right). \tag{3.D.8}$$

We note that Eqs. (3.D.5) and (3.D.8) are valid for $L \geq 3.480$ and that Eq. (3.D.8) gives the correct first-order correction to the random-coil limit of $\langle S^2 \rangle$,

$$\langle S^2 \rangle = \frac{L}{12}\left(1 - \frac{7}{6L} + \cdots\right) \qquad \text{for } L \gg 1. \tag{3.D.9}$$

In fact, however, Eq. (3.D.8) is applicable only for relatively large L, as is evident from the derivation.

In the rigid-ring limit of $L \to 0$, $\langle S^2 \rangle$ may be directly evaluated to be

$$\lim_{L \to 0} \langle S^2 \rangle = \frac{L^2}{4\pi^2}. \tag{3.D.10}$$

The correction to the rigid-ring limit above must be evaluated in a different way in order to join it to the $\langle S^2 \rangle$ given by Eq. (3.D.8) or (3.D.9) (see Chap. 7).

References

1. O. Kratky and G. Porod: Rec. Trav. Chem. **68**, 1106 (1949).
2. G. Porod: J. Polym. Sci. **10**, 157 (1953).
3. H. Benoit and P. Doty: J. Phys. Chem. **57**, 958 (1953).
4. S. Chandrasekhar: Rev. Mod. Phys. **15**, 1 (1943).
5. S. A. Rica and P. Gray: *Statistical Mechanics of Simple Liquids* (Interscience, New York, 1965).
6. J. J. Hermans and R. Ullman: Physica **18**, 951 (1952).
7. H. E. Daniels: Proc. Roy. Soc. (Edinburgh) **A63**, 290 (1952).

8. R. P. Feynman and A. R. Hibbs: *Quantum Mechanics and Path Integrals* (McGraw-Hill, New York, 1965).
9. K. F. Freed: J. Chem. Phys. **54**, 1453 (1971); Adv. Chem. Phys. **22**, 1 (1972).
10. N. Saito, K. Takahashi, and Y. Yunoki: J. Phys. Soc. Japan **22**, 219 (1967).
11. See, for example, L. D. Landau and E. M. Lifshitz: *Theory of Elasticity* (Addison-Wesley, Reading, 1959).
12. S. E. Bresler and Ya. I. Frenkel: Acta Phys.-Chim. USSR **11**, 485 (1939).
13. L. D. Landau and E. M. Lifshitz: *Statistical Physics* (Addison-Wesley, Reading, 1958).
14. R. A. Harris and J. E. Hearst: J. Chem. Phys. **44**, 2595 (1966).
15. H. Yamakawa: J. Chem. Phys. **59**, 3811 (1973).
16. H. Yamakawa: Pure Appl. Chem. **46**, 135 (1976).
17. See, for example, L. I. Schiff: *Quantum Mechanics* (McGraw-Hill, New York, 1968).
18. S. Heine, O. Kratky, G. Porod, and P. J. Schmitz: Makromol. Chem. **44–46**, 682 (1961).
19. K. Nagai: Polym. J. **4**, 35 (1973).
20. H. Yamakawa: *Modern Theory of Polymer Solutions* (Harper & Row, New York, 1971).
21. H. Yamakawa and M. Fujii: Macromolecules **7**, 649 (1974).
22. W. Gobush, H. Yamakawa, W. H. Stockmayer, and W. S. Magee: J. Chem. Phys. **57**, 2839 (1972).
23. I. Prigogine: *Non-Equilibrium Statistical Mechanics* (Interscience, New York, 1962).
24. See also R. M. Mazo: *Statistical Mechanical Theories of Transport Processes* (Pergamon, Oxford, 1967).
25. J. Shimada, M. Fujii, and H. Yamakawa: J. Polym. Sci., Polym. Phys. Ed. **12**, 2075 (1974).
26. K. Nagai: J. Chem. Phys. **38**, 924 (1963).
27. R. L. Jernigan and P. J. Flory: J. Chem. Phys. **50**, 4185 (1969).
28. P. J. Flory: *Statistical Mechanics of Chain Molecules* (Interscience, New York, 1969).
29. H. Yamakawa and M. Fujii: J. Chem. Phys. **59**, 6641 (1973).
30. J. E. Hearst and W. H. Stockmayer: J. Chem. Phys. **37**, 1425 (1962).
31. T. Norisuye, H. Murakami, and H. Fujita: Macromolecules **11**, 966 (1978).
32. See, for example, A. Messiah: *Quantum Mechanics* (North-Holland, Amsterdam, 1972), Vol. I.
33. P. E. Rouse, Jr.: J. Chem. Phys. **21**, 1272 (1953).
34. I. Noda and J. E. Hearst: J. Chem. Phys. **54**, 2342 (1971).
35. M. Fixman and J. Kovac: J. Chem. Phys. **58**, 1564 (1973).
36. Y. Tagami: Macromolecules **2**, 8 (1969).
37. N. Saito and M. Namiki: Prog. Theo. Phys. (Kyoto) **16**, 71 (1956).
38. H. Hoshikawa, N. Saito, and K. Nagayama: Polym. J. **7**, 79 (1975).
39. K. Soda: J. Phys. Soc. Japan **35**, 866 (1973); J. Chem. Phys. **95**, 9337 (1991).
40. R. G. Winkler, P. Reineker, and L. Harnau: J. Chem. Phys. **101**, 8119 (1994).
41. M. Fujii and H. Yamakawa: Macromolecules **8**, 792 (1975).

4 Chain Statistics – Helical Wormlike Chains

As mentioned in Chap. 3, the KP model [1] may be applicable not only to stiff polymers but also to ordinary flexible polymers if the characteristic ratio C_n increases monotonically to its coil-limiting value C_∞ as the number of skeletal bonds n in the chain is increased. For symmetric chains such as polymethylene, polyoxymethylene, and polyoxyethylene there is indeed good agreement between values of $\langle R^2 \rangle$ as a function of n for the KP and RIS models if the contour length L of the former is properly converted to n [2,3]. However, C_n increases to C_∞ more rapidly than expected from the KP model for poly(dimethylsiloxane) [4], while it decreases to C_∞ with increasing n for poly-DL-alanine [5] or even exhibits a maximum in the case of, for instance, syndiotactic poly(methyl methacrylate) [6], as already seen in Fig. 2.4. Such breakdown of the KP model is probably due to the fact that these real chains with different skeletal bond angles possess locally preferred helical conformations. Further, as anticipated in Chap. 3, it is impossible to assign local vectors and tensors to the KP chain unless they are parallel to and cylindrically symmetric about its contour, respectively.

These circumstances make us recognize a need to extend it to a more general elastic wire model that can resolve them. The HW model is one thus presented [7–10]. It has both bending and torsional energies and its chain contour becomes a regular helix at the minimum zero of its total elastic (potential) energy. In this chapter the foundation of the statistical mechanics of the unperturbed HW chain is presented along with some related topics.

4.1 Formulation of the Model

We consider an elastic wire of fixed length L with both bending and torsional energies and affix a localized Cartesian coordinate system (ξ, η, ζ) to it at the contour point s $(0 \leq s \leq L)$ following Landau and Lifshitz [11], where the unit vector $\mathbf{e}_\zeta$ in the direction of the ζ axis is chosen to coincide with the unit vector $\mathbf{u}(s)$ tangential to the contour with the unit vectors $\mathbf{e}_\xi$ and $\mathbf{e}_\zeta$ being in the directions of the principal axes of inertia of its cross section at s, as depicted in Fig. 4.1. (In its application to a given real chain these axes can be affixed to the latter in a definite manner, as shown later.)

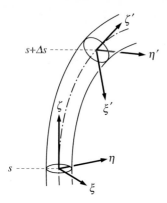

Fig. 4.1. Localized Cartesian coordinate systems (ξ, η, ζ) affixed to the HW chain

The localized coordinate system (ξ', η', ζ') at $s + \Delta s$ is obtained by an infinitesimal rotation $\Delta\overline{\boldsymbol{\Omega}} = (\Delta\overline{\Omega}_\xi, \Delta\overline{\Omega}_\eta, \Delta\overline{\Omega}_\zeta)$ of the (ξ, η, ζ) system at s; that is,

$$\mathbf{e}_{\mu'} = \mathbf{e}_\mu + \Delta\overline{\boldsymbol{\Omega}} \times \mathbf{e}_\mu \qquad (\mu = \xi, \eta, \zeta), \tag{4.1}$$

or in matrix notation,

$$\begin{pmatrix} \mathbf{e}_{\xi'} \\ \mathbf{e}_{\eta'} \\ \mathbf{e}_{\zeta'} \end{pmatrix} = \begin{pmatrix} 1 & \Delta\overline{\Omega}_\zeta & -\Delta\overline{\Omega}_\eta \\ -\Delta\overline{\Omega}_\zeta & 1 & \Delta\overline{\Omega}_\xi \\ \Delta\overline{\Omega}_\eta & -\Delta\overline{\Omega}_\xi & 1 \end{pmatrix} \begin{pmatrix} \mathbf{e}_\xi \\ \mathbf{e}_\eta \\ \mathbf{e}_\zeta \end{pmatrix}. \tag{4.2}$$

The deformed state of the wire may be determined by the "angular velocity" vector $\boldsymbol{\omega}(s) = (\omega_\xi, \omega_\eta, \omega_\zeta)$ defined by

$$\boldsymbol{\omega} = \lim_{\Delta s \to 0} \frac{\Delta\overline{\boldsymbol{\Omega}}}{\Delta s}. \tag{4.3}$$

The HW chain is then defined as the wire whose elastic (potential) energy U per unit contour length is given by

$$U = \frac{1}{2}\alpha\left[\omega_\xi^2 + (\omega_\eta - \kappa_0)^2\right] + \frac{1}{2}\beta(\omega_\zeta - \tau_0)^2, \tag{4.4}$$

where α and β are the bending and torsional force constants, respectively, and are related to each other by the equation

$$\beta = \alpha(1 + \sigma)^{-1} \tag{4.5}$$

with σ being Poisson's ratio ranging from 0 to 0.5, and κ_0 and τ_0 are constants independent of s. The U given by Eq. (4.4) is seen to become a minimum of zero in the *deformed* state $\boldsymbol{\omega} = (0, \kappa_0, \tau_0)$. Then the chain contour as a differentiable space curve becomes a regular helix, as shown below. This is just the requirement for the HW model. However, the definition of the HW model by Eq. (4.4) requires some comments. The fact is that the Bugl–Fujita potential [12] was first adopted [7] as that of the chain having both bending and torsional energies but with relaxation of a certain (unphysical) constraint inherent in it (see Appendix 4.A). The model that resulted was then shown to have eventually the potential given by Eq. (4.4) [13].

Now $\dot{\mathbf{u}}(s)$ is the curvature vector of the chain contour as a differentiable space curve, so that the unit curvature vector $\mathbf{n}(s)$ is given by

$$\mathbf{n} = \frac{\dot{\mathbf{u}}}{|\dot{\mathbf{u}}|}, \tag{4.6}$$

where the over dot denotes the derivative with respect to s as usual. According to differential geometry [14], the form of a space curve is determined by the (*differential-geometrical*) *curvature* $\kappa(s)$ and *torsion* $\tau(s)$ defined by

$$\kappa = |\dot{\mathbf{u}}|, \tag{4.7}$$

$$\tau = (\mathbf{u} \times \mathbf{n}) \cdot \dot{\mathbf{n}}. \tag{4.8}$$

Note that $\mathbf{u} \times \mathbf{n}$ is usually called the unit binormal vector.

For further developments it is convenient to introduce the Euler angles $\Omega = (\theta, \phi, \psi)$ $(0 \le \theta \le \pi, 0 \le \phi \le 2\pi, 0 \le \psi \le 2\pi)$ defining the orientation of the localized coordinate system (ξ, η, ζ) with respect to an external Cartesian coordinate system (x, y, z). That is, the former system is obtained by rotation of the latter by the angles Ω as follows: first rotate the (x, y, z) system by an angle ϕ about the z axis to obtain a system (x', y', z') with $z = z'$, then rotate this system by an angle θ about the y' axis to obtain a system (x'', y'', z'') with $y' = y''$, and finally rotate this system by an angle ψ about the z'' axis to obtain the system $(\xi, \eta, \zeta) = (x''', y''', z''')$ with $z'' = z'''$, as shown in Fig. 4.2. We then have

$$\mathbf{e}_\zeta \equiv \mathbf{u} = (1, \theta, \phi),$$
$$\mathbf{e}_\xi \equiv \mathbf{a} = \mathbf{e}_\theta \cos \psi + \mathbf{e}_\phi \sin \psi, \tag{4.9}$$
$$\mathbf{e}_\eta \equiv \mathbf{b} = -\mathbf{e}_\theta \sin \psi + \mathbf{e}_\theta \cos \psi$$

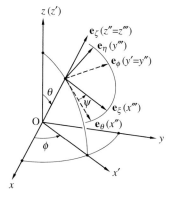

Fig. 4.2. Euler angles $\Omega = (\theta, \phi, \psi)$ defining the orientation of the localized coordinate system (ξ, η, ζ) with respect to an external Cartesian coordinate system (x, y, z)

with $\mathbf{b} = \mathbf{u} \times \mathbf{a}$, where in the first of Eqs. (4.9) $\mathbf{u}$ has been expressed in spherical polar coordinates, and $\mathbf{e}_\theta$ and $\mathbf{e}_\phi$ are the unit vectors in the directions of the x'' and y'' ($= y'$) axes, respectively. The components of $\boldsymbol{\omega}$ may also be expressed in terms of the Euler angles as follows [15],

$$
\begin{aligned}
\omega_\xi &= \dot\theta \sin\psi - \dot\phi \sin\theta \cos\psi\,, \\
\omega_\eta &= \dot\theta \cos\psi + \dot\phi \sin\theta \sin\psi\,, \\
\omega_\zeta &= \dot\phi \cos\theta + \dot\psi\,.
\end{aligned}
\tag{4.10}
$$

In order to express κ and τ readily in terms of these components, we rotate the $(\xi,\,\eta,\,\zeta)$ system by an angle $\psi_0(s)$ about the ζ axis to obtain a system $(\xi_0,\,\eta_0,\,\zeta_0)$ with $\mathbf{e}_\zeta = \mathbf{e}_{\zeta_0}$, as depicted in Fig. 4.3. We then have, from Eqs. (4.10),

$$
\begin{aligned}
\omega_{\xi_0} &= \omega_\xi \cos\psi_0 + \omega_\eta \sin\psi_0\,, \\
\omega_{\eta_0} &= -\omega_\xi \sin\psi_0 + \omega_\eta \cos\psi_0\,, \\
\omega_{\zeta_0} &= \omega_\zeta + \dot\psi_0\,.
\end{aligned}
\tag{4.11}
$$

If ψ_0 is chosen so that $\omega_{\xi_0} = 0$, then $\Delta\mathbf{u}$ is in the direction of $\mathbf{e}_{\xi_0}$ because of no rotation about $\mathbf{e}_{\xi_0}$. We therefore have $\mathbf{e}_{\xi_0} = \mathbf{n}$ and $\mathbf{e}_{\eta_0} = \mathbf{u} \times \mathbf{n}$, and also $\Delta\mathbf{u} = \Delta\overline{\Omega}_{\eta_0}\mathbf{e}_{\xi_0}$ and $\Delta\mathbf{n} \cdot \mathbf{e}_{\eta_0} = \Delta\overline{\Omega}_{\zeta_0}$. Thus we obtain, from Eqs. (4.7) and (4.8), $\kappa = \omega_{\eta_0}$ and $\tau = \omega_{\zeta_0}$, and then, from Eqs. (4.11) with $\omega_{\xi_0} = 0$,

$$
\kappa = (\omega_\xi{}^2 + \omega_\eta{}^2)^{1/2}\,,
\tag{4.12}
$$

$$
\tau = \omega_\zeta - \frac{d}{ds}\tan^{-1}\left(\frac{\omega_\xi}{\omega_\eta}\right).
\tag{4.13}
$$

From Eqs. (4.12) and (4.13), we have $\kappa = \kappa_0$ and $\tau = \tau_0$ at $\boldsymbol{\omega} = (0,\,\kappa_0,\,\tau_0)$. The space curve specified by $\kappa = \kappa_0$ and $\tau = \tau_0$ is a regular helix whose radius ρ and pitch h are given by [14]

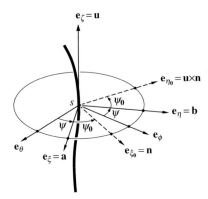

Fig. 4.3. Various unit vectors and rotation angles at s. $\mathbf{n}$ and $\mathbf{u} \times \mathbf{n}$ are the unit curvature and unit binormal vectors, respectively

$$\rho = \frac{\kappa_0}{\kappa_0^2 + \tau_0^2} ,$$

$$h = \frac{2\pi\tau_0}{\kappa_0^2 + \tau_0^2} ,$$

(4.14)

the helix being right-handed for $\tau_0 > 0$ and left-handed for $\tau_0 < 0$. This helix, which is taken by the HW chain contour at the minimum zero of its potential energy, is referred to as the *characteristic helix*. It is schematically depicted in Fig. 4.4a. We note that the HW chain which has the potential U given by Eq. (4.4) is not the only one that becomes a regular helix at the minimum zero of U but that the U given by Eq. (4.4) is of the simplest form of the potentials of those chains (see Appendix 4.A).

In the particular case of $\kappa_0 = 0$, Eq. (4.4) reduces to

$$U = \frac{1}{2}\alpha\dot{\mathbf{u}}^2 + \frac{1}{2}\beta(\omega_\zeta - \tau_0)^2 ,$$

(4.15)

where the first term on the right-hand side is just the bending energy of the KP chain. As seen from Eqs. (4.14), the characteristic helix then becomes a straight line. The chain defined by Eq. (4.15) is referred to as the *generalized KP chain*. It is then convenient to classify it into two types: one (type 1) with $\tau_0 \neq 0$ (KP1) and the other (type 2) with $\tau_0 = 0$ (KP2). Their characteristic helices (rods) are depicted in Figs. 4.4b and c, respectively. The *original KP chain* is defined as the chain with the U given by Eq. (4.15) with $\beta = 0$. All these chains with $\kappa_0 = 0$, both original and generalized (KP1 and KP2), are referred to simply as the KP chain unless necessary to specify.

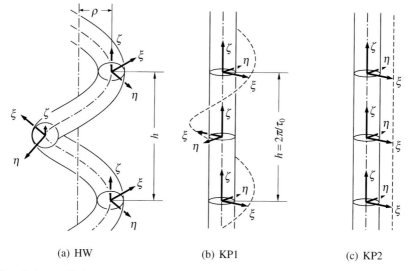

(a) HW   (b) KP1   (c) KP2

Fig. 4.4a–c. Characteristic helix and the localized coordinate systems affixed to it

Finally, it is pertinent to add some discussion of the meaning of the unit vectors $\mathbf{e}_\xi = \mathbf{a}$ and $\mathbf{e}_\eta = \mathbf{b}$. It is true that $\mathbf{e}_{\xi_0}$ and $\mathbf{e}_{\eta_0}$ are the unit curvature and unit binormal vectors, respectively. Now, from Eq. (4.2), we have $\Delta\mathbf{u} = \mathbf{e}_{\zeta'} - \mathbf{e}_\zeta = \Delta\overline{\Omega}_\eta\mathbf{e}_\xi - \Delta\overline{\Omega}_\xi\mathbf{e}_\eta$, so that

$$\dot{\mathbf{u}} = \omega_\eta\mathbf{a} - \omega_\xi\mathbf{b}. \tag{4.16}$$

If we average both sides of Eq. (4.16) (at constant $\mathbf{a}$ and $\mathbf{b}$), we obtain

$$\mathbf{a} = \kappa_0^{-1}\langle\dot{\mathbf{u}}\rangle, \tag{4.17}$$

since we have, from Eq. (4.4), $\langle\omega_\xi\rangle = 0$ and $\langle\omega_\eta\rangle = \kappa_0$. Thus $\mathbf{a}$ has the meaning of the unit *mean* curvature vector, so that $\mathbf{b}$ $(= \mathbf{u} \times \mathbf{a})$ is the unit *mean* binormal vector.

4.2 Diffusion Equations

4.2.1 Path Integrals and Fokker–Planck Equations

We can define the Green function $G(\mathbf{R}, \mathbf{u}, \mathbf{a} \,|\, \mathbf{u}_0, \mathbf{a}_0; L)$, that is, the conditional distribution function of the radius vector $\mathbf{r}(L) = \mathbf{R}$, the unit tangent vector $\mathbf{u}(L) = \mathbf{u}$, and the unit mean curvature vector $\mathbf{a}(L) = \mathbf{a}$ at the terminal end of the HW chain of contour length L when $\mathbf{r}(0) = \mathbf{0}$, $\mathbf{u}(0) = \mathbf{u}_0$, and $\mathbf{a}(0) = \mathbf{a}_0$ at the initial end. We simply denote it by $G(\mathbf{R}, \Omega \,|\, \Omega_0; L)$ with $\Omega = (\theta, \phi, \psi)$ being the Euler angles, since $\mathbf{u}$ and $\mathbf{a}$ uniquely determine the orientation Ω of the system $(\mathbf{u}, \mathbf{a}, \mathbf{b})$ with $\mathbf{b} = \mathbf{u} \times \mathbf{a}$. It is normalized as

$$\int G(\mathbf{R}, \Omega \,|\, \Omega_0; L)d\mathbf{R}d\Omega = 1 \tag{4.18}$$

with $d\Omega = d\mathbf{u}d\mathbf{a} = \sin\theta d\theta d\phi d\psi$. Similarly, we can define the distribution functions $G(\Omega \,|\, \Omega_0; L)$, $G(\mathbf{R}; L)$, and so forth, corresponding to the cases of the KP chain. Further, the characteristic function $I(\mathbf{k}, \Omega \,|\, \Omega_0; L)$ is defined by

$$I(\mathbf{k}, \Omega \,|\, \Omega_0; L) = \int G(\mathbf{R}, \Omega \,|\, \Omega_0; L)\exp(i\mathbf{k} \cdot \mathbf{R})d\mathbf{R}. \tag{4.19}$$

Now, as in Eq. (3.40), $G(\mathbf{R}, \Omega \,|\, \Omega_0; L)$ may be expressed in terms of the path integral over the paths $\mathbf{u}(s)$ and $\mathbf{a}(s)$, which we simply denote by $\Omega(s)$, subject to the condition of Eq. (3.39) as follows,

$$G(\mathbf{R}, \Omega \,|\, \Omega_0; L) = \int_{\Omega(0)=\Omega_0}^{\Omega(L)=\Omega} \delta\left(\mathbf{R} - \int_0^L \mathbf{u}ds\right)$$

$$\times \exp\left(-\frac{1}{k_\mathrm{B}T}\int_0^L Uds\right)\mathcal{D}[\Omega(s)], \tag{4.20}$$

where U is given by Eq. (4.4). Taking the Fourier transform of both sides of Eq. (4.20), we then obtain

$$I(\mathbf{k}, \Omega \,|\, \Omega_0;\; L) = \int_{\Omega(0)=\Omega_0}^{\Omega(L)=\Omega} \exp\left(i \int_0^L \mathcal{L} ds \right) \mathcal{D}\big[\Omega(s)\big] , \qquad (4.21)$$

where $\mathcal{L}$ is the "Lagrangian" given by

$$\mathcal{L} = K - V + \mathbf{k} \cdot \mathbf{u} \qquad (4.22)$$

with

$$K = \frac{i}{4\lambda}\left[\omega_\xi{}^2 + \omega_\eta{}^2 + (1+\sigma)^{-1}\omega_\zeta{}^2 \right] , \qquad (4.23)$$

$$V = \frac{i}{4\lambda}\left[2\kappa_0\omega_\eta + 2(1+\sigma)^{-1}\tau_0\omega_\zeta - \kappa_0{}^2 - (1+\sigma)^{-1}\tau_0{}^2 \right] . \qquad (4.24)$$

In Eqs. (4.23) and (4.24) we have used Eqs. (3.37) and (4.5). It is seen that the I given by Eq. (4.21) with Eqs. (4.22)–(4.24) is just the quantum-mechanical kernel for a symmetric top with the kinetic energy K and the angular-velocity-dependent potential energy V in a gravitational field $\mathbf{k}$. Thus we can derive the "Schrödinger" equation for I.

The "angular momenta" p_μ and the "Hamiltonian" $\mathcal{H}$ are defined by

$$p_\mu = \frac{\partial \mathcal{L}}{\partial \omega_\mu} , \qquad (4.25)$$

$$\mathcal{H} = \sum_\mu \omega_\mu p_\mu - \mathcal{L} . \qquad (4.26)$$

From Eqs. (4.22)–(4.26), we have

$$\mathcal{H} = -i\lambda[p_\xi{}^2 + p_\eta{}^2 + (1+\sigma)p_\zeta{}^2] + \kappa_0 p_\eta + \tau_0 p_\zeta - \mathbf{k} \cdot \mathbf{u} . \qquad (4.27)$$

If we introduce the quantization,

$$p_\mu = -i\frac{\partial}{\partial \overline{\Omega}_\mu} \qquad (4.28)$$

in units of $\hbar$, then we obtain the "Schrödinger" equation,

$$i\frac{\partial I}{\partial L} = \mathcal{H}I , \qquad (4.29)$$

or

$$\left(\frac{\partial}{\partial L} + \mathcal{A} - i\mathbf{k} \cdot \mathbf{u} \right) I(\mathbf{k}, \Omega \,|\, \Omega_0; L) = \delta(L)\delta(\Omega - \Omega_0) , \qquad (4.30)$$

where

$$\mathcal{A} = \kappa_0 L_\eta + \tau_0 L_\zeta - \lambda\sigma L_\zeta{}^2 - \lambda \mathbf{L}^2 \qquad (4.31)$$

with $\mathbf{L} = (L_\xi,\, L_\eta,\, L_\zeta)$ and

$$L_\mu = \frac{\partial}{\partial \overline{\Omega}_\mu} . \tag{4.32}$$

By Fourier inversion of Eq. (4.30), we find the Fokker–Planck equations satisfied by G,

$$\left(\frac{\partial}{\partial L} + \mathcal{A} + \mathbf{u} \cdot \nabla_R \right) G(\mathbf{R}, \Omega \,|\, \Omega_0; L) = \delta(L)\delta(\mathbf{R})\delta(\Omega - \Omega_0), \tag{4.33}$$

$$\left(\frac{\partial}{\partial L} + \mathcal{A} \right) G(\Omega \,|\, \Omega_0; L) = \delta(L)\delta(\Omega - \Omega_0). \tag{4.34}$$

In Eq. (4.34) $G(\Omega \,|\, \Omega_0; L) = I(0, \Omega \,|\, \Omega_0; L)$ is the "free-particle" Green function and is also obtained by integration of $G(\mathbf{R}, \Omega \,|\, \Omega_0; L)$ over $\mathbf{R}$. The components of $\mathbf{L}$, which is the "angular momentum" operator for a rigid body in units of $-i\hbar$ [16, 17], may be expressed in terms of the Euler angles as

$$L_\xi = \sin\psi \frac{\partial}{\partial \theta} - \frac{\cos\psi}{\sin\theta} \frac{\partial}{\partial \phi} + \cot\theta \cos\psi \frac{\partial}{\partial \psi},$$

$$L_\eta = \cos\psi \frac{\partial}{\partial \theta} + \frac{\sin\psi}{\sin\theta} \frac{\partial}{\partial \phi} - \cot\theta \sin\psi \frac{\partial}{\partial \psi}, \tag{4.35}$$

$$L_\zeta = \frac{\partial}{\partial \psi} .$$

We note that the coefficients of the differential operators on the right-hand sides of Eqs. (4.35), e.g., $\partial\theta/\partial\overline{\Omega}_\xi = \sin\psi$, can be obtained by inversion of Eqs. (4.10) with Eq. (4.3).

It is straightforward to generalize Eq. (4.4) for U and hence Eq. (4.31) for the diffusion operator $\mathcal{A}$ in order to consider more general elastic wire models, as shown in Appendix 4.A. However, it is almost impossible to apply them to real chains, since their model parameters are too many to determine unambiguously from experiment. The HW model may be a necessary and sufficient generalization of the KP model. Then there arises an interesting question: from what discrete chains can these continuous chains, in particular, the HW chain, be obtained by the continuous limiting process? This problem is considered in Appendix 4.B.

4.2.2 The Free-Particle Green Function

We solve the diffusion Eq. (4.34) to find the free-particle Green function $G(\Omega \,|\, \Omega_0; L)$ [18]. For this purpose it is convenient to choose as the basis functions the (normalized) *Wigner functions* $\mathcal{D}_l^{mj}(\Omega)$ of the Euler angles Ω [16, 17]. They are the mj elements $\langle lm|\mathcal{R}(\Omega)|lj\rangle$ of the lth rank rotation matrix with $\mathcal{R}(\Omega)$ being the operator of the finite rotation Ω of the coordinate system and are explicity defined in Appendix 4.C. In the remainder of this chapter all lengths are measured in units of λ^{-1} unless otherwise noted, for simplicity. Then, for instance, $\lambda^{-1}\kappa_0$ and $\lambda^{-1}\tau_0$ are replaced by (reduced) κ_0 and τ_0, respectively.

Now the solution of Eq. (4.34) may be expanded in the form

$$G(\Omega \,|\, \Omega_0; L) = \sum_{l=0}^{\infty} \sum_{m=-l}^{l} \sum_{j=-l}^{l} \sum_{j'=-l}^{l} g_l^{jj'}(L)\mathcal{D}_l^{mj}(\Omega)\mathcal{D}_l^{mj'*}(\Omega_0)\,, \quad (4.36)$$

where $g_l^{jj'}(L)$ are the expansion coefficients to be determined, the asterisk indicates the complex conjugate, and the boundary condition is given, from the closure relation of Eq. (4.C.9), by

$$G(\Omega \,|\, \Omega_0; 0) = \delta(\Omega - \Omega_0)$$

$$= \sum_{l=0}^{\infty} \sum_{m=-l}^{l} \sum_{j=-l}^{l} \mathcal{D}_l^{mj}(\Omega)\mathcal{D}_l^{mj*}(\Omega_0)\,. \quad (4.37)$$

The coefficients $g_l^{jj'}$ may be determined by substitution of Eqs. (4.31), (4.36), and (4.37) into Eq. (4.34), noting that the components L_μ of $\mathbf{L}$ operate on $\mathcal{D}_l^{mj}$ as follows [17],

$$L_\xi \mathcal{D}_l^{mj} = \frac{1}{2}ic_l^j \mathcal{D}_l^{m(j+1)} + \frac{1}{2}ic_l^{-j}\mathcal{D}_l^{m(j-1)}\,,$$

$$L_\eta \mathcal{D}_l^{mj} = -\frac{1}{2}c_l^j \mathcal{D}_l^{m(j+1)} + \frac{1}{2}c_l^{-j}\mathcal{D}_l^{m(j-1)}\,, \quad (4.38)$$

$$L_\zeta \mathcal{D}_l^{mj} = ij\mathcal{D}_l^{mj}\,,$$

together with the first of Eqs. (4.C.16), where i is the imaginary unit and c_l^j is defined by

$$c_l^j = \left[(l-j)(l+j+1)\right]^{1/2}\,. \quad (4.39)$$

If $\tilde{g}_l^{jj'}$ is the Laplace transform of $g_l^{jj'}$,

$$\tilde{g}_l^{jj'}(p) = \int_0^{\infty} g_l^{jj'}(L)\exp(-pL)dL\,, \quad (4.40)$$

then $\tilde{g}_l^{jj'}$ satisfies the equations

$$\delta_{jj'} = \left[p + l(l+1) + ij\tau_0 + \sigma j^2\right]\tilde{g}_l^{jj'}$$

$$+\frac{1}{2}\kappa_0 c_l^j \tilde{g}_l^{(j+1)j'} - \frac{1}{2}\kappa_0 c_l^{-j}\tilde{g}_l^{(j-1)j'} \quad (|j|,\ |j'| \leq l)\,. \quad (4.41)$$

If we introduce a $(2l+1) \times (2l+1)$ matrix $\mathbf{A}_l(p)$ whose elements $A_{l,jj'}$ ($|j|$, $|j'| \leq l$) are given by

$$A_{l,jj'} = p + l(l+1) + ij\tau_0 + \sigma j^2 \qquad \text{for } j' = j$$

$$= \frac{1}{2}\kappa_0 c_l^j \qquad \text{for } j' = j+1$$

$$= -\frac{1}{2}\kappa_0 c_l^{-j} \qquad \text{for } j' = j-1$$

$$= 0 \qquad \text{otherwise}, \tag{4.42}$$

then the solution of Eq. (4.41) is

$$\tilde{g}_l^{jj'}(p) = \frac{A_l^{j'j}(p)}{\displaystyle\prod_{j=-l}^{l}(p + z_{l,j})}, \tag{4.43}$$

where $A_l^{j'j}$ is the cofactor of the element $A_{l,j'j}$ and $-z_{l,j}$ are the $2l+1$ roots of the algebraic equation of degree $2l+1$,

$$|\mathbf{A}_l(p)| = 0 \tag{4.44}$$

with $|\mathbf{A}_l|$ being the determinant of $\mathbf{A}_l$. In the particular case of $\sigma = 0$, $z_{l,j}$ is given by [19]

$$z_{l,j} = l(l+1) + ij(\kappa_0^2 + \tau_0^2)^{1/2} \qquad (|j| \le l; \sigma = 0). \tag{4.45}$$

By Laplace inversion of Eq. (4.43), we then find $g_l^{jj'}$ as a sum of residues of $e^{Lp}\tilde{g}_l^{jj'}$.

Now, multiplying both sides of Eq. (4.36) by $\mathcal{D}_l^{mj*}(\Omega)\mathcal{D}_l^{mj'}(\Omega_0)$ and integrating over Ω and Ω_0 with the use of the orthonormality relation of Eq. (4.C.8), we obtain

$$g_l^{jj'}(L) = 8\pi^2 \langle \mathcal{D}_l^{mj*}(\Omega)\mathcal{D}_l^{mj'}(\Omega_0)\rangle, \tag{4.46}$$

where

$$\langle \cdots \rangle = (8\pi^2)^{-1}\int(\cdots)G(\Omega\,|\,\Omega_0; L)d\Omega d\Omega_0. \tag{4.47}$$

Thus $g_l^{jj'}$ have the meaning of the (time-independent) *angular correlation functions*. As seen later, all kinds of equilibrium moments or properties may in principle be expressed in terms of them, so that they are the fundamental quantities in the equilibrium statistical mechanics of the HW chain. Their behavior is examined in detail in Sect. 4.4.

Finally, in the particular case of the KP chain ($\kappa_0 = 0$), we readily have

$$g_l^{jj'}(L) = \delta_{jj'}\exp\{-[l(l+1) + ij\tau_0 + \sigma j^2]L\} \qquad \text{(KP)}. \tag{4.48}$$

Then, if integration over ψ is carried out, Eq. (4.33) reduces to Eq. (3.31), the Fokker–Planck equation for $G(\mathbf{R},\mathbf{u}\,|\,\mathbf{u}_0; L)$, and Eq. (4.36) becomes Eq. (3.42), the expansion of $G(\mathbf{u}\,|\,\mathbf{u}_0; L)$ in terms of Y_l^m with the expansion coefficients $g_l^{00}(L) = \exp[-l(l+1)L]$. Thus the HW chain with $\kappa_0 = 0$ is just identical with the original KP chain only as far as the behavior of the chain contour is concerned.

4.2.3 Formal Solutions

An integral equation for the characteristic function $I(\mathbf{k}, \Omega \,|\, \Omega_0; L)$ may be derived in the same manner as in the derivation of Eq. (3.41) for the KP chain. The result is

$$I(\mathbf{k}, \Omega \,|\, \Omega_0; L) = G(\Omega \,|\, \Omega_0; L) + i\mathbf{k} \cdot \int_0^L \int \mathbf{u}_1 G(\Omega \,|\, \Omega_1; L - s_1)$$
$$\times I(\mathbf{k}, \Omega_1 \,|\, \Omega_0; s_1) ds_1 d\Omega_1 \,. \tag{4.49}$$

Integration of both sides of Eq. (4.49) over $\mathbf{a}_0$ and division by 2π leads to

$$I(\mathbf{k}, \Omega \,|\, \mathbf{u}_0; L) = G(\Omega \,|\, \mathbf{u}_0; L) + i\mathbf{k} \cdot \int_0^L \int \mathbf{u}_1 G(\Omega \,|\, \Omega_1; L - s_1)$$
$$\times I(\mathbf{k}, \Omega_1 \,|\, \mathbf{u}_0; s_1) ds_1 d\Omega_1 \,. \tag{4.50}$$

Further integration over $\mathbf{u}_0$ and division by 4π leads to

$$I(\mathbf{k}, \Omega; L) = G(\Omega; L) + i\mathbf{k} \cdot \int_0^L \int \mathbf{u}_1 G(\Omega \,|\, \Omega_1; L - s_1)$$
$$\times I(\mathbf{k}, \Omega_1; s_1) ds_1 d\Omega_1 \,. \tag{4.51}$$

We find here the formal solutions of Eqs. (4.50) and (4.51) [18] to derive operational expressions for the moments $\langle (\mathbf{R} \cdot \mathbf{u}_0)^n \rangle$ and $\langle R^{2m} \rangle$ in the next section. If $\mathbf{u}_0$ is chosen to be in the direction of the z axis of an external Cartesian coordinate system ($\mathbf{u}_0 = \mathbf{e}_z$), the known parts of the integral Eqs. (4.50) and (4.51) may be written as

$$G(\Omega \,|\, \mathbf{u}_0; L) = \sum_{l=0}^{\infty} \sum_{j=-l}^{l} c_l g_l^{j0} \mathcal{D}_l^{0j}(\Omega) \,, \tag{4.52}$$

$$G(\Omega; L) = c_0 g_0^{00} \mathcal{D}_0^{00}(\Omega) \,, \tag{4.53}$$

where

$$c_l = \left(\frac{2l + 1}{8\pi^2} \right)^{1/2} \,. \tag{4.54}$$

Now both $I(\mathbf{k}, \Omega \,|\, \mathbf{u}_0; L)$ and $I(\mathbf{k}, \Omega; L)$, which we simply denote by $I(L)$, may be expanded in the form

$$I(L) = \sum_{l,m,j} K_l^{mj}(L) \mathcal{D}_l^{mj}(\Omega) \,, \tag{4.55}$$

where the sums over $\mathcal{D}_l^{mj}$ are taken over $l \geq 0$, $|m| \leq l$, and $|j| \leq l$ unless otherwise specified, and $K_l^{mj}(L)$ stands for $K_l^{mj}(\mathbf{k} \,|\, \mathbf{u}_0; L)$ or $K_l^{mj}(\mathbf{k}; L)$, as the case may be. We express $\mathbf{e}_k = \mathbf{k}/k$ and $\mathbf{u}_1$ as $\mathbf{e}_k = (1, \chi, \omega)$ and $\mathbf{u}_1 =$

$(1, \theta_1, \phi_1)$ in spherical polar coordinates. It is then convenient to rewrite Eq. (3.44) as

$$\mathbf{e}_k \cdot \mathbf{u}_1 = \frac{8\pi^2}{3} \sum_m \mathcal{D}_1^{m0}(\tilde{\Omega}) \mathcal{D}_1^{m0*}(\Omega_1), \tag{4.56}$$

where $\tilde{\Omega} = (\chi, \omega, 0)$, $\Omega_1 = (\theta_1, \phi_1, \psi_1)$, and we have used Eq. (4.C.4). Note that $\mathbf{e}_k \cdot \mathbf{u}_1$ is independent of ψ_1.

Substitution of Eqs. (4.55) and (4.56) into Eqs. (4.50) and (4.51) with Eqs. (4.36), (4.52), and (4.53) and integration over Ω_1 with the use of Eqs. (4.C.7) and (4.C.11) leads to the integral equation for $K_l^{mj}(L)$,

$$K_l^{mj} = f_l^{mj} + i\bar{k} \sum_{j'} g_l^{jj'} * \mathcal{L}_{j'} K_l^{mj}, \tag{4.57}$$

where the asterisk indicates the convolution integration defined by Eq. (3.46), $\bar{k}$ is given by

$$\bar{k} = \left(\frac{4\pi}{3}\right)^{1/2} k \tag{4.58}$$

instead of by Eq. (3.47), and f_l^{mj} are given by

$$\begin{aligned} f_l^{mj} &= c_l g_l^{j0} \delta_{m0} && \text{for } K_l^{mj} = K_l^{mj}(\mathbf{k} \,|\, \mathbf{u}_0; L) \\ &= c_0 g_0^{00} \delta_{l0} \delta_{m0} \delta_{j0} && \text{for } K_l^{mj} = K_l^{mj}(\mathbf{k}; L) \end{aligned} \tag{4.59}$$

with c_l being given by Eq. (4.54). In Eq. (4.57) $\mathcal{L}_{j'}$ is an operator defined by

$$\mathcal{L}_{j'} = \sum_{\nu=-1}^{1} [2h(\nu) - 1] Y_1^{\nu}(\chi, \omega) \sum_{\mu=-1}^{1} a_\mu^{\nu j'}, \tag{4.60}$$

where h is the same unit step function as that in the second of Eqs. (3.52), and $a_\mu^{\nu j}$ $(\mu, \nu, j = 0, \pm 1)$ are generalized creation and annihilation operators which operate on $g_l^{jj'}$ as

$$a_\mu^{\nu j''} g_l^{jj'} = g_{l+\mu}^{j'' j'}, \tag{4.61}$$

and on K_l^{mj} and f_l^{mj} as

$$a_\mu^{\nu j'} K_l^{mj} = (-1)^{m-j'} [(2l+1)(2l+2\mu+1)]^{1/2}$$
$$\times \begin{pmatrix} l & 1 & l+\mu \\ -m & -\nu & m+\nu \end{pmatrix} \begin{pmatrix} l & 1 & l+\mu \\ -j' & 0 & j' \end{pmatrix} K_{l+\mu}^{(m+\nu)j'} \tag{4.62}$$

with (:::) being the *Wigner 3-j symbol*. Its definition and properties are given in Appendix 4.C.

Thus the solution for K_l^{mj} may be expressed as

$$K_l^{mj} = \sum_{n=0}^{\infty} (i\bar{k})^n \left(\sum_{j'} g_l^{jj'} * \mathcal{L}_{j'}\right)^n f_l^{mj}. \tag{4.63}$$

Then Eq. (4.55) with Eq. (4.63) gives the desired formal solutions. Integration of the results over Ω leads to

$$I(\mathbf{k}\,|\,\mathbf{u}_0; L) = 2^{3/2}\pi \sum_{n=0}^{\infty}(i\bar{k})^n \left(\sum_j g_0^{0j} * \mathcal{L}_j\right)^n f_0^{00}, \qquad (4.64)$$

$$I(\mathbf{k}; L) = 2^{3/2}\pi \sum_{m=0}^{\infty}(-1)^m \bar{k}^{2m} \left(\sum_j g_0^{0j} * \mathcal{L}_j\right)^{2m} f_0^{00}, \qquad (4.65)$$

where the prime on j has been omitted, and the range of summation over j is explicitly shown in the next subsection.

4.3 Moments

The moments $\langle \mathbf{R} \cdot \mathbf{u}_0 \rangle$ and $\langle R^2 \rangle$ (and hence also $\langle S^2 \rangle$) can readily be obtained [7] from Eq. (4.33) by the procedure of Hermans and Ullman [20]. In general, however, the moments $\langle (\mathbf{R} \cdot \mathbf{u}_0)^n \rangle$ and $\langle R^{2m} \rangle$ may be more efficiently evaluated from operational expressions for them [18] as in the case of the KP chain.

4.3.1 $\langle (\mathbf{R} \cdot \mathbf{u}_0)^n \rangle$

Expanding $(\ )^n f_0^{00}$ in Eq. (4.64), we obtain

$$I(\mathbf{k}\,|\,\mathbf{u}_0; L) = \sum_{n=0}^{\infty}(ik)^n \sum_{q\leq n}(2q+1)^{1/2} \sum_{\substack{\text{paths} \\ (0\to q)}} \sum_{\nu} \sum_{j}(-1)^{(1/2)(n-n^0)}$$
$$\times 2^{-(1/2)(n-n^0)} C_\mu^{\nu j} \Gamma_{0\cdots q}^{j}(L)\cos^{n^0}\chi\,\sin^{(n-n^0)}\chi, \qquad (4.66)$$

where

$$C_\mu^{\nu j} = (f_q^{0j_n})^{-1} a_{\mu_n}^{\nu_n j_n} a_{\mu_{n-1}}^{\nu_{n-1}j_{n-1}} \cdots a_{\mu_2}^{\nu_2 j_2} a_{\mu_1}^{\nu_1 0} f_0^{00}, \qquad (4.67)$$

$$\Gamma_{0\cdots q}^{j}(L) = g_0^{00} * g_{l_1}^{0j_2} * g_{l_2}^{j_2 j_3} * \cdots * g_{l_{n-1}}^{j_{n-1}j_n} * g_q^{j_n 0} \qquad (4.68)$$

with n^0 the number of a_μ^{0j} in $C_\mu^{\nu j}$ and with

$$l_r = \sum_{i=1}^{r}\mu_i \geq 0 \qquad (l_0 = 0,\ l_n = q), \qquad (4.69)$$

$$\sum_{i=1}^{n}\nu_i = 0, \qquad (4.70)$$

$$|j_r| \leq \min(l_r, l_{r-1}) \qquad (j_0 = j_1 = j_{n+1} = 0). \qquad (4.71)$$

Note that $C_\mu^{\nu j}$ is a constant independent of L, that Eq. (4.70) holds because of the first line of Eqs. (4.59), and that $\min(a,b)$ denotes the smaller of a and b. $\Gamma_{0\cdots q}^j(L)$ may be expressed as a sum of residues p_i of the function $Q(p)$,

$$Q(p) = \frac{e^{Lp} \prod\limits_{r=1}^{n} A_{l_r}^{j_{r+1}j_r}(p)}{\prod\limits_{r=0}^{n} \prod\limits_{k=-l_r}^{l_r} (p + z_{l_r,k})}, \qquad (4.72)$$

where $A_l^{j'j}(p)$ and $z_{l,k}$ have been defined in Eq. (4.43). In Eq. (4.66), the third sum is taken over all possible paths $(0l_1l_2\cdots l_{n-1}q)$ from 0 to q, which are different from those for the KP chain since the case of $\mu_i = 0$ may occur in the present case, the fourth sum is taken over $\nu_1, \cdots, \nu_n$ compatible with Eq. (4.70), and the fifth sum is taken over $j_1, \cdots, j_n$ compatible with Eq. (4.71).

If we choose $\mathbf{e}_k = \mathbf{e}_z(= \mathbf{u}_0)$, that is, $\chi = 0$, and compare Eq. (4.66) with Eq. (3.57), then we obtain

$$\langle (\mathbf{R}\cdot\mathbf{u}_0)^n \rangle = n! \sum_{q\leq n}(2q+1)^{1/2} \sum_{\substack{\text{paths}\\(0\to q)}} \sum_j C_\mu^{0j} \Gamma_{0\cdots q}^j(L), \qquad (4.73)$$

where only the terms with $n = n^0$, that is, $\nu_i = 0$ for all i make contribution, so that μ_i is nonzero and the paths are the same as those for the KP chain.

In particular, $\langle \mathbf{R}\cdot\mathbf{u}_0 \rangle$ for $\sigma = 0$ is given by

$$\langle \mathbf{R}\cdot\mathbf{u}_0 \rangle = \frac{1}{2}c_\infty - \frac{1}{\nu^2}e^{-2L}\left\{\frac{1}{2}\tau_0{}^2 + \frac{\kappa_0{}^2}{r^2}[2\cos(\nu L) - \nu\sin(\nu L)]\right\}, \qquad (4.74)$$

where

$$c_\infty = \frac{4 + \tau_0{}^2}{4 + \kappa_0{}^2 + \tau_0{}^2}, \qquad (4.75)$$

$$\nu = (\kappa_0{}^2 + \tau_0{}^2)^{1/2}, \qquad (4.76)$$

$$r = (4 + \nu^2)^{1/2}. \qquad (4.77)$$

4.3.2 $\langle R^{2m} \rangle$ and $\langle S^2 \rangle$

By expanding $(\quad)^{2m} f_0^{00}$, Eq. (4.65) may be reduced to Eq. (3.67) with

$$\langle R^{2m} \rangle = (2m+1)! \sum_{\substack{\text{paths}\\(0\to 0)}} \sum_j C_\mu^{0j} \Gamma_{0\cdots 0}^j(L), \qquad (4.78)$$

where the paths are again the same as those for the KP chain. We note that when $\kappa_0 = 0$, only the terms with $j_i = 0$ for all i make contribution and Eqs. (4.73) and (4.78) reduce to Eqs. (3.64) and (3.69), respectively. In the calculation of $\langle R^{2m} \rangle$, Eq. (4.72) reduces to

$$
Q(p) = \frac{e^{Lp} \prod\limits_{r=1}^{2m-1} A_{l_r}^{j_{r+1} j_r}(p)}{\prod\limits_{j=0}^{m} \left[\prod\limits_{k=-j}^{j} (p + z_{j,k}) \right]^{x_j}}, \tag{4.79}
$$

where x_j is the number of the factors with $l_r = j$ for a given k in the denominator of Eq. (4.72). If we assume that all $z_{j,k}$ are different and that the right-hand side of Eq. (4.79) is already a simple fraction, the formula for residues gives

$$
\Gamma_{0\cdots0}^{j}(L) = \sum_{\substack{j=0 \\ x_j \neq 0}}^{m} \frac{1}{(x_j - 1)!} \sum_{k=-j}^{j} \left[\frac{d^{x_j - 1}}{dp^{x_j - 1}} (p + z_{j,k})^{x_j} Q(p) \right]_{p=-z_{j,k}} . \tag{4.80}
$$

Thus, on recalling that $x_j \leq m - j + 1$, $\langle R^{2m} \rangle$ may be written in the form

$$
\langle R^{2m} \rangle = \sum_{j=0}^{m} \sum_{i=j}^{m} \sum_{k=-j}^{j} A_{ijk}^{(m)} L^{i-j} \exp(-z_{j,k} L), \tag{4.81}
$$

where $A_{ijk}^{(m)}$ are coefficients independent of L but dependent on κ_0, τ_0, and σ.

It has been found that $\langle R^{2m} \rangle$ are rather insensitive to change in σ for flexible chains [18]. Thus, in the remainder of this book, we set $\sigma = 0$ for these chains, for simplicity. (In the case of, for instance, circular DNA, we consider the KP1 chain with $\sigma \neq 0$.) In the case of $\sigma = 0$, in Eq. (4.81) $z_{j,k}$ is given by Eq. (4.45) and the coefficients $A_{ijk}^{(m)}$ ($m = 1, 2$) as functions of κ_0 and τ_0 are given in Appendix II. In particular, $\langle R^2 \rangle$ is given by

$$
\langle R^2 \rangle = c_\infty L - \frac{\tau_0^2}{2\nu^2} - \frac{2\kappa_0^2 (4 - \nu^2)}{\nu^2 r^4}
$$
$$
+ \frac{e^{-2L}}{\nu^2} \left\{ \frac{\tau_0^2}{2} + \frac{2\kappa_0^2}{r^4} \left[(4 - \nu^2) \cos(\nu L) - 4\nu \sin(\nu L) \right] \right\}. \tag{4.82}
$$

From Eqs. (2.25) and (4.82), we then obtain for the mean-square radius of gyration $\langle S^2 \rangle$

$$
\langle S^2 \rangle = \frac{\tau_0^2}{\nu^2} \langle S^2 \rangle_{KP} + \frac{\kappa_0^2}{\nu^2 r^2} \left[\frac{rL}{3} - \cos(2\varphi) + \frac{2}{rL} (\cos 3\varphi) \right.
$$
$$
\left. - \frac{2}{r^2 L^2} \cos(4\varphi) + \frac{2e^{-2L}}{r^2 L^2} \cos(\nu L + 4\varphi) \right], \tag{4.83}
$$

where

$$\varphi = \cos^{-1}\left(\frac{2}{r}\right), \tag{4.84}$$

and $\langle S^2\rangle_{\mathrm{KP}}$ is the $\langle S^2\rangle$ for the KP chain of the same contour length and is given by Eq. (3.6), that is,

$$\langle S^2\rangle_{\mathrm{KP}} = \frac{L}{6} - \frac{1}{4} + \frac{1}{4L} - \frac{1}{8L^2}(1 - e^{-2L}). \tag{4.85}$$

It is seen from Eqs. (4.82) and (4.83) that in the limit of $L \to 0$ (at finite $\kappa_0 < \infty$) the rod limits of Eqs. (3.7) are still obtained, while in the random-coil limit of $L \to \infty$ there hold the relations

$$\langle R^2\rangle = 6\langle S^2\rangle = c_\infty L \qquad \text{for } L \to \infty. \tag{4.86}$$

If lengths are unreduced, we obtain, from Eqs. (2.27), (2.28), (4.74), and (4.82), for the Kuhn segment length A_{K} and the persistence length q,

$$A_{\mathrm{K}} = 2q = c_\infty \lambda^{-1} \le \lambda^{-1}, \tag{4.87}$$

where the third inequality holds since $c_\infty \le 1$, as seen from Eq. (4.75), the third equality holding in the case of the KP chain ($\kappa_0 = 0$ and $c_\infty = 1$).

In order to apply the continuous model to a real chain, the total contour length L of the former must be converted to the number of repeat units (degree of polymerization) x or the molecular weight M. This is done conveniently by introducing a *shift factor* M_{L} as defined by $M_{\mathrm{L}} = M/(\text{unreduced})L$. Thus, in the case of the HW model for flexible polymers, $\kappa_0(\text{reduced})$, $\tau_0(\text{reduced})$, λ^{-1}, and M_{L} may be chosen as the basic model parameters (with $\sigma = 0$). In this subsection we compare HW values with RIS values for $\langle R^2\rangle$, which is in general experimentally unobservable. (A comparison with experiment is made for $\langle S^2\rangle$ in the next chapter.)

Now we equate the $\langle R^2\rangle$ for the HW and RIS models to each other, so that the characteristic ratio C_n of the latter ($n = 2x$) may be related to the $\langle R^2\rangle$ of the former, which we denote by $\langle R^2\rangle_{\mathrm{HW}} = f_R(L; \kappa_0, \tau_0)$, by the equation

$$C_n = \frac{1}{2}(\lambda l)^{-2}\delta^{-1}L^{-1}f_R(L; \kappa_0, \tau_0) \tag{4.88}$$

with

$$\log x = \log L + \log \delta, \tag{4.89}$$

where l is the bond length and $\delta = \lambda^{-1}M_0/M_{\mathrm{L}}$ with M_0 the molecular weight of the repeat unit. The quantities $(\lambda l)^{-2}\delta^{-1}$ and δ may then be determined from a best fit of a plot of the quantity on the right-hand side of Eq. (4.88) against $\log L$ for properly chosen values of κ_0 and τ_0 to that of C_n against $\log x$. Thus we may determine the HW model parameters κ_0, τ_0, λ^{-1}, and M_{L} for a given RIS chain. Figure 4.5, which corresponds to Fig. 2.4, shows such plots, where the points represent the RIS values for PM [21], PDMS [21,22],

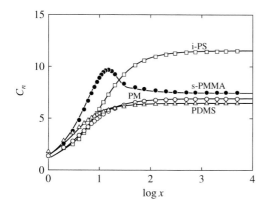

Fig. 4.5. Characteristic ratio C_n plotted against $\log x$ with x the number of repeat units. The points represent the RIS values [6, 21–23], and the curves represent the corresponding best-fit HW values

Table 4.1. Values of the HW model parameters from RIS values of C_n

Polymer	Temp. (°C)	κ_0	$\lvert\tau_0\rvert$	λ^{-1} (Å)	M_L (Å^{-1})
PM	140	0.6	0	14.5	11.5
PDMS	110	1.2	0	16.0	25.0
i-PS	27	8.5	15.0	29.4	43.0
s-PMMA	27	4.2	1.0	60.0	38.0

i-PS [23], and s-PMMA [6], and the curves represent the corresponding best-fit HW values. The values of the HW model parameters so determined are given in Table 4.1. A discussion of the results is deferred to Sect. 4.4.4. We only note that it is difficult to determine unambiguously these parameters from C_n for the first three polymers in Table 4.1.

4.3.3 Persistence Vector

The *persistence vector* $\mathbf{A}$ of the HW chain is defined as the average of the end-to-end vector $\mathbf{R}$ with the orientation Ω_0 of the initial localized Cartesian coordinate system $(\mathbf{e}_{\xi_0}, \mathbf{e}_{\eta_0}, \mathbf{e}_{\zeta_0})$ fixed [24],

$$\mathbf{A} = \langle \mathbf{R} \rangle_{\Omega_0} . \tag{4.90}$$

In what follows, we omit the subscript 0 which refers to the initial localized system. We then express $\mathbf{A}$ as

$$\mathbf{A} = \langle \xi \rangle \mathbf{e}_\xi + \langle \eta \rangle \mathbf{e}_\eta + \langle \zeta \rangle \mathbf{e}_\zeta , \tag{4.91}$$

where it is evident that $\langle \xi \rangle = \langle \mathbf{R} \cdot \mathbf{a}_0 \rangle = -\langle \mathbf{R} \cdot \mathbf{a} \rangle$, $\langle \eta \rangle = \langle \mathbf{R} \cdot \mathbf{b}_0 \rangle = \langle \mathbf{R} \cdot \mathbf{b} \rangle$, and $\langle \zeta \rangle = \langle \mathbf{R} \cdot \mathbf{u}_0 \rangle = \langle \mathbf{R} \cdot \mathbf{u} \rangle$. The reason for the nomenclature of $\mathbf{A}$ is that the persistence length q may also be expressed in terms of $\mathbf{A}$ as

$$q = \lim_{L \to \infty} \mathbf{A} \cdot \mathbf{u}_0 , \qquad (4.92)$$

as seen from Eq. (4.91).

For flexible chains ($\sigma = 0$) the above components of $\mathbf{A}$ (in the initial localized system) may readily be evaluated to be [24, 25]

$$\langle \xi \rangle = \frac{\kappa_0}{r^2} - \frac{\kappa_0}{\nu r^2} e^{-2L} \left[\nu \cos(\nu L) + 2 \sin(\nu L) \right] ,$$

$$\langle \eta \rangle = \frac{\kappa_0 \tau_0}{2r^2} - \frac{\kappa_0 \tau_0}{\nu^2} e^{-2L} \left\{ \frac{1}{2} - \frac{1}{r^2} \left[2 \cos(\nu L) - \nu \sin(\nu L) \right] \right\} , \qquad (4.93)$$

$$\langle \zeta \rangle = \langle \mathbf{R} \cdot \mathbf{u}_0 \rangle ,$$

where $\langle \mathbf{R} \cdot \mathbf{u}_0 \rangle$ is given by Eq. (4.74).

For the RIS model Flory [26, 27] has defined its persistence vector $\mathbf{A}$ ($\mathbf{a}$ in his notation) as the mean end-to-end vector $\langle \mathbf{R} \rangle_{1,2}$ with the first and second bonds fixed, and used a molecular Cartesian coordinate system ($\mathbf{e}_x$, $\mathbf{e}_y$, $\mathbf{e}_z$) such that the x axis is taken along the first bond, the y axis is in the plane of the first and second bonds with its direction chosen at an acute angle with the second bond, and the z axis completes the right-handed system, as depicted in Fig. 4.6. Thus we express this $\mathbf{A}$ as

$$\mathbf{A} = \langle \mathbf{R} \rangle_{1,2} = \langle x \rangle \mathbf{e}_x + \langle y \rangle \mathbf{e}_y + \langle z \rangle \mathbf{e}_z . \qquad (4.94)$$

Now we wish to equate the $\mathbf{A}$ for the two models in order to compare them. It should then be noted that the vector $\mathbf{u}_0$ of the HW chain is not necessarily in the direction of the first bond of the RIS chain (see Fig. 4.6). Therefore, suppose that the (initial) HW localized coordinate system ($\mathbf{e}_\xi$, $\mathbf{e}_\eta$, $\mathbf{e}_\zeta$) is obtained by rotation of the (initial) RIS model coordinate system ($\mathbf{e}_x$, $\mathbf{e}_y$, $\mathbf{e}_z$) by the Euler angles $\tilde{\Omega} = (\tilde{\theta}, \tilde{\phi}, \tilde{\psi})$ [not to be confused with $\tilde{\Omega}$ in Eq. (4.56)]. If $\mathbf{A}(\xi, \eta, \zeta)$ and $\mathbf{A}(x, y, z)$ are the column forms of $\mathbf{A}$ of the HW chain in the two systems, we have

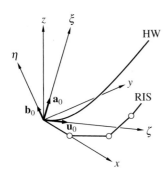

Fig. 4.6. Initial localized Cartesian coordinate systems (ξ, η, ζ) and (x, y, z) of the HW chain and the RIS model, respectively

$$\mathbf{A}(\xi, \eta, \zeta) = \mathbf{Q}(\tilde{\Omega}) \cdot \mathbf{A}(x, y, z) \,, \qquad (4.95)$$

where $\mathbf{Q}$ is the rotational transformation matrix and is given by

$$\mathbf{Q} = \begin{pmatrix} c_{\tilde{\theta}} c_{\tilde{\phi}} c_{\tilde{\psi}} - s_{\tilde{\phi}} s_{\tilde{\psi}} & c_{\tilde{\theta}} s_{\tilde{\phi}} c_{\tilde{\psi}} + c_{\tilde{\phi}} s_{\tilde{\psi}} & -s_{\tilde{\theta}} c_{\tilde{\psi}} \\ -c_{\tilde{\theta}} c_{\tilde{\phi}} s_{\tilde{\psi}} - s_{\tilde{\phi}} c_{\tilde{\psi}} & -c_{\tilde{\theta}} s_{\tilde{\phi}} s_{\tilde{\psi}} + c_{\tilde{\phi}} c_{\tilde{\psi}} & s_{\tilde{\theta}} s_{\tilde{\psi}} \\ s_{\tilde{\theta}} c_{\tilde{\phi}} & s_{\tilde{\theta}} s_{\tilde{\phi}} & c_{\tilde{\theta}} \end{pmatrix} \,. \qquad (4.96)$$

with $s_{\tilde{\theta}} = \sin\tilde{\theta}$, $c_{\tilde{\theta}} = \cos\tilde{\theta}$, and so on. Thus the components of $\mathbf{A}(x, y, z)$ calculated from the inverse of Eq. (4.95) by assigning proper values to $\tilde{\Omega}$ may be equated to $\langle x \rangle$, $\langle y \rangle$, and $\langle z \rangle$ (of the RIS model).

The comparison is made as follows. The parameters to be determined are κ_0, τ_0, λ^{-1}, M_L, and $\tilde{\Omega}$. Let $\mathbf{A}_\infty$ be $\mathbf{A}$ for an infinitely long chain. We first equate $A_\infty = |\mathbf{A}_\infty|$ of the HW chain to that of the RIS model, that is,

$$\lambda^{-1} A_{\infty,\mathrm{HW}} = A_{\infty,\mathrm{RIS}} \,, \qquad (4.97)$$

where we note that the $\mathbf{A}$ of the latter has not been reduced by λ^{-1}. $A_{\infty,\mathrm{HW}}$ may be computed from Eqs. (4.93) for properly chosen values of κ_0 and τ_0, so that λ^{-1} may be determined from Eq. (4.97) with the value of $A_{\infty,\mathrm{RIS}}$. With these values of κ_0, τ_0, λ^{-1}, and A_∞, $\tilde{\Omega}$ may then be determined to give the coincidence between the directions of $\mathbf{A}_\infty$ of the two models and also a best fit of values of the components of $\lambda^{-1}\mathbf{A}(x, y, z)$ of the HW chain as a function of L to those of $\langle x \rangle$, $\langle y \rangle$, and $\langle z \rangle$ of the RIS model. Finally, M_L may be determined, by the use of Eq. (4.89), from a best fit of values of $\lambda^{-1} A_{\mathrm{HW}}$ as a function of L to those of A_{RIS} as a function of x (number of repeat units), where $A = |\mathbf{A}|$.

In Table 4.2 are given the results of such an analysis made [24] using the RIS values for PM [28], PDMS [22], i-PS [29], and s-PMMA [29]. For illustration, Figs. 4.7 and 4.8 show plots of $\langle y \rangle$ and $\langle z \rangle$ against $\langle x \rangle$ for PM and s-PMMA, respectively. The filled and unfilled circles represent the RIS values of $\langle y \rangle$ and $\langle z \rangle$, respectively, the attached numbers indicating the values of x, and the curves represent the corresponding best-fit HW values.

For the KP chain, the components $\langle \xi \rangle$ and $\langle \eta \rangle$ vanish, so that both $\langle y \rangle$ and $\langle z \rangle$ are directly proportional to $\langle x \rangle$. It is seen from Fig. 4.7 that for the PM chain, whose κ_0 is rather small, $\langle y \rangle$ and $\langle z \rangle$ are nearly proportional to

Table 4.2. Values of the HW model parameters from RIS values of $\mathbf{A}$

Polymer	$\tilde{\theta}$ (deg)	$\tilde{\phi}$ (deg)	$\tilde{\psi}$ (deg)	κ_0	τ_0	λ^{-1} (Å)	M_L (Å^{-1})
PM	85.3	35.8	340.0	0.5	4.0	15.2	10.1
PDMS	89.4	23.0	270.0	0.8	0.05	15.8	19.6
i-PS	120.1	41.2	230.0	13.5	−16.5	33.5	41.2
s-PMMA	69.6	28.4	67.0	3.7	0.3	54.0	35.7

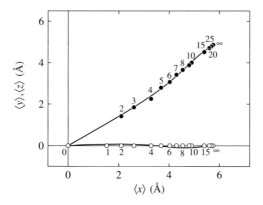

Fig. 4.7. The components $\langle y \rangle$ and $\langle z \rangle$ of the persistence vector **A** plotted against the component $\langle x \rangle$ for PM. The *filled and unfilled circles* represent the RIS values [28] of $\langle y \rangle$ and $\langle z \rangle$, respectively, *the attached numbers* indicating the values of x, and the curves represent the corresponding best-fit HW values

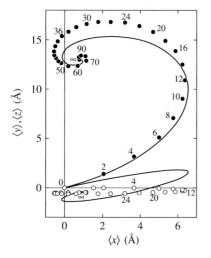

Fig. 4.8. Persistence vector **A** for s-PMMA; see legend for Fig. 4.7. The RIS values have been taken from [29]

$\langle x \rangle$, it being indeed close to the KP chain. On the other hand, it is seen from Fig. 4.8 that the s-PMMA chain has locally helical conformations. Thus it is a typical HW chain, whose C_n as a function of x (or n) also exhibits salient behavior, it passing through a maximum at some value of x before reaching C_∞ (see Fig. 4.5).

4.4 Angular Correlation Functions

In this section we examine in detail the behavior of the angular correlation functions $g_l^{jj'}(L)$ for flexible chains ($\sigma = 0$) and compare them with the corresponding functions $g_l^{jj'}(x)$ properly defined as functions of the number of repeat units x for the RIS model [19]. Baram and Gelbart [30] also considered an "angular correlation function" for the RIS model, but it rather corresponds to the free-particle Green function, its moments corresponding to the present $g_l^{jj'}(x)$ defined in the Flory localized coordinate system.

4.4.1 Explicit Expressions for $\sigma = 0$

In the particular case of $\sigma = 0$ we derive an explicit expression for $g_l^{jj'}(L)$ by expanding $G(\Omega \,|\, \Omega_0; L)$ in terms of the eigenfunctions of the operator $\mathcal{A}$ given by Eq. (4.31), which we denote by $\Psi_l^{mj}(\Omega)$, instead of $\mathcal{D}_l^{mj}(\Omega)$ as follows,

$$G(\Omega \,|\, \Omega_0; L) = \sum_{l,m,j} h_l^j(L) \Psi_l^{mj}(\Omega) \Psi_l^{mj*}(\Omega_0). \qquad (4.98)$$

Thus we first consider the eigenvalue problem for $\mathcal{A}$ [31].

We rotate the localized coordinate system (ξ, η, ζ) by an angle α about the ξ axis to obtain a system (ξ', η', ζ'). Let $\Delta\overline{\Omega}_\mu$ $(\mu = \xi, \eta, \zeta)$ and $\Delta\overline{\Omega}_{\mu'}$ be the components of the infinitesimal rotation $\Delta\overline{\Omega}$ in the two systems, respectively. We then have

$$\begin{pmatrix} \Delta\overline{\Omega}_{\xi'} \\ \Delta\overline{\Omega}_{\eta'} \\ \Delta\overline{\Omega}_{\zeta'} \end{pmatrix} = \begin{pmatrix} 1 & 0 & 0 \\ 0 & \cos\alpha & \sin\alpha \\ 0 & -\sin\alpha & \cos\alpha \end{pmatrix} \begin{pmatrix} \Delta\overline{\Omega}_\xi \\ \Delta\overline{\Omega}_\eta \\ \Delta\overline{\Omega}_\zeta \end{pmatrix}. \qquad (4.99)$$

The components of the angular momentum operator $\mathbf{L}$ in the two systems may therefore be related to each other as

$$\begin{aligned} L_\xi &= L_{\xi'}, \\ L_\eta &= \cos\alpha \, L_{\eta'} - \sin\alpha \, L_{\zeta'}, \qquad (4.100) \\ L_\zeta &= \sin\alpha \, L_{\eta'} + \cos\alpha \, L_{\zeta'}, \end{aligned}$$

so that if we set

$$\alpha = -\tan^{-1}\left(\frac{\kappa_0}{\tau_0}\right) \qquad (-\pi \le \alpha \le 0), \qquad (4.101)$$

then Eq. (4.31) reduces to

$$\mathcal{A} = \nu L_{\zeta'} - \mathbf{L}^2, \qquad (4.102)$$

where ν is given by Eq. (4.75).

Let $\Omega' = (\theta', \phi', \psi')$ be the Euler angles defining the orientation of the system (ξ', η', ζ') with respect to an external coordinate system. As seen from Eqs. (4.102) and (4.C.16), the eigenfunctions of $\mathcal{A}$ are just $\mathcal{D}_l^{mj}(\Omega')$ and its eigenvalues are found to be the $z_{l,j}$ given by Eq. (4.45); that is,

$$\mathcal{A}\Psi_l^{mj} = z_{l,j} \Psi_l^{mj} \qquad (4.103)$$

with $\Psi_l^{mj} = \mathcal{D}_l^{mj}(\Omega')$ and with

$$z_{l,j} = l(l+1) + ij\nu\,. \tag{4.104}$$

Since the rotation Ω' is equal to the resultant of the two successive rotations $\Omega = (\theta,\ \phi,\ \psi)$ and $\Omega_\alpha = (\alpha,\ -\pi/2,\ \pi/2)$, we use Eq. (4.C.13) to obtain

$$\Psi_l^{mj}(\Omega) = c_l^{-1} \sum_k \mathcal{D}_l^{mk}(\Omega)\mathcal{D}_l^{kj}(\Omega_\alpha)\,, \tag{4.105}$$

where c_l is given by Eq. (4.54).

Thus Eq. (4.98) may be rewritten as

$$G(\Omega\,|\,\Omega_0;\,L) = \sum_{l,m,k} c_l^{-2} h_l^k(L) \sum_j \mathcal{D}_l^{mj}(\Omega)\mathcal{D}_l^{jk}(\Omega_\alpha)$$
$$\times \sum_{j'} \mathcal{D}_l^{mj'*}(\Omega_0)\mathcal{D}_l^{j'k*}(\Omega_\alpha)\,. \tag{4.106}$$

From Eq. (4.34) with Eqs. (4.98) and (4.103), we find

$$h_l^j(L) = \exp(-z_{l,j}L)\,. \tag{4.107}$$

By comparing Eq. (4.106) with Eq. (4.36), we obtain

$$g_l^{jj'}(L) = c_l^{-2} \sum_k \exp(-z_{l,k}L)\mathcal{D}_l^{jk}(\Omega_\alpha)\mathcal{D}_l^{j'k*}(\Omega_\alpha)\,. \tag{4.108}$$

This is the desired expression for $g_l^{jj'}(L)$ [32].

4.4.2 The Rotational Isomeric State Model

For the RIS model we can affix a localized Cartesian coordinate system to its rigid body part (the "monomer" unit) composed of two adjacent skeletal bonds, the pth system to the part composed of the $(p-1)$th and pth bonds. Let $\Omega_0 = (\theta_0,\ \phi_0,\ \psi_0)$ and $\Omega = (\theta,\ \phi,\ \psi)$ be the orientations of the pth and $q(=p+n)$th systems, respectively, with respect to an external coordinate system. If we assume that $p \gg 1$ and $N - q \gg 1$ with N the total number of skeletal bonds in the chain, we may ignore end effects to define the Green function $G(\Omega\,|\,\Omega_0;\,n)$.

If Ω_1 is the orientation of the qth system with respect to the pth one, $G(\Omega\,|\,\Omega_0;\,n)$ may be expanded in the form

$$G(\Omega\,|\,\Omega_0;\,n) = G(\Omega_1\,|\,0;\,n)$$
$$= \sum_{l,m,j} f_l^{mj}(n)\mathcal{D}_l^{mj}(\Omega_1)\,, \tag{4.109}$$

where $\Omega = 0$ denotes $\theta = \phi = \psi = 0$, and the expansion coefficients f_l^{mj} depend only on n (and the model parameters). Note that G is invariant to rotation of the external coordinate system. Since the rotation Ω_1 is equal to the resultant of the two successive rotations Ω_0^{-1} and Ω, where Ω_0^{-1} is the inverse of the rotation Ω_0, we use Eq. (4.C.13) to have

$$\mathcal{D}_l^{mj}(\Omega_1) = c_l^{-1} \sum_{m'} \mathcal{D}_l^{mm'}(\Omega_0^{-1}) \mathcal{D}_l^{m'j}(\Omega), \qquad (4.110)$$

where we have, from Eq. (4.C.12),

$$\mathcal{D}_l^{mm'}(\Omega_0^{-1}) = \mathcal{D}_l^{m'm*}(\Omega_0). \qquad (4.111)$$

Substitution of Eq. (4.100) with Eq. (4.111) into Eq. (4.109) leads to

$$G(\Omega \,|\, \Omega_0; n) = \sum_{l,m,j,j'} g_l^{jj'}(n) \mathcal{D}_l^{mj}(\Omega) \mathcal{D}_l^{mj'*}(\Omega_0) \qquad (4.112)$$

with

$$g_l^{jj'}(n) = c_l^{-1} f_l^{j'j}(n). \qquad (4.113)$$

Thus the $g_l^{jj'}(n)$ given by Eq. (4.113) are the angular correlation functions for the RIS model corresponding to the $g_l^{jj'}(L)$ for the HW chain, and Eq. (4.46) holds for $g_l^{jj'}(n)$. However, it is important to note that both $g_l^{jj'}(L)$ and $g_l^{jj'}(n)$ are invariant to rotation of the external coordinate system but that the latter depends on the orientation of the localized coordinate system affixed to the monomer unit with respect to that monomer unit.

For the evaluation of $g_l^{jj'}(n)$ it is convenient to use Eq. (4.113), where $f_l^{mj}(n)$ may be expressed as

$$f_l^{mj}(n) = \langle \mathcal{D}_l^{mj*}(\Omega_q) \rangle_{\Omega_p=0} \qquad (4.114)$$

with

$$\langle \cdots \rangle_{\Omega_p=0} = \int (\cdots) G(\Omega_q \,|\, \Omega_p = 0; n) d\Omega_q. \qquad (4.115)$$

We first evaluate $f_l^{mj}(n)$ in the Flory localized coordinate system as defined below and denote the $g_l^{jj'}(n)$ thus evaluated in this system by $h_l^{mj}(n)$, for convenience. Then we transform $h_l^{mj}(n)$ to $g_l^{jj'}(n)$ expressed in a different localized coordinate system appropriate for a comparison with $g_l^{jj'}(L)$ for the HW chain. We note that the $f_l^{mj*}(n)$ in the Flory localized system is equivalent to the quantity studied by Baram and Gelbart [30].

Now, in order to evaluate the average in Eq. (4.114), we define explicity the kth localized coordinate system $(\mathbf{e}_{x_k}, \mathbf{e}_{y_k}, \mathbf{e}_{z_k})$ as follows. The z_k axis is taken along the kth bond vector $\mathbf{l}_k$, the x_k axis is in the plane of $\mathbf{l}_{k-1}$ and $\mathbf{l}_k$ with its direction chosen at an acute angle with $\mathbf{l}_{k-1}$, and the y_k axis completes the right-handed system, as depicted in Fig. 4.9. Let $\hat{\theta}_k$ be the angle between $\mathbf{l}_k$ and $\mathbf{l}_{k+1}$ (supplement of the bond angle) and let $\hat{\phi}_k$ be the internal rotation

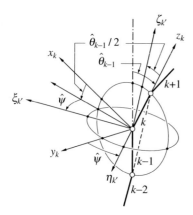

Fig. 4.9. The Flory localized coordinate system $(x_k,\ y_k,\ z_k)$ and the system $(\xi_{k'},\ \eta_{k'},\ \zeta_{k'})$ associated with the HW chain

angle about the kth bond with $\hat{\phi}_k = 0$ in the *trans* conformation, where we distinguish them from the Euler angles by the over caret. This system is essentially the same as that employed by Flory and co-workers [21] except that their $(x,\ y,\ z)$ is replaced by $(z,\ x,\ y)$. This minor change facilitates the use of the Wigner $\mathcal{D}$ functions; the kth system is obtained by rotation of the $(k-1)$th one by the Euler angles $(\hat{\theta}_{k-1},\ \hat{\phi}_{k-1},\ \pi)$. The present system is referred to as the Flory system, for convenience.

The qth system may then be obtained from the pth system by the n successive rotations $(\hat{\theta}_k,\ \hat{\phi}_k,\ \pi)$. We therefore use successively Eq. (4.C.13) and recall the relation, Eq. (4.C.6),

$$\mathcal{D}_l^{mj}(0) = c_l \delta_{mj} \tag{4.116}$$

to find for $\mathcal{D}_l^{mj}(\Omega_q)$

$$\mathcal{D}_l^{mj}(\Omega_q) = c_l^{-n+1} \sum_{\{m\}} \prod_{k=p}^{q-1} \mathcal{D}_l^{m_k m_{k+1}}(\hat{\theta}_k,\ \hat{\phi}_k,\ \pi)$$
$$(m_p = m,\ m_q = j), \tag{4.117}$$

where the sums are taken over $m_{p+1},\ m_{p+2},\ \cdots,\ m_{q-1} = \{m\}$. Thus the $f_l^{mj}(n)$ given by Eq. (4.114) (or its complex conjugate) may be expressed as

$$f_l^{mj*}(n) = Z^{-1} \sum_{\{\hat{\phi}_{N-2}\}} \mathcal{D}_l^{mj}(\Omega_q) \exp[-E(\{\hat{\phi}_{N-2}\})/k_B T], \tag{4.118}$$

where $\mathcal{D}_l^{mj}(\Omega_q)$ is given by Eq. (4.117) and Z is the partition function given by Eq. (2.17), that is,

$$Z = \sum_{\{\hat{\phi}_{N-2}\}} \exp[-E(\{\hat{\phi}_{N-2}\})/k_B T]. \tag{4.119}$$

As shown in Appendix 4.B, the HW chain is a continuous limit of a hypothetical (discrete) chain of "monomer units," each composed of two adjacent
skeletal bonds, one localized system $(\mathbf{e}_\xi, \mathbf{e}_\eta, \mathbf{e}_\zeta)$ being affixed to one unit. In
order to make a comparison of the RIS model (or a given real chain) with
the HW chain, one localized system should be assigned to one monomer unit
defined by two adjacent skeletal bonds of the former so that the total number of systems affixed to the chain of N bonds (with N even) is $N/2$. Then,
the Flory system does not necessarily coincide with the localized system of
the HW chain as yet. Indeed, in the analysis of the persistence vector $\mathbf{A}$ the
orientation of one with respect to the other has been determined to give best
agreement between its components as functions of chain length for the two
models. In the present case, however, the orientation is, to some extent, re-
stricted to preserve certain symmetry relations for the HW chain in the RIS
model. The localized system of the RIS model thus determined to coincide
with that of the HW chain is described below, the symmetry relations being
derived in the next subsection.

The N bonds in the RIS model are numbered $1, 2, \cdots, N$ or $0, 1, 2, \cdots,$
$N - 1$ so that k, p, q, and n are always even; that is, $k = 2k'$, $p = 2p'$,
$q = 2q'$, and $n = 2x$, for simplicity. The k'th localized system $(\mathbf{e}_{\xi_{k'}}, \mathbf{e}_{\eta_{k'}},$
$\mathbf{e}_{\zeta_{k'}})$ $(k' = 1, 2, \cdots)$ of the RIS model *corresponding to* the system $(\mathbf{e}_\xi, \mathbf{e}_\eta,$
$\mathbf{e}_\zeta)$ of the HW chain is obtained by rotation of the kth Flory system $(\mathbf{e}_{x_k},$
$\mathbf{e}_{y_k}, \mathbf{e}_{z_k})$ by the Euler angles $\hat{\Omega} = (\frac{1}{2}\hat{\theta}_{k-1}, 0, \hat{\psi})$, assuming that $|\mathbf{l}_k| = l$ for
all k (see Fig. 4.9 and also Fig. 4.B.1). In other words, $\mathbf{e}_{\zeta_{k'}}$ must be parallel
to $\mathbf{l}_{k-1} + \mathbf{l}_k$. Let $g_l^{jj'}(x, \hat{\psi})$ be the angular correlation functions between the
p'th and $q'(= p' + x)$th systems thus obtained. It may be expressed in terms
of $h_l^{jj'}(n)$ as follows. If $\Omega_{k'} = (\theta_{k'}, \phi_{k'}, \psi_{k'})$ is the orientation of the k'th
system with respect to the external system, we have, from Eq. (4.C.13),

$$\mathcal{D}_l^{mj}(\Omega_{k'}) = c_l^{-1} \sum_{m'} \mathcal{D}_l^{mm'}(\Omega_k) \mathcal{D}_l^{m'j}(\hat{\Omega}). \qquad (4.120)$$

From Eq. (4.46) for $g_l^{jj'}(n)$ and Eq. (4.120), we find

$$g_l^{jj'}(x, \hat{\psi}) = \exp\left[-i(j-j')\hat{\psi}\right] \sum_m \sum_{m'} d_l^{mj}\left(\frac{\hat{\theta}}{2}\right) d_l^{m'j'}\left(\frac{\hat{\theta}}{2}\right) h_l^{mm'}(n), \qquad (4.121)$$

where $\hat{\theta} = \hat{\theta}_{p-1} = \hat{\theta}_{q-1}$ for the chain under consideration, and $d_l^{mj}(\theta)$ is the
θ-dependent part of $\mathcal{D}_l^{mj}(\Omega)$ and is defined by Eq. (4.C.2). The parameter $\hat{\psi}$
is determined from a comparison of the RIS model with the HW chain.

In what follows, for simplicity, the argument $\hat{\psi}$ of $g_l^{jj'}(x, \hat{\psi})$ is omitted unless necessary to specify, and the localized system affixed to the kth monomer
(repeat) unit is called the kth system (without the prime on k), so that $g_l^{jj'}(x)$
are the angular correlation functions between the pth and qth monomer units
and are dependent on p and q as $x = q - p$ (for large p and $N - q$ in the chain

of N repeat units). Further, we note that the restriction of the orientation of the localized system to be affixed to the RIS model (or the real chain) depends on the physical property to be considered; it may be somewhat relaxed for the persistence vector $\mathbf{A}$ and the orientation need not be considered for the moments $\langle R^{2m} \rangle$.

4.4.3 Symmetry Relations

The angular correlation functions $g_l^{jj'}(L)$ and $g_l^{jj'}(x)$ have two kinds of symmetry; one arises from the reality of the Green function, and the other from its invariance to reversal of the initial and terminal ends of the chain or the numbering of the bonds.

The first symmetry relation may readily be obtained. The Green function is real, so that

$$G^* = G .\tag{4.122}$$

Substitution of Eq. (4.36) or Eq. (4.112) into Eq. (4.122) and use of Eq. (4.C.7) leads to

$$g_l^{jj'} = (-1)^{j+j'} g_l^{(-j)(-j')*} ,\tag{4.123}$$

which is valid for both the HW and RIS models.

Next we consider the second symmetry. The contour length of the HW chain may be measured from either end, and the bonds in the RIS model may be numbered from either end. We use the superscripts $(+)$ and $(-)$ to indicate the two senses of measuring chain length, one being the reverse of the other. The localized system or orientation $(\mathbf{a}, \mathbf{b}, \mathbf{u})$ of the HW chain measured in the $(+)$ sense becomes $(\mathbf{a}, -\mathbf{b}, -\mathbf{u})$ when measured in the $(-)$ sense. A similar relation also holds for the RIS model irrespective of the value of $\hat{\psi}$. In other words, the Euler angles $\Omega = (\theta, \phi, \psi)$ of the localized systems of both the HW and RIS models measured in the $(+)$ sense become $\Omega^{(-)} = (\pi - \theta, \phi + \pi, -\psi)$ when measured in the $(-)$ sense. Then the distribution function is invariant to change of the sense of measuring chain length, the distribution of the initial orientation being uniform. Therefore, this is also the case with the conditional distribution function, that is, the Green function, so that we have the relation for the HW chain,

$$G^{(-)}(\Omega_0^{(-)} \,|\, \Omega^{(-)}; L) = G^{(+)}(\Omega \,|\, \Omega_0; L) ,\tag{4.124}$$

and the equivalent relation for $G(\Omega \,|\, \Omega_0; x)$ for the RIS model. Let $g_l^{(\pm)jj'}(L)$ be the expansion coefficients of $G^{(\pm)}(\Omega \,|\, \Omega_0; L)$. Since we have, by the use of Eqs. (4.C.19) and (4.C.20), the relation

$$\mathcal{D}_l^{mj}(\Omega^{(-)}) = (-1)^l \mathcal{D}_l^{m(-j)}(\Omega) ,\tag{4.125}$$

we find, from Eq. (4.124) for the HW chain and from the equivalent for the RIS model,

$$g_l^{(+)jj'}(L) = (-1)^{j+j'} g_l^{(-)j'j}(L) \qquad \text{(HW)}, \qquad (4.126)$$

$$g_l^{(+)jj'}(x,\hat{\psi}) = (-1)^{j+j'} g_l^{(-)j'j}(x,-\hat{\psi}) \qquad \text{(RIS)}. \qquad (4.127)$$

In Eq. (4.127), note that the system $(e_{\xi_k}, -e_{\eta_k}, -e_{\zeta_k})$ assigned in the $(-)$ sense is obtained by rotation $\hat{\Omega}^{(-)}(-\hat{\psi})$ of the Flory system assigned in the $(-)$ sense when the system $(e_{\xi_k}, e_{\eta_k}, e_{\zeta_k})$ in the $(+)$ sense is obtained by rotation $\hat{\Omega}(\hat{\psi})$ of the Flory system in the $(+)$ sense.

For the HW chain, $g_l^{(+)jj'}$ is identical with $g_l^{(-)jj'}$, and therefore we obtain, from Eqs. (4.123) and (4.126), [with suppression of the superscripts $(+)$ and $(-)$]

$$g_l^{jj'}(L) = g_l^{(-j')(-j)*}(L)$$
$$= (-1)^{j+j'} g_l^{j'j}(L) \qquad \text{(HW)} \qquad (4.128)$$

as the desired two symmetry relations. As seen from Eqs. (4.128), a consideration of the ranges $-l \leq j' \leq 0$ and $j' \leq j \leq -j'$ suffices, so that the number of independent jj' components of the lth order angular correlation function $g_l^{jj'}$ is equal to $(l+1)^2$. Further, if $\bar{g}_l^{jj'}$ and $\bar{\bar{g}}_l^{jj'}$ are the real and imaginary parts of $g_l^{jj'}$, respectively, that is,

$$g_l^{jj'} = \bar{g}_l^{jj'} + i\bar{\bar{g}}_l^{jj'}, \qquad (4.129)$$

then we have

$$\bar{\bar{g}}_l^{jj'} = 0 \qquad \text{for } \tau_0 = 0, \qquad (4.130)$$

as can easily be shown from Eqs. (4.42) and (4.43) with $\sigma = 0$.

For the RIS model, the further deduction from Eqs. (4.123) and (4.127) requires a consideration of the stereochemical configuration. Here summarize only the results [19]. The symmetry relations for $g_l^{jj'}(x)$ are the same as Eqs. (4.128) except for $\bar{\bar{g}}_l^{jj'}(x)$ for certain stereochemical sequences of asymmetric chains, provided that the localized system is defined as above with assignment of a proper value of $\hat{\psi}$ ranging from 0 to 2π. Fortunately, however, $\bar{\bar{g}}_l^{jj'}(x)$, which are related to the asymmetry of the chain, have been found numerically to be very small and of minor importance. In particular, for symmetric chains the symmetry relations are completely the same as Eqs. (4.128) if we take $\hat{\psi} = 0$ or π, $\bar{\bar{g}}_l^{jj'}(x)$ vanishing as in the case of the HW chain with $\tau_0 = 0$.

4.4.4 Numerical Results

For the HW model all components $g_l^{jj'}(L)$ $(l \geq 1)$ vanish in the limit of $L \to \infty$, as seen from Eq. (4.108). On the other hand, for the RIS model some of the components $g_l^{jj'}(x)$ for $l \geq 2$ approach finite values, or zero very slowly, as x becomes infinity, as pointed out first by Baram and Gelbart [30]. This behavior of the RIS model is unphysical and is again discussed in the next chapter. Thus we examine numerically the behavior of only $\bar{g}_1^{jj'}$, for which the symmetry relations for the two models may be the same. (Note that $g_l^{00} = 1$.) We then choose as its four independent components $\bar{g}_1^{00}$, $\bar{g}_1^{0(-1)}$, $\bar{g}_1^{1(-1)}$, and $\bar{g}_1^{(-1)(-1)}$. For the HW chain they are explicitly given, from Eq. (4.108), by [25]

$$\bar{g}_1^{00}(L) = \frac{1}{\nu^2} e^{-2L} \left[\kappa_0^{\,2} \cos(\nu L) + \tau_0^{\,2} \right] ,$$

$$\bar{g}_1^{0(-1)}(L) = \frac{\kappa_0}{\sqrt{2}\nu} e^{-2L} \sin(\nu L) ,$$

$$\bar{g}_1^{1(-1)}(L) = \frac{\kappa_0^{\,2}}{2\nu^2} e^{-2L} \left[1 - \cos(\nu L) \right] ,$$

$$\bar{g}_1^{(-1)(-1)}(L) = \frac{1}{2\nu^2} e^{-2L} \left[\kappa_0^{\,2} + (\nu^2 + \tau_0^{\,2}) \cos(\nu L) \right] .$$

(4.131)

For the RIS model the corresponding components $\bar{g}_1^{jj'}(x)$ may be calculated from Eq. (4.121) with Eqs. (4.113) and (4.118).

Now we equate the $\bar{g}_1^{jj'}$ for the two models,

$$\bar{g}_1^{jj'}(L) = \bar{g}_1^{jj'}(x) ,$$

(4.132)

where L is related to x by Eq. (4.89). Thus we may determine κ_0, τ_0, $\hat{\psi}$, and δ from best fits of HW values of $\bar{g}_1^{jj'}$ plotted against $\log L$ to the RIS values plotted against $\log x$. We note that for both the HW and RIS models the mean-square end-to-end distance $\langle R^2 \rangle$ may be expressed in terms of g_1^{00}, and the persistence vector $\mathbf{A}$ in terms of g_1^{00} and $g_1^{0(-1)}$. Since $g_l^{jj'}$ is a dimensionless quantity, λ^{-1} and M_L cannot be determined separately from δ. We therefore assume here the values of M_L determined from $\mathbf{A}$ [24].

In Table 4.3 are given the values of the HW model parameters so determined for PM, PDMS, i-PS, s-PS, i-PMMA, and s-PMMA. Among them,

Table 4.3. Values of the HW model parameters from RIS values of $\bar{g}_1^{jj'}$

Polymer	$\hat{\psi}$ (rad)	κ_0	$|\tau_0|$	λ^{-1} (Å)	M_L (Å^{-1})[a]	Ref. (RIS parameters)
PM	π	0.3	0	14.5	10.1	[21]
PDMS	0	0.8	0	15.3	19.6	[21, 22]
i-PS	$\frac{3}{2}\pi$	11	15	26.4	41.2	[33]
s-PS	0	0.8	2.3	40.4	38.9	[33]
i-PMMA	π	1.7	1.4	32.7	33.5	[34]
s-PMMA	π	4.4	0.8	65.6	35.7	[34]

[a] From RIS values of $\mathbf{A}$

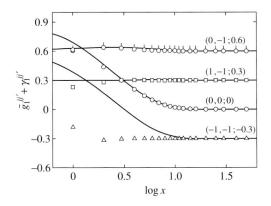

Fig. 4.10. $\bar{g}_1^{jj'} + \gamma_1^{jj'}$ plotted against $\log x$ for PM with x the number of repeat units. The points and curves represent the RIS and HW values, respectively, the numbers in parentheses indicating the values of $(j, j'; \gamma_1^{jj'})$

the first two are symmetric chains, and for PDMS the part containing the Si−O and O−Si bonds has been chosen as the monomer unit. The remaining are asymmetric chains and the part containing the C−C$^\alpha$ and C$^\alpha$−C bonds (with C$^\alpha$ the α carbon) has been chosen as the monomer unit. The values of the RIS parameters necessary for the calculation of $\bar{g}_1^{jj'}(x)$ have been taken from [21, 22, 33, 34] (see Table 4.3). We note that for the above polymers the effect of chain ends on $\bar{g}_1^{jj'}(x)$ may be neglected for $p = N - q \geq 20$.

For illustration, values of $\bar{g}_1^{jj'} + \gamma_1^{jj'}$ with $\gamma_1^{jj'}$ being constants are plotted against $\log x$ in Figs. 4.10 and 4.11 for PM and s-PMMA, respectively. The points and curves represent the RIS and HW values calculated as mentioned above, respectively. There is good agreement between the values for the two models except for $\bar{g}_1^{(-1)(-1)}$, especially for PM. The reason for this is the following. For a straight rod which is permitted to undergo torsional deformation about its axis, the component $g_1^{(-1)(-1)}$ is given, from Eq. (4.46), by

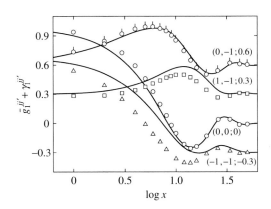

Fig. 4.11. $\bar{g}_1^{jj'} + \gamma_1^{jj'}$ plotted against $\log x$ for s-PMMA; see legend for Fig. 4.10

$$\bar{g}_1^{(-1)(-1)}(L) = \langle \cos(\psi - \psi_0) \rangle \,,$$
$$\tilde{\bar{g}}_1^{(-1)(-1)}(L) = \langle \sin(\psi - \psi_0) \rangle \,,$$
$$\text{(rod)}, \qquad (4.133)$$

where $\theta = \theta_0$ and $\phi = \phi_0$. Thus the $g_1^{(-1)(-1)}$ for both models may be regarded as being closely related to the torsional correlation between two monomer units. Therefore, the slower decay of this component for the HW chain implies that its torsional correlation is of rather long range. This is understandable and may probably be a defect of the elastic wire model.

Now we are in a position to discuss the results in Table 4.3 along with those in Tables 4.1 and 4.2. We first consider the meaning of the results obtained for $\hat{\psi}$. This angle determines the direction of the ξ_k axis of the localized system $(\mathbf{e}_{\xi_k}, \mathbf{e}_{\eta_k}, \mathbf{e}_{\zeta_k})$ of the RIS model and therefore the direction of the unit mean curvature vector $\mathbf{a} = \mathbf{e}_\xi$ of the corresponding HW chain at that point when $\mathbf{e}_{\zeta_k}$ is chosen to coincide with the unit tangent vector $\mathbf{u} = \mathbf{e}_\zeta$ of the latter. In the case of PDMS and PMMA the chain is given a local curvature by the inequality of the two skeletal bond angles with a sequence of successive *trans* conformations being preferred [6,21,22,34]. For example, the α carbons in the local all-*trans* sequence of bonds in the PMMA chain are located on a circle (not a straight line), and the direction of $\mathbf{e}_\xi = \mathbf{a}$ coincides with that of the curvature vector of this circle toward its center if $\hat{\psi} = \pi$, $\mathbf{u}$ being in the direction of the local axis of the chain fully extended, as depicted in Fig. 4.12. We note that in general, $\hat{\psi} = 3\pi/2$ and 0 for isotactic and syndiotactic monosubstituted asymmetric chains, respectively, while $\hat{\psi} = \pi$ for both isotactic and syndiotactic disubstituted asymmetric chains. Thus, in the adaptation of the HW model to a given real chain, we can determine the orientation of the localized coordinate system of the former with respect to the monomer unit, and therefore also express definitely, for instance, the electric dipole moment vector and optical polarizability tensor of the monomer unit in that system.

As for the HW model parameters κ_0, τ_0, λ^{-1}, and M_L, their values obtained from different properties (C_n, $\mathbf{A}$, and $g_l^{jj'}$) are seen to be rather in good agreement with each other. This is also the case with other properties,

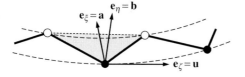

Fig. 4.12. Localized coordinate system affixed to the monomer unit in the local all-*trans* sequence of bonds in a PMMA chain with $\hat{\psi} = \pi$. The *filled circles* represent the α carbons, and $\mathbf{u}$ is in the direction of the local axis of the chain fully extended

provided that the HW model is adapted to a real chain on the bond length or somewhat longer scales. In general, the parameters κ_0 and τ_0 describe the preferred local chain conformation, and the parameter λ^{-1} represents the chain stiffness, as already mentioned (λ represents the degree of thermal fluctuation). Of the polymers listed in Table 4.3, the s-PMMA chain is the most stiff, while the PM chain is the most flexible, and moreover, close to the KP chain. The parameter M_L is related to the chemical structure of the chain. The significance of these parameters is further discussed in the next subsection, and in more detail in later chapters, giving a picture of the chain conformation on the basis of their values determined from experiment for a given polymer.

4.5 Helical Nature of the Chain

The HW model may be characterized simply by the behavior of the ratio $\langle R^2 \rangle / c_\infty L$ ($= C_n/C_\infty$) as a function of the contour length L (or the number of repeat units x). It is therefore worth while to establish relations between the model parameters κ_0 and τ_0 and the behavior of this ratio, assuming that $\sigma = 0$.

Figure 4.13 shows a (κ_0, τ_0)-plane, where we consider only its first quadrant since κ_0 is nonnegative and $\langle R^2 \rangle$ is an even function of τ_0. The diameter 2ρ and pitch h of the characteristic helix are equal to each other on the straight line,

$$\frac{\tau_0}{\pi^{-1}\kappa_0} = 1 \left(= \frac{h}{2\rho} \right),\qquad(4.134)$$

passing through the origin, as seen from Eqs. (4.14); and $2\rho > h$ and $2\rho < h$ below and above it, respectively. If κ_0 and τ_0 (and hence the stiffness parameter λ^{-1}) become very large, the chain approaches the characteristic

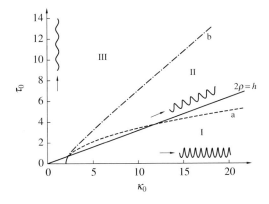

Fig. 4.13. Characteristics of a (κ_0, τ_0)-plane. It is divided into three domains I, II, and III according to the type of the first maximum of $\langle R^2 \rangle / c_\infty L$ as a function of L (see the text)

Table 4.4. Typical values of the HW model parameters and their domains

Code	Domain	κ_0	τ_0	c_∞
1	I	2.5	0.5	0.40476
2	I	5.0	1.0	0.16667
3	III	1.0	1.0	0.83333
4	III	3.0	6.0	0.81633
5	II	30.0	8.0	0.07025

helix, so that its *helical nature* becomes strong, as illustrated in the figure. In other words, the thermal fluctuation in the chain conformation from the characteristic helix is small for very large κ_0 and τ_0. (Recall that κ_0 and τ_0 are reduced by λ.) If κ_0 becomes very small, the chain approaches the KP chain irrespective of the value of τ_0. In the limit of the KP chain with $\kappa_0 = 0$, the characteristic helix degenerates into the straight line (rod), the type-1 or -2 rod according to the KP1 or KP2 chain (see Fig. 4.4).

The (κ_0, τ_0)-plane may be divided into three domains I, II, and III, as shown in Fig. 4.13, where the dashed curve a is the boundary between the domains I and II, and the dot-dashed curve b is the boundary between the domains II and III. The ratio $\langle R^2 \rangle / c_\infty L$ as a function of L exhibits at least one maximum in the domains I and II, and the first peak (occuring as L is increased) is higher and lower than the coil-limiting value of unity in I and II, respectively. In the domain III the ratio is an increasing function of L but exhibits inflection in some cases.

In Table 4.4 are given five sets of values of κ_0 and τ_0 as typical examples along with the domain of each code and the values of c_∞. Values of $\langle R^2 \rangle / c_\infty L$ are plotted against $\log L$ in Fig. 4.14 for these codes. It also includes the values for the KP chain and the random coil (C). We note that typical HW chains belong to the domain I and that the cases belonging to the domain II rarely occur for real chains.

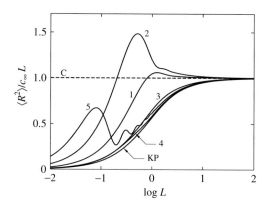

Fig. 4.14. $\langle R^2 \rangle / c_\infty L$ plotted against $\log L$ for the HW codes of Table 4.4, and also for the KP chain and the random coil (C)

4.6 Distribution Functions

The most general distribution function for the HW chain of contour length L is the Green function $G(\mathbf{R}, \Omega \,|\, \Omega_0; L)$, which is obtained by Fourier inversion of the characteristic function $I(\mathbf{k}, \Omega \,|\, \Omega_0; L)$. In Sect. 4.2.3 we have already obtained the formal solutions for $I(\mathbf{k} \,|\, \mathbf{u}_0; L)$ and $I(\mathbf{k}; L)$ as its special cases. In this section we generalize the procedure to find the formal solution for $I(\mathbf{k}, \Omega \,|\, \Omega_0; L)$ and then general developments for $G(\mathbf{R}, \Omega \,|\, \Omega_0; L)$ for the case of $\sigma \neq 0$. For the latter we explore two types of asymptotic expansions: the Daniels-type distributions and the moment-based distributions [35], as in the case of the KP chain. We note that $G(\mathbf{R} \,|\, \Omega_0; L)$ corresponds to the moment-based distribution function considered by Flory [26, 27] for the RIS model.

4.6.1 General Developments

The starting equation is the integral equation (4.49) for $I(\mathbf{k}, \Omega \,|\, \Omega_0; L)$, where $\mathbf{u}_0$ is chosen to be in an arbitrary direction ($\mathbf{u}_0 \neq \mathbf{e}_z$). The solution for I may then be expanded in the form

$$I(\mathbf{k}, \Omega \,|\, \Omega_0; L) = \sum K_{ll'}^{mm',jj'}(\mathbf{k}; L) \mathcal{D}_l^{mj}(\Omega) \mathcal{D}_{l'}^{m'j'*}(\Omega_0) \qquad (4.135)$$

with

$$K_{ll'}^{mm',jj'} = \sum_{n=0}^{\infty} (i\bar{k})^n \left(\sum_{j''} g_l^{jj''} * \mathcal{L}_{j''} \right)^n f_{ll'}^{mm',jj'}, \qquad (4.136)$$

where $\bar{k}$ is given by Eq. (4.58) and $f_{ll'}^{mm',jj'}$ is given by

$$f_{ll'}^{mm',jj'} = g_l^{jj'} \delta_{ll'} \delta_{mm'}. \qquad (4.137)$$

The operator $\mathcal{L}_{j''}$ is defined by Eq. (4.60), where the operators $a_\mu^{\nu j''}$ operate on $g_l^{jj'}$ in the same way as in Eq. (4.61) and also operate on $K_{ll'}^{mm',jj'}$ and $f_{ll'}^{mm',jj'}$ to change only the indices l, m, and j (not l', m', and j') in the same way as in Eq. (4.62).

   Now, in order to simplify the operational equation, we rotate the external coordinate system by the Euler angles $\tilde{\Omega} = (\chi, \omega, 0)$ to take $\mathbf{e}_k = \mathbf{e}_z$, that is, $\chi = 0 \ (= \omega)$. Then only the terms with $\nu = 0$ make a contribution since $Y_1^\nu(0, \omega) = (3/4\pi)^{1/2} \delta_{\nu 0}$. In what follows, let $K_{ll'}^{mm',jj'}$ denote its value in this new system. It becomes a spherically symmetric function of $\mathbf{k}$ such that

$$K_{ll'}^{mm',jj'}(\mathbf{k}; L) = K_{ll'}^{mm,jj'}(k; L) \delta_{mm'} \qquad (4.138)$$

with

$$K_{ll'}^{mm,jj'}(k; L) = \sum_{n \geq |l-l'|} (ik)^n \sum_{\substack{\text{paths} \\ (l \to l')}} \sum_j C_\mu^{0j}(l, m) \Gamma_{l \cdots l'}^j(L), \qquad (4.139)$$

where

$$C_\mu^{0j}(l, m) = (f_{l'l'}^{mm,jnj'})^{-1} a_{\mu_n}^{0j_n} a_{\mu_{n-1}}^{0j_{n-1}} \cdots a_{\mu_1}^{0j_1} f_{ll'}^{mm,jj'}, \qquad (4.140)$$

$$\Gamma_{l \cdots l'}^j(L) = g_l^{jj_1} * g_{l_1}^{j_1 j_2} * g_{l_2}^{j_2 j_3} * \cdots * g_{l_{n-1}}^{j_{n-1} j_n} * g_{l'}^{j_n j'} \qquad (4.141)$$

with

$$l_r = l + \sum_{i=1}^r \mu_i \geq |m| \qquad (l_0 = l,\ l_n = l'), \qquad (4.142)$$

$$\nu_i = 0, \qquad (4.143)$$

$$|j_r| \leq \min(l_r, l_{r-1}). \qquad (4.144)$$

In Eq. (4.139), the second sum is taken over all possible paths $(l l_1 l_2 \cdots l_{n-1} l')$ from l to l' (for which the case of $\mu_i = 0$ may occur), and the third sum over $j_1, \cdots, j_n$ compatible with Eq. (4.144).

In the present case of $\nu = 0$, Eq. (4.62) reduces to

$$a_\mu^{0j''} K_{ll'}^{mm,jj'} = E_{l+(1/2)(\mu-1)}^{mj''} K_{(l+\mu)l'}^{mm,j''j'} \qquad \text{for } \mu = \pm 1$$

$$= F_l^{mj''} K_{ll'}^{mm,j''j'} \qquad \text{for } \mu = 0 \qquad (4.145)$$

with

$$E_l^{mj} = \left[\frac{(l+1+m)(l+1-m)(l+1+j)(l+1-j)}{(l+1)^2(2l+1)(2l+3)} \right]^{1/2}, \qquad (4.146)$$

$$F_l^{mj} = \frac{mj}{l(l+1)} \qquad \text{for } l \neq 0$$

$$= 0 \qquad \text{for } l = 0. \qquad (4.147)$$

Thus the operation (on a computer) becomes easy in the system with $\mathbf{e}_k = \mathbf{e}_z$.

We perform transformation back into the original system by rotating the $\mathbf{e}_k = \mathbf{e}_z$ system by the Euler angles $\tilde{\Omega}^{-1}$. Let $\Omega' = (\theta', \phi', \psi')$ be the Euler angles defining the orientation of the localized coordinate system $(\mathbf{e}_\xi, \mathbf{e}_\eta, \mathbf{e}_\zeta)$ in the latter system, and we have, from Eq. (4.C.13),

$$\mathcal{D}_l^{mj}(\Omega') = c_l^{-1} \sum_{m'} \mathcal{D}_l^{mm'}(\tilde{\Omega}^{-1}) \mathcal{D}_l^{m'j}(\Omega). \qquad (4.148)$$

Equation (4.135) may therefore be rewritten in the form

$$I(\mathbf{k}, \Omega \,|\, \Omega_0;\ L) = \sum_{l_i, m_i, m, j_i} (c_{l_1} c_{l_2})^{-1} K_{l_1 l_2}^{mm,j_1 j_2}(k; L)$$

$$\times \mathcal{D}_{l_1}^{mm_1}(\tilde{\Omega}^{-1}) \mathcal{D}_{l_2}^{mm_2*}(\tilde{\Omega}^{-1}) \mathcal{D}_{l_1}^{m_1 j_1}(\Omega) \mathcal{D}_{l_2}^{m_2 j_2*}(\Omega_0), \qquad (4.149)$$

where the sum over m is taken for $|m| \leq \min(l_1, l_2)$, and $K_{l_1 l_2}^{mm, j_1 j_2}$ is given by Eq. (4.139) and is invariant to rotation [l_1, l_2, j_1, and j_2 not to be confused with those in Eqs. (4.142) and (4.144)].

By the use of Eq. (4.C.10), the product of the first two $\mathcal{D}$ functions may be expanded in terms of $\mathcal{D}_{l_3}^{m_3 j_3}$. Further, we have, from Eq. (4.C.12) with Eq. (4.C.4),

$$\mathcal{D}_l^{0j}(\tilde{\Omega}^{-1}) = (2\pi)^{-1/2}(-1)^{(j+|j|)/2} Y_l^{-j}(\chi, \omega). \qquad (4.150)$$

Equation (4.149) may then be further rewritten in the form

$$I(\mathbf{k}, \Omega \,|\, \Omega_0; L) = \sum_{l_i, m_i, j_i} \mathcal{I}_{l_1 l_2 l_3}^{m_1 m_2, j_1 j_2}(k; \ L) \mathcal{D}_{l_1}^{m_1 j}(\Omega)$$

$$\times \mathcal{D}_{l_2}^{m_2 j_2 *}(\Omega_0) Y_{l_3}^{m_2 - m_1}(\chi, \omega) \qquad (4.151)$$

with

$$\mathcal{I}_{l_1 l_2 l_3}^{m_1 m_2, j_1 j_2}(k; L) = \sum_m (-1)^{(1/2)(|m_1 - m_2| + m_1 - m_2) + m - m_1} \left[4\pi(2l_3 + 1)\right]^{1/2}$$

$$\times \begin{pmatrix} l_1 & l_2 & l_3 \\ m & -m & 0 \end{pmatrix} \begin{pmatrix} l_1 & l_2 & l_3 \\ m_1 & -m_2 & m_2 - m_1 \end{pmatrix} K_{l_1 l_2}^{mm, j_1 j_2}(k; L), \quad (4.152)$$

where we have used the selection rules for the 3-j symbol given by Eqs. (4.C.26) and (4.C.27), so that $j_3 = m_1 - m_2$, and l_1, l_2, and l_3 satisfy the triangular inequalities. Equation (4.151) with Eq. (4.152) is the desired general development of the characteristic function.

The Green function $G(\mathbf{R}, \Omega \,|\, \Omega_0; L)$ is obtained by Fourier inversion of Eq. (4.151). If we express $\mathbf{R}$ as $\mathbf{R} = (R, \Theta, \Phi)$ in spherical polar coordinates and use Eq. (3.B.16) with Eq. (3.B.15), that is,

$$\exp(-i\mathbf{k} \cdot \mathbf{R}) = 4\pi \sum_{l,m} (-i)^l j_l(kR) Y_l^m(\Theta, \Phi) Y_l^{-m}(\chi, \omega), \qquad (4.153)$$

where $j_l(x)$ is the spherical Bessel function of the first kind defined by

$$j_l(x) = (2x)^l \sum_{r=0}^{\infty} \frac{(-1)^r (l+r)!}{r!(2l + 2r + 1)!} x^{2r}, \qquad (4.154)$$

then we find

$$G(\mathbf{R}, \Omega \,|\, \Omega_0; L) = \sum_{l_i, m_i, j_i} \mathcal{G}_{l_1 l_2 l_3}^{m_1 m_2, j_1 j_2}(R; L) \mathcal{D}_{l_1}^{m_1 j_1}(\Omega)$$

$$\times \mathcal{D}_{l_2}^{m_2 j_2 *}(\Omega_0) Y_{l_3}^{m_2 - m_1}(\Theta, \Phi) \qquad (4.155)$$

with

$$\mathcal{G}_{l_1 l_2 l_3}^{m_1 m_2, j_1 j_2}(R; L) = (2\pi^2)^{-1}(-i)^{l_3} \int_0^{\infty} k^2 j_{l_3}(kR) \mathcal{I}_{l_1 l_2 l_3}^{m_1 m_2, j_1 j_2}(k; L) dk. \qquad (4.156)$$

Equation(4.155) with Eq. (4.156) is the desired general development of the distribution fuction.

In particular, we have

$$G(\mathbf{R}\,|\,\mathbf{u}_0 = \mathbf{e}_z, \mathbf{a}_0 = \mathbf{e}_x; L) = \sum_{l,m}(2l+1)^{1/2}$$
$$\times \mathcal{G}_{0ll}^{0m,0m}(R;L)Y_l^m(\Theta,\Phi)\,, \qquad (4.157)$$

$$G(\mathbf{R}\,|\,\mathbf{u}_0 = \mathbf{e}_z; L) = \sum_{l}(2l+1)^{1/2}\mathcal{G}_{0ll}^{00,00}(R;L)Y_l^0(\Theta,\Phi)\,, \qquad (4.158)$$

$$G(\mathbf{R}; L) = (4\pi)^{-1/2}\mathcal{G}_{000}^{00,00}(R;L)\,. \qquad (4.159)$$

4.6.2 Daniels-Type Distributions

The Gobush operator $\delta\mathcal{L}$ [36] introduced in the evaluation of the Daniels-type distribution function [37] for the KP chain operates on the basis functions, while the creation and annihilation operators a_μ^ν operate on the coefficients of expansion in terms of the basis functions. As shown in Sect. 3.4.1, however, $\delta\mathcal{L}$ may be written in terms of a_μ^ν and the two representations have one-to-one correspondence. Thus the present formulation [35] follows the latter procedure but is closely related to the development [38] leading to Eq. (3.83) by the use of $\delta\mathcal{L}$ written in terms of a_μ^ν.

The problem is to expand the $\mathcal{G}_{\cdots}^{\cdots}(R;L)$ given by Eq. (4.156) in inverse powers of L, suppressing all exponential terms of order $\exp(-\text{const}.L)$. This is the Daniels approximation to $G(\mathbf{R},\Omega\,|\,\Omega_0;L)$. For this purpose, we first consider the expansion of $K_{ll'}^{mm,jj'}(k;L)$. As seen from Eqs. (4.43), (4.139), and (4.141), those paths for which all l_rs are positive lead to only the exponential terms in $K_{ll'}^{mm,jj'}$. In the Daniels approximation it is therefore sufficient to consider contributions from those paths for which at least one of the l_rs is zero. Then we need only to consider $K_{ll'}^{00,jj'}$ with $m=0$ since according to Eq. (4.142) $l_r > 0$ for all r if $|m| > 0$, so that only the operators $a_\mu^{0j''}$ with $\mu = \pm 1\,(\neq 0)$ appear, as seen from Eqs. (4.145) and (4.147).

Now we consider the Laplace transforms $\tilde{K}_{ll'}^{00,jj'}(k;p)$ and $\tilde{\Gamma}_{l\ldots l'}^{j}(p)$ of $K_{ll'}^{00,jj'}(k;L)$ and $\Gamma_{l\ldots l'}^{j}(L)$, respectively, in Eq. (4.139). The paths from l to l' may be conveniently represented by the stone-fence diagrams in an (i,l_i)-plane as in Fig. 3.3. Under the conditions above, each of these paths may be decomposed into $irreducible$ paths $(ll_1\cdots 0)$, $(0l_i\cdots 0)$, $\cdots$, $(0l_j\cdots 0)$, and $(0l_k\cdots l_{n-1}l')$. By the term irreducible, we mean that all of the indices specifying such a path are positive except the initial and terminal ones. Then $\tilde{K}_{ll'}^{00,jj'}$ may be written in the form

$$\tilde{K}_{ll'}^{00,jj'} = \sum_{n\geq|l-l'|}\sum_{\text{paths}}\Delta_0\Delta^{k_0}\Delta_1\Delta^{k_1}\Delta_2\cdots\Delta^{k_{m-1}}\Delta_m\Delta^{k_m}\Delta_{\mathrm{T}}\,, \qquad (4.160)$$

where k_α are nonnegative integers, $\Delta_{\mathrm{T}} \equiv \Delta_{m+1}$, and

$$\Delta_\alpha = \left[\delta_{\alpha,m+1} + (1 - \delta_{\alpha,m+1})p\right](ik)^{s_\alpha} \sum_{\mathbf{j}_\alpha} C_{\boldsymbol{\mu}_\alpha}^{0\mathbf{j}_\alpha} \tilde{I}_{\mathbf{l}_\alpha}^{\mathbf{j}_\alpha}(p)$$

$$(0 \le \alpha \le m + 1) \qquad (4.161)$$

with $s_\alpha \ge 4$ for $1 \le \alpha \le m$; and

$$\Delta = p(ik)^2 C_{1(-1)}^{00} \tilde{I}_{010}^0(p)$$

$$= -\frac{k^2}{3pf(p)}\left[(p + 2 + \sigma)^2 + \tau_0{}^2\right] \qquad (4.162)$$

with

$$f(p) = (p + 2)\left[(p + 2 + \sigma)^2 + \tau_0{}^2\right] + \kappa_0{}^2(p + 2 + \sigma). \qquad (4.163)$$

In Eq. (4.161), the sets $\boldsymbol{\mu}_\alpha$, $\mathbf{j}_\alpha$, and $\mathbf{l}_\alpha$ are associated with the irreducible subpath Δ_α, and the arguments of $C_{\boldsymbol{\mu}_\alpha}^{0\mathbf{j}_\alpha}$ have been omitted, for simplicity. The factor Δ_0 corresponds to an initial s_0-step subpath ($l \to 0$), Δ_α ($1 \le \alpha \le m$) to an $s_\alpha(\ge 4)$-step subpath ($0 \to 0$), Δ_{T} to a terminal s_{m+1}-step subpath ($0 \to l'$), and Δ to a two-step subpath (010). Note that $\Delta_0 = 1$ and $s_0 = 0$ when $l = 0$, that $\Delta_{\mathrm{T}} = p^{-1}$ and $s_{m+1} = 0$ when $l' = 0$, and that

$$n = 2\sum_{\alpha=0}^{m} k_\alpha + \sum_{\alpha=0}^{m+1} s_\alpha. \qquad (4.164)$$

In Fig. 4.15 is shown a diagram corresponding to a term in Eq. (4.160) with $n = 25$, $m = 2$, $k_0 = 1$, $k_1 = 0$, $k_2 = 2$, $l = 2$, $l' = 3$, $\mathbf{l}_0 = (210)$, $\mathbf{l}_1 = (01232343210)$, $\mathbf{l}_2 = (01210)$, and $\mathbf{l}_3 = (0123)$, for illustration.

The sums over n and over paths in Eq. (4.160) may be converted to sums over m, k_0, k_1, $\cdots$, k_m (each from 0 to ∞) and over all possible subpaths Δ_0, Δ_1, $\cdots$, Δ_m, and Δ_{T}. We then obtain

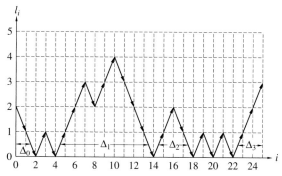

Fig. 4.15. Stone-fence diagram corresponding to a Daniels term in Eq. (4.160) with $n = 25$, $m = 2$, $k_0 = 1$, $k_1 = 0$, $k_2 = 2$, $l = 2$, $l' = 3$, $\mathbf{l}_0 = (210)$, $\mathbf{l}_1 = (01232343210)$, $\mathbf{l}_2 = (01210)$, and $\mathbf{l}_3 = (0123)$

$$\tilde{K}_{ll'}^{00,jj'} = \sum_{m=0}^{\infty}\sum_{r=0}^{\infty}\binom{m+r}{m}\Delta^r \sum_{\{\Delta_\alpha\}}\Delta_0\Delta_1\cdots\Delta_m\Delta_{\mathrm{T}}$$

$$= (\underbrace{\sum\Delta_0}_{\Delta_0})(\underbrace{\sum\Delta_{\mathrm{T}}}_{\Delta_{\mathrm{T}}})\sum_{m=0}^{\infty}(1-\Delta)^{-m-1}\prod_{\alpha=1}^{m}(\underbrace{\sum\Delta_\alpha}_{\Delta_\alpha}),\quad (4.165)$$

where $\prod_\alpha(\ )\equiv 1$ when $m=0$. Let $\overline{\Delta}_\alpha(s_\alpha)$ be the sum of Δ_α over all possible subpaths Δ_α of s_α steps. It is independent of α for $1\le\alpha\le m$ and is denoted by $\overline{\Delta}$. Further, note that $s_0 = l+2r$, $s_{m+1}=l'+2r$, and $s_\alpha = 2r+4$ for $1\le\alpha\le m$, where r is a nonnegative integer. Thus Eq. (4.165) may be rewritten as

$$\tilde{K}_{ll'}^{00,jj'} = \left[\sum_{r=0}^{\infty}\overline{\Delta}_0(l+2r)\right]\left[\sum_{r=0}^{\infty}\overline{\Delta}_{\mathrm{T}}(l'+2r)\right]$$

$$\times\sum_{m=0}^{\infty}(1-\Delta)^{-m-1}\left[\sum_{r=0}^{\infty}\overline{\Delta}(2r+4)\right]^m.\qquad (4.166)$$

Now $(1-\Delta)^{-m-1}$ may be expanded in powers of p and $(p+c_\infty k^2/6)^{-1}$, where

$$c_\infty = \frac{2}{f(0)}\left[(2+\sigma)^2+\tau_0{}^2\right],\qquad (4.167)$$

which reduces to Eq. (4.74) if $\sigma=0$. The other factors in Eq. (4.166) may be expanded in powers of p and ik. Then, on retaining terms up to $\mathcal{O}(L^{-s})$, where s is a positive integer, $\tilde{K}_{ll'}^{00,jj'}$ may be expanded in the form

$$\tilde{K}_{ll'}^{00,jj'}(k;p) = \sum_{n_1=0}^{2s'}\sum_{n_2=0}^{[s']}\sum_{n_3=0}^{2s'}A_{ll',n_1n_2n_3}^{jj'}$$

$$\times\frac{p^{n_1}(ik)^{l+l'+2n_3}}{\left(p+\frac{1}{6}c_\infty k^2\right)^{n_2+1}}+\mathcal{O}(L^{-s-1/2}),\qquad (4.168)$$

where $A_{\cdots}^{\cdots}$ are coefficients independent of p and k, $[x]$ is Gauss' symbol indicating the largest integer $\le x$, and the sums are taken under the restriction

$$0\le n_1-n_2+n_3\le s'\equiv s-\frac{1}{2}(l+l').\qquad (4.169)$$

Recall that $k^2=\mathcal{O}(L^{-1})$, $p=\mathcal{O}(L^{-1})$, and $(p+\mathrm{const.}\,k^2)^{-n}=\mathcal{O}(L^{n-1})$ [36].

From Eqs. (4.152) and (4.168), we find $\mathcal{I}_{l_1l_2l_3}^{m_1m_2,j_1j_2}(k;L)$. We then note that it is nonvanishing if l_1, l_2, and l_3 satisfy the triangular inequalities with $l_1+l_2+l_3$ even, because of the selection rules for the first 3-j symbol with $m=0$ in Eq. (4.152) (see Appendix 4.C). Therefore, l_1+l_2 is even (odd) when l_3 is even (odd). Thus, replacing again l and l' by l_1 and l_2, respectively, in Eq. (4.168), we may put $l_1 + l_2 + 2n_3 = l_3 + 2n_3'$ to change indices from n_3 to n_3'. After dropping the prime on n_3, we obtain

$$
\tilde{\mathcal{I}}_{l_1 l_2 l_3}^{m_1 m_2, j_1 j_2}(k; p) = \sum_{n_1=0}^{2s'} \sum_{n_2=0}^{[s']} \sum_{n_3=0}^{2s'} B_{l_1 l_2 l_3, n_1 n_2 n_3}^{m_1 m_2, j_1 j_2}
$$
$$
\times \frac{p^{n_1} (ik)^{l_3 + 2n_3}}{\left(p + \frac{1}{6} c_\infty k^2\right)^{n_2 + 1}} + \mathcal{O}(L^{-s-1/2}), \qquad (4.170)
$$

where $B_{...}^{...}$ are coefficients independent of p and k, and the sums are taken under the restriction

$$
\frac{1}{2}(l_1 + l_2 - l_3) \le n_1 - n_2 + n_3 \le s' \equiv s - \frac{1}{2} l_3. \qquad (4.171)
$$

$\mathcal{I}_{...}^{...}$ is found by Laplace inversion of Eq. (4.170); it is given by a sum of residues of the right-hand side of Eq. (4.170) multiplied by e^{Lp},

$$
\mathcal{I}_{l_1 l_2 l_3}^{m_1 m_2, j_1 j_2}(k; L) = \exp\left(-\frac{1}{6} c_\infty L k^2\right) \sum_{n_1=0}^{2s'} \sum_{n_2=0}^{[s']} C_{l_1 l_2 l_3, n_1 n_2}^{m_1 m_2, j_1 j_2}
$$
$$
\times (ik)^{l_3 + 2n_1} L^{n_2} + \mathcal{O}(L^{-s-1/2}), \qquad (4.172)
$$

where $C_{...}^{...}$ are coefficients independent of k and L, n_1 and n_2 have been redefined, and the sums are taken under the restriction

$$
\frac{1}{2}(l_1 + l_2 - l_3) \le n_1 - n_2 \le s' \equiv s - \frac{1}{2} l_3. \qquad (4.173)
$$

Finally, we find $\mathcal{G}_{l_1 l_2 l_3}^{m_1 m_2, j_1 j_2}(R; L)$ from Eqs. (4.156) and (4.172). Then a useful formula is

$$
\int_0^\infty k^{\mu-1} J_\nu(Rk) \exp(-a^2 k^2) dk = \frac{1}{2a^\mu} \left[\frac{1}{2}(\mu - \nu) - 1 \right]! \left(\frac{R}{2a} \right)^\nu
$$
$$
\times L_{(1/2)(\mu-\nu)-1}^{(\nu)} \left(\frac{R^2}{4a^2} \right) \exp\left(-\frac{R^2}{4a^2} \right), \qquad (4.174)
$$

where $\mu - \nu$ is even, $J_\nu(x)$ is the (ordinary) Bessel function of the first kind defined by

$$
j_l(x) = \left(\frac{\pi}{2x} \right)^{1/2} J_{l+1/2}(x), \qquad (4.175)
$$

and $L_n^{(\nu)}(x)$ is the Laguerre polynomial defined by

$$
L_n^{(\nu)}(x) = \sum_{r=0}^n \binom{n+\nu}{n-r} \frac{(-x)^r}{r!}. \qquad (4.176)
$$

Thus the final result is

$$\mathcal{G}_{l_1 l_2 l_3}^{m_1 m_2, j_1 j_2}(\mathbf{R}; L) = \left(\frac{3}{2\pi c_\infty L}\right)^{3/2} \exp\left(-\frac{3R^2}{2c_\infty L}\right)$$

$$\times \left[\left(\frac{R}{c_\infty L}\right)^{l_3} \sum_{n_1=(l_1+l_2-l_3)/2}^{[s-l_3/2]} \sum_{n_2=0}^{2n_1} D_{l_1 l_2 l_3, n_1 n_2}^{m_1 m_2, j_1 j_2} \right.$$

$$\left. \times \left(\frac{1}{c_\infty L}\right)^{n_1} \left(\frac{R^2}{c_\infty L}\right)^{n_2} + \mathcal{O}(L^{-s-1/2}) \right], \qquad (4.177)$$

where $D_{\cdots}^{\cdots}$ are coefficients independent of R and L, and n_1 and n_2 have been redefined. In the sth Daniels approximation to the $\mathcal{G}_{\cdots}^{\cdots}$ given by Eq. (4.177), l_1, l_2, and l_3 satisfy

$$|l_1 - l_2| \le l_3 \le l_1 + l_2 \le 2s \qquad (4.178)$$

with $l_1 + l_2 + l_3$ even. The coefficients $D_{\cdots}^{\cdots}$ may be computed for given κ_0, τ_0, and σ efficiently by the use of a computer.

4.6.3 Moment-Based Distributions

The moment-based distribution function is obtained as a straightforward consequence of the general development [35]. Its leading term is spherical Gaussian, while that of the moment-based distribution function of the Flory type [26, 27] is generalized (or ellipsoidal) Gaussian. Of course, both become the Hermite polynomial expansion [21, 39, 40] when reduced to $G(\mathbf{R}; L)$ as in the case of the KP chain.

As seen from Eq. (4.156) with Eq. (4.154), $R^{-l_3} \mathcal{G}_{l_1 l_2 l_3}^{m_1 m_2, j_1 j_2}(R; L)$ may be expanded in even powers of R. By the use of this fact and Eq. (4.155), $\mathcal{G}_{\cdots}^{\cdots}(R; L)$ may therefore be expanded in terms of Laguerre polynomials as follows,

$$\mathcal{G}_{l_1 l_2 l_3}^{m_1 m_2, j_1 j_2}(R; L) = \left(\frac{3}{2\langle R^2\rangle}\right)^{3/2} \exp(-\rho^2)$$

$$\times \sum_{n=0}^{\infty} M_{l_1 l_2 l_3, n}^{m_1 m_2, j_1 j_2}(L) \rho^{l_3} L_n^{(l_3+1/2)}(\rho^2) \qquad (4.179)$$

with

$$\rho = \left(\frac{3}{2\langle R^2\rangle}\right)^{1/2} R. \qquad (4.180)$$

$L_n^{(\nu)}$ have the "orthonormality" property

$$\int_0^\infty L_n^{(l+1/2)}(\rho^2) L_m^{(l+1/2)}(\rho^2) \rho^{2l+2} e^{-\rho^2} d\rho = N_n^{(l)} \delta_{nm} \qquad (4.181)$$

with

$$N_n^{(l)} = \frac{\pi^{1/2}(2n + 2l + 1)!!}{2^{n+l+2} n!}. \qquad (4.182)$$

By the use of Eq. (4.181) and also the orthonormality properties of Y_l^m and $\mathcal{D}_l^{mj}$, Eqs. (3.B.5) and (4.C.8), we find for the expansion coefficients in Eq. (4.179)

$$M_{l_1 l_2 l_3, n}^{m_1 m_2, j_1 j_2}(L) = \left(\frac{8\pi^2}{N_n^{(l_3)}}\right) \langle \rho^{l_3} L_n^{(l_3+1/2)}(\rho^2)$$
$$\times \mathcal{D}_{l_1}^{m_1 j_1 *}(\Omega) \mathcal{D}_{l_2}^{m_2 j_2}(\Omega_0) Y_{l_3}^{m_1-m_2}(\Theta, \Phi)\rangle, \quad (4.183)$$

where

$$\langle \cdots \rangle = (8\pi^2)^{-1} \int (\cdots) G d\mathbf{R} d\Omega d\Omega_0. \quad (4.184)$$

Now the problem is to evaluate the average in Eq. (4.183). From Eqs. (4.176) and (4.183), it is seen to be a sum of terms of the form

$$M = \langle R^{l_3+2r} \mathcal{D}_{l_1}^{m_1 j_1 *}(\Omega) \mathcal{D}_{l_2}^{m_2 j_2}(\Omega_0) Y_{l_3}^{m_1-m_2}(\Theta, \Phi)\rangle$$
$$\equiv \langle R\mathcal{D}^*\mathcal{D}Y\rangle. \quad (4.185)$$

In order to evaluate this average, we consider the average,

$$M' \equiv \langle \exp(i\mathbf{k}\cdot\mathbf{R})\mathcal{D}_{l_1}^{m_1 j_1 *}(\Omega)\mathcal{D}_{l_2}^{m_2 j_2}(\Omega_0)\rangle \quad (4.186)$$
$$= (8\pi^2)^{-1}\int \mathcal{D}_{l_1}^{m_1 j_1 *}\mathcal{D}_{l_2}^{m_2 j_2} I d\Omega d\Omega_0. \quad (4.187)$$

We have, from Eq. (4.186) with Eq. (4.154),

$$M' = 4\pi \sum_{l_3, m_3} i^{l_3} \langle j_{l_3}(kR)\mathcal{D}_{l_1}^{m_1 j_1 *}\mathcal{D}_{l_2}^{m_2 j_2} Y_{l_3}^{-m_3}(\Theta, \Phi)\rangle Y_{l_3}^{m_3}(\chi, \omega), \quad (4.188)$$

and from Eq. (4.187) with Eq. (4.151),

$$M' = (8\pi^2)^{-1}\sum_{l_3} \mathcal{I}_{l_1 l_2 l_3}^{m_1 m_2, j_1 j_2}(k; L) Y_{l_3}^{m_2-m_1}(\chi, \omega). \quad (4.189)$$

Equating the coefficients of $Y_{l_3}^{m_3}(\chi, \omega)$ in Eqs. (4.188) and (4.189), we obtain for the average on the right-hand side of Eq. (4.188)

$$\langle j_{l_3}\mathcal{D}_{l_1}^{m_1 j_1 *}\mathcal{D}_{l_2}^{m_2 j_2}Y_{l_3}^{-m_3}\rangle = (32\pi^3 i^{l_3})^{-1}\mathcal{I}_{l_1 l_2 l_3}^{m_1 m_2, j_1 j_2}(k; L)$$
$$\text{for } m_3 = m_2 - m_1 \quad (4.190)$$
$$= 0 \qquad \text{for } m_3 \neq m_2 - m_1,$$

where l_1, l_2, and l_3 satisfy the triangular inequalities.

Thus substitution of Eq. (4.154) into the first line of Eqs. (4.190) leads to

$$\mathcal{I}_{l_1 l_2 l_3}^{m_1 m_2, j_1 j_2}(k; L) = 32\pi^3 \cdot 2^{l_3}\sum_{r=0}^{\infty}\frac{(l_3+r)!}{r!(2l_3+2r+1)!}\langle R\mathcal{D}^*\mathcal{D}Y\rangle(ik)^{l_3+2r}.$$
$$(4.191)$$

From Eqs. (4.139), (4.152), and (4.191), the evaluation of the moments $M = \langle RD^*DY \rangle$ is seen to be similar to that of $\langle R^{2m} \rangle$. It may be eventually written in the form

$$\langle RD^*DY \rangle = \sum_{i,j,k}{}' A^{m_1 m_2, j_1 j_2}_{l_1 l_2 l_3, r, ijk} L^i \exp(-z_{j,k} L)\,, \qquad (4.192)$$

where $-z_{j,k}$ are the roots of the algebraic Eq. (4.44), and the numerical coefficients $A^{\cdots}_{\cdots}$ may be calculated for given κ_0, τ_0, and σ by the use of a computer. Note that $g^{jj'}_l$, $\langle (\mathbf{R} \cdot \mathbf{u}_0)^n \rangle$, and $\langle R^{2m} \rangle$ are special cases of $\langle RD^*DY \rangle$, which are the generalized moments of the distribution function $G(\mathbf{R}, \Omega \,|\, \Omega_0; L)$.

Finally, we briefly mention the moment-based distribution function $G(\mathbf{R}, \Omega \,|\, \Omega_0; L)$ of the Flory type for the HW model [35]. Its asymptotic form is a generalized Gaussian function of $\mathbf{R} - \mathbf{A}$ with $\mathbf{A}$ the persistence vector and it may be expanded in terms of Hermite polynomials. In this case there is no efficient method of calculating the expansion coefficients. However, the distribution functions $G(\mathbf{R}; L)$ derived from these moment-based $G(\mathbf{R}, \Omega \,|\, \Omega_0; L)$ are the same, that is, the well-known Hermite polynomial expansion [21, 39, 40]. It is pertinent to reproduce here the result. It reads

$$G(\mathbf{R}; L) = \left(\frac{3}{2\pi \langle R^2 \rangle} \right)^{3/2} \exp(-\rho^2) \sum_{\nu=0}^{\infty} h_{2\nu} \rho^{-1} H_{2\nu+1}(\rho)\,, \qquad (4.193)$$

where

$$h_{2\nu} = \frac{1}{2^{2(\nu+1)}(2\nu+1)!} \langle \rho^{-1} H_{2\nu+1}(\rho) \rangle\,, \qquad (4.194)$$

ρ is defined by Eq. (4.180), and H_ν is the Hermite polynomial defined by

$$H_\nu(x) = (-1)^\nu e^{x^2} \frac{d^\nu (e^{-x^2})}{dx^\nu}\,. \qquad (4.195)$$

Note that the $G(\mathbf{R}; L)$ truncated at $\nu = s$ (the sth Hermite polynomial approximation) involves the moments $\langle R^{2m} \rangle$ with $m \le s$.

4.6.4 Convergence

We examine the convergence of the two types of asymptotic expansions of the distribution function, in particular, with respect to the ring-closure probability $G(\mathbf{0}; L)$ and the mean reciprocal of the end-to-end distance $\langle R^{-1} \rangle$, that is, the convergence of the Daniels and Hermite polynomial expansions of $G(\mathbf{R}; L)$ as in the case of the KP chain.

The Daniels and Hermite values of $G(\mathbf{0}; L)$ are plotted against L in Figs. 4.16 and 4.17, respectively, for $\kappa_0 = 5$, $\tau_0 = 1$, and $\sigma = 0$ (Code 2 of Table 4.4), which is a typical HW case (of the strong helical nature). The numbers attached to the curves indicate the values of the degree s of approximation, and the dashed curve represents the coil-limiting values $(3/2\pi c_\infty L)^{3/2}$.

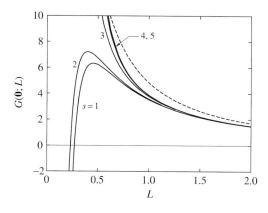

Fig. 4.16. Daniels values of the ring-closure probability $G(\mathbf{0}; L)$ plotted against L for the HW chain with $\kappa_0 = 5$, $\tau_0 = 1$, and $\sigma = 0$ (Code 2). The *dashed curve* represents the coil-limiting values $(3/2\pi c_\infty L)^{3/2}$

It is seen that both are convergent for $L \gtrsim 1.5$ and divergent for smaller L; $G(\mathbf{0}; L)$ must vanish at $L = 0$. It is then interesting to recall that for the KP chain, the Daniels approximation is convergent for $L \gtrsim 3$, while the convergence of the Hermite polynomial approximation is much worse. Thus it may be concluded that as the helical nature is increased, the convergence of the Daniels and Hermite polynomial approximations to $G(\mathbf{0}; L)$ becomes better (in particular for the latter) and their radii L of convergence become almost the same.

Figure 4.18 shows plots of $L\langle R^{-1}\rangle$ against L for the same code. The solid and dashed curves represent the Daniels and Hermite values, respectively, and the numbers attached to the curves indicate again the values of the degree s of approximation. Figure 4.19 shows similar plots for $\kappa_0 = \tau_0 = 1$ and $\sigma = 0$ (Code 3 of Table 4.4), which is rather close to the KP chain. It is again seen that for the code of the strong helical nature both approximations are convergent for $L \gtrsim 1.5$ and that for the code close to the KP chain the convergence of the Hermite polynomial approximation is worse. Although

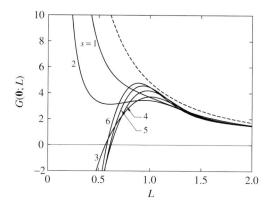

Fig. 4.17. Hermite values of $G(\mathbf{0}; L)$ for the HW chain with $\kappa_0 = 5$, $\tau_0 = 1$, and $\sigma = 0$ (Code 2); see legend to Fig. 4.16

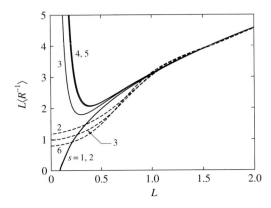

Fig. 4.18. $L\langle R^{-1}\rangle$ plotted against L for the HW chain with $\kappa_0 = 5$, $\tau_0 = 1$, and $\sigma = 0$ (Code 2). The *solid and dashed curves* represent the Daniels and Hermite values, respectively

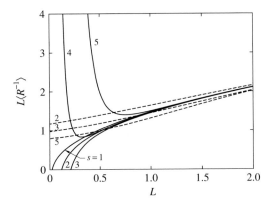

Fig. 4.19. $L\langle R^{-1}\rangle$ for the HW chain with $\kappa_0 = \tau_0 = 1$ and $\sigma = 0$ (Code 3); see legend for Fig. 4.18

both are, of course, divergent for smaller L, it is interesting to see that as L approaches zero, the Hermite value of $L\langle R^{-1}\rangle$ becomes finite; in particular, its third approximation gives $L\langle R^{-1}\rangle \simeq 1$ at $L = 0$.

4.7 Approximations

Necessarily, the convergence of the asymptotic expansions from the coil limit considered in the last section is very slow as the rod limit is approached. From the practical point of view it is therefore necessary to establish more efficient and useful approximation methods. Thus, in this section, we consider two such methods. One [41] is a modification of the Hermite (or Laguerre) polynomial approximations to the distribution functions and also a generalization of the procedure of Fixman and Skolnick [42]. This is called the *weighting function method*. The other [41,43] is a simple method which gives expansions of, for instance, $\langle R^{2m}\rangle$ in terms of the relative deviation ϵ of R^2 near the rod limit; and thus any moment may be expanded in powers of contour length L if desired. This is referred to as the ϵ *method*, for convenience. In later chapters these two approximation methods along with the second Daniels approximation are used to give values or construct interpolation formulas for various properties which are valid in a good approximation over the whole range of L.

4.7.1 Weighting Function Method

The right-hand side of Eq. (4.179) for $\mathcal{G}_{l_1l_2l_3}^{m_1m_2,j_1j_2}(R;L)$ may be regarded as its expansion in terms of Laguerre polynomials with a Gaussian weighting function, and therefore it may be generalized to an orthogonal polynomial expansion with an arbitrary given weighting function w,

$$\mathcal{G}_{l_1l_2l_3}^{m_1m_2,j_1j_2}(R;L) = \left(\frac{3}{2\langle R^2\rangle}\right)^{3/2} w(\rho)\sum_{n=0}^{\infty} M_{l_1l_2l_3,n}^{m_1m_2,j_1j_2}(L)\rho^{l_3}h_n^{(l_3)}(\rho^2),$$

$$(4.196)$$

where ρ is defined by Eq. (4.180) and $h_n^{(l_3)}$ are certain orthogonal polynomials of degree n.

A recurrence formula for $h_n^{(l)}$ may be derived by a standard method of constructing orthogonal polynomials [44]. The result is

$$h_n^{(l)}(\rho^2) = (-n^{-1}\rho^2 + \beta_n)h_{n-1}^{(l)}(\rho^2) - \gamma_n h_{n-2}^{(l)}(\rho^2) \qquad \text{for } n \geq 1 \quad (4.197)$$

with $h_0^{(l)} \equiv 1$ and $h_{-1}^{(l)} \equiv 0$ and with

$$\beta_n = \frac{1}{nN_{n-1}^{(l)}}\int_0^{\infty}\left[h_{n-1}^{(l)}(\rho^2)\right]^2\rho^{2l+4}w(\rho)d\rho, \qquad (4.198)$$

$$\gamma_n = \frac{(n-1)N_{n-1}^{(l)}}{nN_{n-2}^{(l)}} \qquad (\gamma_1 \equiv 0), \qquad (4.199)$$

$$N_n^{(l)}\delta_{nm} = \int_0^{\infty} h_n^{(l)}(\rho^2)h_m^{(l)}(\rho^2)\rho^{2l+2}w(\rho)d\rho, \qquad (4.200)$$

where the coefficient of the highest power ρ^{2n} of $h_n^{(l)}(\rho^2)$ has been chosen to be $(-1)^n/n!$, for convenience, and Eq. (4.200) gives the "orthonormality" property. Note that if $w(\rho) = \exp(-\rho^2)$, $h_n^{(l)}(\rho^2)$ is the Laguerre polynomial $L_n^{(l+1/2)}(\rho^2)$. By the use of Eq. (4.200) and the orthonormality properties of Y_l^m and $\mathcal{D}_l^{mj}$, Eqs. (3.B.5) and (4.C.8), we find for the expansion coefficient in Eq. (4.196)

$$M_{l_1l_2l_3,n}^{m_1m_2,j_1j_2}(L) = \left(\frac{8\pi^2}{N_n^{(l_3)}}\right)\langle\rho^{l_3}h_n^{(l_3)}(\rho^2)$$

$$\times\mathcal{D}_{l_1}^{m_1j_1*}(\Omega)\mathcal{D}_{l_2}^{m_2j_2}(\Omega_0)Y_{l_3}^{m_1-m_2}(\Theta,\Phi)\rangle. \quad (4.201)$$

The moment in Eq. (4.201) may be evaluated in the same manner as that used in the evaluation of the moment in Eq. (4.183).

Now we truncate the series in Eq. (4.196) to derive successive approximations. Suppose that we retain terms of the characteristic function $I(\mathbf{k}, \Omega \,|\, \Omega_0; L)$ up to $\mathcal{O}(k^{2s})$. From Eq. (4.191), it is seen that the terms of I up to $\mathcal{O}(k^{l_3+2n_1})$ can give exactly the coefficients $M_{\cdots}^{\cdots}(L)$ for $0 \leq n \leq n_1$. The desired approximation may therefore be obtained by truncating the series in Eq. (4.196) at $n = [s - l_3/2]$, so that $l_3 \leq 2s$, with $[x]$ being Gauss' symbol. In the particular case of $G(\mathbf{R}; L)$ with $l_i = m_i = j_i = 0$, this gives an orthogonal polynomial expansion truncated at $n = s$, involving $\langle R^{2m} \rangle$ with $m \leq s$.

By a theorem regarding least-squares polynomial approximations [44], the coefficients $M_{\cdots}^{\cdots}(L)$ so determined for $n \leq [s - l_3/2]$ minimize the weighted mean-square error,

$$
e^2 = \int_0^\infty \left\{ \left(\frac{2\langle R^2 \rangle}{3} \right)^{3/2} \mathcal{G}_{l_1 l_2 l_3}^{m_1 m_2, j_1 j_2}(R; L) \left[\rho^{l_3} w(\rho) \right]^{-1} \right.
$$
$$
\left. - \sum_{n=0}^{[s-l_3/2]} M_{l_1 l_2 l_3, n}^{m_1 m_2, j_1 j_2}(L) h_n^{l_3}(\rho^2) \right\}^2 \rho^{2l_3+2} w(\rho) d\rho . \quad (4.202)
$$

In the particular case of $l_i = m_i = j_i = 0$, this corresponds to Eq. (4.2) of [42]. It is then convenient to rewrite the sth approximation to $\mathcal{G}_{\cdots}^{\cdots}(L)$ in the form

$$
\mathcal{G}_{l_1 l_2 l_3}^{m_1 m_2, j_1 j_2}(R; L) = \left(\frac{3}{2\langle R^2 \rangle} \right)^{3/2} w(\rho) \sum_{n=0}^{[s-l_3/2]} \mathcal{F}_{l_1 l_2 l_3, n}^{m_1 m_2, j_1 j_2}(L) \rho^{l_3+2n} . \quad (4.203)
$$

The coefficients $\mathcal{F}_{\cdots}^{\cdots}(L)$ may be determined by the use of the least-squares theorem, instead of constructing the polynomials from Eq. (4.197); that is, they are the solutions of the linear simultaneous equations

$$
8\pi^2 \langle \rho \mathcal{D}^* \mathcal{D} Y \rangle = \sum_{n=0}^{[s-l_3/2]} \mathcal{F}_{l_1 l_2 l_3, n}^{m_1 m_2, j_1 j_2}(L) \int_0^\infty \rho^{2(l_3+n+n'+1)} w(\rho) d\rho
$$
$$
\text{for } 0 \leq n' \leq [s - l_3/2] , \quad (4.204)
$$

where $\langle \rho \mathcal{D}^* \mathcal{D} Y \rangle = \langle \rho^{l_3+2n'} \mathcal{D}_{l_1}^{m_1 j_1 *} \mathcal{D}_{l_2}^{m_2 j_2} Y_{l_3}^{m_1 - m_2} \rangle$ is equivalent to $\langle R \mathcal{D}^* \mathcal{D} Y \rangle$ and may be evaluated from Eq. (4.192). Equation (4.155) with Eq. (4.203) gives the distribution function approximated by the weighting function method. We note that the corresponding characteristic function cannot in general be found analytically from the former.

The problem that remains is to choose a suitable weighting function $w(\rho)$. Fixman and Skolnick [42] have chosen the function

$$
w_{\mathrm{FS}}(\rho) = \exp\left[-a\rho^2 - (b\rho^2)^\nu \right] , \quad (4.205)
$$

where the parameters a and b as functions of L (in its application to the HW model) are determined so that the 0th approximation to $G(\mathbf{R}; L)$, that is, the normalized weighting function gives the exact $\langle R^2 \rangle$ and $\langle R^4 \rangle$, and ν is an integer ranging from 2 to 5. Koyama [45] has also approximated $G(\mathbf{R}; L)$ by its 0th approximation with

$$w_{\mathrm{K}}(\rho) = \rho^{-1} \exp(-a\rho^2)\sinh(b\rho)\,, \qquad (4.206)$$

where the parameters a and b (as functions of L) are determined in the same way as in Eq. (4.205). For these weighting functions, however, the solutions for a and b do not exist in the range of $\langle R^4 \rangle / \langle R^2 \rangle^2 > 5/3$. Indeed, such cases can occur for the HW model. For example, values of $\langle R^6 \rangle / \langle R^2 \rangle^3$ are plotted against those of $\langle R^4 \rangle / \langle R^2 \rangle^2$ in Fig. 4.20 for the HW chain with $\kappa_0 = 10$, $\tau_0 = 2$, and $\sigma = 0$. The numbers attached to the curves indicate the values of L as an auxiliary variable, and the vertical line segment is drawn at $\langle R^4 \rangle / \langle R^2 \rangle^2 = 5/3$.

Thus, in order to remove the difficulty and also to obtain better approximations, we choose the following two weighting functions

$$w_{\mathrm{I}}(\rho) = \exp\left[-a_1\rho^2 - a_2\rho - (b\rho^2)^\nu\right]\,, \qquad (4.207)$$

$$w_{\mathrm{II}}(\rho) = \exp\left[-a_1\rho^2 - a_2\rho^4 - (b\rho^2)^\nu\right]\,, \qquad (4.208)$$

where in both cases ν is set equal to 5, and a_1, a_2, and b as functions of L are determined so that the normalized weighting function gives the exact

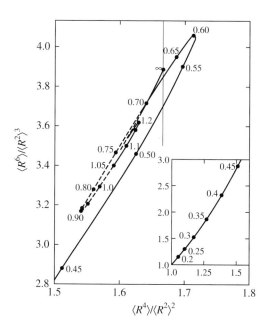

Fig. 4.20. $\langle R^6 \rangle / \langle R^2 \rangle^3$ plotted against $\langle R^4 \rangle / \langle R^2 \rangle^2$ for the HW chain with $\kappa_0 = 10$, $\tau_0 = 2$, and $\sigma = 0$. The *numbers attached to the curves* indicate the values of L, and the vertical line segment is drawn at $\langle R^4 \rangle / \langle R^2 \rangle^2 = 5/3$. The solutions for the weighting function w_{II} do not exist in the ranges of L indicated by the *dashed curves*

$\langle R^2 \rangle$, $\langle R^4 \rangle$, and $\langle R^6 \rangle$. When $a_2 = 0$, both w_{I} and w_{II} agree with w_{FS}. For w_{I}, the solutions for a_1, a_2, and b exist over the whole range of L except for the cases in which the pitch h of the characteristic helix is much smaller than its radius ρ. The distribution function G with w_{I} does not fulfil the requirement

$$\left. \frac{\partial G(\mathbf{R}; L)}{\partial R} \right|_{\mathbf{R}=0} = 0 , \tag{4.209}$$

which arises from the fact that $G(\mathbf{R}; L)$ is a spherically symmetric function of $\mathbf{R}$ and is analytic at $\mathbf{R} = \mathbf{0}$. However, the effect of this defect may be regarded as small except on $G(\mathbf{0}; L)$ for some cases and also on the properties related to $\partial G / \partial R$. On the other hand, w_{II} satisfies Eq. (4.209) but the solutions for its a_1, a_2, and b exist only in the limited range of L in some cases. For example, in the case of Fig. 4.20 the solutions for w_{II} do not exist for $0.7 \lesssim L \lesssim 1.05$ and $1.35 \lesssim L \lesssim 5$, which ranges are indicated by the dashed curves in the figure. Note that in the case of $G(\mathbf{R}; L)$ we take $s \geq 3$ since there are no correction polynomials for $s \leq 3$ if w_{I} or w_{II} is used.

In practice, we determine the constants in w and evaluate the integral on the right-hand side of Eq. (4.204) following the procedure of Fixman and co-workers [42, 46]. The required integrals are of the form

$$\int_0^\infty \rho^m w(\rho) d\rho = b^{-(m+1)/2} I_m(c_1, c_2) \tag{4.210}$$

with

$$I_m = \int_0^\infty x^m \exp(-c_1 x^2 - c_2 x^\alpha - x^{2\nu}) dx , \tag{4.211}$$

$$c_1 = a_1/b , $$
$$c_2 = a_2/b^{\alpha/2} , \tag{4.212}$$

where $\alpha = 1$ or 4, and $\nu = 5$. If I_m are evaluated by numerical integrations for $0 \leq m \leq 2\nu - 1$, we find I_m for $m \geq 2\nu$ from the recurrence relation

$$2\nu I_{m+2\nu} = (m+1)I_m - c_2\alpha I_{m+\alpha} - 2c_1 I_{m+2} \quad \text{for } m \geq 0 . \tag{4.213}$$

Then the even moments $\langle \rho^{2m} \rangle_w$ of the normalized weighting function are given by

$$\langle \rho^{2m} \rangle_w = \frac{\displaystyle\int_0^\infty \rho^{2m+2} w(\rho) d\rho}{\displaystyle\int_0^\infty \rho^2 w(\rho) d\rho} = \frac{I_{2m+2}}{b^m I_2} . \tag{4.214}$$

The conditions $\langle \rho^{2m} \rangle = \langle \rho^{2m} \rangle_w$ for $m = 1$–3 that determine a_1, a_2, and b may be rewritten as

$$\langle \rho^2 \rangle = 3/2 = I_4/bI_2 \,, \tag{4.215}$$

$$\langle \rho^4 \rangle / \langle \rho^2 \rangle^2 = I_6 I_2 / I_4{}^2 \,, \tag{4.216}$$

$$\langle \rho^6 \rangle / \langle \rho^2 \rangle^3 = I_8 I_2{}^2 / I_4{}^3 \,. \tag{4.217}$$

Thus a_1, a_2, and b may be determined as follows: (1) we first determine c_1 and c_2 as the solutions of the nonlinear simultaneous Eqs. (4.216) and (4.217) (which are found by the Newtonian method), (2) we then determine b from Eq. (4.215), and (3) finally we determine a_1 and a_2 from Eqs. (4.212). However, it must be noted that these parameters cannot be determined accurately for such small L that $\langle R^4 \rangle / \langle R^2 \rangle^2$ is smaller than about 1.03.

4.7.2 Epsilon Method

We define the relative deviations ϵ and $\delta_{l_1 l_2 l_3}^{m_1 m_2, j_1 j_2}$ of R^2 and $R^{l_3} \mathcal{D}_{l_1}^{m_1 j_1 *}(\Omega)$ $\times \mathcal{D}_{l_2}^{m_2 j_2}(\Omega_0) Y_{l_3}^{m_1 - m_2}(\Theta, \Phi)$ by

$$R^2 = \langle R^2 \rangle_0 (1 + \epsilon) \,, \tag{4.218}$$

$$R^{l_3} \mathcal{D}_{l_1}^{m_1 j_1 *} \mathcal{D}_{l_2}^{m_2 j_2} Y_{l_3}^{m_1 - m_2} = \langle R^{l_3} \mathcal{D}_{l_1}^{m_1 j_1 *} \mathcal{D}_{l_2}^{m_2 j_2} Y_{l_3}^{m_1 - m_2} \rangle_0 (1 + \delta_{l_1 l_2 l_3}^{m_1 m_2, j_1 j_2}) \,, \tag{4.219}$$

respectively, where $\langle A \rangle_0$ is set equal either to $\langle A \rangle$ or to its rod-limiting value according to the convergence of the quantity to be considered.

We then have, from Eqs. (4.218) and (4.219),

$$\langle \epsilon^m \rangle = \frac{\langle R^{2m} \rangle}{\langle R^2 \rangle_0{}^m} - \sum_{r=0}^{m-1} \binom{m}{r} \langle \epsilon^r \rangle \qquad (m \geq 1) \,, \tag{4.220}$$

$$\langle \delta \rangle = \frac{\langle R^{l_3} X \rangle}{\langle R^{l_3} X \rangle_0} - 1 \,, \tag{4.221}$$

$$\langle \delta \epsilon^m \rangle = \frac{\langle R^{l_3 + 2m} X \rangle}{\langle R^{l_3} X \rangle_0 \langle R^2 \rangle_0{}^m} - \sum_{r=0}^{m} \binom{m}{r} \langle \epsilon^r \rangle$$

$$- \sum_{r=0}^{m-1} \binom{m}{r} \langle \delta \epsilon^r \rangle \qquad (m \geq 1) \,, \tag{4.222}$$

where we have abbreviated $\delta_{...}^{...}$ and $\mathcal{D}^* \mathcal{D} Y$ to δ and X, respectively, so that $\langle R^{l_3 + 2m} X \rangle = \langle R \mathcal{D}^* \mathcal{D} Y \rangle$. Thus $\langle \epsilon^m \rangle$ ($m \geq 1$) and $\langle \delta \epsilon^m \rangle$ ($m \geq 0$) may be expressed successively in terms of $\langle R^{2m} \rangle$ and $\langle R \mathcal{D}^* \mathcal{D} Y \rangle$. We note that $\langle \epsilon \rangle = \langle \delta \rangle = 0$ for $\langle A \rangle_0 = \langle A \rangle$, while $\langle \epsilon \rangle = \mathcal{O}(L)$ and $\langle \delta \rangle = \mathcal{O}(L)$ in the case of the rod-limiting values for $\langle A \rangle_0$, but that $\langle \epsilon^m \rangle = \mathcal{O}(L^m)$ for $m \geq 2$ and $\langle \delta \epsilon^m \rangle = \mathcal{O}(L^{m+1})$ for $m \geq 1$ in both cases.

If we retain terms up to $\mathcal{O}(L^s)$, the generalized moments $\langle RD^*DY \rangle$ may be expanded in terms of $\langle \epsilon^r \rangle$ and $\langle \delta \epsilon^r \rangle$ as follows,

$$\langle RD^*DY \rangle = \langle R^{l_3} X \rangle_0 \langle R^2 \rangle_0^m \left[\sum_{r=0}^{m} \binom{m}{r} \langle (1+\delta)\epsilon^r \rangle \right] \qquad \text{for } m \leq s$$

$$= \langle R^{l_3} X \rangle_0 \langle R^2 \rangle_0^m \left[\sum_{r=0}^{s-1} \binom{m}{r} \langle (1+\delta)\epsilon^r \rangle \right.$$

$$\left. + \binom{m}{s} \langle \epsilon^s \rangle + \mathcal{O}(L^{s+1}) \right] \qquad \text{for } m > s. \quad (4.223)$$

A similar expansion of $\langle R^{l_3+n} X \rangle$ with n being a positive or negative integer may easily be obtained. Whichever values of $\langle A \rangle_0$ are assigned, we may expand these generalized moments along with $\langle \epsilon^m \rangle$ and $\langle \delta \epsilon^m \rangle$ in powers of L if we want. In particular, we have

$$\langle R^{2m} \rangle = \langle R^2 \rangle_0^m \sum_{r=0}^{m} \binom{m}{r} \langle \epsilon^r \rangle \qquad \text{for } m \leq s$$

$$= \langle R^2 \rangle_0^m \left[\sum_{r=0}^{s} \binom{m}{r} \langle \epsilon^r \rangle + \mathcal{O}(L^{s+1}) \right] \qquad \text{for } m > s, \quad (4.224)$$

$$\langle R^{-1} \rangle = \langle R^2 \rangle_0^{-1/2} \left[\sum_{r=0}^{s} \frac{(-1)^r (2r-1)!!}{2^r r!} \langle \epsilon^r \rangle + \mathcal{O}(L^{s+1}) \right]. \quad (4.225)$$

Substitution of Eqs. (4.223) into Eq. (4.191) leads to the sth-order expansion of $\mathcal{I}_{\cdots}^{\cdots}$,

$$\mathcal{I}_{l_1 l_2 l_3}^{m_1 m_2, j_1 j_2}(k; L) = 32\pi^3 i^{l_3} \langle R^{l_3} X \rangle_0 \langle R^2 \rangle_0^{-l_3/2}$$

$$\times \left[\sum_{r=0}^{s-1} \frac{(-x)^r}{2^r r!} \langle (1+\delta)\epsilon^r \rangle j_{l_3+r}(x) + \frac{(-x)^s}{2^s s!} \langle \epsilon^s \rangle j_{l_3+s}(x) \right] \quad (4.226)$$

with

$$x = \langle R^2 \rangle_0^{1/2} k, \quad (4.227)$$

where we have used Eq. (4.154). In particular, we have

$$I(\mathbf{k}; L) = (4\pi)^{-1/2} \mathcal{I}_{000}^{00,00}(k; L)$$

$$= \sum_{r=0}^{s} \frac{(-x)^r}{2^r r!} \langle \epsilon^r \rangle j_r(x). \quad (4.228)$$

Now we derive the sth-order expansions in powers of L for the case of $\sigma = 0$. The averages $\langle \epsilon^m \rangle$ and $\langle (1+\delta)\epsilon^m \rangle$ may be evaluated from Eqs. (4.220)–(4.222) with the moments $\langle R^{2m} \rangle$ and $\langle RD^*DY \rangle$. The results may then be written in the form,

$$\langle \epsilon^m \rangle = \sum_{n=m}^{s} E_{mn}(\kappa_0, \tau_0)L^n \,, \qquad (4.229)$$

$$\langle (1+\delta)\epsilon^m \rangle = \sum_{n=m}^{s} D_{l_1 l_2 l_3, mn}^{m_1 m_2, j_1 j_2}(\kappa_0, \tau_0)L^n \,, \qquad (4.230)$$

where we note that $E_{00} = 1$ and that $D_{\cdots}^{\cdots}$ are not to be confused with the $D_{\cdots}^{\cdots}$ in Eq. (4.177). For convenience, we consider here only $\langle R^{2m} \rangle$, $\langle R^{-1} \rangle$, and $I(\mathbf{k}; L)$. Substitution of Eq. (4.229) into Eqs. (4.224), (4.225), and (4.228) leads to

$$L^{-2m}\langle R^{2m} \rangle = 1 + \sum_{r=1}^{m}\sum_{n=r}^{s} \binom{m}{r} E_{rn}L^n \qquad \text{for } m \le s$$

$$= 1 + \sum_{n=1}^{s}\sum_{r=1}^{n} \binom{m}{r} E_{rn}L^n \qquad \text{for } m > s\,, \qquad (4.231)$$

$$L\langle R^{-1} \rangle = 1 + \sum_{n=1}^{s}\sum_{r=1}^{n} \frac{(-1)^r(2r-1)!!}{2^r r!} E_{rn}L^n \,, \qquad (4.232)$$

$$I(\mathbf{k}; L) = j_0(Lk) + \sum_{n=1}^{s}\sum_{r=1}^{n} \frac{(-1)^r}{2^r r!} E_{rn}L^n (Lk)^r j_r(Lk) \,, \qquad (4.233)$$

where in Eq. (4.233) we have assumed $\langle R^2 \rangle_0 = L^2$ (the rod-limiting value). Note that the above derivation of Eq. (4.231) is trivial since Eq. (4.229) has in fact been obtained from Eq. (4.231). The coefficients E_{mn} ($1 \le m \le n \le 5$) and also those $D_{l_1 l_2 l_3, mn}^{00,00}$ ($0 \le m < n \le 5$), which are required later, are given as functions of κ_0 and τ_0 in Appendix III.

As seen from Eqs. (4.231)–(4.233) with the E_{mn} given in Appendix III, the coefficients of terms linear in L of such quantities are constants independent of κ_0 and τ_0, those of square and cubic terms are functions only of κ_0, and those of higher terms are functions of κ_0 and τ_0. Thus they include as special cases the WKB approximations (first-order terms) as given by Eqs. (3.96) [and also by Eq. (3.97)] and also the expansions given by Eq. (3.97) and Eq. (3.103) with the first of Eqs. (3.100) for the KP chain. We note that the convergence of the expansion of I given by Eq. (4.228) with $\langle R^2 \rangle_0 = \langle R^2 \rangle$ is better than that of the expansion with $\langle R^2 \rangle_0 = L^2$ or of the expansion given by Eq. (4.233).

4.7.3 Convergence

We examine the convergence of the weighting function method with respect to the ring-closure probability $G(\mathbf{0}; L)$ and that of the ϵ method with respect to the mean reciprocal of the end-to-end distance $\langle R^{-1} \rangle$ and the characteristic function $I(\mathbf{k}; L)$. For simplicity, the approximations with the weighting functions w_I, w_{II}, and w_{FS} (with $\nu = 5$) are indicated by WIs, WIIs, and FSs,

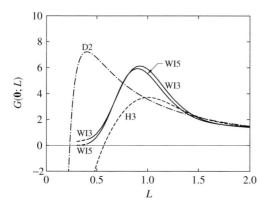

Fig. 4.21. WI3, WI5, D2, and H3 values of $G(0; L)$ plotted against L for the HW chain with $\kappa_0 = 5$, $\tau_0 = 1$, and $\sigma = 0$ (Code 2)

respectively, and the ϵ approximations by ϵs, where the number s indicates the degree of approximation. We also consider the Daniels approximations (Ds) and the Hermite polynomial approximations (Hs) in some cases.

Values of $G(0; L)$ are plotted against L in Fig. 4.21 for $\kappa_0 = 5$, $\tau_0 = 1$, and $\sigma = 0$ (Code 2 of Table 4.4) and in Fig. 4.22 for $\kappa_0 = \tau_0 = 1$ and $\sigma = 0$ (Code 3 of Table 4.4), where the D2 and H3 values have already been plotted in Figs. 4.16 and 4.17, respectively. It is seen that the convergence of the weighting function method is in general much better than that of the Daniels and Hermite polynomial approximations, and also becomes better as the helical nature is increased (better for Code 2 than for Code 3). In particular, it is important to note that the weighting function method can in general give

$$G(0; L) = 0 \qquad \text{for } L \ll 1, \tag{4.234}$$

as shown in Fig. 4.21, except for codes close to the KP chain such as Code 3.

The $\epsilon 1$ to $\epsilon 5$ values of $L\langle R^{-1}\rangle$ calculated from Eq. (4.232) are plotted

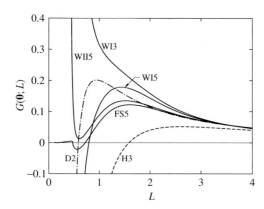

Fig. 4.22. WI3, WI5, WII5, FS5, D2, and H3 values of $G(0; L)$ plotted against L for the HW chain with $\kappa_0 = \tau_0 = 1$ and $\sigma = 0$ (Code 3)

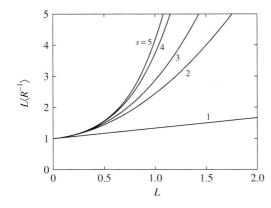

Fig. 4.23. ϵs values of $L\langle R^{-1}\rangle$ plotted against L for the HW chain with $\kappa_0 = 5$, $\tau_0 = 1$, and $\sigma = 0$ (Code 2)

against L in Fig. 4.23 for Code 2 as an example. Although for this code the convergence is good for $L \lesssim 0.5$ (radius of convergence), it cannot be improved appreciably even if s is increased, since we have assumed $|\epsilon| < 1$. In general, the convergence becomes poorer as the helical nature is increased. The $\epsilon 1$, $\epsilon 4$, and $\epsilon 5$ values of $I(\mathbf{k}; L)$ calculated from the second line of Eqs. (4.228) with $\langle R^2\rangle_0 = \langle R^2\rangle$ at $k = 30$ for the same code are plotted against L in Fig. 4.24. For comparison, the H3 values and the rod-limiting values (R) are also plotted. For this case the radius L of convergence is about 0.4, and in general it becomes smaller as k is increased and as the helical nature is increased. In any case, however, we can join the ϵ values to the WI, WII, or FS values, and then conveniently to the D2 values in order to obtain good approximations valid over the whole range of L.

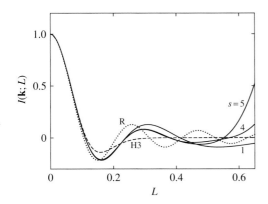

Fig. 4.24. $\epsilon 1$, $\epsilon 4$, $\epsilon 5$, and H3 values of $I(\mathbf{k}; L)$ plotted against L for the HW chain with $\kappa_0 = 5$, $\tau_0 = 1$, and $\sigma = 0$ (Code 2) at $k = 30$. The *dotted curve* R represents the rod-limiting values

4.8 Some Other Topics

4.8.1 Multivariate Distribution Functions, Etc.

We can in general evaluate the multivariate distribution function $P(\{\mathbf{R}_p\}, \Omega,$ $\Omega_0; L)$ of $\{\mathbf{R}_p\} = \mathbf{R}_1, \mathbf{R}_2, \cdots, \mathbf{R}_p$, $\Omega(L) = \Omega$, and $\Omega(0) = \Omega_0$ for the HW chain of contour length L, where $\mathbf{R}_j$ is the vector distance between the contour points s_j and $s_{j'}$ ($0 \leq s_j < s_{j'} \leq L$; $j = 1, 2, \cdots, p$) [31, 47]. This distribution function may be used to evaluate the moments $\langle S^{2m} \rangle$ of the radius of gyration and the moments of inertia tensor of linear [48] and branched [47] chains. However, a comparison with experiment for these quantities cannot directly be made (except for $\langle S^2 \rangle$), nor are they used in later chapters. Thus we do not reproduce the results for them.

4.8.2 Temperature Coefficients of $\langle R^2 \rangle$

We consider the temperature coefficient of (unperturbed) $\langle R^2 \rangle$ in the coil limit [49]. We denote it by $\langle R^2 \rangle_{(C)}$. When unreduced, it is given, from Eqs. (4.86) with Eqs. (4.75) and (4.76), by

$$\langle R^2 \rangle_{(C)} = \frac{(4 + \tau_0^2)L}{(4 + \nu^2)\lambda^2}, \qquad (4.235)$$

where λ is related to the bending force constant α by Eq. (3.37), and the reduced quantities L, κ_0, and τ_0 on the right-hand side are related to the respective unreduced quantities (primed) by $L = \lambda L'$, $\kappa_0 = \lambda^{-1}\kappa_0'$, and $\tau_0 = \lambda^{-1}\tau_0'$.

Now, for the elastic wire model the temperature coefficients of L', $\kappa_0'^{-1}$, and $\tau_0'^{-1}$ must be of the same order of magnitude as linear thermal expansion coefficients of ordinary solids (10^{-6}–10^{-5} deg^{-1}), so that their dependence on temperature T may be ignored. Further, we assume that α is independent of T. Then, the only quantity that depends on T is λ, which is proportional to T, and we have

$$\frac{d\ln\langle R^2 \rangle_{(C)}}{dT} = -\frac{16 - 4\kappa_0^2 + (8 + \nu^2)\tau_0^2}{(4 + \tau_0^2)(4 + \nu^2)}T^{-1}. \qquad (4.236)$$

For the KP chain ($\kappa_0 = 0$), therefore, the temperature coefficient defined by the left-hand side of Eq. (4.236) is always equal to $-T^{-1}$.

A contour map of the temperature coefficient in a (κ_0, τ_0)-plane calculated from Eq. (4.236) is shown in Fig. 4.25, where the solid and dashed curves are the contour lines at $T = 300$ and 400 K, respectively, the attached numbers indicating the values of $10^3 d\ln\langle R^2 \rangle_{(C)}/dT$ (in deg^{-1}). Along the heavy solid curve 0, it vanishes at all temperatures. It is interesting to see from a comparison of Fig. 4.25 with Fig. 4.13 that the temperature coefficient becomes

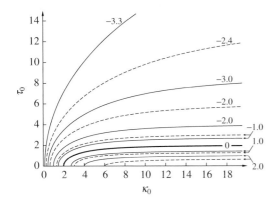

Fig. 4.25. Contour map of $10^3 d\ln\langle R^2\rangle_{(C)}/dT$ (in deg^{-1}) in a (κ_0, τ_0)-plane. The *solid and dashed curves* are the contour lines at $T = 300$ and 400 K, respectively

positive for typical HW chains. Here only note that the observed temperature coefficient is in fact positive for PDMS [50] and s-PMMA [34,51], which may be regarded as typical HW chains from their model parameters determined from experiment, as shown in the next chapter.

Appendix 4.A Generalization and Other Related Models

Equation (4.4) for the potential energy U per unit contour length may be generalized to [13]

$$U = (4D_\xi)^{-1}(\omega_\xi - c_\xi)^2 + (4D_\eta)^{-1}(\omega_\eta - c_\eta)^2 + (4D_\zeta)^{-1}(\omega_\zeta - c_\zeta)^2 , \quad (4.A.1)$$

where D_μ and c_μ ($\mu = \xi, \eta, \zeta$) are constants independent of s. This potential becomes a minimum in the state $\boldsymbol{\omega} = (c_\xi, c_\eta, c_\zeta)$, in which the chain contour is a regular helix specified by

$$\kappa_0 = (c_\xi^2 + c_\eta^2)^{1/2} ,$$
$$\tau_0 = c_\zeta , \qquad\qquad (4.A.2)$$

as seen from Eqs. (4.12) and (4.13). For this chain we can also define the Green function $G(\mathbf{R}, \Omega \,|\, \Omega_0; L) = G(\mathbf{R}, \mathbf{u}, \mathbf{a} \,|\, \mathbf{u}_0, \mathbf{a}_0; L)$, where $\mathbf{u}$ and $\mathbf{a}$ are defined by the first and second of Eqs. (4.9), respectively, but $\mathbf{a}$ and $\mathbf{b} = \mathbf{u} \times \mathbf{a}$ are not necessarily the mean unit curvature and mean unit binormal vectors. The Fokker–planck equation for G is still given by Eq. (4.33) but with the diffusion operator

$$\mathcal{A} = c_\xi L_\xi + c_\eta L_\eta + c_\zeta L_\zeta - k_\mathrm{B}T(D_\xi L_\xi^2 + D_\eta L_\eta^2 + D_\zeta L_\zeta^2) \quad (4.A.3)$$

with L_μ the angular momentum operators given by Eqs. (4.32) and (4.35).

Now Eq. (4.34) with Eq. (4.A.3) is just the general, standard equation of anisotropic diffusion in a convective field (c_ξ, c_η, c_ζ) in an Ω space. The moments of the infinitesimal rotation vector $\Delta\overline{\boldsymbol{\Omega}}$ may therefore be readily found to be [52]

$$\langle\Delta\overline{\Omega}_\mu\rangle = c_\mu\Delta s\,,$$
$$\langle\Delta\overline{\Omega}_\mu\Delta\overline{\Omega}_\nu\rangle = 2k_\mathrm{B}TD_\mu\delta_{\mu\nu}\Delta s \quad (\mu,\nu=\xi,\eta,\zeta)\,. \tag{4.A.4}$$

This generalized chain reduces to the HW chain if $c_\xi = 0$, $c_\eta = \kappa_0$, $c_\zeta = \tau_0$, $D_\xi = D_\eta = \lambda/k_\mathrm{B}T$, and $D_\zeta = \lambda(1+\sigma)/k_\mathrm{B}T$, and the latter reduces to the Bugl–Fujita (BF) chain [12] if $D_\xi \to 0$ with the other constants remaining unchanged. Note that to let the bending force constant about the ξ axis approach infinity ($D_\xi \to 0$) with $c_\xi = 0$ is equivalent to $\omega_\xi = 0$. This constraint ($\omega_\xi = 0$) in the BF chain is unphysical. The moments of $\Delta\overline{\boldsymbol{\Omega}}$ for these chains are then obtained as

$$\begin{aligned}
\langle\Delta\overline{\Omega}_\xi\rangle &= 0\,,\\
\langle\Delta\overline{\Omega}_\eta\rangle &= \kappa_0\Delta s\,,\\
\langle\Delta\overline{\Omega}_\zeta\rangle &= \tau_0\Delta s\,,\\
\langle(\Delta\overline{\Omega}_\eta)^2\rangle &= 2\lambda\Delta s\,,\\
\langle(\Delta\overline{\Omega}_\zeta)^2\rangle &= 2\lambda(1+\sigma)\Delta s\,,\\
\langle\Delta\overline{\Omega}_\mu\Delta\overline{\Omega}_\nu\rangle &= 0 \quad \text{for } \mu\neq\nu\,,
\end{aligned} \qquad \text{(HW and BF)} \tag{4.A.5}$$

and

$$\begin{aligned}
\langle(\Delta\overline{\Omega}_\xi)^2\rangle &= 2\lambda\Delta s && \text{(HW)}\\
&= 0 && \text{(BF)}\,.
\end{aligned} \tag{4.A.6}$$

The moments given by Eqs. (4.A.4)–(4.A.6) are used in Appendix 4.B, where the continuous limits of discrete chains are considered.

However, the HW chain as a special case of Eq. (4.A.1) requires some comments. The necessary and sufficient condition under which the above generalized chain reduces to the HW chain is $D_\xi = D_\eta$ (an isotropically bending wire or a symmetric top), the condition $c_\xi = 0$ being unnecessary. If $D_\xi = D_\eta$ and $c_\xi \neq 0$, we rotate the (ξ, η, ζ) system by a constant angle $\psi_0 = -\tan^{-1}(c_\xi/c_\eta)$ about the ζ axis at every point s to transform Eq. (4.A.1) into

$$U = (4D_\xi)^{-1}\left[\omega_{\xi_0}^2 + (\omega_{\eta_0} - \kappa_0)^2\right] + (4D_\zeta)^{-1}(\omega_{\zeta_0} - \tau_0)^2 \tag{4.A.7}$$

in the new system (ξ_0, η_0, ζ_0), where κ_0 and τ_0 are given by Eqs. (4.A.2). This is just the standard form of U given by Eq. (4.4) for the HW chain.

Equation (4.A.1) may be further generalized, although formally, to its most general from [53], as done by Miyake and Hoshino [54, 55], in which $\mathbf{e}_\zeta$, one of the principal axes of inertia, does not necessarily coincide with the unit tangent vector $\mathbf{u}$, so that

$$\mathbf{u} = l_1\mathbf{e}_\xi + l_2\mathbf{e}_\eta + l_3\mathbf{e}_\zeta \tag{4.A.8}$$

with l_i the direction cosines of $\mathbf{u}$ in its localized coordinate system (ξ, η, ζ). However, the energy of their original chain [54] becomes a minimum in the state $\boldsymbol{\omega} = (\omega l_1, \omega l_2, \omega l_3) = \omega \mathbf{u}$, in which the contour is a straight line. Therefore, it has the KP case but not the regular helix extreme, and is not a helical wormlike chain.

Appendix 4.B Corresponding Discrete Chains

We find a hypothetical discrete chain which tends to the generalized continuous chain defined by Eq. (4.A.1) and therefore also to the HW and KP chains in the continuous limit. For this purpose we start from a discrete chain with coupled rotations. Suppose that its kth monomer unit ($k = 1, 2, \cdots, x$) is composed of the $(i-1)$th and ith skeletal bonds with $i = 2k$ ($i = 1, 2, \cdots, n = 2x$). We affix a localized Cartesian coordinate system $(\mathbf{e}_{\xi_k}, \mathbf{e}_{\eta_k}, \mathbf{e}_{\zeta_k})$ to each monomer unit in such a way that $\mathbf{e}_{\zeta_k}$ is in the direction of $\mathbf{l}_{i-1} + \mathbf{l}_i \equiv \bar{\mathbf{l}}_k$ with $\mathbf{l}_i$ the ith bond vector, $\mathbf{e}_{\xi_k}$ is in the plane of $\mathbf{l}_{i-1}$ and $\mathbf{l}_i$ with an acute angle between $\mathbf{e}_{\xi_k}$ and $\mathbf{l}_{i-1}$, and $\mathbf{e}_{\eta_k}$ completes the right-handed system, as depicted in Fig. 4.B.1. Suppose that an external coordinate system agrees with the first localized system, and let the Euler angles $\overline{\Omega}_k = (\bar{\theta}_k, \bar{\phi}_k, \bar{\psi}_k)$ define the orientation of the kth system with respect to the $(k-1)$th system. We then assume that the total potential energy E of the chain is of the form

$$E = \sum_{k=2}^{x} E_k(\overline{\Omega}_k), \qquad (4.B.1)$$

where E_k are the same for all k, and also $\bar{l}_k = \bar{l}$, $\bar{\theta}_k = \bar{\theta}$, $\bar{\phi}_k = \bar{\phi}$, and $\bar{\psi}_k = \bar{\psi}$ for all k. Thus the chain defined by Eq. (4.B.1) may be regarded as a coarse-grained discrete chain with coupled rotations (composed of x bonds of length $\bar{l}$), and it is the hypothetical chain whose continuous limit is taken.

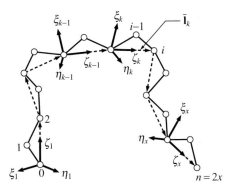

Fig. 4.B.1. Coarse-grained discrete chain with coupled rotations composed of x bonds of length $\bar{l}$, which is a hypothetical chain whose continuous limit is taken

Now we take the continuous limit by letting $\bar{l} \to 0$, $\bar{\theta} \to 0$, and $\bar{\phi} + \bar{\psi} \to 0$ at constant $x\bar{l} = L$ and under certain additional conditions [13]. Note that the rotation with $\bar{\phi} + \bar{\psi} = 0$ is an identity transformation at $\bar{\theta} = 0$. Those conditions are found as follows. If we retain terms of the first order in $\bar{\theta}$ and $\bar{\phi} + \bar{\psi}$, the transformation from the $(k-1)$th system to the kth system is given by

$$\begin{pmatrix} \mathbf{e}_{\xi_k} \\ \mathbf{e}_{\eta_k} \\ \mathbf{e}_{\zeta_k} \end{pmatrix} = \begin{pmatrix} 1 & \bar{\phi} + \bar{\psi} & -\bar{\theta} \cos \bar{\psi} \\ -\bar{\phi} - \bar{\psi} & 1 & \bar{\theta} \sin \bar{\psi} \\ \bar{\theta} \cos \bar{\psi} & -\bar{\theta} \sin \bar{\psi} & 1 \end{pmatrix} \begin{pmatrix} \mathbf{e}_{\xi_{k-1}} \\ \mathbf{e}_{\eta_{k-1}} \\ \mathbf{e}_{\zeta_{k-1}} \end{pmatrix}. \quad (4.B.2)$$

By a comparison of Eq. (4.B.2) with Eq. (4.2), we may relate $\bar{\theta}$, $\bar{\phi}$, and $\bar{\psi}$ to the infinitesimal rotation vector $\Delta \overline{\mathbf{\Omega}}$ of the continuous chain for $\Delta s = \bar{l}$ by the associations

$$\begin{aligned} \Delta \overline{\Omega}_\xi &\longleftrightarrow \bar{\theta} \sin \bar{\psi} \equiv \alpha_\xi \,, \\ \Delta \overline{\Omega}_\eta &\longleftrightarrow \bar{\theta} \cos \bar{\psi} \equiv \alpha_\eta \,, \\ \Delta \overline{\Omega}_\zeta &\longleftrightarrow \bar{\phi} + \bar{\psi} \equiv \alpha_\zeta \,. \end{aligned} \quad (4.B.3)$$

Thus we obtain, from Eqs. (4.A.4) and (4.B.3), the desired additional conditions

$$\begin{aligned} \langle \alpha_\mu \rangle &= c_\mu \bar{l} \,, \\ \langle \alpha_\mu \alpha_\nu \rangle &= 2k_B T D_\mu \delta_{\mu\nu} \bar{l} \quad (\mu, \nu = \xi, \eta, \zeta) \,. \end{aligned} \quad (4.B.4)$$

In particular, the HW chain may be obtained under the additional conditions,

$$\begin{aligned} \langle \alpha_\xi \rangle &= 0 \,, \\ \langle \alpha_\eta \rangle &= \kappa_0 l \,, \\ \langle \alpha_\zeta \rangle &= \tau_0 l \,, \\ \langle \alpha_\xi^2 \rangle &= \langle \alpha_\eta^2 \rangle = 2\lambda l \,, \\ \langle \alpha_\zeta^2 \rangle &= 2\lambda(1 + \sigma)l \,, \\ \langle \alpha_\mu \alpha_\nu \rangle &= 0 \quad \text{for } \mu \neq \nu \,, \end{aligned} \quad (4.B.5)$$

corresponding to Eqs. (4.A.5) and (4.A.6). From the fourth of Eqs. (4.B.5) with Eqs. (4.B.3), we also have for the fluctuation in $\bar{\theta}$

$$\langle \bar{\theta}^2 \rangle = 4\lambda l \,. \quad (4.B.6)$$

The HW chain may be characterized by the fourth of Eqs. (4.B.5), that is, the special correlation $\langle \bar{\theta}^2 \sin^2 \bar{\psi} \rangle = \langle \bar{\theta}^2 \cos^2 \bar{\psi} \rangle$.

In the particular case of the KP chain that is the HW chain with $\kappa_0 = 0$, the first, second, and fourth of Eqs. (4.B.5) require that the distribution of $\bar{\psi}$ be uniform. Then the distribution of $\bar{\phi}$ must also be uniform, although there is coupling between $\bar{\phi}$ and $\bar{\psi}$, as seen from the third and fifth of Eqs. (4.B.5).

Thus the discrete chain corresponding to the KP chain is a freely rotating chain (with bond length $\bar{l}$ and bond angle $\pi - \bar{\theta}$) having the same fluctuation in $\bar{\theta}$ as that given by Eq. (4.B.6) only as far as the chain contour is concerned (see also Sect. 4.2.2). This is the well-known result given in Chap. 3.

Finally, we note that Miyake and Hoshino [56] also derived their general continuous chain as the continuous limit of a discrete chain with independent rotations (without coupling). This is very interesting, but rather surprising since for such a discrete chain the characteristic ratio C_n is only a monotonically increasing function of n [21].

Appendix 4.C Wigner $\mathcal{D}$ Functions and 3-j Symbols

The normalized Wigner function $\mathcal{D}_l^{mj}$ used in this book is related to the unnormalized function $\bar{\mathcal{D}}_l^{mj}$ by the relation $\mathcal{D}_l^{mj} = c_l \bar{\mathcal{D}}_l^{mj}$ with c_l being given by Eq. (4.54), and $\bar{\mathcal{D}}_l^{mj}$ corresponds to Edmonds' $\mathcal{D}_{jm}^{(l)}$ [16] and Davydov's D_{mj}^l [17].

Now $\mathcal{D}_l^{mj}(\Omega) = \mathcal{D}_l^{mj}(\theta, \phi, \psi)$ is defined by

$$\mathcal{D}_l^{mj}(\Omega) = c_l e^{im\phi} d_l^{mj}(\theta) e^{ij\psi} \tag{4.C.1}$$

with

$$d_l^{mj}(\theta) = \left[\frac{(l+j)!(l-j)!}{(l+m)!(l-m)!} \right]^{1/2}$$
$$\times \left(\cos \frac{1}{2}\theta \right)^{j+m} \left(\sin \frac{1}{2}\theta \right)^{j-m} P_{l-j}^{(j-m,j+m)}(\cos \theta), \tag{4.C.2}$$

where $P_n^{(\alpha,\beta)}(x)$ is the Jacobi polynomial defined by

$$P_n^{(\alpha,\beta)}(x) = \frac{(-1)^n}{2^n n!}(1-x)^{-\alpha}(1+x)^{-\beta}$$
$$\times \frac{d^n}{dx^n}\left[(1-x)^{\alpha+n}(1+x)^{\beta+n} \right] \quad (|x| \le 1). \tag{4.C.3}$$

In particular, we have

$$\mathcal{D}_l^{m0}(\theta, \phi, \psi) = (2\pi)^{-1/2}(-1)^{(m+|m|)/2} Y_l^m(\theta, \phi), \tag{4.C.4}$$

$$\mathcal{D}_l^{0j}(\theta, \phi, \psi) = (2\pi)^{-1/2}(-1)^{(j-|j|)/2} Y_l^j(\theta, \psi), \tag{4.C.5}$$

$$\mathcal{D}_l^{mj}(0, 0, 0) = c_l \delta_{mj}. \tag{4.C.6}$$

We have the complex conjugation

$$\mathcal{D}_l^{mj*} = (-1)^{m-j} \mathcal{D}_l^{(-m)(-j)}, \tag{4.C.7}$$

and also the orthonormality and closure relations

$$\int \mathcal{D}_l^{mj*} \mathcal{D}_{l'}^{m'j'} \, d\Omega = \delta_{ll'} \delta_{mm'} \delta_{jj'} , \tag{4.C.8}$$

$$\sum_{l=0}^{\infty} \sum_{m=-l}^{l} \sum_{j=-l}^{l} \mathcal{D}_l^{mj*}(\Omega) \mathcal{D}_l^{mj}(\Omega') = \frac{1}{\sin\theta} \delta(\theta - \theta') \delta(\phi - \phi') \delta(\psi - \psi')$$

$$= \delta(\Omega - \Omega') . \tag{4.C.9}$$

The product of two $\mathcal{D}$ functions may be expanded in terms of single $\mathcal{D}$ functions as follows,

$$\mathcal{D}_{l_1}^{m_1 j_1} \mathcal{D}_{l_2}^{m_2 j_2} = 8\pi^2 \sum_{l_3=|l_1-l_2|}^{l_1+l_2} \sum_{m_3=-l_3}^{l_3} \sum_{j_3=-l_3}^{l_3} c_{l_1} c_{l_2} c_{l_3}$$

$$\times \begin{pmatrix} l_1 & l_2 & l_3 \\ m_1 & m_2 & m_3 \end{pmatrix} \begin{pmatrix} l_1 & l_2 & l_3 \\ j_1 & j_2 & j_3 \end{pmatrix} \mathcal{D}_{l_3}^{m_3 j_3 *} , \tag{4.C.10}$$

where $(:::)$ is the Wigner $3-j$ symbol, which is defined below. The integral of the product of three $\mathcal{D}$ functions is then found from Eqs. (4.C.8) and (4.C.10) to be

$$\int \mathcal{D}_{l_1}^{m_1 j_1} \mathcal{D}_{l_2}^{m_2 j_2} \mathcal{D}_{l_3}^{m_3 j_3} \, d\Omega = 8\pi^2 c_{l_1} c_{l_2} c_{l_3}$$

$$\times \begin{pmatrix} l_1 & l_2 & l_3 \\ m_1 & m_2 & m_3 \end{pmatrix} \begin{pmatrix} l_1 & l_2 & l_3 \\ j_1 & j_2 & j_3 \end{pmatrix} . \tag{4.C.11}$$

If Ω^{-1} is the inverse of the rotation Ω, we have $\Omega^{-1} = (-\theta, -\psi, -\phi)$ or $(\theta, \pi - \psi, \pi - \phi)$, and thus

$$\mathcal{D}_l^{mj}(\Omega^{-1}) = \mathcal{D}_l^{jm*}(\Omega) . \tag{4.C.12}$$

If Ω is the resultant of two successive rotations Ω_1 and Ω_2 in this order, we have

$$\mathcal{D}_l^{mj}(\Omega) = c_l^{-1} \sum_k \mathcal{D}_l^{mk}(\Omega_1) \mathcal{D}_l^{kj}(\Omega_2) . \tag{4.C.13}$$

Thus, when the coordinate system is rotated by Ω_1, the $\mathcal{D}$ function transforms the spherical harmonics from $Y_l^m(\theta, \phi)$ to $Y_l^j(\theta', \phi')$ in the new system as follows,

$$\tilde{Y}_l^m(\theta, \phi) = c_l^{-1} \sum_j \mathcal{D}_l^{mj}(\Omega_1) \tilde{Y}_l^j(\theta', \phi') , \tag{4.C.14}$$

where $\tilde{Y}_l^m = (-1)^{(m+|m|)/2} Y_l^m$ is the spherical harmonics with the phase factor $(-1)^m$ for $m > 0$. When $m = 0$, Eq. (4.C.14) reduces to Eq. (3.B.15). Further, we note that the matrix $\mathcal{D}_l^{mj}(\Omega)$ is unitary, that is,

$$\sum_{m=-l}^{l} \mathcal{D}_l^{mj*} \mathcal{D}_l^{mj'} = c_l^2 \delta_{jj'} . \tag{4.C.15}$$

The spherical harmonics Y_l^m are the eigenfunctions of ∇_u^2 (squared angular momentum operator) [see Eq. (3.B.10)], while $\mathcal{D}_l^{mj}$ are the simultaneous eigenfunctions of $\mathbf{L}^2$, $L_z = \partial/\partial\phi$, and L_ζ,

$$\mathbf{L}^2 \mathcal{D}_l^{mj} = -l(l+1)\mathcal{D}_l^{mj} ,$$
$$L_z \mathcal{D}_l^{mj} = im\mathcal{D}_l^{mj} , \tag{4.C.16}$$
$$L_\zeta \mathcal{D}_l^{mj} = ij\mathcal{D}_l^{mj} .$$

We have the symmetries of $d_l^{mj}(\theta)$,

$$d_l^{mj}(-\theta) = d_l^{jm}(\theta) , \tag{4.C.17}$$

$$d_l^{mj}(\pi+\theta) = (-1)^{l-j} d_l^{m(-j)}(\theta) , \tag{4.C.18}$$

$$d_l^{mj}(\pi-\theta) = (-1)^{l-j} d_l^{(-j)m}(\theta) , \tag{4.C.19}$$

$$d_l^{mj}(\theta) = (-1)^{j-m} d_l^{jm}(\theta) = (-1)^{j-m} d_l^{(-m)(-j)}(\theta) . \tag{4.C.20}$$

The orthonormality of $d_l^{mj}(\theta)$ is found from Eq. (4.C.15) to be

$$\sum_{m=-l}^{l} d_l^{mj}(\theta) d_l^{mj'}(\theta) = \delta_{jj'} . \tag{4.C.21}$$

The explicit expressions for $d_l^{mj}(\theta)$ with $l=1$ and 2 are: for $l=1$,

$$d_1^{(-1)(-1)}(\theta) = d_1^{11}(\theta) = \tfrac{1}{2}(1+\cos\theta) ,$$
$$d_1^{(-1)1}(\theta) = d_1^{1(-1)}(\theta) = \tfrac{1}{2}(1-\cos\theta) ,$$
$$d_1^{(-1)0}(\theta) = -d_1^{0(-1)}(\theta) = d_1^{01}(\theta) = -d_1^{10}(\theta) = \tfrac{1}{\sqrt{2}}\sin\theta , \tag{4.C.22}$$
$$d_1^{00}(\theta) = \cos\theta ;$$

for $l=2$,

$$d_2^{(-2)(-2)}(\theta) = d_2^{22}(\theta) = \tfrac{1}{4}(1+\cos\theta)^2 ,$$
$$d_2^{(-2)2}(\theta) = d_2^{2(-2)}(\theta) = \tfrac{1}{4}(1-\cos\theta)^2 ,$$
$$d_2^{(-2)(-1)}(\theta) = -d_2^{(-1)(-2)}(\theta) = d_2^{12}(\theta) = -d_2^{21}(\theta) = \tfrac{1}{2}\sin\theta\,(1+\cos\theta) ,$$
$$d_2^{(-2)1}(\theta) = -d_2^{1(-2)}(\theta) = d_2^{(-1)2}(\theta) = -d_2^{2(-1)}(\theta) = \tfrac{1}{2}\sin\theta\,(1-\cos\theta) ,$$
$$d_2^{(-2)0}(\theta) = d_2^{0(-2)}(\theta) = d_2^{02}(\theta) = d_2^{20}(\theta) = \tfrac{\sqrt{6}}{4}\sin^2\theta , \tag{4.C.23}$$
$$d_2^{(-1)(-1)}(\theta) = d_2^{11}(\theta) = -\tfrac{1}{2}(1+\cos\theta)(1-2\cos\theta) ,$$

$$d_2^{(-1)1}(\theta) = d_2^{1(-1)}(\theta) = \tfrac{1}{2}(1 - \cos\theta)(1 + 2\cos\theta)\,,$$

$$d_2^{(-1)0}(\theta) = -d_2^{0(-1)}(\theta) = d_2^{01}(\theta) = -d_2^{10}(\theta) = \tfrac{1}{\sqrt{2}}\sin\theta\cos\theta\,,$$

$$d_2^{00}(\theta) = -\tfrac{1}{2}(1 - 3\cos^2\theta)\,.$$

Now the 3-j symbol is defined by

$$\begin{pmatrix} l_1 & l_2 & l_3 \\ m_1 & m_2 & m_3 \end{pmatrix} = (-1)^{l_1 - l_2 - m_3}(2l_3 + 1)^{-1/2}$$

$$\times\, (l_1\, m_1\, l_2\, m_2 \,|\, l_1\, l_2\, l_3\, -m_3)\,, \qquad (4.\text{C}.24)$$

where $(\cdots|\cdots)$ is the vector-coupling (Clebsh–Gordan) coefficient defined by

$$(l_1\, m_1\, l_2\, m_2 \,|\, l_1\, l_2\, l\, m) = \delta_{m,m_1+m_2}$$

$$\times \left[\frac{(2l+1)(l_1 + l_2 - l)!(l_1 - m_1)!(l_2 - m_2)!(l + m)!(l - m)!}{(l_1 + l_2 + l + 1)!(l_1 - l_2 + l)!(-l_1 + l_2 + l)!(l_1 + m_1)!(l_2 + m_2)!} \right]^{1/2}$$

$$\times \sum_n (-1)^{n+l_1-m_1} \frac{(l_1 + m_1 + n)!(l_2 + l - m_1 - n)!}{n!(l_1 - m_1 - n)!(l - m - n)!(l_2 - l + m_1 + n)!}\,,$$

$$(4.\text{C}.25)$$

where the sum over n is taken so that the arguments in the denominator are nonnegative. This coefficient has the orthogonality and unitarity properties

$$\sum_{l,m} (l_1\, m_1'\, l_2\, m_2' \,|\, l_1\, l_2\, l\, m)(l_1\, l_2\, l\, m \,|\, l_1\, m_1\, l_2\, m_2) = \delta_{m_1' m_1}\delta_{m_2' m_2}\,, \quad (4.\text{C}.26)$$

$$\sum_{m_1,m_2} (l_1\, l_2\, l'\, m' \,|\, l_1\, m_1\, l_2\, m_2)(l_1\, m_1\, l_2\, m_2 \,|\, l_1\, l_2\, l\, m) = \delta_{l'l}\delta_{m'm}\delta(l_1\, l_2\, l)\,,$$

$$(4.\text{C}.27)$$

where $\delta(l_1\, l_2\, l) = 1$ if $l = |l_1 - l_2|, |l_1 - l_2| + 1, \cdots, l_1 + l_2 - 1, l_1 + l_2$ (triangular inequalities) and is zero otherwise.

The 3-j symbol is real, and we have the selection rules: the 3-j symbol vanishes if the following two conditions are not satisfied at the same time,

$$m_1 + m_2 + m_3 = 0\,, \qquad (4.\text{C}.28)$$

$$|l_1 - l_2| \le l_3 \le l_1 + l_2\,, \qquad (4.\text{C}.29)$$

where Eq. (4.C.29) is called the triangular inequalities. In other words, Eqs. (4.C.28) and (4.C.29) are the necessary conditions for the nonvanishing of the 3-j symbol. In the particular case of $m_1 = m_2 = m_3 = 0$, the 3-j symbol does not vanish if the triangular inequalities hold with $l_1 + l_2 + l_3$ being even.

We have the symmetry,

$$\begin{pmatrix} l_1 & l_2 & l_3 \\ m_1 & m_2 & m_3 \end{pmatrix} = (-1)^{l_1+l_2+l_3} \begin{pmatrix} l_1 & l_2 & l_3 \\ -m_1 & -m_2 & -m_3 \end{pmatrix}. \quad (4.C.30)$$

We have as special cases

$$\begin{pmatrix} l & l & 0 \\ m & -m & 0 \end{pmatrix} = (-1)^{l-m}(2l+1)^{-1/2}, \quad (4.C.31)$$

$$\begin{pmatrix} l_1 & l_2 & l_1+l_2 \\ m_1 & m_2 & -m_1-m_2 \end{pmatrix} = (-1)^{l_1-l_2+m_1+m_2}$$
$$\times \left[\frac{(2l_1)!(2l_2)!(l_1+l_2+m_1+m_2)!(l_1+l_2-m_1-m_2)!}{(2l_1+2l_2+1)!(l_1+m_1)!(l_1-m_1)!(l_2+m_2)!(l_2-m_2)!} \right]^{1/2},$$
$$(4.C.32)$$

$$\begin{pmatrix} l_1 & l_2 & l_3 \\ l_1 & -l_1-m & m \end{pmatrix} = (-1)^{-l_1+l_2+m}$$
$$\times \left[\frac{(2l_1)!(-l_1+l_2+l_3)!(l_1+l_2+m)!(l_3-m)!}{(l_1+l_2+l_3+1)!(l_1-l_2+l_3)!(l_1+l_2-l_3)!(-l_1+l_2-m)!(l_3+m)!} \right]^{1/2}.$$
$$(4.C.33)$$

The recurrence relations, Eqs. (3.7.12) and (3.7.13) of Edmonds [16], are also useful.

References

1. O. Kratky and G. Porod: Rec. Trav. Chem. **68**, 1106 (1949).
2. H. Maeda, N. Saito, and W. H. Stockmayer: Polym. J. **2**, 94 (1971).
3. See also M. Fixman: J. Chem. Phys. **58**, 1559 (1973).
4. P. J. Flory and J. A. Semlyen: J. Am. Chem. Soc. **88**, 3209 (1966).
5. W. G. Miller, D. A. Brant, and P. J. Flory: J. Mol. Biol. **23**, 67 (1967).
6. D. Y. Yoon and P. J. Flory: Polymer **16**, 645 (1975).
7. H. Yamakawa and M. Fujii: J. Chem. Phys. **64**, 5222 (1976).
8. H. Yamakawa: Macromolecules **10**, 692 (1977).
9. H. Yamakawa: Ann. Rev. Phys. Chem. **35**, 23 (1984).
10. H. Yamakawa: In *Molecular Conformation and Dynamics of Macromolecules in Condensed Systems*, M. Nagasawa, ed. (Elsevier, Amsterdam, 1988), p. 21.
11. L. D. Landau and E. M. Lifshitz: *Theory of Elasticity* (Addison-Wesley, Reading, 1959).
12. P. Bugl and S. Fujita: J. Chem. Phys. **50**, 3137 (1969).
13. H. Yamakawa and J. Shimada: J. Chem. Phys. **68**, 4722 (1978).
14. D. J. Struik: *Differential Geometry* (Addison-Wesley, Reading, 1950).
15. See, for example, E. T. Whittaker: *A Treatise on the Analytical Dynamics of Particles and Rigid Bodies* (Cambridge Univ., London, 1970).

16. A. R. Edmonds: *Angular Momentum in Quantum Mechanics* (Princeton Univ., Princeton, 1974).
17. A. S. Davydov: *Quantum Mechanics* (Pergamon, Oxford, 1965).
18. H. Yamakawa, M. Fujii, and J. Shimada: J. Chem. Phys. **65**, 2371 (1976).
19. H. Yamakawa and J. Shimada: J. Chem. Phys. **70**, 609 (1979).
20. J. J. Hermans and R. Ullman: Physica **18**, 951 (1952).
21. P. J. Flory: *Statistical Mechanics of Chain Molecules* (Interscience, New York, 1969).
22. P. J. Flory and V. W. C. Chang: Macromolecules **9**, 33 (1976).
23. D. Y. Yoon and P. J. Flory: Macromolecules **9**, 294 (1975).
24. H. Yamakawa and M. Fujii: J. Chem. Phys. **66**, 2584 (1977).
25. M. Fujii, K. Nagasaka, J. Shimada, and H. Yamakawa: Macromolecules **16**, 1613 (1983).
26. P. J. Flory: Proc. Nat. Acad. Sci. USA **70**, 1819 (1973).
27. P. J. Flory and D. Y. Yoon: J. Chem. Phys. **61**, 5358 (1974).
28. D. Y. Yoon and P. J. Flory: J. Chem. Phys. **61**, 5366 (1974).
29. D. Y. Yoon and P. J. Flory: J. Polym. Sci., Polym. Phys. Ed. **14**, 1425 (1976).
30. A. Baram and W. M. Gelbart: J. Chem. Phys. **66**, 617 (1977).
31. J. Shimada and H. Yamakawa: J. Chem. Phys. **73**, 4037 (1980).
32. H. Yamakawa and T. Yoshizaki: J. Chem. Phys. **75**, 1016 (1981).
33. D. Y. Yoon, P. R. Sundararajan, and P. J. Flory: Macromolecules **8**, 776 (1975).
34. P. R. Sundararajan and P. J. Flory: J. Am. Chem. Soc. **96**, 5025 (1974).
35. J. Shimada and H. Yamakawa: J. Chem. Phys. **67**, 344 (1977).
36. W. Gobush, H. Yamakawa, W. H. Stockmayer, and W. S. Magee: J. Chem. Phys. **57**, 2839 (1972).
37. H. E. Daniels: Proc. Roy. Soc. (Edinburgh) **A63**, 290 (1952).
38. J. Shimada, M. Fujii, and H. Yamakawa: J. Polym. Sci., Polym. Phys. Ed. **12**, 2075 (1974).
39. K. Nagai: J. Chem. Phys. **38**, 924 (1963).
40. R. L. Jernigan and P. J. Flory: J. Chem. Phys. **50**, 4185 (1969).
41. H. Yamakawa, J. Shimada, and M. Fujii: J. Chem. Phys. **68**, 2140 (1978).
42. M. Fixman and J. Skolnick: J. Chem. Phys. **65**, 1700 (1976).
43. M. Fujii and H. Yamakawa: J. Chem. Phys. **72**, 6005 (1980).
44. G. Szegö: *Orthogonal Polynomials* (American Mathematical Society, Providence, 1967).
45. R. Koyama: J. Phys. Soc. Japan **34**, 1029 (1973).
46. M. Fixman and R. Alben: J. Chem. Phys. **58**, 1553 (1973).
47. M. Fujii, K. Nagasaka, J. Shimada, and H. Yamakawa: J. Chem. Phys. **77**, 986 (1982).
48. J. Shimada, K. Nagasaka, and H. Yamakawa: J. Chem. Phys. **75**, 469 (1981).
49. H. Yamakawa and T. Yoshizaki: Macromolecules **15**, 1444 (1982).
50. J. E. Mark and P. J. Flory: J. Am. Chem. Soc. **86**, 138 (1964).
51. I. Sakurada, A. Nakajima, O. Yoshizaki, and K. Nakamae: Kolloid Z. **186**, 41 (1962).
52. S. Chandrasekhar: Rev. Mod. Phys. **15**, 1 (1943).
53. A. E. H. Love: *A Treatise on the Mathematical Theory of Elasticity* (Dover, New York, 1927).
54. A. Miyake and Y. Hoshino: Rep. Prog. Polym. Phys. Japan **18**, 69 (1975); **19**, 47 (1976).
55. A. Miyake and Y. Hoshino: J. Phys. Soc. Japan **47**, 942 (1979).
56. A. Miyake and Y. Hoshino: J. Phys. Soc. Japan **46**, 1324 (1979).

5 Equilibrium Properties

This chapter presents the statistical-mechanical treatments of equilibrium conformational or static properties, such as the mean-square radius of gyration, scattering function, mean-square optical anisotropy, and mean-square electric dipole moment, of the unperturbed HW chain, including the KP wormlike chain as a special case, by an application of its chain statistics developed in Chap. 4. A comparison of theory with experiment is made with experimental data obtained for several flexible polymers in the Θ state over a wide range of molecular weight, including the oligomer region, and also for typical semiflexible polymers (without excluded volume) in some cases. It must be noted that well-characterized samples have recently been used for measurements of dilute-solution properties of the former; they are sufficiently narrow in molecular weight distribution, and have a fixed stereochemical composition independent of the molecular weight in the case of asymmetric polymers.

5.1 Mean-Square Radius of Gyration

5.1.1 Basic Equations and Model Parameters

We begin by making a comparison of theory with experiment with respect to the mean-square radius of gyration $\langle S^2 \rangle$ for several flexible and semiflexible polymers to determine their HW model parameters κ_0, τ_0, λ^{-1}, and M_L (with Poisson's ratio $\sigma = 0$). For this purpose, it is convenient to use the number of repeat units in the chain (or the degree of polymerization) x instead of its total contour length L.

Equation (4.83) with Eq. (4.89) may then be rewritten as

$$\frac{\langle S^2 \rangle}{x} = \frac{M_0 \lambda^{-1}}{M_L} \left[\frac{f_S(\lambda L; \lambda^{-1}\kappa_0, \lambda^{-1}\tau_0)}{\lambda L} \right] \tag{5.1}$$

with

$$\log x = \log(\lambda L) + \log\left(\frac{\lambda^{-1} M_L}{M_0}\right), \tag{5.2}$$

where M_0 is the molecular weight of the repeat unit and the function f_S is given by

$$f_S(L; \kappa_0, \tau_0) = \frac{\tau_0{}^2}{\nu^2} f_{S,\text{KP}}(L) + \frac{\kappa_0{}^2}{\nu^2 r^2} \left[\frac{rL}{3} - \cos(2\varphi) + \frac{2}{rL} \cos(3\varphi) \right.$$

$$\left. - \frac{2}{r^2 L^2} \cos(4\varphi) + \frac{2}{r^2 L^2} e^{-2L} \cos(\nu L + 4\varphi) \right] \qquad (5.3)$$

with

$$\nu = (\kappa_0{}^2 + \tau_0{}^2)^{1/2} , \qquad (5.4)$$

$$r = (4 + \nu^2)^{1/2} , \qquad (5.5)$$

$$\varphi = \cos^{-1} \left(\frac{2}{r} \right) , \qquad (5.6)$$

and with $f_{S,\text{KP}}$ being the function f_S for the KP chain given by

$$f_{S,\text{KP}}(L) = \frac{L}{6} - \frac{1}{4} + \frac{1}{4L} - \frac{1}{8L^2}(1 - e^{-2L}) . \qquad (5.7)$$

In the limit of $\lambda L \to \infty$, we have

$$\lim_{\lambda L \to \infty} \left[\frac{f_S(\lambda L)}{\lambda L} \right] = \frac{1}{6} c_\infty , \qquad (5.8)$$

so that

$$\lim_{x \to \infty} \left(\frac{\langle S^2 \rangle}{x} \right) \equiv \left(\frac{\langle S^2 \rangle}{x} \right)_\infty = \frac{M_0 \lambda^{-1} c_\infty}{6 M_\text{L}} , \qquad (5.9)$$

where

$$c_\infty = \frac{4 + (\lambda^{-1} \tau_0)^2}{4 + (\lambda^{-1} \kappa_0)^2 + (\lambda^{-1} \tau_0)^2} . \qquad (5.10)$$

Recall that for the KP chain $\kappa_0 = 0$ and $c_\infty = 1$.

Figure 5.1 shows double-logarithmic plots of $\langle S^2 \rangle / x$ (in Å^2) against x for atactic (a-) PS with the fraction of racemic diads $f_\text{r} = 0.59$ in cyclohexane at 34.5°C (Θ) [1, 2], a-PMMA with $f_\text{r} = 0.79$ in acetonitrile at 44.0°C (Θ) [3], i-PMMA with $f_\text{r} = 0.01$ in acetonitrile at 28.0°C (Θ) [4], poly(n-butyl isocyanate) (PBIC) in tetrahydrofuran (THF) at 40°C [5], DNA in 0.2 mol/l NaCl at 25°C [6], and schizophyllan in 0.01 N NaCl at 25°C [7], where for DNA x has been chosen to be the number of base pairs. Among these polymers, the first three are flexible and the other three are semiflexible. The data have been obtained from light scattering measurements except for fractions of the flexible polymers with $\langle S^2 \rangle^{1/2} \lesssim 80\text{Å}$, for which those have been obtained from small-angle X-ray scattering (SAXS) measurements. We note that proper corrections for chain thickness (the spatial distribution of electrons around the chain contour) have been made to the values of $\langle S^2 \rangle$ from the SAXS measurements following the procedure given in Appendix 5.A.

In the figure the solid curves represent the best-fit HW (or KP) theoretical values calculated from Eq. (5.1) with Eq. (5.2) with the values of the model parameters listed in Table 5.1, where we note that the values of

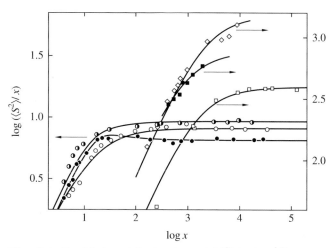

Fig. 5.1. Double-logarithmic plots of $\langle S^2 \rangle/x$ (in Å^2) against x for a-PS in cyclohexane at 34.5°C ($\circ$) [1,2], a-PMMA in acetonitrile at 44.0°C ($\bullet$) [3], i-PMMA in acetonitrile at 28.0°C ($\circleddash$) [4], PBIC in THF at 40°C ($\square$) [5], DNA in 0.2 mol/l NaCl at 25°C ($\blacksquare$) [6], and schizophyllan in 0.01 N NaOH at 25°C ($\diamond$) [7]. The *solid curves* represent the best-fit HW (or KP) theoretical values

Table 5.1. Values of the HW model parameters for typical flexible and semiflexible polymers from $\langle S^2 \rangle$

Polymer (f_r)	Solvent	Temp. (°C)	$\lambda^{-1}\kappa_0$	$\lambda^{-1}\tau_0$	λ^{-1} (Å)	M_L (Å^{-1})	Ref. (Obs.)
a-PS (0.59)	Cyclohexane	34.5	3.0	6.0	20.6	35.8	[1,2]
a-PMMA (0.79)	Acetonitrile	44.0	4.0	1.1	57.9	36.3	[3]
i-PMMA (0.01)	Acetonitrile	28.0	2.5	1.3	38.0	32.5	[4]
PBIC	THF	40	0	$\cdots$	1320	55.1	[5]
DNA	0.2 mol/l NaCl	25	0	$\cdots$	1360	195	[6]
Schizophyllan	0.01 N NaOH	25	0	$\cdots$	3000	217	[7]

$\lambda^{-1}\kappa_0$ and $\lambda^{-1}\tau_0$ for a-PS have been determined from the mean-square optical anisotropy (see Sect. 5.3.3). It is seen that for both flexible and semiflexible polymers the behavior of $\langle S^2 \rangle$ may be well explained by the HW (or KP) continuous model. The reader is also referred to the review article by Norisuye [8], in which the values of the KP model parameters are summarized for a wide variety of semiflexible polymers.

5.1.2 Chain Stiffness and Local Chain Conformations

In general, the (static) stiffness parameter λ^{-1} may be considered smaller and larger than about 100Å for flexible and semiflexible (or stiff) polymers, respectively. This is rather the definition of the chain stiffness from the point of view of the continuous model. It may also be defined by the behavior of the (unperturbed) ratio $\langle S^2 \rangle / x$; that is, this ratio becomes independent of x for $x \gtrsim 300$ for flexible polymers but levels off only at much larger x for semiflexible polymers, as seen from Fig. 5.1.

We first give a brief discussion of the results for the model parameters for the semiflexible polymers given in Table 5.1, for convenience. The value 55.1 Å^{-1} of M_L for PBIC is close to the values 54.5 and 51.1 Å^{-1} corresponding to the Troxell–Scheraga [9] and Schmueli–Traub–Rosenheck [10] 8_3 helices of PBIC, respectively, indicating that its chain takes preferentially such a helical form in dilute solution. The values of M_L for DNA and schizophyllan correspond to those for their double and triple helices, respectively. The structures of all these typical semiflexible or stiff polymer chains, whose λ^{-1} are greater than hundreds of angstroms, are in general very symmetric about their helix axes, so that they may be well represented by the KP chain whose contour coincides with the helix axis. The schizophyllan chain has the very large value of λ^{-1} and is the most stiff of the polymers studied so far [8].

Now we discuss the results for the flexible polymers. The asymptotic ratio $(\langle S^2 \rangle / x)_\infty$ in Eq. (5.9) is equal to 8.1_3, 6.5_7, and 9.3_1 Å^2 for a-PS, a-PMMA, and i-PMMA, respectively, and cannot be directly correlated to the chain stiffness λ^{-1}, as seen from Table 5.1. From Eqs. (4.87) and (5.9), it is seen to be related to the Kuhn segment length A_K and the persistence length q by the equations

$$A_K = 2q = c_\infty \lambda^{-1}$$

$$= \frac{6 M_L}{M_0} \left(\frac{\langle S^2 \rangle}{x} \right)_\infty . \tag{5.11}$$

For polymer chains having the same ratio M_L / M_0 the asymptotic ratio $(\langle S^2 \rangle / x)_\infty$ is then proportional to A_K and q. (Note that the values of M_L / M_0 are close to each other for these three flexible polymers.) Thus neither A_K nor q is a measure of chain stiffness except for the KP chain for which $c_\infty = 1$. This is also the case with the characteristic ratio C_∞, which is given by

$$C_\infty = \frac{3}{l^2} \left(\frac{\langle S^2 \rangle}{x} \right)_\infty \tag{5.12}$$

with l the bond length if the number of skeletal bonds is equal to $2x$. It must be emphasized that the a-PMMA chain is much stiffer than the i-PMMA chain, which is stiffer than the a-PS chain.

It is seen from Fig. 4.13 and the values of $\lambda^{-1}\kappa_0$ and $\lambda^{-1}\tau_0$ in Table 5.1 that the a-PMMA chain is of the strongest helical nature of the above three flexible polymers and the a-PS chain is of the weakest. Indeed, for a-PMMA, the ratio $\langle S^2 \rangle / x$ as a function of x passes through a maximum at $x \simeq 50$ before reaching its asymptotic value for large x, as seen from Fig. 5.1. We

Table 5.2. Values of the characteristic helix parameters

Polymer (f_r)	Solvent	Temp. (°C)	ρ (Å)	h (Å)
a-PS (0.59)	Cyclohexane	34.5	1.3_7	$17._3$
a-PMMA (0.79)	Acetonitrile	44.0	$13._5$	$23._3$
i-PMMA (0.01)	Acetonitrile	28.0	$12._0$	$39._1$

note that this maximum cannot be explained by any type of RIS models for a-PMMA with $f_r = 0.79$ (having hydrogen atoms at both terminal ends) [3], although the RIS values (of C_n) exhibit it for s-PMMA, as shown in Fig. 4.5 (see also Sect. 5.2.2).

According to the HW model, a flexible polymer chain in dilute solution may be pictured as a regular helix (that is, the characteristic helix) disturbed (or destroyed) by thermal fluctuations or a random coil retaining more or less helical portions. The shape of the characteristic helix may be determined as a space curve by the radius ρ and pitch h, which are given by Eqs. (4.14),

$$\rho = \left[\frac{\lambda^{-1}\kappa_0}{(\lambda^{-1}\kappa_0)^2 + (\lambda^{-1}\tau_0)^2} \right] \lambda^{-1},$$

$$h = 2\pi \left[\frac{\lambda^{-1}\tau_0}{(\lambda^{-1}\kappa_0)^2 + (\lambda^{-1}\tau_0)^2} \right] \lambda^{-1}, \tag{5.13}$$

and the degree of disturbance (thermal fluctuation) may be represented by the parameter λ, so that the regular helical structure is destroyed to a lesser extent in the chain with larger stiffness λ^{-1}. In general, the chain of strong helical nature has large ρ (compared to h) and large λ^{-1}, and thus retains rather large and clearly distinguishable helical portions in dilute solution. Note that the chain with vanishing ρ (the KP chain) has no helical nature irrespective of the value of λ^{-1} and that the chain with small λ^{-1} is not of strong helical nature irrespective of the shape of its characteristic helix.

In Table 5.2 are given the values of ρ and h calculated for the above three flexible polymers from Eqs. (5.13) with the values of $\lambda^{-1}\kappa_0$, $\lambda^{-1}\tau_0$, and λ^{-1} given in Table 5.1. With those values, their characteristic helices are illustratively drawn in Fig. 5.2. It is seen that the characteristic helix is most and least extended for the a-PS and a-PMMA chains, respectively. These shapes and the values of λ^{-1} make us easily understand their degrees of helical nature.

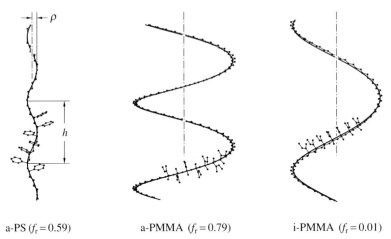

a-PS $(f_r = 0.59)$ a-PMMA $(f_r = 0.79)$ i-PMMA $(f_r = 0.01)$

Fig. 5.2. Illustration of the characteristic helices for a-PS, a-PMMA, and i-PMMA

5.1.3 HW Monte Carlo Chains

The difference in local chain conformation between flexible polymers may be visualized by generating instantaneous configurations of the contour of the HW chain, that is, HW Monte Carlo chains [3]. For this purpose, we divide the HW chain of contour length L into N identical parts, each of contour length $\Delta s = L/N$, to consider its discrete analog. For the case of $\lambda \Delta s \ll 1$, the bond vector $\mathbf{a}_p$ ($p = 1, 2, \cdots, N$) as defined as the vector distance between the contour points $(p-1)\Delta s$ and $p\Delta s$ may be assumed to be of length Δs and in the direction of the vector tangential to the contour or the ζ axis of the localized Cartesian coordinate system (ξ, η, ζ) affixed at the contour point $(p-1)\Delta s$. Let $\Delta \tilde{\boldsymbol{\Omega}}_p$ ($p = 1, 2, \cdots, N-1$) be the infinitesimal rotation vector by which the localized coordinate system at the contour point $p\Delta s$ is obtained from the one at the contour point $(p-1)\Delta s$, let $\Delta \tilde{\Omega}_{p\xi}$, $\Delta \tilde{\Omega}_{p\eta}$, and $\Delta \tilde{\Omega}_{p\zeta}$ be the Cartesian components of $\Delta \tilde{\boldsymbol{\Omega}}_p$ expressed in the latter system, and let $\Delta \tilde{\boldsymbol{\Omega}}_p = (|\Delta \tilde{\boldsymbol{\Omega}}_p|, \theta_p, \phi_p)$ in spherical polar coordinates in that system.

The transformation from the latter to the former system may then be represented by the transformation matrix $\mathbf{T}_p$ defined by

$$\mathbf{T}_p = \mathbf{A}^{-1}(\theta_p, \phi_p) \cdot \mathbf{R}(|\Delta \tilde{\boldsymbol{\Omega}}_p|) \cdot \mathbf{A}(\theta_p, \phi_p), \tag{5.14}$$

where the rotation matrices $\mathbf{A}(\theta, \phi)$ and $\mathbf{R}(\psi)$ are given by

$$\mathbf{A}(\theta, \phi) = \begin{pmatrix} \cos\theta\cos\phi & \cos\theta\sin\phi & -\sin\theta \\ -\sin\theta & \cos\phi & 0 \\ \sin\theta\cos\phi & \sin\theta\sin\phi & \cos\theta \end{pmatrix}, \tag{5.15}$$

$$\mathbf{R}(\psi) = \begin{pmatrix} \cos\psi & \sin\psi & 0 \\ -\sin\psi & \cos\psi & 0 \\ 0 & 0 & 1 \end{pmatrix}. \qquad (5.16)$$

By the use of the transformation matrices $\mathbf{T}_p$ thus defined, the pth bond vector $\mathbf{a}_p$ ($p = 2, 3, \cdots, N$) expressed in the localized system at the contour point 0 may be written as

$$\mathbf{a}_p = \mathbf{T}_1^{-1} \cdot \mathbf{T}_2^{-1} \cdots \mathbf{T}_{p-1}^{-1} \cdot \mathbf{a}, \qquad (5.17)$$

where $\mathbf{a}$ is the pth bond vector $\mathbf{a}_p$ in the $(p-1)$th system, so that $\mathbf{a}$ is $\mathbf{a}_1 = (0, 0, \Delta s)$ in the 0th system.

An instantaneous configuration of this entire (discrete) HW chain may be generated by joining the bond vectors $\mathbf{a}_p$ successively, so that we have only to generate a set of $N-1$ infinitesimal rotation vectors $\Delta\tilde{\mathbf{\Omega}}_p$ ($p = 1, 2, \cdots, N-1$). The potential energy U of the (continuous) HW chain per unit contour length is given by Eq. (4.4) with $\alpha = \beta$ when $\sigma = 0$. Therefore, if we assume that the "angular velocity" vector $\boldsymbol{\omega}$ takes the constant value $\Delta\tilde{\mathbf{\Omega}}_p/\Delta s$ between the contour points $(p-1)\Delta s$ and $p\Delta s$ for $\lambda\Delta s \ll 1$, the potential energy U_p for the rotation $\Delta\tilde{\mathbf{\Omega}}_p$ may be written in the form

$$U_p = \frac{k_B T}{4\lambda\Delta s}\left[(\Delta\tilde{\Omega}_{p\xi})^2 + (\Delta\tilde{\Omega}_{p\eta} - \kappa_0\Delta s)^2 + (\Delta\tilde{\Omega}_{p\zeta} - \tau_0\Delta s)^2\right], \quad (5.18)$$

where Eq. (3.37) has been used. We can then readily generate $\Delta\tilde{\mathbf{\Omega}}_p$ by the introduction of the Boltzmann factor $\exp(-U_p/k_B T)$ as the equilibrium probability distribution function of $\Delta\tilde{\mathbf{\Omega}}_p$. For the generation of the chain, $\lambda\Delta s$ must be taken to be as small as possible, say 0.02.

In Fig. 5.3 are depicted representative instantaneous contours of the a-PS, a-PMMA, and i-PMMA chains so obtained for $x = 500$, where their

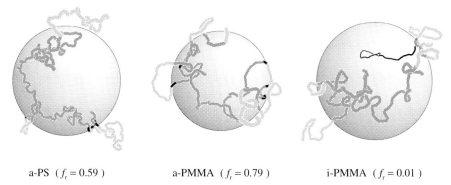

a-PS ($f_r = 0.59$)   a-PMMA ($f_r = 0.79$)   i-PMMA ($f_r = 0.01$)

Fig. 5.3. Representative instantaneous contours of HW Monte Carlo chains corresponding to a-PS, a-PMMA, and i-PMMA with $x = 500$ such that their radii of gyration S are just equal to their respective $\langle S^2\rangle^{1/2}$

radii of gyration S are just equal to the respective values of $\langle S^2 \rangle^{1/2}$. The shaded sphere has the radius S, which is equal to 63.2, 57.4, and 67.8 Å for a-PS, a-PMMA, and i-PMMA, respectively, and is nearly proportional to $(\langle S^2 \rangle / x)_\infty^{1/2}$. The a-PS chain seems just the random-flight chain. On the other hand, several helical portions are clearly observed in the picture for a-PMMA, for example, in its bottom-right part, as was expected from the discussion in the last subsection, while such portions do not appear for i-PMMA. Thus the i-PMMA chain tends to take more extended conformations than the a-PMMA chain, so that the ratio $(\langle S^2 \rangle / x)_\infty$ is larger for the former despite the fact that λ^{-1} is smaller for the former. However, the chain contour of i-PMMA is still rather smooth compared to that of a-PS. This is due to the fact that λ^{-1} is larger for the former than for the latter. It is because of this chain stiffness that the ratio $(\langle S^2 \rangle / x)_\infty$ is even larger for i-PMMA than for a-PS. (Note that the ratio is smaller for a-PMMA than for a-PS because of the strong helical nature of the former.)

The HW chain takes account of both chain stiffness and local chain conformations in a satisfactory manner.

5.2 Scattering Function

5.2.1 Scattering Function for the Chain Contour

We first evaluate the *scattering function* $P(k; L)$ for the HW chain of contour length L in the continuous-point-scatterer approximation such that the scatterers are uniformly and continuously distributed on the chain contour. From Eq. (5.A.2) with Eq. (5.A.4), it is then given by

$$P(k; L) = L^{-2} \left\langle \left| \int_0^L \exp\left[i\mathbf{k} \cdot \mathbf{r}(s) \right] ds \right|^2 \right\rangle$$
$$= 2L^{-2} \int_0^L (L - s) I(\mathbf{k}; s) ds, \qquad (5.19)$$

where i is the imaginary unit, $\mathbf{r}(s)$ is the radius vector of the contour point s, $I(\mathbf{k}; s)$ is the characteristic function for the chain of contour length s, and $\mathbf{k}$ is the scattering vector whose magnitude k is given by

$$k = (4\pi / \tilde{\lambda}) \sin(\theta / 2) \qquad (5.20)$$

with $\tilde{\lambda}$ the wavelength in the medium and θ the scattering angle. Note that $\tilde{\lambda}$ is equal to the wavelength λ_0 in vacuum in the case of SAXS and small-angle neutron scattering (SANS). In what follows, all lengths are measured in units of λ^{-1} unless otherwise noted.

In order to evaluate $P(\mathbf{k}; L)$ from the second line of Eqs. (5.19), we adopt two approximation methods for the evaluation of I, that is, the weighting function method and the ϵ method given in Sects. 4.7.1 and 4.7.2, respectively. In the first method, the characteristic function $I(\mathbf{k}; L)$ may be evaluated, although numerically, from the Fourier transform of the distribution function $G(\mathbf{R}; L)$ of the end-to-end vector $\mathbf{R}$,

$$
I(\mathbf{k}; L) = \int G(\mathbf{R}; L) \exp(i\mathbf{k} \cdot \mathbf{R}) d\mathbf{R}
$$

$$
= 4\pi k^{-1} \int_0^\infty R \sin(kR) G(\mathbf{R}; L) dR , \tag{5.21}
$$

where $G(\mathbf{R}; L)$ is approximated by

$$
G(\mathbf{R}; L) = \left(\frac{3}{2\langle R^2 \rangle}\right)^{3/2} w(\rho) \sum_{n=0}^s M_n(L) \rho^{2n} \tag{5.22}
$$

with

$$
\rho = \left(\frac{3}{2\langle R^2 \rangle}\right)^{1/2} R . \tag{5.23}
$$

For the weighting function $w(\rho)$, we adopt the function $w_I(\rho)$ given by Eq. (4.207),

$$
w(\rho) = \exp\left[-a_1\rho^2 - a_2\rho - (b\rho^2)^5\right] . \tag{5.24}
$$

The coefficients a_1, a_2, and b in Eq. (5.24) are first determined so that the normalized weighting function gives the exact moments $\langle R^2 \rangle$, $\langle R^4 \rangle$, and $\langle R^6 \rangle$, and then $M_n = (4\pi)^{-1} \mathcal{F}_{000,n}^{00,00}$ ($n = 0$–s; $s \geq 3$) in Eq. (5.22) are determined from Eq. (4.204), that is, in such a way that the $G(\mathbf{R}; L)$ given by Eq. (5.22) with this w gives the exact moments $\langle R^{2m} \rangle$ ($m = 0$–s). Note that when $a_2 = b = 0$, Eq. (5.22) gives the Hermite polynomial expansion of $G(\mathbf{R}; L)$.

For very small L, the second method is used, and then $I(\mathbf{k}; L)$ is approximately given by Eq. (4.228),

$$
I(\mathbf{k}; L) = \sum_{m=0}^s \frac{(-x)^m}{2^m m!} \langle \epsilon^m \rangle j_m(x) , \tag{5.25}
$$

where $j_m(x)$ is the spherical Bessel function of the first kind. If $\langle R^2 \rangle_0$ is chosen to be $\langle R^2 \rangle$ in Eq. (4.220), we have

$$
x = \langle R^2 \rangle^{1/2} k , \tag{5.26}
$$

$$
\langle \epsilon^m \rangle = \frac{\langle R^{2m} \rangle}{\langle R^2 \rangle^m} - \sum_{r=0}^{m-1} \binom{m}{r} \langle \epsilon^r \rangle \tag{5.27}
$$

with $\epsilon = R^2/\langle R^2 \rangle - 1$, so that $\langle \epsilon^m \rangle$ ($m \geq 1$) may be expressed successively in terms of $\langle R^{2r} \rangle$ ($r = 1$–m). The required moments $\langle R^{2m} \rangle$ may be evaluated from Eq. (4.81) (for $\sigma = 0$).

It has been found that if k is not very large, accurate values of $I(\mathbf{k}; L)$ may be obtained over the whole range of L using the values from the ϵ method for

L smaller than some small value and from the weighting function method for L larger than that value, both for the degree of approximation s equal to 5, for all those values of κ_0 and τ_0 for which the latter method has the solution [11]. The integration in the second line of Eqs. (5.19) must then be carried out numerically to find $P(k; L)$.

For the particular case of the KP chain, we give an interpolation formula for $P(k; L)$ constructed on the basis of the numerical results [11]. It may be well approximated by

$$P(k; L) = P_0(k; L)\Gamma(k; L). \tag{5.28}$$

$P_0(k; L)$ is given by

$$P_0(k; L) = \left[1 - \chi(k;\ L)\right]P_{(C^*)}(k; L) + \chi(k; L)P_{(R)}(k; L), \tag{5.29}$$

where $P_{(C^*)}(k; L)$ is the Debye scattering function for the random coil (Gaussian chain) [12] having the same mean-square radius of gyration $\langle S^2 \rangle$ as that of the KP chain under consideration,

$$P_{(C^*)}(k; L) = 2u^{-2}(e^{-u} + u - 1) \tag{5.30}$$

with

$$u = \langle S^2 \rangle k^2, \tag{5.31}$$

$P_{(R)}(k; L)$ is the scattering function for the rod [12, 13],

$$P_{(R)}(k;\ L) = 2v^{-2}\left[v\text{Si}(v) + \cos v - 1\right] \tag{5.32}$$

with

$$v = Lk, \tag{5.33}$$

and with $\text{Si}(v)$ being the sine integral

$$\text{Si}(v) = \int_0^v t^{-1} \sin t\, dt, \tag{5.34}$$

and $\chi(k; L)$ is defined by

$$\chi = \exp(-\xi^{-5}) \tag{5.35}$$

with

$$\xi = \pi\langle S^2 \rangle k/2L. \tag{5.36}$$

In Eq. (5.28), $\Gamma(k; L)$ is given by

$$\Gamma(k; L) = 1 + (1 - \chi)\sum_{i=2}^{5} A_i\xi^i + \chi\sum_{i=0}^{2} B_i\xi^{-i} \tag{5.37}$$

with

Table 5.3. Values of $a_{k,ij}$ and $b_{k,ij}$ in Eqs. (5.38) and (5.39)

i	j	$a_{1,ij}$	$a_{2,ij}$	$b_{1,ij}$	$b_{2,ij}$
0	0	$\cdots$	$\cdots$	1.3489	$\cdots$
0	1	$\cdots$	$\cdots$	1.6527 (1)	1.3544 (1)
0	2	$\cdots$	$\cdots$	-6.5909 (1)	6.0772 (1)
1	0	$\cdots$	$\cdots$	-2.0350	$\cdots$
1	1	$\cdots$	$\cdots$	-3.0016 (1)	3.2504 (1)
1	2	$\cdots$	$\cdots$	1.1290 (2)	-1.3836 (2)
2	0	1.7207 $(-1)^{\mathrm{a}}$	$\cdots$	1.3744	$\cdots$
2	1	-7.0881	3.3157 (-1)	1.2268 (1)	-5.1258 (1)
2	2	1.9577 (1)	-1.0692	-4.6316 (1)	7.2212 (1)
3	0	7.7459 (-2)	$\cdots$	$\cdots$	$\cdots$
3	1	4.8101	-3.9383	$\cdots$	$\cdots$
3	2	-2.0099 (2)	1.1279 (1)	$\cdots$	$\cdots$
4	0	9.6330 (-1)	$\cdots$	$\cdots$	$\cdots$
4	1	2.6450 (1)	1.2608 (1)	$\cdots$	$\cdots$
4	2	4.0647 (2)	-3.8021 (1)	$\cdots$	$\cdots$
5	0	-1.1307	$\cdots$	$\cdots$	$\cdots$
5	1	-2.3971 (1)	-9.7252	$\cdots$	$\cdots$
5	2	-2.2471 (2)	3.3515 (1)	$\cdots$	$\cdots$

^{a} $a(n)$ means $a \times 10^n$

$$A_i = \sum_{j=0}^{2} a_{1,ij} L^{-j} e^{-10/L} + \sum_{j=1}^{2} a_{2,ij} L^{-j} e^{-2L} , \tag{5.38}$$

$$B_i = \sum_{j=0}^{2} b_{1,ij} L^{-j} + \sum_{j=1}^{2} b_{2,ij} L^{j} e^{-2L} , \tag{5.39}$$

where $a_{1,ij}$, $a_{2,ij}$, $b_{1,ij}$, and $b_{2,ij}$ are numerical coefficients and their values are given in Table 5.3. The application of Eq. (5.28) with Eqs. (5.29)–(5.39) for the KP chain is limited to the range of $k \lesssim 10$.

For the HW chain, including the KP chain, of very small L, $P(k; L)$ may be evaluated analytically, although in a series form, as given in Sect. 5.3.5.

For convenience, we define a function $F(k; L)$ by

$$F(k; L) = Lk^2 P(k; L) . \tag{5.40}$$

Note that $F(k; L)$ corresponds to the quantity (*Kratky function*) often plotted in SAXS and SANS experiments. In the following we examine the behavior of the scattering function $F(k; L)$ thus calculated as a function of k.

Values of $F(k; L)$ for two typical HW chains with $\kappa_0 = 2.5$ and $\tau_0 = 0.5$ (Code 1 of Table 4.4) and with $\kappa_0 = 5.0$ and $\tau_0 = 1.0$ (Code 2 of Table 4.4) and also the KP chain in the range of convergence are represented by the solid curves 1, 2, and KP in Figs. 5.4 and 5.5 for $L = 80$ and 10^4, respectively. The dashed curves C(2) and C(KP) represent the values of $F_{(C^*)}$ calculated from Eq. (5.40) with Eq. (5.30) for the random coils having the same $\langle S^2 \rangle$

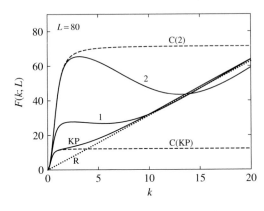

Fig. 5.4. Plots of $F(k; L)$ against k for $L = 80$. The *solid curves* 1, 2, and KP represent the values for HW Code 1 ($\kappa_0 = 2.5$ and $\tau_0 = 0.5$), HW Code 2 ($\kappa_0 = 5.0$ and $\tau_0 = 1.0$), and the KP chain, respectively. The *dashed curves* C(2) and C(KP) represent the values of $F_{(C*)}$ for the random coils having the same $\langle S^2 \rangle$ as those of Code 2 and the KP chain, respectively, and the *dotted curve* R the values of $F_{(R)}$ for the rod

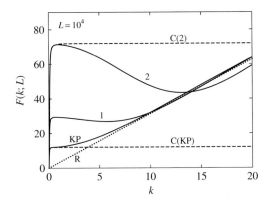

Fig. 5.5. Plots of $F(k; L)$ against k for $L = 10^4$; see legend for Fig. 5.4

as those of Code 2 and the KP chain, respectively, and the dotted curves R represent the values of $F_{(R)}$ calculated from Eq. (5.40) with Eq. (5.32) with the respective values of L. For the typical HW chains, $F(k; L)$ exhibits a maximum and a minimum.

All the solid curves in Figs. 5.4 and 5.5 are seen to approach straight lines asymptotically as k is increased. Indeed, we have

$$F(k; L) \longrightarrow \pi k + C(L) \qquad \text{for } k \to \infty \qquad (5.41)$$

with

$$C(L) = -2L^{-1} \qquad \text{(rod)}, \qquad (5.42)$$

$$C(\infty) = \frac{2}{3} \qquad \text{(KP)}, \qquad (5.43)$$

where Eq. (5.43) is due to des Cloizeaux [14]. We note that it is impossible to evaluate $C(L)$ analytically for the KP and HW chains of finite L and that for the random coil we have

$$\lim_{k\to\infty} F_{(C)}(k;L) = \lim_{L\to\infty} F_{(C)}(k;L) = 12c_\infty^{-1}, \qquad (5.44)$$

where $F_{(C)}$ is equal to $F_{(C^*)}$ with $\langle S^2 \rangle = c_\infty L/6$. On the other hand, we have [15]

$$F(0;\infty) \equiv \lim_{k\to 0}\left[\lim_{L\to\infty} F(k;L)\right] = 12c_\infty^{-1}, \qquad (5.45)$$

so that $F(0;\infty) = F_{(C)}(\infty;L)$, and $F(0;\infty) = 12$ for the KP chain.

Finally, mention must be made of earlier investigations of the scattering function for the KP chain. The KP chain statistics have been introduced approximately by Peterlin [16], Heine et al. [17], and Koyama [18]. The evaluation has been carried out numerically by des Cloizeaux [14] for the infinitely long chain for $k \leq 8$. His results for F almost agree with the corresponding values shown in Fig. 5.5 for $L = 10^4$ in the range of $k \geq 0.5$. For the KP chain, it may therefore be concluded that $F(k;L)$ increases monotonically with increasing k for all values of L (see Figs. 5.4 and 5.5). In other words, it does not exhibit even a plateau in the transition range of k from random coil to rod. Some earlier theories [17, 18] happen to predict the existence of the plateau region for the KP chain because of the approximations involved. If it is observed experimentally, it should be explained by the HW model. The exact evaluation has also been carried out for limited values of L in the light-scattering range [18–21] (see also Sect. 5.3.5). In particular, we note that Sharp and Bloomfield [19] have derived the first Daniels approximation to the scattering function [11].

5.2.2 Comparison with the RIS Model

As seen in the last subsection, for typical HW codes the theory can predict the first maximum and minimum but not the second ones in the Kratky function such as observed by Kirste and Wunderlich [22–24] in their SAXS experiment for s-PMMA. However, Yoon and Flory [25] have carried out Monte Carlo calculations on the basis of the RIS model to show the existence of such oscillation for s-PMMA, taking the α-carbon atoms as the scatterers. As already noted, on the other hand, the RIS model can predict the maximum in $\langle S^2 \rangle/x$ for s-PMMA but not for a-PMMA with $f_r = 0.79$ [3]. Thus we examine the dependence on f_r of the Kratky function on the basis of the RIS model.

For this purpose, we adopt only the three-state RIS model [26] for PMMA chains, assuming as before that both terminal ends are hydrogen atoms [3], for convenience. Following the procedure of Yoon and Flory [25], we evaluate the characteristic function for a part of the RIS chain by the Monte Carlo method if the number of repeat units x in that part is smaller than or equal to 30, and by the eighth-order Hermite polynomial approximation otherwise. The scattering function $P(k;x)$ is then given by

$$P(k;x) = P_1(k;x) + P_2(k;x) \qquad (5.46)$$

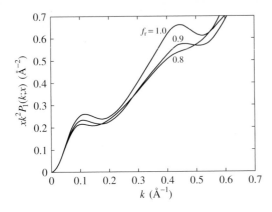

Fig. 5.6. Plots of $xk^2P_1(k;x)$ against k for the three-state RIS model for PMMA with $x = 1000$ at 300 K for the indicated values of f_r

with

$$P_1(k;x) = x^{-1} + 2x^{-2}\sum_{j=1}^{30}(x-j)\left\langle\frac{\sin(kr_j)}{kr_j}\right\rangle_{\mathrm{MC}}, \tag{5.47}$$

$$P_2(k;x) = 2x^{-2}\sum_{j=31}^{x-1}(x-j)I(\mathbf{k};j), \tag{5.48}$$

where r_j is the distance between two α-carbon atoms connected by $2j$ successive skeletal bonds, $\langle\cdots\rangle_{\mathrm{MC}}$ denotes a Monte Carlo average, and $I(\mathbf{k};j)$ is the characteristic function. We note that P_2 makes no contribution to P in the range of large k. Thus, for the examination of the behavior of P in such a k region, we consider only P_1, for simplicity.

Figure 5.6 shows plots of xk^2P_1 against k for the three-state model for PMMA with $x = 1000$ for $f_r = 1.0, 0.9$, and 0.8 at 300 K [27]. Note that $f_r = 1$ for s-PMMA. It is seen that the amplitude of the oscillation of the (Kratky) plot in the range of large k becomes small as f_r is decreased but that the weak oscillation still exists even for $f_r = 0.8$ in contrast to the corresponding case of the HW chain (close to Code 2 in Figs. 5.4 and 5.5). The second maximum and minimum (oscillation) in the Kratky function indicate the "crystal-like" behavior of the chain, and their occurrence should rather be regarded as a defect of the RIS model, for which some components of the angular correlation function $g_l^{jj'}(x)$ do not vanish in the limit of $x \to \infty$, as pointed out by Baram and Gelbart [28] and mentioned in Sect. 4.4.4. It is believed that both the RIS theory prediction and the experiment by Kirste and Wunderlich are wrong. Their experiment is further discussed in Sect. 5.2.4. We note that the oscillation does not appear even if the point scatterers are discretely arrayed on the HW chain contour [27].

5.2.3 Effects of Chain Thickness

In SAXS and SANS experiments, the scatterers are atomic electrons and hydrogen nuclei distributed around the chain contour, respectively, and then the scattering function directly observed, which we denote by $P_s(k; L)$, contains effects of the spatial distribution of scatterers, that is, effects of chain thickness (see Appendix 5.A). We evaluate P_s, assuming two types of scatterer distribution [27]. One is a uniform scatterer distribution within a flexible cylinder of contour length L having a uniform circular cross section of diameter d whose center is on the HW chain contour (cylinder model), and the other is an assembly of N identical (touched) oblate spheroids of principal diameters d_b and γd_b ($0 < \gamma \le 1$) in which the scatterers are uniformly distributed (touched-spheroid model). All lengths are measured in units of λ^{-1}.

(a) Cylinder Model

For this case, $P_s(k; L)$ is given, from Eq. (5.A.2) with Eq. (5.A.5), by

$$P_s(k; L) = 2L^{-2} \int_0^L (L - s) I_s(k; s) ds \tag{5.49}$$

with

$$I_s(k; L) = \left(\frac{4}{\pi d^2}\right)^2 \left\langle \left| \int_{C_0} d\bar{r}_0 \int_C d\bar{r} \exp\left[i k \cdot (\mathbf{R} + \bar{r} - \bar{r}_0)\right] \right| \right\rangle, \tag{5.50}$$

where $\mathbf{R} = \mathbf{R}(L)$ is the end-to-end vector distance of the chain of contour length L, $\bar{r}_0$ (or $\bar{r}$) is the vector distance from the initial (or terminal) contour point to an arbitrary point in the normal cross section at that point, and the integrations are carried out over the respective cross sections. The equilibrium average $\langle \cdots \rangle$ is given by Eq. (4.184),

$$\langle \cdots \rangle = (8\pi^2)^{-1} \int (\cdots) G(\mathbf{R}, \Omega \mid \Omega_0; L) d\mathbf{R} d\Omega d\Omega_0, \tag{5.51}$$

where G is the Green function defined in Sect. 4.2.1.

Before making further developments, it is convenient to consider the case of rigid rods. In this case $I_s(k; L)$ may be evaluated to be

$$I_s(k; L) = j_0(Lk) \sum_{n_0=0}^{\infty} \left[F_{n_0}(dk)\right]^2 + \sum_{n=1}^{\infty} (4n + 1) j_{2n}(Lk)$$

$$\times \sum_{n_0, n_L=0}^{\infty} (-1)^{n_0+n_L} \left[(4n_0 + 1)(4n_L + 1)\right]^{1/2}$$

$$\times \begin{pmatrix} 2n & 2n_0 & 2n_L \\ 0 & 0 & 0 \end{pmatrix}^2 F_{n_0}(dk) F_{n_L}(dk) \quad \text{(rod)}, \tag{5.52}$$

where $(::::)$ is the Wigner 3-j symbol, and $F_n(dk)$ is a function of dk and is defined by

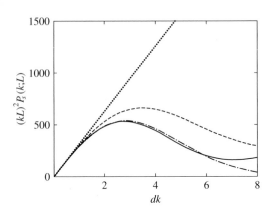

Fig. 5.7. Plots of $(Lk)^2 P_s(k; L)$ against dk for the rigid rod. The *solid curve* represents the exact values for the rod of $L/d = 100$ having finite thickness, and the *dotted curve* for the corresponding rod with vanishing thickness. The *dot-dashed and dashed curves* represent the approximate values for the same rod with finite thickness calculated from Eq. (5.54) for the conventional method and from Eq. (5.49) with Eq. (5.52) only with the j_0 term, respectively

$$F_n(dk) = (-1)^n \left(\frac{8}{\pi^{1/2} d^2} \right) \int_{C_0} j_{2n}(\bar{r}_0 k) Y_{2n}^0 \left(\frac{\pi}{2}, \phi_0 \right) d\bar{r}_0$$

$$= \frac{1}{2} (4n + 1)^{1/2} \frac{(2n - 1)!!}{(2n)!!}$$

$$\times \sum_{m=0}^{\infty} \frac{(-1)^m \Gamma(\frac{1}{2})}{(m + n + 1) m! (m + 2n + \frac{3}{2})} \left(\frac{dk}{4} \right)^{2(m+n)} \tag{5.53}$$

with Γ the gamma function and with $\bar{r}_0 = (\bar{r}_0, \pi/2, \phi_0)$ in spherical polar coordinates in the localized coordinate system at the contour point 0. In Eq. (5.52), we note that $j_0(Lk)$ is just equal to the characteristic function $I(\mathbf{k}; L)$ for the rigid rod of length L and that the sum of $\left[F_{n_0}(dk) \right]^2$ over n_0 is equal to the scattering function for the circular disk of diameter d.

Values of $(Lk)^2 P_s(k; L)$ calculated as a function of dk for the rigid rod from Eq. (5.49) with Eq. (5.52) for $L/d = 100$ are represented by the solid curve in Fig. 5.7. [Note that the dimensionless quantity $(Lk)^2 P_s$ is a function of dk and L/d.] The dotted curve represents the values for the rigid rod with vanishing d, that is, its contour. [Note that $(Lk)^2 P_s$ for this rod with $d = 0$ is a function of $Lk = 100 dk$, that is, dk.] It is seen that the scattering function for the rod with finite d becomes much smaller than that for the rod with vanishing d in the range of large dk because of the additional interference due to the spatial distribution of scatterers around the contour.

We also examine the behavior of the scattering function P_s obtained by the conventional method for the correction for chain thickness. It is given by [24, 29]

$$P_s(k; L) = P(k; L) \exp\left(-\frac{1}{16} d^2 k^2 \right). \tag{5.54}$$

The factor $\exp(-d^2 k^2/16)$ represents approximately the additional interference mentioned above. In Fig. 5.7, the dot-dashed curve represents the values calculated from Eq. (5.54) with Eq. (5.32) for the contour scattering function $P(k; L)$ for the rod. For the rigid rod Eq. (5.54) is seen to give a good approximation to P_s in the range of $dk \lesssim 6$.

For the flexible chain, however, the orientational correlation between the two normal cross sections at two contour points diminishes rapidly as the contour distance between them is increased, so that the effect of the additional interference may be considered to become smaller than that for the rigid rod. Thus we neglect this correlation, for simplicity. In the case of the rigid rod this approximation gives the values represented by the dashed curve in Fig. 5.7, which have been calculated from Eqs. (5.49) and (5.52) with neglect of all terms other than the first (j_0) term on the right-hand side of Eq. (5.52). This is, of course, not a good approximation for the rigid rod, for which the orientational correlation never vanishes even if it is infinitely long. We further neglect the correlation between the end-to-end distance R and the orientation Ω. This approximation causes no serious errors for flexible chains except for very small L. Then we have for the desired expression for $I_s(\mathbf{k}; L)$

$$I_s(\mathbf{k}; L) = I(\mathbf{k}; L) \sum_{n=0}^{\infty} g_{2n}^{00}(L) \left[F_n(dk) \right]^2, \tag{5.55}$$

where $I(\mathbf{k}; L)$ is the characteristic function, $g_l^{jj'}(L)$ is the angular correlation function given by Eq. (4.108) (for $\sigma = 0$), and we note that $g_0^{00} = 1$.

We have examined the convergence of the sum on the right-hand side of Eq. (5.55) and found that the summands with $n \geq 2$ may be neglected in the ordinary range of k in which SAXS and SANS measurements are carried out. Thus we have for the final explicit expression for $I_s(\mathbf{k}; L)$

$$I_s(\mathbf{k}; L) = I(\mathbf{k}; L) \left\{ \left[F_0(dk) \right]^2 + g_2^{00}(L) \left[F_1(dk) \right]^2 \right\}, \tag{5.56}$$

where

$$g_2^{00}(L) = e^{-6L} \left[\frac{3\kappa_0^4}{4\nu^4} \cos(2\nu L) + \frac{3\kappa_0^2 \tau_0^2}{\nu^4} \cos(\nu L) + \frac{1}{4} \left(\frac{3\tau_0^2}{\nu^2} - 1 \right)^2 \right], \tag{5.57}$$

$$F_0(x) = 4x^{-2} \left[1 - \cos(x/2) \right], \tag{5.58}$$

$$F_1(x) = 4x^{-3} \left\{ x \left[\cos(x/2) - 1 \right] - 6 \left[\sin(x/2) - 1 \right] \right\}. \tag{5.59}$$

(b) *Touched-Spheroid Model*

For this case, $P_s(k; L)$ is given, from Eq. (5.A.2) with Eq. (5.A.6), by

$$P_s(k; L) = N^{-1} I_s(\mathbf{k}; 0) + 2N^{-2} \sum_{j=1}^{N-1} (N - j) I_s(\mathbf{k}; j\gamma d_b) \tag{5.60}$$

with

$$I_s(\mathbf{k}; L) = \left(\frac{6}{\pi\gamma d_b^{\,3}}\right)^2 \left\langle \int_{V_1} d\bar{\mathbf{r}}_1 \int_{V_N} d\bar{\mathbf{r}}_N \exp\left[i\mathbf{k}\cdot(\mathbf{R}+\bar{\mathbf{r}}_N-\bar{\mathbf{r}}_1)\right]\right\rangle, \quad (5.61)$$

where $\mathbf{R} = \mathbf{R}(L)$ with $L = (N-1)\gamma d_b$, $\bar{\mathbf{r}}_j$ $(j = 1, N)$ is the vector distance from the center of the jth spheroid to an arbitrary point within it, and the integrations are carried out within the respective spheroids.

In the same approximations as those in the case of the cylinder model, $I_s(\mathbf{k}; L)$ may be given by Eq. (5.56) with $F_n(d_b k)$ in place of $F_n(dk)$, where $F_n(d_b k)$ is defined by

$$F_n(d_b k) = (-1)^n \left(\frac{12}{\pi^{1/2}\gamma d_b^{\,3}}\right) \int_{V_1} j_{2n}(k\bar{r}_1)Y_{2n}^0(\theta_1, \phi_1)d\bar{\mathbf{r}}_1 \quad (5.62)$$

with $\bar{\mathbf{r}}_1 = (\bar{r}_1, \theta_1, \phi_1)$ in spherical polar coordinates in the localized coordinate system at the contour point 0. The required F_n are explicitly given by

$$F_0(x) = \frac{24}{\gamma x^3}\int_0^1 \left[xf(y)\right]^2 j_1\left[xf(y)\right]dy, \quad (5.63)$$

$$F_1(x) = -\frac{12\sqrt{5}}{\gamma x^3}\int_0^1 (3y^2-1)\{-4\sin\left[xf(y)\right] + xf(y)\cos\left[xf(y)\right]$$
$$+3\mathrm{Si}\left[xf(y)\right]\}dy, \quad (5.64)$$

where $\mathrm{Si}(v)$ is the sine integral given by Eq. (5.34) and $f(y)$ is given by

$$f(y) = \frac{1}{2}\left[1 + (\gamma^{-2}-1)y^2\right]^{-1/2}. \quad (5.65)$$

The integrations in Eqs. (5.63) and (5.64) must be carried out numerically.

In the particular case of $\gamma = 1$, F_n vanish for $n \geq 1$ and F_0 is given by

$$F_0(d_b k) = 12(d_b k)^{-3}\left[2\sin(d_b k/2) - d_b k\cos(d_b k/2)\right] \quad (\gamma = 1). \quad (5.66)$$

Note that the square of this F_0 is just the scattering function for the sphere of diameter d_b. Then Eq. (5.60) is the exact expression for the scattering function for the touched-sphere (bead) model, and $P_s(k; L)$ is simply factored into $(F_0)^2$ and the contour scattering function $P(k; L)$ given by Eq. (5.60) with $I_s(\mathbf{k}; 0) = 1$ and $I_s = I$ for $j \geq 1$; that is,

$$P_s(k; L) = P(k; L)\left[F_0(d_b k)\right]^2 \quad (\gamma = 1). \quad (5.67)$$

This is the relation derived by Burchard and Kajiwara [30].

(c) Numerical Results

Finally, we give numerical results for the Kratky function $F_s(k; L)$ defined by Eq. (5.40) with P_s in place of P for the two models. Before doing this, we consider the relation between them. If we introduce the requirement that the coefficients of k^2 in the expansions of P_s given by Eq. (5.A.7) for them be identical with each other (for very large L), the squared radius of gyration $(2 + \gamma^2)d_b^2/20$ of the spheroid is identical with the one $d^2/8$ of the circular cross section of the cylinder. Then d_b may be related to d by the equation

$$d_b = \left[\frac{5}{2(2 + \gamma^2)}\right]^{1/2} d \,. \qquad (5.68)$$

In what follows, we use instead of d_b the diameter d from this relation for the touched-spheroid model, for convenience. In an application of this model to a real polymer chain, we replace the repeat unit of the latter by one spheroid such that its principal diameter γd_b is identical with the contour length per repeat unit, the number of spheroids N being equal to that of repeat units x. With the value of γd_b evaluated and that of d properly assigned for a given real polymer, the parameter γ is then calculated from Eq. (5.68), and therefore d_b is also determined. For comparison, we also consider the touched-sphere (bead) model of (approximately) the same contour length such that its bead diameter d_b is given by Eq. (5.68) with $\gamma = 1$ for a given value of d.

Figure 5.8 shows plots of $F_s(k; L)$ against k for the a-PMMA chain with $f_r = 0.79$ and $N = 1000$ for the indicated values of d. With the values of its λ^{-1} and M_L given in Table 5.1, the value of γd_b is evaluated to be 0.0476, and then we have $L = 47.55$. The solid curves represent the values for the cylinder model, and the dashed and dotted curves represent the values for the corresponding touched-spheroid and touched-sphere (bead) models, respectively. Numerical results have also been obtained for other values of N and for the a-PS chain [27]. From these results, the following two rather

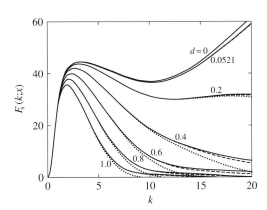

Fig. 5.8. Plots of $F_s(k; L)$ against k for a-PMMA ($\kappa_0 = 4.0$ and $\tau_0 = 1.1$) with $N = 1000$ and $\gamma d_b = 0.0476$ for the indicated values of d. The *solid, dashed, and dotted curves* represent the values for the cylinder, touched-spheroid, and touched-sphere models, respectively

obvious but important facts may be pointed out. First, in the range of small and intermediate k ($dk \lesssim 7$), the directly observed Kratky function F_s (with finite chain thickness) is almost independent of the model for the scatterer distribution but depends on the local conformation of the chain contour. Second, for larger k ($dk \gtrsim 7$), F_s (or its decay) depends strongly on the local scatterer distribution, so that it is dangerous to construct the contour Kratky function F from F_s there, as has often been done by the use of an approximate formula for the chain-thickness correction.

5.2.4 Comparison with Experiment

In this subsection we make a comparison of theory with experiment with respect to the Kratky function $F_s(k)$ mainly with SAXS data recently obtained for a-PS, a-PMMA, i-PMMA, and s-PMMA by the use of a point-focussing camera (together with a Kratky U-slit camera in the range of small k) [31–34]. In this case, F_s is defined by

$$F_s(k) = Mk^2 P_s(k) \tag{5.69}$$

with M the polymer molecular weight. For convenience, the theoretical values for the above polymers (except for s-PMMA) are calculated from Eq. (5.69) with Eqs. (5.49) and (5.56) for the cylinder model, using the values of the model parameters given in Table 5.1. The diameter d as an adjustable parameter is determined from a best fit of the theoretical values to the data.

Figure 5.9 shows plots of $F_s(k)$ against k for four fractions of a-PS ($f_r = 0.59$) with the indicated values of M in cyclohexane at 34.5°C [31]. For comparison, it also includes SANS data (filled circles) obtained by Huber et al. [35] for a fraction of a-PS with $M = 1.07 \times 10^4$ in cyclohexane-d_{12} at 35°C. The solid curves represent the best-fit HW theoretical values. The values of d thus determined by the curve fitting to the SAXS data are 6.8, 12.2, 13.7, and 13.9 Å for the fractions with the lowest to highest molecular weights, respectively, and its value from the SANS data is 9.9 Å. Agreement between theory and experiment is rather good in the range of $k \lesssim 0.25$ Å^{-1} but becomes poor for larger k, indicating that the details of distribution of electrons or hydrogen nuclei as the scatterers around the chain contour must there be taken into account. The diameter d from the SAXS data increases somewhat with increasing M for $M \lesssim 1.0 \times 10^4$, but the reason for this is not clear. We note that the value of d from the apparent mean-square radius of gyration $\langle S^2 \rangle_s$ is 9.4 Å (see Appendix 5.A). Further, note that the value 9.9 Å above from the SANS data is smaller than the value 13.7 Å from the SAXS data for the fraction with almost the same M ($= 1.01 \times 10^4$). This disagreement may be regarded as arising from the fact that for PS the distribution of electrons as the scatterers around the chain contour in SAXS is broader than that of hydrogen nuclei as those in SANS.

Figure 5.10 shows similar plots for four fractions of a-PMMA ($f_r = 0.79$)

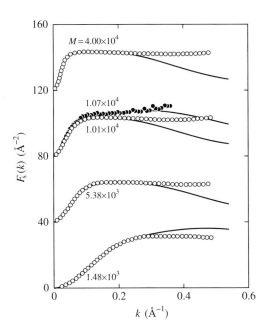

Fig. 5.9. Plots of $F_s(k)$ against k for a-PS with the indicated values of M; ○, SAXS data in cyclohexane at 34.5°C [31]; ●, SANS data in cyclohexane-d$_{12}$ at 35°C [35]. The *solid curves* represent the best-fit HW theoretical values calculated with the values of the model parameters given in Table 5.1 and proper values of d (see the text)

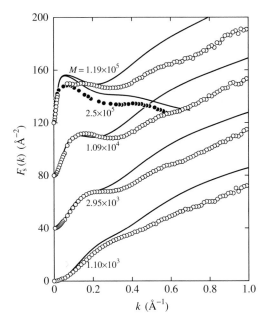

Fig. 5.10. Plots of $F_s(k)$ against k for a-PMMA with the indicated values of M; ○, SAXS data in acetonitrile at 44.0°C [32]; ●, SANS data in the bulk [36]; see legend for Fig. 5.9

in acetonitrile at 44.0°C [32]. It also includes SANS data (filled circles) obtained by Dettenmaier et al. [36] for a fraction of a-PMMA ($f_r = 0.78$) with

$M \simeq 2.5 \times 10^5$ in the bulk. The values of d from the SAXS and SANS data are 2.8 and 11.3 Å, respectively. In this case, agreement between theory and experiment is only semiquantitative. It is however important to see that for $M \gtrsim 3 \times 10^3$, the observed Kratky plot exhibits the maximum and minimum but not the second ones (or oscillation) such as observed by Kirste and Wunderlich [22–24], being consistent with the HW theory prediction. We believe that the desmeared SAXS data obtained by them (with a Kratky U-slit camera) are not correct for large k. The value 11.3 Å of d from the SANS data is remarkably larger than the value 2.8 Å from the SAXS data and even the value 8.2 Å from $\langle S^2 \rangle_s$ (see Appendix 5.A). This is due to the fact that the scatterers (electrons) are distributed in rather small regions around the α carbon atoms and the ester groups in SAXS, while they are the hydrogen nuclei in SANS.

Figure 5.11 also shows similar plots for four fractions of i-PMMA ($f_r = 0.01$) in acetonitrile at 28.0°C [33]. It also includes SANS data (filled circles) obtained by O'Reilly et al. [37] for a fraction of i-PMMA ($f_r = 0.03$) with $M = 1.20 \times 10^5$ in the bulk and corrected by Vacatello et al. [38]. The value of d from the SAXS data is 3.0 Å, which is close to the corresponding value 2.8 Å above for a-PMMA in the same Θ solvent, and the value from the SANS data is 13.7 Å, which is also close to the corresponding value 11.3 Å above for a-PMMA in the bulk. In this case the theory may accidentally well explain the experimental results over the whole range of k examined. As was expected, the SAXS data do not exhibit the maximum and minimum since the helical nature of the i-PMMA chain is weaker than that of the a-PMMA chain.

For comparison, the SAXS data and the corresponding theoretical values for the above a-PS, a-PMMA, and i-PMMA fractions with $M \simeq 10^4$ are plotted in Fig. 5.12. It also includes data for a fraction of s-PMMA ($f_r = 0.92$) with $M = 3.76 \times 10^4$ in acetonitrile at 44.0°C (Θ) [34], although the theoretical values have not been calculated since the values of its model parameters have not been determined. It is important to see that the Kratky plot does not exhibit the second maximum and minimum even for s-PMMA. The behavior of F_s in the range of $k \lesssim 0.2$ Å^{-1} may be considered to reflect the local chain conformation since the effect of electron distribution is rather small there. It may therefore be concluded that the HW theory may in fact well explain the observed differences in F_s in such a range of k. As for this range, we note that the difference in the observed height of the so-called plateau in the Kratky plot, which strictly cannot be observed for a-PMMA, between a-PS and a-PMMA cannot be explained by the Gaussian chain model. For this model, the height is equal to $2M/\langle S^2 \rangle$, as seen from Eq. (5.30) or Eqs. (5.44). However, these a-PS and a-PMMA fractions have almost the same $\langle S^2 \rangle/M$ [1,3].

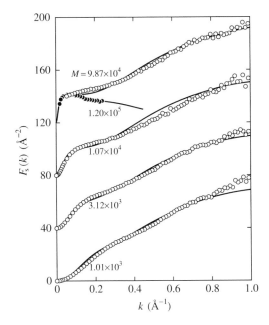

Fig. 5.11. Plots of $F_s(k)$ against k for i-PMMA with the indicated values of M; $\circ$, SAXS data in acetonitrile at 28.0°C [33]; $\bullet$, SANS data in the bulk [37]; see legend for Fig. 5.9

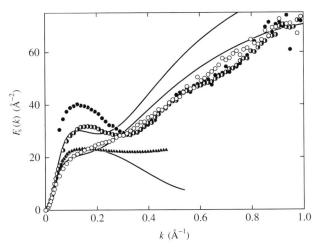

Fig. 5.12. Plots of $F_s(k)$ against k with SAXS data for a-PS ($\blacktriangle$), a-PMMA ($\bullet$), and i-PMMA ($\circ$) with $M \simeq 10^4$ in the respective Θ solvents [31–33], and s-PMMA ($f_r = 0.92$) with $M = 3.76 \times 10^4$ in acetonitrile at 44.0°C ($\bullet$) [34]. The *solid curves* represent the best-fit HW theoretical values for a-PS, a-PMMA, and i-PMMA

5.3 Anisotropic Light Scattering –
Mean-Square Optical Anisotropy

5.3.1 Basic Equations

The chain we have considered in the last section is the one having optically isotropic scatterers in the light-scattering case. In this section we treat
anisotropic light scattering on the basis of the HW chain which has the ex-
cess local polarizability tensors $\boldsymbol{\alpha}(s)$ and $\tilde{\boldsymbol{\alpha}}(s)$ (over the mean polarizability
of the solvent) per unit contour length at the contour point s ($0 \leq s \leq L$),
expressed in the localized and external Cartesian coordinate systems, respectively, where $\boldsymbol{\alpha}(s)$ is assumed to be independent of s [39]. All lengths are
measured in units of λ^{-1} unless otherwise noted.

Now we consider the excess intensity I_{fi} of scattered light with wave vector
$\mathbf{k}_{\mathrm{f}}$ and polarization $\mathbf{n}_{\mathrm{f}}$ for the case of monochromatic plane-polarized incident light with the intensity I_{i}^0, wave vector $\mathbf{k}_{\mathrm{i}}$, and polarization $\mathbf{n}_{\mathrm{i}}$, where
the subscripts i and f refer to "initial" (incident) and "final" (scattered), re-
spectively, and the polarization is defined as the unit vector in the direction
of the electric field of light. If λ_0 is the wavelength of light in vacuum and r
is the distance from the center of the system (the single HW chain) to the
detector, the ratio of I_{fi} to I_{i}^0 is given by [40]

$$\frac{I_{\mathrm{fi}}}{I_{\mathrm{i}}^0} = \frac{16\pi^4(\bar{\alpha}L)^2 F_{\mathrm{fi}}}{\lambda_0{}^4 r^2} , \tag{5.70}$$

where $\bar{\alpha}$ is the (excess) mean local polarizability per unit contour length and
is given by

$$\bar{\alpha} = \frac{1}{3}\mathrm{Tr}\,\boldsymbol{\alpha} , \tag{5.71}$$

and F_{fi} is given by

$$F_{\mathrm{fi}} = (\bar{\alpha}L)^{-2}\left\langle \left| \int_0^L \alpha_{\mathrm{fi}}(s)\exp\left[i\mathbf{k}\cdot\mathbf{r}(s)\right]ds \right|^2 \right\rangle$$
$$= (\bar{\alpha}L)^{-2}\int_0^L ds_1 \int_0^L ds_2 \,\langle\alpha_{\mathrm{fi}}(s_1)\alpha_{\mathrm{fi}}(s_2)\exp\left[i\mathbf{k}\cdot\mathbf{R}(s_1,\,s_2)\right]\rangle \tag{5.72}$$

with

$$\alpha_{\mathrm{fi}}(s) = \mathbf{n}_{\mathrm{f}}\cdot\tilde{\boldsymbol{\alpha}}(s)\cdot\mathbf{n}_{\mathrm{i}} , \tag{5.73}$$

$$\mathbf{k} = \mathbf{k}_{\mathrm{f}} - \mathbf{k}_{\mathrm{i}} , \tag{5.74}$$

and with $\mathbf{R}(s_1, s_2) = \mathbf{r}(s_2) - \mathbf{r}(s_1)$ being the vector distance between the
contour points s_1 and s_2. (Note that $\mathbf{k}$ is the scattering vector as before.)

The average in the second line of Eqs. (5.72) is given by Eq. (5.51) (with
$s = |s_1 - s_2|$ in place of L), so that it may readily be reduced to

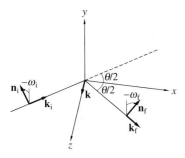

Fig. 5.13. Scattering geometry (see the text)

$$F_{\text{fi}} = (2\pi\bar{\alpha}L)^{-2} \int_0^L ds\,(L-s) \int d\Omega d\Omega_0\, \alpha_{\text{fi}}(s)\alpha_{\text{fi}}(0)I(\mathbf{k},\Omega\,|\,\Omega_0;s)\,, \quad (5.75)$$

where $I(\mathbf{k},\Omega\,|\,\Omega_0;s)$ is the characteristic function for the chain of contour length s.

In order to proceed to carry out evaluation, we adopt a specific scattering geometry, as depicted in Fig. 5.13. Suppose that the scatterer (the chain) is located at the origin of the external Cartesian coordinate system $(\mathbf{e}_x, \mathbf{e}_y, \mathbf{e}_z)$, and take the xz plane as the scattering plane spanned by $\mathbf{k}_i$ and $\mathbf{k}_f$ with $\mathbf{k}_f \cdot \mathbf{e}_x > 0$, $\mathbf{k}$ being in the positive direction of the z axis. This choice of the z axis proves very convenient in later developments. Let $\mathbf{n}_i$ ($\mathbf{n}_f$) be parallel to a vector obtained by rotation of $\mathbf{e}_y$ by an angle $-\omega_i$ ($-\omega_f$) about $\mathbf{k}_i$ ($\mathbf{k}_f$). The scattering system may then be completely determined by the scattering angle θ and the angles ω_i and ω_f. In particular, we use the symbols v (V) and h (H) to indicate that $\mathbf{n}_i$ ($\mathbf{n}_f$) is vertical and horizontal, respectively, with respect to the scattering plane, and also q (Q) to indicate that ω_i (ω_f) $= \pi/4$. For example, I_{Hv} denotes the horizontal component of the scattered intensity for the case of vertically polarized incident light. The important components I_{fi} (with α_{fi}) discussed later are the following five: Vv ($\omega_i = \omega_f = 0$), Hv ($\omega_i = 0$, $\omega_f = \pi/2$), Vh ($\omega_i = \pi/2$, $\omega_f = 0$), Hh ($\omega_i = \omega_f = \pi/2$), and Qq ($\omega_i = \omega_f = \pi/4$).

If $I_{\text{fi}}^{(N)}$ is the excess intensity of light scattered by N molecules in the scattering volume V, the reduced component R_{fi} experimentally determined is defined by

$$R_{\text{fi}} = \frac{I_{\text{fi}}^{(N)} r^2}{I_i^0 V}\,. \quad (5.76)$$

At infinite dilution ($I_{\text{fi}}^{(N)} = N I_{\text{fi}}$), it is related to F_{fi} by the equations

$$R_{\text{fi}} = \frac{16\pi^4 N_A c(\bar{\alpha}L)^2 F_{\text{fi}}}{\lambda_0^4 M}$$
$$= 2KMcF_{\text{fi}} \quad (5.77)$$

with

$$K = \frac{2\pi^2 \tilde{n}_0{}^2}{N_A \lambda_0{}^4} \left(\frac{\partial \tilde{n}}{\partial c} \right)^2 , \tag{5.78}$$

where N_A is the Avogadro constant, c $(= MN/N_A V)$ the polymer mass concentration, $\tilde{n}_0$ the refractive index of the solvent, and $\partial \tilde{n}/\partial c$ the refractive index increment. Thus we have five components R_{fi} corresponding to the five F_{fi}.

Further, we introduce two other R_{fi}, which are often measured. One is the reduced intensity R_{Uv} of the unpolarized scattered light for the case of $I_i^0 = I_v^0$, and the other is the reduced intensity R_{Uu} of the unpolarized scattered light for the case of unpolarized incident light with the intensity $I^0 = I_v^0 + I_h^0$ $(I_v^0 = I_h^0)$. At infinite dilution, we have

$$R_{Uv} = 2KMcF_{Uv} \tag{5.79}$$

with

$$F_{Uv} = F_{Vv} + F_{Hv} , \tag{5.80}$$

and

$$R_{Uu} \equiv R_\theta = \frac{N I_{Uu} r^2}{I^0 V (1 + \cos^2 \theta)}$$
$$= KMcF_{Uu} \tag{5.81}$$

with

$$F_{Uu} = \frac{F_{Vv} + 2F_{Hv} + F_{Hh}}{1 + \cos^2 \theta} . \tag{5.82}$$

We note that for isotropic scatterers F_{Vv}, $F_{Hh}/\cos^2 \theta$, $4F_{Qq}/(1 + \cos \theta)^2$, F_{Uv}, and F_{Uu} become identical with the isotropic scattering function $P(k)$ (at infinite dilution).

5.3.2 Components of the Scattered Intensity

We evaluate the scattered components F_{fi}. The evaluation consists of two main steps: (1) to express α_{fi} in terms of the polarizability $\boldsymbol{\alpha}$ expressed in the localized coordinate system and its orientation Ω with respect to the external coordinate system and (2) to express F_{fi} in terms of the expansion coefficients of the characteristic function.

For the first purpose, we first express α_{fi} in terms of the polarizability tensor $\tilde{\boldsymbol{\alpha}} = (\tilde{\alpha}_{\mu\nu})$ $(\mu, \nu = x, y, z)$ expressed in the external system. The components of $\mathbf{n}_i$ and $\mathbf{n}_f$ in this system are

$$\mathbf{n}_i = (-\hat{s}s_i, c_i, -\hat{c}s_i) ,$$
$$\mathbf{n}_f = (\hat{s}s_f, c_f, -\hat{c}s_f) , \tag{5.83}$$

where $\hat{s} = \sin(\theta/2)$, $\hat{c} = \cos(\theta/2)$, $s_i = \sin \omega_i$, $c_i = \cos \omega_i$, $s_f = \sin \omega_f$, and $c_f = \cos \omega_f$. We then have

$$\alpha_{fi} = \tilde{\boldsymbol{\alpha}} : \mathbf{A} \tag{5.84}$$

with

$$\mathbf{A} = \begin{pmatrix} -\hat{s}^2 s_i s_f & -\hat{s}s_i c_f & \hat{s}\hat{c}s_i s_f \\ \hat{s}c_i s_f & c_i c_f & -\hat{c}c_i s_f \\ -\hat{s}\hat{c}s_i s_f & -\hat{c}s_i c_f & \hat{c}^2 s_i s_f \end{pmatrix} . \tag{5.85}$$

In order to carry out the integrations over Ω and Ω_0 in Eq. (5.75), it is convenient to introduce the spherical (irreducible) components $\tilde{\alpha}_l^m$ (α_l^m) $(l = 0, 1, 2; m = 0, \pm 1, \pm 2)$ of the tensor $\tilde{\boldsymbol{\alpha}}$ $(\boldsymbol{\alpha})$ [40], which are defined in terms of the Cartesian components $\tilde{\alpha}_{\mu\nu}$ $(\alpha_{\mu'\nu'})$ as in Eqs. (5.B.2) of Appendix 5.B and which satisfy the relation given by Eq. (5.B.4), where $\alpha_{\mu'\nu'}$ $(\mu', \nu' = \xi, \eta, \zeta)$ are the components expressed in the localized system $(\mathbf{e}_\xi, \mathbf{e}_\eta, \mathbf{e}_\zeta)$. The reason for this is that $\tilde{\alpha}_l^m$ may be transformed to α_l^m by the same transformation rule as the spherical harmonics, that is, by Eq. (5.B.6). For molecules which do not absorb light and which are optically inactive, the polarizability tensor is real and symmetric in any coordinate system, so that $\tilde{\alpha}_1^m = \alpha_1^m = 0$ $(m = 0, \pm 1)$. The following development is, of course, limited to this case.

With these spherical tensors, the five components α_{fi} introduced in the last subsection may be written, from Eqs. (5.84), (5.85), and (5.B.2), in the matrix form

$$\begin{pmatrix} \alpha_{Vv} \\ \alpha_{Vh} \\ \alpha_{Hv} \\ \alpha_{Hh} \\ \alpha_{Qq} \end{pmatrix} = \begin{pmatrix} \frac{1}{\sqrt{3}} & -\frac{1}{\sqrt{6}} & 0 & 0 & -\frac{1}{2} & -\frac{1}{2} \\ 0 & 0 & -\frac{i\hat{c}}{2} & -\frac{i\hat{c}}{2} & \frac{i\hat{s}}{2} & -\frac{i\hat{s}}{2} \\ 0 & 0 & -\frac{i\hat{c}}{2} & -\frac{i\hat{c}}{2} & -\frac{i\hat{s}}{2} & \frac{i\hat{s}}{2} \\ \frac{2\hat{c}^2-1}{\sqrt{3}} & \frac{1+\hat{c}^2}{\sqrt{6}} & 0 & 0 & -\frac{\hat{s}^2}{2} & -\frac{\hat{s}^2}{2} \\ \frac{\hat{c}^2}{\sqrt{3}} & \frac{\hat{c}^2}{2\sqrt{6}} & -\frac{i\hat{c}}{2} & -\frac{i\hat{c}}{2} & -\frac{1+\hat{s}^2}{4} & -\frac{1+\hat{s}^2}{4} \end{pmatrix} \begin{pmatrix} \tilde{\alpha}_0^0 \\ \tilde{\alpha}_2^0 \\ \tilde{\alpha}_2^1 \\ \tilde{\alpha}_2^{-1} \\ \tilde{\alpha}_2^2 \\ \tilde{\alpha}_2^{-2} \end{pmatrix} . \tag{5.86}$$

For arbitrary ω_i and ω_f, α_{fi} may also be expressed in terms of $\tilde{\alpha}_l^m$ if we use Eq. (5.86) and the relation

$$\alpha_{fi} = \cos\omega_i \cos\omega_f \alpha_{Vv} + \sin\omega_i \cos\omega_f \alpha_{Vh}$$
$$+ \cos\omega_i \sin\omega_f \alpha_{Hv} + \sin\omega_i \sin\omega_f \alpha_{Hh} , \tag{5.87}$$

which can be easily verified from Eqs. (5.84) and (5.85) in the external system. Thus Eq. (5.86) or (5.87) with Eqs. (5.B.2) and (5.B.6) (with α in place of T) is the desired expression for α_{fi} in terms of α_l^m or $\alpha_{\mu'\nu'}$.

Now, according to Nagai's theorem [41], F_{fi} for arbitrary ω_i and ω_f may be expressed generally as a linear combination of four independent scattered components. In order to prove it in a simpler way and to proceed to the second step, it is convenient to introduce "correlation functions" defined by

$$\langle A, B \rangle = \int A^*(s) B(0) I(\mathbf{k}, \Omega \mid \Omega_0; s) d\Omega d\Omega_0 , \tag{5.88}$$

where A and B stand for $\tilde{\alpha}_l^m$ or α_{fi}, and the asterisk indicates the complex conjugate as usual. If we use Eq. (4.151) with Eqs. (4.C.8) and (5.B.6) in

Eq. (5.88) and note that $\mathbf{k} = k\mathbf{e}_z$, so that $\chi = 0$ and $Y_{l_3}^{m_2-m_1}(0,\omega) = [(2l_3+1)/4\pi]^{1/2}\delta_{m_1 m_2}$, then we obtain

$$\langle \tilde{\alpha}_{l_1}^{m_1}, \tilde{\alpha}_{l_2}^{m_2} \rangle = (4\pi)^{-1/2}\delta_{m_1 m_2}(c_{l_1}c_{l_2})^{-1} \sum_{l_3=|l_1-l_2|}^{l_1+l_2} (2l_3+1)^{1/2}$$
$$\times \sum_{|j_1|\le l_1} \sum_{|j_2|\le l_2} \alpha_{l_1}^{j_1*}\alpha_{l_2}^{j_2} \mathcal{I}_{l_1 l_2 l_3}^{m_1 m_2, j_1 j_2}(k; s). \tag{5.89}$$

As shown in Appendix 5.C, we have, by the use of Eqs. (5.86), (5.87), and (5.89),

$$F_{\text{fi}} = \cos\omega_{\text{i}} \cos\omega_{\text{f}} \cos(\omega_{\text{i}}+\omega_{\text{f}})F_{\text{Vv}} + \sin^2(\omega_{\text{i}}-\omega_{\text{f}})F_{\text{Hv}}$$
$$- \sin\omega_{\text{i}} \sin\omega_{\text{f}} \cos(\omega_{\text{i}}+\omega_{\text{f}})F_{\text{Hh}} + \sin 2\omega_{\text{i}} \sin 2\omega_{\text{f}} F_{\text{Qq}}. \tag{5.90}$$

This is Nagai's theorem.

The second problem is to express the four independent components on the right-hand side of Eq. (5.90) in terms of $\mathcal{I}_{\cdots}^{\cdots}(k; s)$. It is seen that these components may be written in terms of $\langle \alpha_{\text{Vv}}, \alpha_{\text{Vv}} \rangle$, $\langle \alpha_{\text{Hv}}, \alpha_{\text{Hv}} \rangle$, $\langle \alpha_{\text{Hh}}, \alpha_{\text{Hh}} \rangle$, and $\langle \alpha_{\text{Qq}}, \alpha_{\text{Qq}} \rangle$, respectively, and therefore of $\langle \tilde{\alpha}_{l_1}^{m_1}, \tilde{\alpha}_{l_2}^{m_2} \rangle$. Then, from Eqs. (5.86) and (5.89), they are seen to contribute to F_{fi} as $\langle \tilde{\alpha}_{l_1}^{m_1}, \tilde{\alpha}_{l_2}^{m_1} \rangle + \langle \tilde{\alpha}_{l_1}^{-m_1}, \tilde{\alpha}_{l_2}^{-m_1} \rangle$. If we use the relations

$$\mathcal{I}_{l_1 l_2 l_3}^{(-m_1)(-m_1), j_1 j_2}(k; L) = (-1)^{l_1+l_2+l_3} \mathcal{I}_{l_1 l_2 l_3}^{m_1 m_1, j_1 j_2}(k; L), \tag{5.91}$$

$$\mathcal{I}_{l_1 l_2 l_3}^{m_1 m_1, j_1 j_2}(k; L) = (-1)^{m_1} \begin{pmatrix} l_1 & l_2 & l_3 \\ m_1 & -m_1 & 0 \end{pmatrix}$$
$$\times \begin{pmatrix} l_1 & l_2 & l_3 \\ 0 & 0 & 0 \end{pmatrix}^{-1} \mathcal{I}_{l_1 l_2 l_3}^{00, j_1 j_2}(k; L), \tag{5.92}$$

where Eq. (5.92) is valid for $l_1 + l_2 + l_3$ even, then we find

$$\langle \tilde{\alpha}_{l_1}^{m_1}, \tilde{\alpha}_{l_2}^{m_1} \rangle + \langle \tilde{\alpha}_{l_1}^{-m_1}, \tilde{\alpha}_{l_2}^{-m_1} \rangle = (-1)^{m_1} \left(\frac{\bar{\alpha}^2}{\pi^{1/2}c_{l_1}c_{l_2}} \right) \sum_{l_3=|l_1-l_2|}^{l_1+l_2} (2l_3+1)^{1/2}$$
$$\times \begin{pmatrix} l_1 & l_2 & l_3 \\ m_1 & -m_1 & 0 \end{pmatrix} \begin{pmatrix} l_1 & l_2 & l_3 \\ 0 & 0 & 0 \end{pmatrix}^{-1} \mathcal{I}_{l_1 l_2 l_3}(k; s), \tag{5.93}$$

where

$$\mathcal{I}_{l_1 l_2 l_3}(k; s) = \sum_{|j_1|\le l_1} \sum_{|j_2|\le l_2} \hat{\alpha}_{l_1}^{j_1*}\hat{\alpha}_{l_2}^{j_2} \mathcal{I}_{l_1 l_2 l_3}^{00, j_1 j_2}(k; s) \tag{5.94}$$

with $\hat{\alpha}_l^m = \alpha_l^m/\bar{\alpha}$. We note that Eqs. (5.91) and (5.92) have been obtained from Eq. (4.152), Eq. (4.C.28) having also been used for the former, and that

Eq. (5.93) with Eq. (5.94) is also valid for any chain other than the HW chain if s is properly interpreted.

Before proceeding further, we make two remarks on $\mathcal{I}_{l_1 l_2 l_3}(k; s)$. First, the only required ones are $\mathcal{I}_{000}$, $\mathcal{I}_{202}$, $\mathcal{I}_{022}$, $\mathcal{I}_{220}$, $\mathcal{I}_{222}$, and $\mathcal{I}_{224}$. Second, these $\mathcal{I}_{l_1 l_2 l_3}$ are real, as seen below. As in Sect. 4.4.3, we can derive a symmetry relation for the function $\mathcal{G}_{\cdots}^{\cdots}(R; L)$ in Eq. (4.155) from the reality of $G(\mathbf{R}, \Omega \,|\, \Omega_0; L)$. The result is

$$\mathcal{G}_{l_1 l_2 l_3}^{m_1 m_2, j_1 j_2}(R; L) = (-1)^{m_1 + m_2 + j_1 + j_2}$$
$$\times \mathcal{G}_{l_1 l_2 l_3}^{(-m_1)(-m_2),(-j_1)(-j_2)*}(R; L). \qquad (5.95)$$

From Eqs. (4.156) and (5.95), we find the corresponding symmetry relation for $\mathcal{I}_{\cdots}^{\cdots}(k; L)$,

$$\mathcal{I}_{l_1 l_2 l_3}^{m_1 m_2, j_1 j_2}(k; L) = (-1)^{m_1 + m_2 + j_1 + j_2 + l_3}$$
$$\times \mathcal{I}_{l_1 l_2 l_3}^{(-m_1)(-m_2),(-j_1)(-j_2)*}(k; L). \qquad (5.96)$$

From Eqs. (5.94), (5.96), and (5.B.4), $\mathcal{I}_{l_1 l_2 l_3}$ are seen to be real.

Now we introduce vectors $\mathbf{F}$ and $\mathbf{Z}$,

$$\mathbf{F}^T = (F_{\mathrm{Vv}}, F_{\mathrm{Hv}}, F_{\mathrm{Hh}}, F_{\mathrm{Qq}}), \qquad (5.97)$$

$$\mathbf{Z}^T = \left(Z_{000}, \sqrt{2}\tilde{Z}_{202}, \frac{1}{5}Z_{220}, \frac{1}{\sqrt{5}}Z_{222}, Z_{224} \right) \qquad (5.98)$$

with

$$\tilde{Z}_{202} = \frac{1}{2}(Z_{202} + Z_{022}), \qquad (5.99)$$

where the superscript T indicates the transpose, and $Z_{l_1 l_2 l_3}$ is defined by

$$Z_{l_1 l_2 l_3}(k; L) = \pi^{-1/2} L^{-2} \int_0^L (L - s) \mathcal{I}_{l_1 l_2 l_3}(k; s) ds. \qquad (5.100)$$

The four components F_{fi} may then be written in the matrix form

$$\mathbf{F} = \mathbf{W} \cdot \mathbf{Z}, \qquad (5.101)$$

where $\mathbf{W}$ is the 4×5 matrix given by

$$\mathbf{W} = \begin{pmatrix} \frac{1}{3} & -\frac{1}{3} & \frac{2}{3} & -\frac{1}{3} & \frac{3}{20} \\ 0 & 0 & \frac{1}{2} & \frac{-1+3c}{8} & -\frac{3+5c}{40} \\ \frac{c^2}{3} & \frac{c(3+c)}{6} & \frac{3+c^2}{6} & \frac{3+6c-c^2}{12} & \frac{19+10c+3c^2}{80} \\ \frac{(1+c)^2}{12} & \frac{(1+c)^2}{24} & \frac{13+2c+c^2}{24} & \frac{-7+16c-c^2}{48} & \frac{-21-34c+3c^2}{320} \end{pmatrix} \qquad (5.102)$$

with $c = \cos\theta$.

Thus the four components F_{fi} may be expressed generally as linear combinations of the five fundamental quantities, that is, the components of $\mathbf{Z}$. Among these, Z_{000} is related to the isotropic scattering function $P(k; L)$ (that is the ordinary scattering function the chain would have if it were optically isotropic) by the equation

$$P(k; L) = \frac{1}{3} Z_{000}(k; L) \,. \tag{5.103}$$

However, it is important to see that we cannot in general determine the five components of $\mathbf{Z}$ inversely from the four observed components of $\mathbf{F}$ at arbitrary θ; it is mathematically impossible unless a specific model is assumed, as pointed out by Nagai [41], although his fundamental quantities are different from $\mathbf{Z}$.

For the HW chain, Eq. (5.94) may be simplified. As in Sect. 4.4.3, we can derive a second symmetry relation for $\mathcal{G}_{\cdots}^{\cdots}(R; L)$. The Green function is invariant to the reversal of the initial and terminal ends of the chain; that is,

$$G(\mathbf{R}, \mathbf{u}, \mathbf{a} \,|\, \mathbf{u}_0, \mathbf{a}_0; L) = G(-\mathbf{R}, -\mathbf{u}_0, \mathbf{a}_0 \,|\, -\mathbf{u}, \mathbf{a}; L) \,, \tag{5.104}$$

so that

$$\mathcal{G}_{l_1 l_2 l_3}^{m_1 m_2, j_1 j_2}(R; L) = (-1)^{m_1 + m_2 + j_1 + j_2 + l_1 + l_2 + l_3}$$
$$\times \mathcal{G}_{l_2 l_1 l_3}^{(-m_2)(-m_1), j_2 j_1}(R; L) \,. \tag{5.105}$$

From Eqs. (4.156) and (5.105), we obtain

$$\mathcal{I}_{l_1 l_2 l_3}^{m_1 m_2, j_1 j_2}(k; L) = (-1)^{m_1 + m_2 + j_1 + j_2 + l_1 + l_2 + l_3}$$
$$\times \mathcal{I}_{l_2 l_1 l_3}^{(-m_2)(-m_1), j_2 j_1}(k; L) \,. \tag{5.106}$$

By the use of Eqs. (5.96), (5.106), and (5.B.4), we find for the required $\mathcal{I}_{l_1 l_2 l_3}(k; s)$ for the HW chain

$$\mathcal{I}_{000} = (\hat{\alpha}_0^0)^2 \mathcal{I}_{000}^{00,00} \,, \tag{5.107}$$

$$\mathcal{I}_{202} = \mathcal{I}_{022}$$
$$= \sum_{j=0,2} f_j \mathrm{Re}(\hat{\alpha}_2^{j*} \hat{\alpha}_0^0) \mathrm{Re}(\mathcal{I}_{202}^{00,j0}) - 2\mathrm{Im}(\hat{\alpha}_2^{1*} \hat{\alpha}_0^0) \mathrm{Im}(\mathcal{I}_{202}^{00,10}) \,, \tag{5.108}$$

$$\mathcal{I}_{22l} = \sum_{\substack{j_2=0 \\ (j_1+j_2=\mathrm{even})}}^{2} \sum_{|j_1| \le j_2} f_{j_1 j_2} \mathrm{Re}(\hat{\alpha}_2^{j_1*} \hat{\alpha}_2^{j_2}) \mathrm{Re}(\mathcal{I}_{22l}^{00,j_1 j_2})$$
$$- \sum_{\substack{j_2=0 \\ (j_1+j_2=\mathrm{odd})}}^{2} \sum_{|j_1| \le j_2} f_{j_1 j_2} \mathrm{Im}(\hat{\alpha}_2^{j_1*} \hat{\alpha}_2^{j_2}) \mathrm{Im}(\mathcal{I}_{22l}^{00,j_1 j_2}) \,, \tag{5.109}$$

where Re and Im indicate the real and imaginary parts, respectively, and f_j and $f_{j_1 j_2}$ are defined by

$$f_j = \frac{2}{1 + \delta_{j0}}, \tag{5.110}$$

$$f_{j_1 j_2} = \frac{4}{(1 + \delta_{j_1 j_2})(1 + \delta_{j_1, -j_2})}. \tag{5.111}$$

Note that $\tilde{Z}_{202} = Z_{202} = Z_{022}$ for the HW chain.

Finally, we consider the two extreme cases of the HW chain, that is, random coils and rods. In the true coil limit, the anisotropic parts of F_{fi} may be neglected compared to the isotropic part (Z_{000}), the former being of $\mathcal{O}(L^{-1})$ in relation to the latter. Indeed, if we make order of magnitude estimates of $Z_{l_1 l_2 l_3}(k; L)$ in the important range of $k = \mathcal{O}(L^{-1/2})$ from Eqs. (5.94) and (5.100) with Eq. (4.191), as in Sect. 4.6.2 (Daniels-type distributions), then we have $Z_{000} = \mathcal{O}(1)$, $Z_{202} = \mathcal{O}(L^{-1})$, $Z_{220} = \mathcal{O}(L^{-1})$, $Z_{222} = \mathcal{O}(L^{-2})$, and $Z_{224} = \mathcal{O}(L^{-2})$. Therefore, the coil limit considered here is that region near the true limit in which the anisotropic correction terms of $\mathcal{O}(L^{-1})$ must be retained. In this region we may neglect Z_{222} and Z_{224} to obtain F_{fi} to terms of $\mathcal{O}(L^{-1})$ from Eq. (5.101). In particular, we have for the polarized and depolarized components

$$F_{\mathrm{Vv}} = \frac{1}{3} Z_{000} - \frac{\sqrt{2}}{3} Z_{202} + \frac{2}{15} Z_{220}, \tag{5.112}$$

$$F_{\mathrm{Hv}} = \frac{1}{10} Z_{220}. \tag{5.113}$$

We note that the results obtained by Horn [42] and by Utiyama and Kurata [43] for the Gaussian chain without correlations between orientations of the scatterers (beads) contain only Z_{000} and Z_{220}, the other components of $\mathbf{Z}$ vanishing, and that Tagami's results [44] for the same model are incorrect, her F_{Hv} being proportional to the Debye function.

As for rods, we have two types of rods (R1 and R2) as the rod limits $(\lambda \to 0)$ of the KP1 and KP2 chains, as depicted in Figs. 4.4b and c, respectively. For optical (and electrical) problems, we may further consider a third type of the KP chain (KP3) and the corresponding rod (R3) such that it has vanishing κ_0 and arbitrary τ_0 and that its local polarizability tensors are cylindrically symmetric about the chain contour $(\mathbf{e}_\zeta)$. The scattered components for the most general R1 rod have been evaluated [39], but the results are not reproduced because of their length. We also note only that a special case of the R2 rod and the R3 rod have been treated by Tagami [44] and by Horn et al. [45], respectively.

5.3.3 Mean-Square Optical Anisotropy

The independent scattered components F_{fi} at $\theta = 0$ are related to the *mean-square optical anisotropy* $\langle \Gamma^2 \rangle$. At $k = 0$ ($\theta = 0$), there is no interference between the scattered waves, and we obtain, from Eqs. (5.72),

$$F_{\mathrm{fi}} = \langle (\mathbf{n}_{\mathrm{f}} \cdot \boldsymbol{\gamma} \cdot \mathbf{n}_{\mathrm{i}})^2 \rangle, \tag{5.114}$$

where $\boldsymbol{\gamma}$ is the polarizability tensor of the entire chain in the external system and is given by

$$\boldsymbol{\gamma} = \int_0^L \tilde{\boldsymbol{\alpha}}(s)ds. \tag{5.115}$$

Now, from Eq. (4.191), it is seen that at $k = 0$, the $\mathcal{I}_{l_1 l_2 l_3}^{00,j_1 j_2}$ that contribute to F_{fi} are only $\mathcal{I}_{000}^{00,00}$ and $\mathcal{I}_{220}^{00,j_1 j_2}$. We then find, from Eqs. (5.94) and (5.100), that the nonvanishing components of $\mathbf{Z}$ are Z_{000} and Z_{220}, and obtain, from Eqs. (5.101) and (5.102) with this $\mathbf{Z}$,

$$F_{\mathrm{Vv}} = F_{\mathrm{Hh}} = F_{\mathrm{Qq}} = \frac{1}{3} Z_{000} + \frac{2}{15} Z_{220}$$

$$= 1 + \frac{4}{45} (\bar{\alpha} L)^{-2} \langle \Gamma^2 \rangle, \tag{5.116}$$

$$F_{\mathrm{Hv}} = \frac{1}{10} Z_{220} \equiv \frac{1}{15} (\bar{\alpha} L)^{-2} \langle \Gamma^2 \rangle. \tag{5.117}$$

The second equality of Eqs. (5.117) comprises the present definition of $\langle \Gamma^2 \rangle$. If γ_i ($i = 1, 2, 3$) are the principal values of $\boldsymbol{\gamma} = (\gamma_{\mu\nu})$ ($\mu, \nu = x, y, z$), it may in general be written, from Eqs. (5.114) and (5.117), in the form

$$\langle \Gamma^2 \rangle = \frac{1}{2} \langle (\gamma_1 - \gamma_2)^2 + (\gamma_2 - \gamma_3)^2 + (\gamma_3 - \gamma_1)^2 \rangle$$

$$= \left\langle \frac{1}{2} \left[(\gamma_{xx} - \gamma_{yy})^2 + (\gamma_{yy} - \gamma_{zz})^2 + (\gamma_{zz} - \gamma_{xx})^2 \right] \right.$$

$$\left. + 3(\gamma_{xy}^2 + \gamma_{yz}^2 + \gamma_{zx}^2) \right\rangle. \tag{5.118}$$

Note that Eqs. (5.116)–(5.118) are valid for any chain.

For the HW chain, we have, from Eqs. (4.36), (4.151), and (5.92),

$$\mathcal{I}_{ll0}^{00,j_1 j_2}(0; L) = (4\pi)^{1/2} g_l^{j_1 j_2}(L). \tag{5.119}$$

Thus we find

$$\langle \Gamma^2 \rangle = 3 \sum_{\substack{j_2=0 \\ (j_1+j_2=\mathrm{even})}}^{2} \sum_{|j_1| \leq j_2} f_{j_1 j_2} \mathrm{Re}(\alpha_2^{j_1*} \alpha_2^{j_2}) \bar{X}_2^{j_1 j_2}(L)$$

$$- 3 \sum_{\substack{j_2=0 \\ (j_1+j_2=\mathrm{odd})}}^{2} \sum_{|j_1| \leq j_2} f_{j_1 j_2} \mathrm{Im}(\alpha_2^{j_1*} \alpha_2^{j_2}) \bar{\bar{X}}_2^{j_1 j_2}(L), \tag{5.120}$$

where $f_{j_1 j_2}$ is given by Eq. (5.111), and $\bar{X}_2^{j_1 j_2}$ and $\bar{\bar{X}}_2^{j_1 j_2}$ are the real and imaginary parts, respectively, of the function $X_2^{j_1 j_2}(L)$ defined by

$$X_l^{j_1 j_2}(L) = \int_0^L (L - s) g_l^{j_1 j_2}(s) ds. \tag{5.121}$$

By the use of Eq. (4.108) for $g_l^{jj'}$ (for the chain with Poisson's ratio $\sigma = 0$), Eq. (5.120) may be reduced to

$$\langle \Gamma^2 \rangle = L \sum_{j=0}^{2} C_j(\boldsymbol{\alpha}, \nu^{-1} \kappa_0, \nu^{-1} \tau_0) f_j(L, \nu), \tag{5.122}$$

where ν is given by Eq. (5.4), and C_j and f_j are given by

$$C_0(\boldsymbol{\alpha}, x, y) = \frac{1}{2} \left[2\alpha_{\zeta\zeta} - \alpha_{\xi\xi} - \alpha_{\eta\eta} + 3x^2(\alpha_{\eta\eta} - \alpha_{\zeta\zeta}) + 6xy\alpha_{\eta\zeta} \right]^2,$$

$$C_1(\boldsymbol{\alpha}, x, y) = 6 \left[xy(\alpha_{\eta\eta} - \alpha_{\zeta\zeta}) + (2y^2 - 1)\alpha_{\eta\zeta} \right]^2$$
$$+ 6(x\alpha_{\xi\eta} + y\alpha_{\zeta\xi})^2, \tag{5.123}$$

$$C_2(\boldsymbol{\alpha}, x, y) = \frac{3}{2}(\alpha_{\xi\xi} - y^2\alpha_{\eta\eta} - x^2\alpha_{\zeta\zeta} + 2xy\alpha_{\eta\zeta})^2$$
$$+ 6(y\alpha_{\xi\eta} + x\alpha_{\zeta\xi})^2,$$

$$f_j(L, \nu) = \frac{1}{(36 + j^2\nu^2)L} \left\{ 6(36 + j^2\nu^2)L - 36 + j^2\nu^2 \right.$$
$$\left. + e^{-6L} \left[(36 - j^2\nu^2) \cos(j\nu L) - 12j\nu \sin(j\nu L) \right] \right\}. \tag{5.124}$$

For the KP2 and KP3 chains, Eq. (5.122) may be further reduced to

$$\langle \Gamma^2 \rangle = \frac{1}{3} \Gamma_L^2 \left(L - \frac{1}{6} + \frac{1}{6} e^{-6L} \right) \quad \text{(KP2, KP3)}, \tag{5.125}$$

where Γ_L^2 is the squared local anisotropy per unit length and is given by

$$\Gamma_L^2 = \lim_{L \to 0} \left(\frac{\langle \Gamma^2 \rangle}{L^2} \right)$$
$$= \frac{1}{2} \left[(\alpha_1 - \alpha_2)^2 + (\alpha_2 - \alpha_3)^2 + (\alpha_3 - \alpha_1)^2 \right] \tag{5.126}$$

with α_i ($i = 1, 2, 3$) being the principal values of $\boldsymbol{\alpha}$. Equation (5.126) with $\alpha_1 = \alpha_2$ is the result derived by Nagai [46] and by Arpin et al. [47] for the KP3 chain.

For a comparison of theory with experiment, it is convenient to use the polarizability tensor $\boldsymbol{\alpha}_0$ of the repeat unit instead of $\boldsymbol{\alpha}$ and also the number of repeat units x instead of L. Then $\boldsymbol{\alpha}$ and $\langle \Gamma^2 \rangle$ (unreduced) are given by

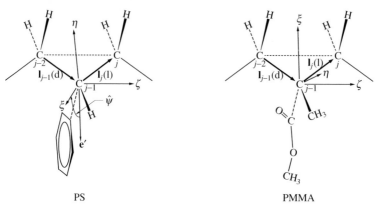

PS   PMMA

Fig. 5.14. Localized Cartesian coordinate systems for PS and PMMA (see the text)

$$\boldsymbol{\alpha} = (M_{\mathrm{L}}/M_0)\boldsymbol{\alpha}_0 , \tag{5.127}$$

$$\frac{\langle \Gamma^2 \rangle}{x} = \frac{\lambda^{-1} M_{\mathrm{L}}}{M_0} \sum_{j=0}^{2} C_j(\boldsymbol{\alpha}_0, \nu^{-1}\kappa_0, \nu^{-1}\tau_0) f_j(\lambda L, \lambda^{-1}\nu) , \tag{5.128}$$

where x is related to L by Eq. (5.2) and M_0 is the molecular weight of the repeat unit.

Now we make a comparison of theory with experiment for $\langle \Gamma^2 \rangle$ for a-PS, a-PMMA, and i-PMMA. We then assume that isotactic and syndiotactic sequences are randomly distributed in the atactic chain, so that $\boldsymbol{\alpha}_0$ for it is given by

$$\boldsymbol{\alpha}_0 = (1 - f_{\mathrm{r}})\boldsymbol{\alpha}_{0,\mathrm{i}} + f_{\mathrm{r}}\boldsymbol{\alpha}_{0,\mathrm{s}} , \tag{5.129}$$

where $\boldsymbol{\alpha}_{0,\mathrm{i}}$ and $\boldsymbol{\alpha}_{0,\mathrm{s}}$ are $\boldsymbol{\alpha}_0$ for the isotactic ($f_{\mathrm{r}} = 0$) and syndiotactic ($f_{\mathrm{r}} = 1$) chains, respectively.

For the calculation of theoretical values of $\langle \Gamma^2 \rangle$ from Eq. (5.128), the components of $\boldsymbol{\alpha}_{0,\sigma}$ ($\sigma = \mathrm{i, s}$) in the localized coordinate system (ξ, η, ζ) must be evaluated. For this purpose, it is necessary to affix this system to the monomer unit of a given chain, corresponding to that of the HW chain. This has already been done for PS and PMMA in Sect. 4.4.3, taking as the monomer unit the part of the chain containing the C–C$^\alpha$ and C$^\alpha$–C bonds, as shown in Fig. 5.14. That is, the localized coordinate system ($\mathbf{e}_{\xi_k}$, $\mathbf{e}_{\eta_k}$, $\mathbf{e}_{\zeta_k}$) affixed to the kth monomer unit containing the $(j-1)$th and jth skeletal bonds ($j = 2k$) corresponds to the system ($\mathbf{e}_\xi$, $\mathbf{e}_\eta$, $\mathbf{e}_\zeta$) of the HW chain as follows: $\mathbf{e}_{\zeta_k}$ is parallel to $\mathbf{l}_{j-1} + \mathbf{l}_j$ with $\mathbf{l}_j$ the jth bond vector, and $\mathbf{e}_{\xi_k}$ is defined by rotation of $\mathbf{e}'$ by the angle $\hat{\psi}$ about the ζ_k axis, where $\mathbf{e}'$ is the unit vector in the plane of $\mathbf{l}_{j-1}$ and $\mathbf{l}_j$ with $\mathbf{e}' \cdot \mathbf{e}_{\zeta_k} = 0$, its positive direction being chosen at an acute angle with $\mathbf{l}_{j-1}$. The values of $\hat{\psi}$ for the i- and s-PS and PMMA chains are given in Table 4.3. For convenience, in Fig. 5.14, the

$(j-1)$th and jth bonds are d- and l-chiral, respectively, according to the Flory convention [48], so that the sequence of bonds in the monomer unit displayed is represented by $d|l$, where the vertical line indicates the location of the α carbon atom.

The components of $\boldsymbol{\alpha}_{0,\sigma}$ may then be evaluated by the use of the values of the bond polarizabilities and group polarizability tensors (and also those for methyl acetate or methyl isobutyrate) determined by Flory and co-workers [49–52] with the use of the procedure of Carlson and Flory [53], assuming their additivity. Then, in the case of PS we average $\boldsymbol{\alpha}_{0,\sigma}$ over internal rotation angles on the basis of the RIS model, while in the case of PMMA we use $\boldsymbol{\alpha}_{0,\sigma}$ for the all-*trans* conformation because of the predominance of the *tt* conformation [54], and assume that the plane of the ester group is perpendicular to the plane of the C–C$^\alpha$ and C$^\alpha$–C bonds and that the ester group occupies the two possible states in the former plane with equal probability, for simplicity. Further, all the $\boldsymbol{\alpha}_{0,\sigma}$ so evaluated are multiplied by the factor $\sqrt{1.94}$ to obtain good agreement between theory and experiment for $\langle \Gamma^2 \rangle$ for a-PS. This means that the polarizabilities determined by the procedure of Carlson and Flory are too small by this factor. Thus the values of the (traceless) $\boldsymbol{\alpha}_{0,\sigma}$ we adopt are

$$\boldsymbol{\alpha}_{0,\sigma} = \begin{pmatrix} 1.64 & \mp 1.96 & 0 \\ \mp 1.96 & 1.84 & 0 \\ 0 & 0 & -3.48 \end{pmatrix} \text{\AA}^3 \qquad \text{for i-PS}$$

$$= \begin{pmatrix} 1.82 & \pm 2.03 & 0 \\ \pm 2.03 & 1.75 & 0 \\ 0 & 0 & -3.57 \end{pmatrix} \text{\AA}^3 \qquad \text{for s-PS}, \qquad (5.130)$$

$$\boldsymbol{\alpha}_{0,\sigma} = \begin{pmatrix} 0.581 & \pm 0.266 & 0 \\ \pm 0.266 & 0.712 & 0 \\ 0 & 0 & -1.293 \end{pmatrix} \text{\AA}^3 \qquad \text{for i-PMMA}$$

$$= \begin{pmatrix} 0.581 & 0 & 0 \\ 0 & 0.712 & 0 \\ 0 & 0 & -1.293 \end{pmatrix} \text{\AA}^3 \qquad \text{for s-PMMA}, \qquad (5.131)$$

where the upper and lower signs of the $\xi\eta$ and $\eta\xi$ components are taken for the bond chiralities $d|l$ and $l|d$, respectively, in Fig. 5.14 and the like, and for s-PMMA these components have been put equal to zero on the average since their sign changes alternately along the chain. Note that for i- and s-PSs and i-PMMA either sign of these components may be taken and that the traceless $\boldsymbol{\alpha}_0$ contributes to $\langle \Gamma^2 \rangle$, as seen from Eqs. (5.123) and (5.127). For the a-PMMA chain with $f_r = 0.79$, we assume $\boldsymbol{\alpha}_0 = \boldsymbol{\alpha}_{0,s}$, for simplicity.

Figure 5.15 shows double-logarithmic plots of $\langle \Gamma^2 \rangle / x$ (in $\text{\AA}^6$) against x with data recently obtained from anisotropic light scattering measurements with a Fabry–Perot interferometer (with corrections for effects of the internal field) for a-PS ($f_r = 0.59$) [55, 56], a-PMMA ($f_r = 0.79$) [56], and i-PMMA

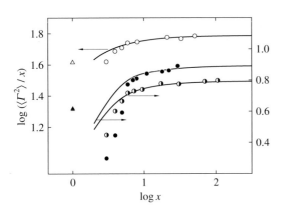

Fig. 5.15. Double-logarithmic plots of $\langle \Gamma^2 \rangle / x$ (in Å^6) against x for a-PS ($\circ$) and cumene ($\triangle$) in cyclohexane at 34.5°C [55,56], a-PMMA ($\bullet$) and methyl isobutyrate ($\blacktriangle$) in acetonitrile at 44.0°C [56], and i-PMMA ($\circleddash$) and methyl isobutyrate ($\blacktriangle$) in acetonitrile at 28.0°C [57]. The *solid curves* represent the HW theoretical values calculated with the values of the model parameters given in Table 5.1 and those of α_0 given by Eq. (5.129) with Eqs. (5.130) and (5.131) (see the text)

($f_r = 0.01$) [57] in the respective Θ solvents given in Table 5.1, and also for cumene (the monomer of PS) and methyl isobutyrate (the monomer of PMMA) in the corresponding solvents. The solid curves represent the respective HW theoretical values calculated from Eq. (5.128) with Eq. (5.129) with the values of the model parameters given in Table 5.1 and those of α_0 given above, where we have used $\lambda^{-1} = 22.7$ Å and $M_L = 37.1$ Å^{-1} for a-PS (somewhat different from those in Table 5.1) and replaced α_0 commonly by $0.73\alpha_0$ for both a- and i-PMMAs to obtain good agreement between theory and experiment for $(\langle \Gamma^2 \rangle / x)_\infty$. The necessity of this replacement of α_0 indicates that the values of the polarizability tensor for the ester group estimated from those for methyl acetate or methyl isobutyrate may be somewhat altered in the PMMA chains. At any rate, there is rather good agreement between theory and experiment in all cases for $x \gtrsim 6$, especially with respect to the rate of increase in $\langle \Gamma^2 \rangle / x$ with increasing x. The disagreement for smaller x may probably be mainly due to effects of chain ends.

5.3.4 Isotropic Scattering Function

We consider the problem of determining the isotropic scattering function $P(\theta)$ [$= P(k)$] as a function of θ from observed independent scattered components [39]. As already mentioned, it is in general impossible to express Z_{000}, and therefore $P(\theta)$, in terms of the four independent scattered components by solving Eq. (5.101) with respect to the five components of **Z**. It is then inevitable to introduce an approximation in order to establish a procedure for the determination of $P(\theta)$. Thus we consider two kinds of those linear combinations $R_{\theta(j)}$ ($j = 1, 2$) of the reduced scattered components R_{fi} which are approximations to R_θ, so that at infinite dilution, the corresponding approximate isotropic scattering functions $P_{(j)}(\theta)$ are given by

$$P_{(j)}(\theta) = R_{\theta(j)}/KMc. \tag{5.132}$$

Now, Z_{224} is of order k^4, and the other components of $\mathbf{Z}$ are of order unity or k^2. [Note that $Z_{l_1 l_2 l_3} = \mathcal{O}(k^{l_3})$.] Therefore, we first neglect Z_{224} in Eq. (5.101) and solve it to find an approximation to Z_{000}. The approximate $P(\theta)$ so obtained from Eq. (5.103) is denoted by $P_{(1)}(\theta)$. Then $R_{\theta(1)}$ is given by

$$R_{\theta(1)} = \mathbf{r} \cdot \tilde{\mathbf{W}} \cdot \tilde{\mathbf{R}}, \tag{5.133}$$

$$\tilde{\mathbf{R}}^T = (R_{\mathrm{Vv}}, R_{\mathrm{Hv}}, R_{\mathrm{Hh}}, R_{\mathrm{Qq}}), \tag{5.134}$$

$$\mathbf{r} = \left(-\frac{1}{6}, -\frac{1}{9}, \frac{1}{2}, \frac{1}{6} \right), \tag{5.135}$$

$$\tilde{\mathbf{W}} = \frac{1}{s^2(1+c)} \begin{pmatrix} -c(1+c)^2 & 2s^2 & -(1+c)^2 & 4c(1+c) \\ 2c(1+c) & -2(1-c)^2 & 2(1+c) & -8c \\ (1+c)^2 & -2s^2 & (1+c)^2 & -4c(1+c) \\ -2(1+c)^2 & -4(1-c)^2 & -2(1+c)^2 & 8(1+c^2) \end{pmatrix} \tag{5.136}$$

with $s = \sin\theta$ and $c = \cos\theta$. Thus $R_{\theta(1)}$ may be expressed as a linear combination of the four independent scattered components R_{fi}. $P_{(1)}(\theta)$ gives the correct coefficient of k^2 or $\sin^2(\theta/2)$ with $P_{(1)}(0) = 1$ *if* R_{fi} are truncated at k^2. Indeed, this is the basis of Nagai's procedure [41] for the determination of $P(\theta)$. However, it is important to note that the k^4 term of the neglected Z_{224} contributes to the coefficient of $\sin^2(\theta/2)$ because of $\tilde{\mathbf{W}}$, so that the corresponding coefficient in $P_{(1)}(\theta)$ is no longer correct. It can be shown that the difference $R_\theta - R_{\theta(1)}$, and therefore $R_{\theta(1)}$ itself, are finite determinate over the whole range of θ. From Eqs. (5.133)–(5.136), however, it is seen that the coefficients of R_{Vv}, R_{Hh}, and R_{Qq} in $R_{\theta(1)}$ are singular at $\theta = 0$, and so are all the coefficients at $\theta = \pi$. Extrapolation to $\theta = 0$ from $R_{\theta(1)}$ thus determined at finite θ may therefore involve appreciable errors, as pointed out by Nagai [41].

A second approximation consists of constructing a linear combination of three of the four independent R_{fi} by neglecting Z_{222} and Z_{224}, which are small compared to the other three components of $\mathbf{Z}$ in the coil limit, as shown in Sect. 5.3.2. Then there are three possible linear combinations of this kind. Among these, a linear combination of R_{Vv}, R_{Hv}, and R_{Qq}, which we denote by $R_{\theta(2)}$, is the only one that is finite determinate and has the nonsingular coefficients at $\theta = 0$. It reads

$$R_{\theta(2)} = \frac{1}{6} R_{\mathrm{Vv}} - \frac{5 + 2\cos\theta + \cos^2\theta}{3(1+\cos\theta)^2} R_{\mathrm{Hv}} + \frac{4}{3(1+\cos\theta)^2} R_{\mathrm{Qq}}. \tag{5.137}$$

This gives correctly $P_{(2)}(0) = 1$, although the coefficient of $\sin^2(\theta/2)$ is, of course, approximate. In this connection, we note that the procedure proposed by Utiyama and Kurata [43] for the Gaussian chain is equivalent to neglecting Z_{202}, Z_{222}, and Z_{224} to obtain

$$R_{\theta(\mathrm{UK})} = \frac{1}{2} R_{\mathrm{Vv}} - \frac{2}{3} R_{\mathrm{Hv}}. \tag{5.138}$$

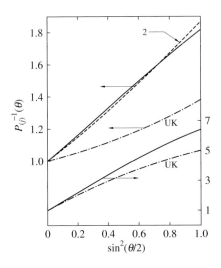

Fig. 5.16. Deviations of $P_{(2)}^{-1}(\theta)$ and $P_{(\mathrm{UK})}^{-1}(\theta)$ from $P^{-1}(\theta)$ as functions of $\sin^2(\theta/2)$ for the KP3 chain with $\epsilon_1 = 5$ and $\epsilon_2 = 0$ for $L = 1$ and $\tilde{\lambda} = 2$ (*upper three curves*) and for $L = 10$ and $\tilde{\lambda} = 4$ (*lower two curves*). The *solid curves* represent the values of $P^{-1}(\theta)$

This also gives $P_{(\mathrm{UK})}(0) = 1$, the coefficient of $\sin^2(\theta/2)$ being approximate except for the Gaussian chain.

In the numerical examination that follows, we assume that $\boldsymbol{\alpha}$ is diagonal, so that the spherical tensors α_l^m are real, for simplicity. It is then convenient to use the dimensionless parameters ϵ_1 and ϵ_2 defined by

$$\epsilon_1 = \left[\alpha_{\zeta\zeta} - \frac{1}{2}(\alpha_{\xi\xi} + \alpha_{\eta\eta})\right] \Big/ \bar{\alpha}\,,$$

$$\epsilon_2 = (\alpha_{\xi\xi} - \alpha_{\eta\eta})/\bar{\alpha} \tag{5.139}$$

with $\alpha_{\xi\xi} = \alpha_1$, $\alpha_{\eta\eta} = \alpha_2$, and $\alpha_{\zeta\zeta} = \alpha_3$. If $\boldsymbol{\alpha}$ is cylindrically symmetric about $\mathbf{e}_\zeta = \mathbf{u}$ ($\alpha_1 = \alpha_2$), we have $\epsilon_2 = 0$ and need only $\mathcal{I}_{l_1 l_2 l_3}^{00,00}$ in Eqs. (5.107)–(5.109). These $\mathcal{I}_{l_1 l_2 l_3}^{00,00}$ may be evaluated by the Laguerre polynomial expansion method rather than the weighting function method since k is rather small for light scattering [39].

We first make brief mention of the theoretical error in $P_{(j)}(\theta)$ examined. It is in general large for the KP chain, or codes close to it, with large ϵ_1 and ϵ_2, and for small L and $\tilde{\lambda}$. In the experimentally important ranges $P^{-1}(\pi) \gtrsim 1.1$ and $\sin^2(\theta/2) \lesssim 0.75$, the error in $P_{(j)}^{-1} - 1$ ($j = 1, 2$) does not exceed 1% except for the KP chain; for the KP chain it does not exceed 1% for $L \gtrsim 1$, $\tilde{\lambda} \gtrsim 1$, and $\Gamma_L^2/\bar{\alpha}^2 \lesssim 4$, and 2% for $L \gtrsim 2$, $\tilde{\lambda} \gtrsim 4$, and $\Gamma_L^2/\bar{\alpha}^2 \lesssim 25$. As an example of the cases of large error, values of $P_{(2)}^{-1}(\theta)$ and $P^{-1}(\theta)$ as functions of $\sin^2(\theta/2)$ for the KP3 chain with $\epsilon_1 = 5$ and $\epsilon_2 = 0$ for $L = 1$ and $\tilde{\lambda} = 2$ are represented by the dashed curve 2 and the solid curve close to it, respectively, in Fig. 5.16, values of $P_{(1)}^{-1}$ being intermediate between them. Necessarily, for this case the deviation of $P_{(\mathrm{UK})}^{-1}$ (upper dot-dashed curve UK) from P^{-1} is

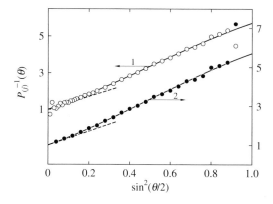

Fig. 5.17. Simulation of experimental values of $P_{(1)}^{-1}(\theta)$ and $P_{(2)}^{-1}(\theta)$ for the KP3 chain with $\epsilon_1 = 5$ and $\epsilon_2 = 0$ for $L = 8$ and $\tilde{\lambda} = 4$. The *solid curves* represent the theoretical values of $P^{-1}(\theta)$, and the *dashed lines* indicate their initial slopes

very large. The lower solid and dot-dashed curves represent the corresponding values for the same chain for $L = 10$ and $\tilde{\lambda} = 4$. In this case the values of $P_{(j)}^{-1}$ ($j = 1, 2$) agree with P^{-1} within the thickness of the curve, while $P_{(UK)}^{-1}$ still differs appreciably from P^{-1}. For larger L, the difference between $P_{(UK)}^{-1}$ and P^{-1} becomes, of course, small. However, the effects themselves are negligibly small for large L or in the coil limit.

Finally, we give results of an examination of the amplification of experimental errors in R_{fi} which is caused by the singularities of their coefficients in $R_{\theta(j)}$ ($j = 1, 2$). For this purpose, "experimental" values of R_{fi} have been simulated by multiplying the theoretical R_{fi} by random numbers normally distributed with a mean of unity and a standard deviation of 0.05. "Experimental" values of $P_{(1)}^{-1}$ and $P_{(2)}^{-1}$ so obtained are represented by the unfilled and filled circles, respectively, in Fig. 5.17 for the same KP3 chain as above for $L = 8$ and $\tilde{\lambda} = 4$. The solid curves represent the theoretical values of P^{-1}, their initial slopes being indicated by the dashed lines. As was expected, the amplification of experimental errors is remarkable near $\theta = 0$ in the first procedure, and near $\theta = \pi$ in both; there is no defect for $\sin^2(\theta/2) \lesssim 0.75$ in the second procedure. Note that these results are independent of ϵ_1 and ϵ_2.

5.3.5 Near the Rod Limit

The procedure presented in the last subsection is not very accurate near the rod limit, especially for the KP chain. Thus we derive analytical expressions for the scattered components near this limit to obtain accurate numerical results which suggest a more accurate method of analyzing experimental data for very stiff polymers [58]. All lengths are measured in units of λ^{-1} as

For simplicity, we consider the chain as having cylindrically symmetric polarizabilities ($\epsilon_2 = 0$), for which only the $\mathcal{I}_{l_1 l_2 l_3}^{00,00}$ contribute to the scattered components, as already mentioned. Then the only required $\mathcal{I}_{...}^{...}$ are $\mathcal{I}_{000}^{00,00}$, $\mathcal{I}_{202}^{00,00}$, and $\mathcal{I}_{22l}^{00,00}$ ($l = 0, 2, 4$). All lengths are measured in units of λ^{-1} as before. Evaluation is carried out by an application of the ϵ method given in Sect. 4.7.2, assuming that $\sigma = 0$, and assigning the rod-limiting values to $\langle R^2 \rangle_0$ and $\langle R^{l_3} X \rangle_0$ in Eqs. (4.220)–(4.222); that is, $\langle R^2 \rangle_0 = L^2$ and

$$\langle R^{l_3} \mathcal{D}_{l_1}^{00*} \mathcal{D}_{l_2}^{00} Y_{l_3}^0 \rangle_0 = (2\pi)^{1/2} c_{l_1} c_{l_2} c_{l_3} \begin{pmatrix} l_1 & l_2 & l_3 \\ 0 & 0 & 0 \end{pmatrix}^2 L^{l_3}, \qquad (5.140)$$

where c_l is given by Eq. (4.54), and Eq. (5.140) has been derived from Eqs. (4.152) and (4.191). Then the sth-order expansion of $I(\mathbf{k}; L)$ is given by Eq. (4.233), and that of $\mathcal{I}_{l_1 l_2 l_3}^{00,00}(\mathbf{k}; L)$ may be written, from Eq. (4.226) with Eqs. (4.229) and (4.230), as

$$\mathcal{I}_{l_1 l_2 l_3}^{00,00}(\mathbf{k}; L) = (-1)^{l_3/2} \left[4\pi (2l_1 + 1)(2l_2 + 1)(2l_3 + 1) \right]^{1/2}$$

$$\times \begin{pmatrix} l_1 & l_2 & l_3 \\ 0 & 0 & 0 \end{pmatrix}^2 \sum_{n=0}^{s} \sum_{r=0}^{n} \frac{(-1)^r}{2^r r!} D_{l_1 l_2 l_3, rn}^{00,00} L^n (Lk)^r j_{l_3+r}(Lk), \qquad (5.141)$$

where $D_{\cdots}^{\cdots}$ are given in Appendix III. Note that $D_{l_1 l_2 l_3, mm}^{00,00} = E_{mm}$ with $E_{00} = 1$ and that $D_{000, mn}^{00,00} = E_{mn}$, where E_{mn} are also given in Appendix III. Equation (5.94) also reduces to

$$\mathcal{I}_{l_1 l_2 l_3}(k; s) = \hat{\alpha}_{l_1}^0 \hat{\alpha}_{l_2}^0 \mathcal{I}_{l_1 l_2 l_3}^{00,00}(k; s). \qquad (5.142)$$

Now we consider only the three components F_{Vv}, F_{Hv}, and F_{Hh}. They may be written in the form

$$\mathbf{F} = \mathbf{W}' \cdot \mathbf{Z}, \qquad (5.143)$$

where

$$\mathbf{F}^T = (F_{Vv}, F_{Hv}, F_{Hh}), \qquad (5.144)$$

$\mathbf{W}'$ is the 3×5 matrix given by Eq. (5.102) without the fourth row, and $\mathbf{Z}$ is given by Eq. (5.98). After the integration over s in Eq. (5.100) with Eq. (5.142), the components of $\mathbf{Z}$ are given by

$$Z_{l_1 l_2 l_3}(k; L) = \hat{\alpha}_{l_1}^0 \hat{\alpha}_{l_2}^0 \sum_{n=0}^{s} L^n Z_{l_1 l_2 l_3, n}(Lk), \qquad (5.145)$$

where

$$Z_{l_1 l_2 l_3, n}(x) = (-1)^{l_3/2} \left[(2l_1 + 1)(2l_2 + 1)(2l_3 + 1) \right]^{1/2} \begin{pmatrix} l_1 & l_2 & l_3 \\ 0 & 0 & 0 \end{pmatrix}^2$$

$$\times x^{-(n+2)} \sum_{r=0}^{n} \frac{(-1)^r}{2^{r-1} r!} D_{l_1 l_2 l_3, rn}^{00,00} J_{l_3+r}^{n+r}(x) \qquad (5.146)$$

with

$$J_l^m(x) = \int_0^x (x - v) v^m j_l(v) dv. \qquad (5.147)$$

The function $J_l^m(x)$ may be evaluated analytically but we do not reproduce the results [58] because of their length.

It has been shown that the convergence of Eq. (5.145) is not very good. Its better alternative is obtained by expanding the reciprocal of the sum over n in Eq. (5.145) in powers of L as follows,

$$Z_{l_1 l_2 l_3}(k; L) = \hat{\alpha}_{l_1}^0 \hat{\alpha}_{l_2}^0 L^2 Z_{l_1 l_2 l_3, 0}(Lk) \left[1 + \sum_{n=1}^{s} L^n \bar{Z}_{l_1 l_2 l_3, n}(Lk) \right]^{-1} \quad (5.148)$$

with

$$\begin{aligned}
\bar{Z}_1 &= -Z_0^{-1} Z_1 , \\
\bar{Z}_2 &= -Z_0^{-1} Z_2 + (Z_0^{-1} Z_1)^2 , \\
\bar{Z}_3 &= -Z_0^{-1} Z_3 + 2Z_0^{-2} Z_1 Z_2 - (Z_0^{-1} Z_1)^3 , \\
\bar{Z}_4 &= -Z_0^{-1} Z_4 + Z_0^{-2}(Z_2^2 + 2Z_1 Z_3) - 3Z_0^{-3} Z_1^2 Z_2 + (Z_0^{-1} Z_1)^4 , \\
\bar{Z}_5 &= -Z_0^{-1} Z_5 + 2Z_0^{-2}(Z_1 Z_4 + Z_2 Z_3) - 3Z_0^{-3} Z_1(Z_2^2 + Z_1 Z_3) \\
&\quad + 4Z_0^{-4} Z_1^3 Z_2 - (Z_0^{-1} Z_1)^5 ,
\end{aligned} \quad (5.149)$$

where we have abbreviated $Z_{l_1 l_2 l_3, n}$ and $\bar{Z}_{l_1 l_2 l_3, n}$ to Z_n and $\bar{Z}_n$, respectively.

In particular, the isotropic scattering function $P(k; L)$ is obtained, from Eqs. (5.103) and (5.145), as

$$P(k; L) = \sum_{n=0}^{s} L^n P_n(Lk) , \quad (5.150)$$

and, from Eqs. (5.103) and (5.148), as

$$P(k; L) = P_0(Lk) \left[1 + \sum_{n=1}^{s} L^n \bar{P}_n(Lk) \right]^{-1} \quad (5.151)$$

(with $s = 5$), where $P_n(x)$ in Eq. (5.150) (not to be confused with the Legendre polynomial) are given by

$$\begin{aligned}
P_n(x) &= 2x^{-2} J_0^0(x) && \text{for } n = 0 \\
&= x^{-(n+2)} \sum_{r=1}^{n} \frac{(-1)^r}{2^{r-1} r!} E_{rn} J_r^{n+r}(x) && \text{for } n \geq 1 , \quad (5.152)
\end{aligned}$$

and $\bar{P}_n(x)$ in Eq. (5.151) are given by Eqs. (5.149) with P_n and $\bar{P}_n$ in place of Z_n and $\bar{Z}_n$, respectively. Note that $P_0(Lk)$ in Eqs. (5.150) and (5.151) is just the $P(k; L)$ for the rod. Koyama [18] and Norisuye et al. [21] have evaluated $P(k; L)$ to terms of $\mathcal{O}(L)$ and $\mathcal{O}(L^5)$, respectively, for the KP chain. Equations (5.150) and (5.151) include their results as special cases. [The term -256 of P_3 in Eqs. (26) of [21] should be replaced by $-256x$.]

The numerical results [58] show that the difference between the values of R_{Vv} for the HW chain and the R3 rod is very small for small L and k, and is also smaller (for R_{Vv}) than for R_{Uv}. This suggests that in order to determine $\langle S^2 \rangle$ experimentally, we should measure the Vv component rather than the Uv so that we may use approximately the equation for the R3 rod [58].

5.4 Electrical Properties

5.4.1 Mean-Square Electric Dipole Moment

The *mean-square electric dipole moment* $\langle \mu^2 \rangle$ is one of the electrical properties closely related to the equilibrium conformational behavior of polymer chains, in particular, to the mean-square end-to-end distance $\langle R^2 \rangle$. We evaluate it by affixing local permanent electric dipole moment vectors to the HW chain [59]. Let $\mathbf{m}(s)$ and $\tilde{\mathbf{m}}(s)$ be those vectors per unit length at the contour point s $(0 \leq s \leq L)$, expressed in the localized and external Cartesian coordinate systems, respectively. We assume that $\mathbf{m}(s)$ is also independent of s. All lengths are measured in units of λ^{-1} unless otherwise noted.

Now the instantaneous dipole moment $\boldsymbol{\mu}$ of the entire chain in the external system is given by

$$\boldsymbol{\mu} = \int_0^L \tilde{\mathbf{m}}(s) ds \,, \tag{5.153}$$

so that $\langle \mu^2 \rangle$ is given by

$$\langle \mu^2 \rangle = 2 \int_0^L (L - s)\langle \tilde{\mathbf{m}}(s) \cdot \tilde{\mathbf{m}}(0) \rangle ds \,, \tag{5.154}$$

where the average in the integrand may be evaluated with the Green function $G(\Omega \,|\, \Omega_0; s)$ for the chain of contour length s as

$$\langle \tilde{\mathbf{m}}(s) \cdot \tilde{\mathbf{m}}(0) \rangle = (8\pi^2)^{-1} \int \tilde{\mathbf{m}}(s) \cdot \tilde{\mathbf{m}}(0) G(\Omega \,|\, \Omega_0; s) d\Omega d\Omega_0 \,. \tag{5.155}$$

As in the case of the polarizability tensor, it is convenient to introduce the spherical components $\tilde{m}^{(j)}$ $(m^{(j)})$ $(j = 0, \pm 1)$ of the vector $\tilde{\mathbf{m}}$ $(\mathbf{m})$ [60], which are defined in terms of the Cartesian components $\tilde{m}_\mu$ $(m_{\mu'})$ $(\mu = x,$ $y, z;$ $\mu' = \xi, \eta, \zeta)$ as in Eqs. (5.B.1) of Appendix 5.B and which satisfy the relation given by Eq. (5.B.3), since $\tilde{m}^{(j)}$ may be transformed to $m^{(j)}$ by Eq. (5.B.5). With $\tilde{m}^{(j)}$, the scalar product in the integrand of Eq. (5.155) may then be written in the form

$$\tilde{\mathbf{m}}(s) \cdot \tilde{\mathbf{m}}(0) = \sum_{j=-1}^{1} \tilde{m}^{(j)*}(s)\, \tilde{m}^{(j)}(0) \,. \tag{5.156}$$

Thus we obtain, from Eqs. (5.154) and (5.155) with Eqs. (4.36), (5.156), and (5.B.5),

$$\langle \mu^2 \rangle = 2 \big[\bar{m}^{00} \bar{X}_1^{00}(L) - 2\bar{m}^{11} \bar{X}_1^{(-1)1}(L)$$
$$- 2\bar{m}^{(-1)1} \bar{X}_1^{11}(L) - 4\bar{\bar{m}}^{01} \bar{\bar{X}}_1^{01}(L) \big] , \tag{5.157}$$

where $\bar{m}^{j_1 j_2}$ and $\bar{\bar{m}}^{j_1 j_2}$ are the real and imaginary parts, respectively, of the quantity $m^{j_1 j_2}$ defined by

$$m^{j_1 j_2} = m^{(j_1)} m^{(j_2)} , \tag{5.158}$$

$\bar{X}_1^{j_1 j_2}$ and $\bar{\bar{X}}_1^{j_1 j_2}$ are those of the function $X_1^{j_1 j_2}(L)$ defined by Eq. (5.121), and we have used Eqs. (4.128) for the symmetry relations for $g_l^{jj'}$.

By the use of Eq. (4.108) for $g_l^{jj'}$ (for the chain with Poisson's ratio $\sigma = 0$), Eq. (5.157) may be reduced to

$$\langle \mu^2 \rangle = 2m^2 \nu^{-2} \big[\hat{\tau}_0{}^2 \bar{v}_{10}(L) + \hat{\kappa}_0{}^2 \bar{v}_{11}(L) \big] , \tag{5.159}$$

where ν is given by Eq. (5.4), $\hat{\kappa}_0$ and $\hat{\tau}_0$ are defined by

$$\hat{\kappa}_0 = (\nu^2 - \hat{\tau}_0{}^2)^{1/2} , \tag{5.160}$$

$$\hat{\tau}_0 = m^{-1}(\kappa_0 m_\eta + \tau_0 m_\zeta) \tag{5.161}$$

with $m = |\mathbf{m}|$, so that $\nu = (\hat{\kappa}_0{}^2 + \hat{\tau}_0{}^2)^{1/2}$, and $\bar{v}_{lk}$ is the real part of the function $v_{lk}(L)$ defined by

$$v_{lk}(L) = z^{-2}(zL - 1 + e^{-zL}) \tag{5.162}$$

with

$$z = l(l+1) + ik\nu , \tag{5.163}$$

that is,

$$\bar{v}_{1j}(L) = \frac{1}{(4 + j^2\nu^2)^2} \big\{ 2(4 + j^2\nu^2)L - 4 + j^2\nu^2$$
$$+ e^{-2L} \big[(4 - j^2\nu^2)\cos(j\nu L) - 4j\nu \sin(j\nu L) \big] \big\} . \tag{5.164}$$

Equation (4.82) for $\langle R^2 \rangle$ may then be rewritten in the form

$$\langle R^2 \rangle = 2\nu^{-2} \big[\tau_0{}^2 \bar{v}_{10}(L) + \kappa_0{}^2 \bar{v}_{11}(L) \big] . \tag{5.165}$$

Comparing Eq. (5.159) with Eq. (5.165), we find

$$\langle \mu^2 \rangle = m^2 \langle \hat{R}^2 \rangle = m^2 f_R(L; \hat{\kappa}_0, \hat{\tau}_0) , \tag{5.166}$$

where $\langle \hat{R}^2 \rangle$ is the mean-square end-to-end distance of the HW chain of contour length L such that the curvature and torsion of its characteristic helix are equal to $\hat{\kappa}_0$ and $\hat{\tau}_0$, respectively. Note that for the chain having type-A dipoles [61] along its contour ($m_\xi = m_\eta = 0$), we have $\hat{\kappa}_0 = \kappa_0$ and $\hat{\tau}_0 = \tau_0$, and therefore $\langle \hat{R}^2 \rangle = \langle R^2 \rangle$. For the KP2 chain and R2 rod, Eq. (5.166) may be further reduced to

$$\langle\mu^2\rangle = m^2\langle R^2\rangle = m^2\left(L - \frac{1}{2} + \frac{1}{2}e^{-2L}\right) \quad \text{(KP2)}, \qquad (5.167)$$

$$\langle\mu^2\rangle = m^2 L^2 \qquad \text{(R2)}. \qquad (5.168)$$

It is important to note that Eqs. (5.167) and (5.168) are valid even for the chain whose dipoles are not of type A.

For a comparison of theory with experiment, it is convenient to write $\langle\mu^2\rangle$ (unreduced), from Eq. (5.166), as

$$\frac{\langle\mu^2\rangle}{x} = (\lambda^{-1}m)^2\left(\frac{\lambda^{-1}M_{\mathrm{L}}}{M_0}\right)^{-1}\left[\frac{f_\mu(\lambda L; \lambda^{-1}\kappa_0, \lambda^{-1}\tau_0, \lambda^{-1}\mathbf{m})}{\lambda L}\right], \quad (5.169)$$

where x is the number of repeat units, M_0 is its molecular weight, and f_μ is given by

$$f_\mu(\lambda L; \lambda^{-1}\kappa_0, \lambda^{-1}\tau_0, \lambda^{-1}\mathbf{m}) = f_R(\lambda L; \lambda^{-1}\hat{\kappa}_0, \lambda^{-1}\hat{\tau}_0) \qquad (5.170)$$

with f_R being given by Eq. (4.82),

$$f_R(L; \kappa_0, \tau_0) = c_\infty L - \frac{\tau_0^2}{2\nu^2} - \frac{2\kappa_0^2(4 - \nu^2)}{\nu^2 r^4} + \frac{1}{\nu^2}e^{-2L}\left\{\frac{1}{2}\tau_0^2\right.$$
$$\left. + \frac{2\kappa_0^2}{r^4}\left[(4 - \nu^2)\cos(\nu L) - 4\nu\sin(\nu L)\right]\right\}. \qquad (5.171)$$

Now we make a comparison of theory with experiment, taking as examples PDMS, PBIC, and poly(n-hexyl isocyanate) (PHIC). The PDMS chain has type-B dipoles [61] perpendicular to its contour and this local dipole moment vector $\mathbf{m}$ may be attached unambiguously in the localized coordinate system, which is affixed to the monomer unit containing the Si–O and O–Si bonds, as mentioned in Sect. 4.4.3 and depicted in Fig. 5.18. That is, the ζ axis is taken along a line passing through the two successive Si atoms, the ξ axis is in the plane of the Si–O and O–Si bonds with its positive direction chosen at an acute angle with the Si→O bond ($\hat{\psi} = 0$), and the η axis completes the right-handed system. Then the vector $\mathbf{m}$ is in the negative direction of the ξ axis, so that $m_\xi = -m$ and $m_\eta = m_\zeta = 0$. On the other hand, the PBIC and PHIC chains, which are typical semiflexible chains, may be treated as the KP chain having type-A dipoles.

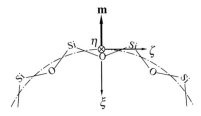

Fig. 5.18. Localized Cartesian coordinate system and the local electric dipole moment vector $\mathbf{m}$ for PDMS (see the text)

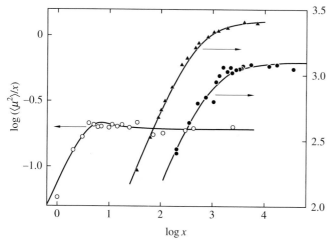

Fig. 5.19. Double-logarithmic plots of $\langle\mu^2\rangle/x$ (in D^2) against x for PDMS in cyclohexane at $25.0°C$ (○) [62], PBIC in CCl_4 at $22.9°C$ (●) [63], and PHIC in toluene at $25.0°C$ (▲) [64]. The *solid curves* represent the best-fit HW (or KP) theoretical values calculated with the values of the model parameters given in Table 5.4

Figure 5.19 shows double-logarithmic plots of $\langle\mu^2\rangle/x$ (in D^2) against x for PDMS in cyclohexane at $25.0°C$ [62], PBIC in carbon tetrachloride (CCl_4) at $22.9°C$ [63], and PHIC in toluene at $25.0°C$ [64]. The excluded-volume effect on $\langle\mu^2\rangle$ of the chain having type-B dipoles may be regarded as negligibly small if any. Thus theoretical values of $\langle\mu^2\rangle$ for all these three polymers may be calculated from Eq. (5.169) with values of $\lambda^{-1}\kappa_0$, $\lambda^{-1}\tau_0$, $\lambda^{-1}M_L$, and $\lambda^{-1}m = m_0(\lambda^{-1}M_L/M_0)$, where m_0 is the permanent electric dipole moment of the repeat unit. Note therefore that λ^{-1} and M_L cannot be separately determined from $\langle\mu^2\rangle$, although it is possible for m_0. The solid curves in Fig. 5.19 represent the best-fit HW (or KP) theoretical values thus calculated with the values of the model parameters given in Table 5.4, where we have put $\lambda^{-1}\tau_0 = 0$ for PDMS, corresponding to Table 4.3, and assumed the values of M_L as noted in order to determine λ^{-1}, for convenience. From the

Table 5.4. Values of the HW model parameters from electrical properties

Polymer	Solvent	Temp. (°C)	$\lambda^{-1}\kappa_0$	$\lambda^{-1}\tau_0$	λ^{-1} (Å)	M_L (Å^{-1})	m_0 (D)	Obs. (Ref.)
PDMS	Cyclohexane	25.0	2.6	0	18.0	$(25.0)^a$	0.29	$\langle\mu^2\rangle$ ([62])
PBIC	CCl_4	22.9	0	⋯	1440	$(55.1)^b$	1.25	$\langle\mu^2\rangle$ ([63])
	CCl_4	room temp.	0	⋯	1440	$(55.1)^b$	⋯	A_{ED} ([9])
PHIC	Toluene	25.0	0	⋯	740	$(74.0)^b$	2.46	$\langle\mu^2\rangle$ ([64])

[a] From RIS values of C_n (see Table 4.1)
[b] From $\langle S^2\rangle$

obtained values of κ_0 and τ_0 (reduced) in the (κ_0, τ_0)-plane of Fig. 4.13, the PDMS chain is seen to be a typical HW chain. Indeed, both experimental and theoretical values of $\langle \mu^2 \rangle / x$ exhibit a maximum, and also the temperature coefficient of $\langle R^2 \rangle$ is positive, as mentioned in Sect. 4.8.2. As for the semi-flexible polymer PBIC, the value of λ^{-1} from $\langle \mu^2 \rangle$ is rather consistent with that from $\langle S^2 \rangle$ in Table 5.1, although it may depend somewhat on solvent.

5.4.2 Electric Birefringence

A second electrical property we consider is *electric birefringence* [59]. Let $\boldsymbol{\alpha}(s)$ and $\tilde{\boldsymbol{\alpha}}(s)$ be the local optical polarizability tensors per unit contour length of the chain expressed in the localized and external Cartesian coordinate systems, respectively, as before, and let $\boldsymbol{\alpha}'(s)$ and $\tilde{\boldsymbol{\alpha}}'(s)$ be the corresponding polarizability tensors. An equation similar to Eq. (5.115) for the optical polarizability tensor $\boldsymbol{\gamma}$ of the entire chain holds for the corresponding static polarizability tensor $\boldsymbol{\gamma}'$ in the external system $(\mathbf{e}_x, \mathbf{e}_y, \mathbf{e}_z)$.

Then the molecular distribution function P under the influence of an applied static electric field $\mathbf{E} = E\mathbf{e}_z$ is given by

$$P = C(E) \exp\left[-(U - \mu_z E - \frac{1}{2}\gamma'_{zz} E^2)/k_B T\right], \qquad (5.172)$$

where U is the intramolecular potential energy and μ_z is the z component of the electric dipole moment $\boldsymbol{\mu}$. The principal axes of the $\boldsymbol{\gamma}$ averaged with P, which we denote by $\langle \boldsymbol{\gamma} \rangle_E$, are in the directions of $\mathbf{e}_x$, $\mathbf{e}_y$, and $\mathbf{e}_z$, its x and y principal values being equal to each other. The required quantity is the difference $\Delta\Gamma$ between the z and x principal values. It is given by [65,66]

$$\Delta\Gamma = \langle \gamma_{zz} - \gamma_{xx} \rangle_E$$
$$= \frac{E^2}{2k_B T} \left(\frac{\Delta^{(D)}}{k_B T} + \Delta^{(P)} \right) + \mathcal{O}(E^3) \qquad (5.173)$$

with

$$\Delta^{(D)} = \left\langle (\gamma_{zz} - \gamma_{xx})\mu_z^2 \right\rangle, \qquad (5.174)$$

$$\Delta^{(P)} = \left\langle (\gamma_{zz} - \gamma_{xx})\gamma'_{zz} \right\rangle, \qquad (5.175)$$

where $\langle\ \rangle$ denotes an equilibrium average (at $E = 0$) as before. Nagai and Ishikawa [65] and Flory [66] have further reduced Eqs. (5.174) and (5.175) to

$$\Delta^{(D)} = \frac{1}{15}\left[3\langle \boldsymbol{\mu} \cdot \boldsymbol{\gamma} \cdot \boldsymbol{\mu} \rangle - \langle \mu^2 \mathrm{Tr}\,\boldsymbol{\gamma} \rangle\right], \qquad (5.176)$$

$$\Delta^{(P)} = \frac{1}{15}\left[3\langle \mathrm{Tr}\,(\boldsymbol{\gamma} \cdot \boldsymbol{\gamma}') \rangle - \langle (\mathrm{Tr}\,\boldsymbol{\gamma})(\mathrm{Tr}\,\boldsymbol{\gamma}') \rangle\right]. \qquad (5.177)$$

For the HW chain, however, it is more efficient to start from Eqs. (5.174) and (5.175) with the use of the spherical tensors.

From Eqs. (5.115), (5.153), (5.174), and (5.175), $\Delta^{(D)}$ and $\Delta^{(P)}$ may be written as

$$\Delta^{(D)} = 2 \int_0^L ds(L-s) \int_0^s dt \langle \Delta\tilde{\alpha}(0)\tilde{m}_z(t)\tilde{m}_z(s) \rangle$$
$$+ \tilde{m}_z(0)\Delta\tilde{\alpha}(t)\tilde{m}_z(s) + \tilde{m}_z(0)\tilde{m}_z(t)\Delta\tilde{\alpha}(s) \rangle, \qquad (5.178)$$

$$\Delta^{(P)} = \int_0^L (L-s) \left[\langle \Delta\tilde{\alpha}(s)\tilde{\alpha}'_{zz}(0) \rangle + \langle \Delta\tilde{\alpha}(0)\tilde{\alpha}'_{zz}(s) \rangle \right] ds \qquad (5.179)$$

with

$$\Delta\tilde{\alpha}(s) = \tilde{\alpha}_{zz}(s) - \tilde{\alpha}_{xx}(s). \qquad (5.180)$$

$\Delta^{(P)}$ is closely related to the mean-square optical anisotropy $\langle \Gamma^2 \rangle$; indeed, when $\alpha = \alpha'$, we have

$$\Delta^{(P)} = \frac{2}{15} \langle \Gamma^2 \rangle. \qquad (5.181)$$

We omit the details of the evaluation of the integrals in Eqs. (5.178) and (5.179), which is straightforward, and give only some of the final results [59], for simplicity. In the particular case of the HW chain having type-A dipoles and polarizability tensors $\boldsymbol{\alpha}$ and $\boldsymbol{\alpha}'$ cylindrically symmetric about $\mathbf{e}_\zeta = \mathbf{u}$ (HWA′), we have

$$\Delta^{(D)} = \frac{(\Delta\alpha)m^2}{15\nu^4(16+\nu^2)} \{ \nu^2 \tau_0^2 (32 - \kappa_0^2 + 2\tau_0^2)\bar{v}_{10}$$
$$+ \nu^2 \kappa_0^2 (32 + 3\tau_0^2)\bar{v}_{11}$$
$$+ \frac{\kappa_0^2}{\nu}(\nu^4 - 9\nu^2\tau_0^2 - 48\nu^2 - 144\tau_0^2)\bar{\bar{v}}_{11}$$
$$- (\kappa_0^2 - 2\tau_0^2)(\nu^2\tau_0^2 - 8\kappa_0^2 + 16\tau_0^2)\bar{v}_{20}$$
$$- 3\kappa_0^2\tau_0^2 \left[(32+\nu^2)\bar{v}_{21} + 4\nu\bar{\bar{v}}_{21} \right]$$
$$- 6\kappa_0^4(4\bar{v}_{22} + \nu\bar{\bar{v}}_{22})$$
$$- (16+\nu^2)(\kappa_0^2 - 2\tau_0^2)(2\tau_0^2\bar{v}_{10}^1 - \kappa_0^2\bar{v}_{11}^1) \} \quad (\text{HWA}'), \quad (5.182)$$

$$\Delta^{(P)} = \frac{(\Delta\alpha)(\Delta\alpha')}{15\nu^4} \left[(\kappa_0^2 - 2\tau_0^2)^2\bar{v}_{20} \right.$$
$$\left. + 12\kappa_0^2\tau_0^2\bar{v}_{21} + 3\kappa_0^4\bar{v}_{22} \right] \qquad (\text{HWA}'), \quad (5.183)$$

where $\Delta\alpha = \alpha_{\zeta\zeta} - \alpha_{\xi\xi}$ ($\alpha_{\xi\xi} = \alpha_{\eta\eta}$) and $\Delta\alpha' = \alpha'_{\zeta\zeta} - \alpha'_{\xi\xi}$ ($\alpha'_{\xi\xi} = \alpha'_{\eta\eta}$); and $\bar{v}_{lk}$ and $\bar{\bar{v}}_{lk}$ are the real and imaginary parts of the function $v_{lk}(L)$ defined by Eq. (5.162), and $\bar{v}_{lk}^1$ and $\bar{\bar{v}}_{lk}^1$ are those of the function $v_{lk}^1(L)$ defined by

$$v_{lk}^1(L) = z^{-3} \left[zL - 2 + (zL + 2)e^{-zL} \right] \qquad (5.184)$$

with z being given by Eq. (5.163).

For the KP2 chain, we have for $\Delta^{(D)}$

$$\Delta^{(D)} = \Delta_L^{(D)} \left(\frac{6}{5}L - \frac{13}{18} + \frac{3}{4}e^{-2L} + \frac{1}{2}Le^{-2L} - \frac{1}{36}e^{-6L} \right) \quad \text{(KP2)},$$

(5.185)

where $\Delta_L^{(D)}$ is the local $\Delta^{(D)}$ per unit (reduced) length and is in general given by

$$
\begin{aligned}
\Delta_L^{(D)} &= \lim_{L \to 0} \left(\frac{\Delta^{(D)}}{L^3} \right) \\
&= \frac{1}{15} \{ \sqrt{6}\alpha_2^0 m^{(0)} m^{(0)} + \sqrt{6}\alpha_2^0 m^{(1)} m^{(-1)} \\
&\quad - 6\sqrt{2}\mathrm{Re}[\alpha_2^1 m^{(0)} m^{(-1)}] + 6\mathrm{Re}[\alpha_2^2 m^{(-1)} m^{(-1)}] \} \\
&= \frac{1}{15} [3(\mathbf{m} \cdot \boldsymbol{\alpha} \cdot \mathbf{m}) - m^2 \mathrm{Tr}\,\boldsymbol{\alpha}] .
\end{aligned}
$$

(5.186)

We note that the third equality of Eqs. (5.186) has been obtained from Eq. (5.176) and that in the case of type-A dipoles the second equality of Eqs. (5.186) reduces to

$$\Delta_L^{(D)} = \frac{\sqrt{6}}{15}\alpha_2^0 m^2 \quad \text{(A)} .$$

(5.187)

For the KP3 chain, we have for $\Delta^{(D)}$

$$
\Delta^{(D)} = \frac{2\Delta\alpha}{15} \left\{ m_\zeta^2 (\bar{v}_{10} - \bar{v}_{20} + 2\bar{v}_{10}^1) - (m_\xi^2 + m_\eta^2) \right. \\
\left. \times \left[\frac{2(4\bar{v}_{11} - 4\bar{v}_{20} - |\tau_0|\bar{\bar{v}}_{11})}{16 + \tau_0^2} + \bar{v}_{11}^1 \right] \right\} \quad \text{(KP3)},
$$

(5.188)

where $\Delta\alpha$ is the same as that in Eq. (5.182). Recall that the optical anisotropy for the KP3 chain is independent of τ_0.

For the KPj chain ($j = 1, 2, 3$) having type-A dipoles (KPA), $\Delta^{(D)}$ is given by Eq. (5.185) with Eq. (5.187) irrespective of the values of τ_0 and α.

For the KP2 and KP3 chains, we have for $\Delta^{(P)}$

$$\Delta^{(P)} = \frac{1}{3}\Delta_L^{(P)} \left(L - \frac{1}{6} + \frac{1}{6}e^{-6L} \right) \quad \text{(KP2, KP3)},$$

(5.189)

where $\Delta_L^{(P)}$ is the local $\Delta^{(P)}$ per unit (reduced) length and is in general given by

$$
\begin{aligned}
\Delta_L^{(P)} &= \lim_{L \to 0} \left(\frac{\Delta^{(P)}}{L^2} \right) \\
&= \frac{1}{5} [\alpha_2^0 \alpha_2'^0 + 2\mathrm{Re}(\alpha_2^{1*}\alpha_2'^1) + 2\mathrm{Re}(\alpha_2^{2*}\alpha_2'^2)] \\
&= \frac{1}{15} [3\mathrm{Tr}\,(\boldsymbol{\alpha} \cdot \boldsymbol{\alpha}') - (\mathrm{Tr}\,\boldsymbol{\alpha})(\mathrm{Tr}\,\boldsymbol{\alpha}')] .
\end{aligned}
$$

(5.190)

We note that for the KP2 and KP3 chains $\Delta^{(P)}$ and $\langle \Gamma^2 \rangle$ have the same dependence on L, as seen from Eqs. (5.125) and (5.189), and that the third line of Eqs. (5.190) has been obtained from Eq. (5.177).

For the KPA chain, $\Delta^{(P)}$ is given by Eq. (5.189).

We note that for all types of rods having type-A dipoles (RA) we have $\Delta^{(D)} = \Delta_L^{(D)} L^3$ and $\Delta^{(P)} = \Delta_L^{(P)} L^2$, which agree with the results derived by Benoit [67] for the R3A rod, and also that in the coil limit $\Delta^{(D)}$ and $\Delta^{(P)}$ are proportional to L, being consistent with the results obtained by Peterlin and Stuart [68] for the freely jointed chain and by Stockmayer and Baur [69] for the spring-bead model.

Finally, we make brief mention of the *Kerr constant K* experimentally determined. It is defined by

$$K = \lim_{\substack{c \to 0 \\ E \to 0}} \left(\frac{\Delta \tilde{n}}{\tilde{n}_0 c E^2} \right),$$ (5.191)

where $\Delta \tilde{n}$ is the difference between the refractive indices of the solution of concentration c in the z and x directions and is given by

$$\Delta \tilde{n} = \frac{2 \pi N_A c \Delta \Gamma}{\tilde{n}_0 M}.$$ (5.192)

We then have, from Eqs. (5.173), (5.191), and (5.192),

$$K = \frac{Q}{M} \left(\frac{\Delta^{(D)}}{k_B T} + \Delta^{(P)} \right)$$ (5.193)

with

$$Q = \frac{\pi N_A}{\tilde{n}_0^2 k_B T}.$$ (5.194)

The right-hand side of Eq. (5.194) must be multiplied by a proper factor if effects of the internal field are taken into account.

5.4.3 Electric Dichroism

The theory of electric birefringence in the last subsection may be translated into the theory of *electric linear dichroism* by regarding the local optical polarizability tensor $\boldsymbol{\alpha}(s)$ as the local dichroic tensor [70]. Let $\boldsymbol{\mu}_{0j}(s)$ be the local electric dipole transition moment per unit contour length of the chain for the electronic transition $0 \to j$ between the ground and jth excited states. The local dichroic tensor $\boldsymbol{\alpha}(s)$ is defined by

$$\boldsymbol{\alpha}(s) = \boldsymbol{\mu}_{0j}(s) \boldsymbol{\mu}_{0j}(s),$$ (5.195)

so that γ and α_l^m in the last subsection are also to be reinterpreted according to Eq. (5.195). Then the molecular extinction coefficient ϵ_ν for the plane-polarized light with polarization $\mathbf{e}_\nu$ ($\nu = x, y, z$) in the applied static electric field $\mathbf{E}$ is given by

$$\epsilon_\nu = f\langle\gamma_{\nu\nu}\rangle_E\,, \tag{5.196}$$

where f is a proportionality constant and the average is taken with the P given by Eq. (5.172). From Eq. (5.196) with the first line of Eqs. (5.173), we have for the electric dichroism $\Delta\epsilon$

$$\Delta\epsilon = \epsilon_z - \epsilon_x = f\Delta\Gamma\,. \tag{5.197}$$

Since the molecular extinction coefficient $\bar{\epsilon}$ for $E = 0$ is given by

$$\bar{\epsilon} = \frac{1}{3}f\mathrm{Tr}\,\langle\gamma\rangle = f\bar{\alpha}L\,, \tag{5.198}$$

we obtain, from Eqs. (5.197) and (5.198) with the second line of Eqs. (5.173),

$$\frac{\Delta\epsilon}{\bar{\epsilon}} = A_{\mathrm{ED}}E^2 + \mathcal{O}(E^3) \tag{5.199}$$

with

$$A_{\mathrm{ED}} = \frac{1}{2k_{\mathrm{B}}T\bar{\alpha}L}\left(\frac{\Delta^{(\mathrm{D})}}{k_{\mathrm{B}}T} + \Delta^{(\mathrm{P})}\right), \tag{5.200}$$

where $\Delta^{(\mathrm{D})}$ and $\Delta^{(\mathrm{P})}$ are given by the equations in the last subsection with the above reinterpretation. Note that the right-hand side of Eq. (5.200) must be multiplied by a proper factor if effects of the internal field are taken into account.

We make a comparison of theory with experiment for A_{ED} in the case of the KPA chain for which $\Delta^{(\mathrm{P})}$ may be neglected. Equation (5.200) may then be rewritten in the form

$$(k_{\mathrm{B}}T)^2 A_{\mathrm{ED}} = \frac{1}{18}(\lambda^{-1}m)^2\epsilon_1 f_{\mathrm{ED}}(\lambda L)\,, \tag{5.201}$$

where ϵ_1 is given by the first of Eqs. (5.139) and f_{ED} is given, from Eqs. (5.185) and (5.200), by

$$f_{\mathrm{ED}}(L) = 1 - \frac{13}{15L} + \frac{9}{10L}e^{-2L} + \frac{3}{5}e^{-2L} - \frac{1}{30}e^{-6L} \quad \text{(KP2)}\,. \tag{5.202}$$

Figure 5.20 shows double-logarithmic plots of $(k_{\mathrm{B}}T)^2 A_{\mathrm{ED}}$ (in D^2) against x for PBIC in CCl_4 at room temperature, the data being due to Troxell and Scheraga [9]. The solid curve represents the best-fit KPA theoretical values calculated from Eq. (5.201) with Eq. (5.202) with the values of the model parameters given in Table 5.4 along with $m_0 = 1.25$ D and $\epsilon_1 = 2.60$. It is interesting to see that the values of λ^{-1} obtained from A_{ED} and $\langle\mu^2\rangle$ agree with each other.

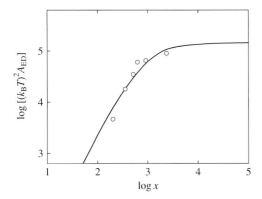

Fig. 5.20. **Fig. 5.20.** Double-logarithmic plots of $(k_BT)^2 A_{ED}$ (in D^2) against x for PBIC in CCl$_4$ at room temperature [9]. The *solid curve* represents the best-fit KPA theoretical values calculated with the values of the model parameters given in Table 5.4 along with $m_0 = 1.25$ D and $\epsilon_1 = 2.60$

Appendix 5.A Chain-Thickness Correction for the Apparent Mean-Square Radius of Gyration

The reciprocal of the excess reduced scattered intensity R_θ for dilute solutions of mass concentration c may be expanded in the form [31, 71]

$$\frac{Kc}{R_\theta} = \frac{1}{MP_s(k)} + 2A_2 Q(k)c + \cdots ,\qquad (5.A.1)$$

where K is the optical constant, M is the polymer molecular weight, $P_s(k)$ is the scattering function as a function of the magnitude k of the scattering vector $\mathbf{k}$ given by Eq. (5.20), A_2 is the second virial coefficient, and $Q(k)$ represents the intermolecular interference. The function P_s contains effects of the spatial distribution of scatterers (electrons or hydrogen nuclei), that is, effects of chain thickness in the case of small-angle X-ray or neutron scattering. In general, it may be written in the form [1]

$$P_s(k; L) = \left\langle \left| \int \rho(\mathbf{r}) \exp(i\mathbf{k} \cdot \mathbf{r}) d\mathbf{r} \right|^2 \right\rangle ,\qquad (5.A.2)$$

where we have explicitly indicated that $P_s(k)$ also depends on the contour length L of the chain, $\langle \ \rangle$ denotes an equilibrium average over chain conformations, i is the imaginary unit, and $\rho(\mathbf{r})$ is the excess scatterer density at vector position $\mathbf{r}$ and is normalized as

$$\int \rho(\mathbf{r}) d\mathbf{r} = 1 .\qquad (5.A.3)$$

In the case for which the scatterers are distributed on the chain contour, $\rho(\mathbf{r})$ is given by

$$\rho(\mathbf{r}) = L^{-1} \int_0^L \delta\left[\mathbf{r} - \mathbf{r}(s)\right] ds ,\qquad (5.A.4)$$

where $\delta(\mathbf{r})$ is a three-dimensional Dirac delta function and $\mathbf{r}(s)$ is the radius vector of the contour point s ($0 \leq s \leq L$) of the chain. Then Eq. (5.A.2) with Eq. (5.A.4) gives the contour scattering function $P(k; L)$ (without effects of chain thickness).

In the case of a cylinder model for which the scatterers are uniformly distributed within a (flexible) cylinder having a uniform cross section of area a_c whose center of mass is on the chain contour, $\rho(\mathbf{r})$ is given by

$$\rho(\mathbf{r}) = (La_c)^{-1} \int_0^L ds \int_{C_S} \delta[\mathbf{r} - \mathbf{r}(s) - \bar{\mathbf{r}}_s]d\bar{\mathbf{r}}_s, \qquad (5.A.5)$$

where $\bar{\mathbf{r}}_s$ is the vector distance from the contour point s to an arbitrary point in the normal cross section at that point and the second integration is carried out over the cross section.

In the case of a touched-subbody model for which the scatterers are uniformly distributed in each of N identical touched subbodies of volume v_s whose centers of mass are on the chain contour, $\rho(\mathbf{r})$ is given by

$$\rho(\mathbf{r}) = (Nv_s)^{-1} \sum_{j=1}^N \int_{V_j} \delta(\mathbf{r} - \mathbf{r}_j - \bar{\mathbf{r}}_j)d\bar{\mathbf{r}}_j, \qquad (5.A.6)$$

where $\mathbf{r}_j$ is the vector position of the center of mass of the jth subbody, $\bar{\mathbf{r}}_j$ is the vector distance from $\mathbf{r}_j$ to an arbitrary point within the jth subbody, and the integration is carried out within it.

The scattering function P_s may be expanded in the form

$$P_s(k; L) = 1 - \frac{1}{3}\langle S^2 \rangle_s k^2 + \mathcal{O}(k^4). \qquad (5.A.7)$$

This is the defining equation for the *apparent mean-square radius of gyration* $\langle S^2 \rangle_s$ for the chain. It is related to the mean-square radius of gyration $\langle S^2 \rangle$ for the chain contour by the equation

$$\langle S^2 \rangle_s = \langle S^2 \rangle + S_c^2, \qquad (5.A.8)$$

where S_c is the radius of gyration for the cross section (cylinder model) or subbody (touched-subbody model) and is given by

$$S_c^2 = \frac{1}{8}d^2 \qquad \text{(cylinder)}, \qquad (5.A.9)$$

$$S_c^2 = \frac{3}{20}d_b^2 \qquad \text{(bead)} \qquad (5.A.10)$$

for the cylinder of diameter d and the sphere (bead) of diameter d_b, respectively. For the cylinder model, d has been calculated to be 9.2, 8.2, and 8.1 Å for a-PS, a-PMMA, and i-PMMA, respectively, from the partial specific volume [1, 3, 4]. For a-PS, however, the value 9.4 Å of d has been adopted in Eq. (5.A.8) [1].

Appendix 5.B Spherical Vectors and Tensors

The spherical (irreducible) components $r^{(j)}$ ($j = 0, \pm 1$) of a vector $\mathbf{r} = (x, y, z)$ are defined in terms of the Cartesian components x, y, and z by [60]

$$r^{(\pm 1)} = \mp \tfrac{1}{\sqrt{2}}(x \pm iy),$$
$$r^{(0)} = z.$$

(5.B.1)

The spherical components T_l^m ($l = 0, 1, 2$; $m = 0, \pm 1, \pm 2$) of a tensor $\mathbf{T} = (T_{\mu\nu})$ ($\mu, \nu = x, y, z$) are defined in terms of the Cartesian components $T_{\mu\nu}$ by

$$T_0^0 = \tfrac{1}{\sqrt{3}}(T_{xx} + T_{yy} + T_{zz}),$$
$$T_1^0 = \tfrac{1}{2}(T_{xy} - T_{yx}),$$
$$T_1^{\pm 1} = \mp \tfrac{1}{2\sqrt{2}}\left[(T_{yz} - T_{zy}) \pm i(T_{zx} - T_{xz})\right],$$
$$T_2^0 = \tfrac{1}{\sqrt{6}}\left[3T_{zz} - (T_{xx} + T_{yy} + T_{zz})\right],$$
$$T_2^{\pm 1} = \mp \tfrac{1}{2}\left[(T_{zx} + T_{xz}) \pm i(T_{zy} + T_{yz})\right],$$
$$T_2^{\pm 2} = \tfrac{1}{2}\left[(T_{xx} - T_{yy}) \pm i(T_{xy} + T_{yx})\right].$$

(5.B.2)

We note that the third and fifth of Eqs. (7.4.1) of [40] and all related equations are incorrect.

We have the symmetry relations

$$r^{(-j)} = (-1)^j r^{(j)*},$$

(5.B.3)

$$T_l^{-m} = (-1)^m T_l^{m*},$$

(5.B.4)

where the asterisk indicates the complex conjugate.

We have the same transformation rule as Eq. (4.C.14), that is,

$$\tilde{r}^{(j)} = c_1^{-1} \sum_{j'=-1}^{1} \mathcal{D}_1^{jj'}(\Omega) r^{(j')},$$

(5.B.5)

$$\tilde{T}_l^m = c_l^{-1} \sum_{j=-l}^{l} \mathcal{D}_l^{mj}(\Omega) T_l^j,$$

(5.B.6)

where the components $\tilde{r}^{(j)}$ and $\tilde{T}_l^m$ are transformed to the components $r^{(j)}$ and T_l^m, respectively, expressed in a new Cartesian coordinate system obtained by rotation Ω of a coordinate system in which the former components are defined, $\mathcal{D}_l^{mj}$ are the normalized Wigner $\mathcal{D}$ functions, and c_l is given by Eq. (4.54).

Appendix 5.C Proof of Nagai's Theorem

We introduce temporarily quantities $\beta^{(\pm)}$ defined by

$$\beta^{(\pm)} = \alpha_{\text{Vh}} \pm \alpha_{\text{Hv}} \,. \tag{5.C.1}$$

We have $\langle \tilde{\alpha}_{l_1}^{m_1}, \tilde{\alpha}_{l_2}^{m_2} \rangle = 0$ for $m_1 \neq m_2$, as seen from Eq. (5.89), and therefore we obtain, from Eqs. (5.86) and (5.C.1), the relations,

$$\langle \alpha_{\text{Vv}}, \beta^{(+)} \rangle = \langle \alpha_{\text{Hh}}, \beta^{(+)} \rangle = \langle \beta^{(-)}, \beta^{(+)} \rangle = 0 \,,$$
$$\langle \beta^{(+)}, \alpha_{\text{Vv}} \rangle = \langle \beta^{(+)}, \alpha_{\text{Hh}} \rangle = \langle \beta^{(+)}, \beta^{(-)} \rangle = 0 \,, \tag{5.C.2}$$

and also

$$\langle \alpha_{\text{Vv}}, \beta^{(-)} \rangle + \langle \beta^{(-)}, \alpha_{\text{Vv}} \rangle = 0 \,,$$
$$\langle \alpha_{\text{Hh}}, \beta^{(-)} \rangle + \langle \beta^{(-)}, \alpha_{\text{Hh}} \rangle = 0 \,. \tag{5.C.3}$$

We may express $\langle \alpha_{\text{fi}}, \alpha_{\text{fi}} \rangle$ in terms of its components $\langle \alpha_{\text{Vv}}, \alpha_{\text{Vv}} \rangle$ and so on by the use of Eq. (5.87) with $\beta^{(+)}$ and $\beta^{(-)}$ instead of α_{Vh} and α_{Hv}. Then, if we use Eqs. (5.C.2) and (5.C.3) and change $\beta^{(+)}$ and $\beta^{(-)}$ back to α_{Vh} and α_{Hv}, we find

$$\langle \alpha_{\text{fi}}, \alpha_{\text{fi}} \rangle = c_{\text{i}}{}^2 c_{\text{f}}{}^2 \langle \alpha_{\text{Vv}}, \alpha_{\text{Vv}} \rangle + s_{\text{i}}{}^2 s_{\text{f}}{}^2 \langle \alpha_{\text{Hh}}, \alpha_{\text{Hh}} \rangle + \frac{1}{2}(c_{\text{i}}{}^2 s_{\text{f}}{}^2 + s_{\text{i}}{}^2 c_{\text{f}}{}^2)$$
$$\times \big(\langle \alpha_{\text{Vh}}, \alpha_{\text{Vh}} \rangle + \langle \alpha_{\text{Hv}}, \alpha_{\text{Hv}} \rangle \big) + c_{\text{i}} s_{\text{i}} c_{\text{f}} s_{\text{f}} \big(\langle \alpha_{\text{Vv}}, \alpha_{\text{Hh}} \rangle$$
$$+ \langle \alpha_{\text{Hh}}, \alpha_{\text{Vv}} \rangle + \langle \alpha_{\text{Vh}}, \alpha_{\text{Hv}} \rangle + \langle \alpha_{\text{Hv}}, \alpha_{\text{Vh}} \rangle \big) \,. \tag{5.C.4}$$

If we set $\omega_{\text{i}} = 0$ and $\omega_{\text{f}} = \pi/2$ in Eq. (5.C.4), we obtain the relation

$$\langle \alpha_{\text{Vh}}, \alpha_{\text{Vh}} \rangle = \langle \alpha_{\text{Hv}}, \alpha_{\text{Hv}} \rangle \,. \tag{5.C.5}$$

If $\omega_{\text{i}'}$ and $\omega_{\text{f}'}$ are certain values of ω_{i} and ω_{f} for which the last term on the right-hand side of Eq. (5.C.4) does not vanish, this term may be expressed as a linear combination of $\langle \alpha_{\text{Vv}}, \alpha_{\text{Vv}} \rangle$, $\langle \alpha_{\text{Hv}}, \alpha_{\text{Hv}} \rangle$ ($= \langle \alpha_{\text{Vh}}, \alpha_{\text{Vh}} \rangle$), $\langle \alpha_{\text{Hh}}, \alpha_{\text{Hh}} \rangle$, and $\langle \alpha_{\text{f}'\text{i}'}, \alpha_{\text{f}'\text{i}'} \rangle$. Therefore, it turns out that $\langle \alpha_{\text{fi}}, \alpha_{\text{fi}} \rangle$ for arbitrary ω_{i} and ω_{f} may be expressed as a linear combination of $\langle \alpha_{\text{Vv}}, \alpha_{\text{Vv}} \rangle$, $\langle \alpha_{\text{Hv}}, \alpha_{\text{Hv}} \rangle$, $\langle \alpha_{\text{Hh}}, \alpha_{\text{Hh}} \rangle$, and $\langle \alpha_{\text{f}'\text{i}'}, \alpha_{\text{f}'\text{i}'} \rangle$. Thus F_{fi} may be expressed as a linear combination of F_{Vv}, $F_{\text{Hv}} (= F_{\text{Vh}})$, F_{Hh}, and $F_{\text{f}'\text{i}'}$. If we choose as the fourth component $F_{\text{f}'\text{i}'} = F_{\text{Qq}}$ with $\omega_{\text{i}'} = \omega_{\text{f}'} = \pi/4$, we obtain Eq. (5.90).

References

1. T. Konishi, T. Yoshizaki, T. Saito, Y. Einaga, and H. Yamakawa: Macromolecules **23**, 290 (1990).
2. T. Konishi, T. Yoshizaki, and H. Yamakawa: Macromolecules **24**, 5614 (1991).
3. Y. Tamai, T. Konishi, Y. Einaga, M. Fujii, and H. Yamakawa: Macromolecules **23**, 4067 (1990).
4. M. Kamijo, N. Sawatari, T. Konishi, T. Yoshizaki, and H. Yamakawa: Macromolecules **27**, 5697 (1994).
5. M. R. Ambler, D. McIntyre, and L. J. Fetters: Macromolecules **11**, 300 (1978).
6. J. E. Godfrey and H. Eisenberg: Biophys. Chem. **5**, 301 (1976).
7. Y. Kashiwagi, T. Norisuye, and H. Fujita: Macromolecules **14**, 1220 (1981).
8. T. Norisuye: Prog. Polym. Sci. **18**, 543 (1993).
9. T. C. Troxell and H. A. Scheraga: Macromolecules **4**, 528 (1971).
10. U. Schmueli, W. Traub, and K. Rosenheck: J. Polym. Sci. Part A-2 **7**, 515 (1969).
11. T. Yoshizaki and H. Yamakawa: Macromolecules **13**, 1518 (1980).
12. P. Debye: J. Phys. Coll. Chem. **51**, 18 (1947).
13. T. Neugebauer: Ann. Phys. **42**, 509 (1943).
14. J. des Cloizeaux: Macromolecules **6**, 403 (1973).
15. M. Fujii and H. Yamakawa: J. Chem. Phys. **66**, 2578 (1977).
16. A. Peterlin: J. Polym. Sci. **47**, 403 (1960).
17. S. Heine, O. Kratky, G. Porod, and P. J. Schmitz: Makromol. Chem. **44**, 682 (1961).
18. R. Koyama: J. Phys. Soc. Japan **34**, 1029 (1973).
19. P. Sharp and V. A. Bloomfield: Biopolymers **6**, 1201 (1968).
20. H. Yamakawa and M. Fujii: Macromolecules **7**, 649 (1974).
21. T. Norisuye, H. Murakami, and H. Fujita: Macromolecules **11**, 966 (1978).
22. R. G. Kirste and W. Wunderlich: Makromol. Chem. **73**, 240 (1964).
23. W. Wunderlich and R. G. Kirste: Ber. Bunsen-Ges. Phys. Chem. **68**, 646 (1964).
24. R. G. Kirste and R. C. Oberthür: In *Small Angle X-ray Scattering*, O. Glatter and O. Kratky, ed. (Academic Press, New York, 1982), p.387.
25. D. Y. Yoon and P. J. Flory: Macromolecules **9**, 299 (1976).
26. P. R. Sundararajan: Macromolecules **19**, 415 (1986).
27. K. Nagasaka, T. Yoshizaki, J. Shimada, and H. Yamakawa: Macromolecules **24**, 924 (1991).
28. A. Baram and W. M. Gelbart: J. Chem. Phys. **66**, 617 (1977).
29. G. Porod: In *Small Angle X-ray Scattering*, O. Glatter and O. Kratky, ed. (Academic Press, New York, 1982), p.17.
30. W. Burchard and K. Kajiwara: Proc. Roy. Soc. London **A316**, 185 (1970).
31. H. Koyama, T. Yoshizaki, Y. Einaga, H. Hayashi, and H. Yamakawa: Macromolecules **24**, 932 (1991).
32. T. Yoshizaki, H. Hayashi, and H. Yamakawa: Macromolecules **26**, 4037 (1993).
33. K. Horita, T. Yoshizaki, H. Hayashi, and H. Yamakawa: Macromolecules **27**, 6492 (1994).
34. T. Yoshizaki, H. Hayashi, and H. Yamakawa: Macromolecules **27**, 4259 (1994).
35. K. Huber, W. Burchard, and S. Bantle: Polymer **28**, 863 (1987).
36. A. Dettenmaier, A. Maconnachie, J. S. Higgins, H. H. Kausch, and T. Q. Nguyen: Macromolecules **19**, 773 (1986).
37. J. M. O'Reilly, D. M. Teegarden, and G. D. Wignall: Macromolecules **18**, 2747 (1985).
38. M. Vacatello, D. Y. Yoon, and P. J. Flory: Macromolecules **23**, 1993 (1990).

39. H. Yamakawa, M. Fujii, and J. Shimada: J. Chem. Phys. **71**, 1611 (1979).
40. B. J. Berne and R. Pecora: *Dynamic Light Scattering* (Interscience, New York, 1976).
41. K. Nagai: Polym. J. **3**, 563 (1972).
42. P. Horn: Ann. Phys. (Paris) **10**, 386 (1955).
43. H. Utiyama and M. Kurata: Bull. Inst. Chem. Res. Kyoto Univ. **42**, 128 (1964); H. Utiyama: J. Phys. Chem. **69**, 4138 (1965).
44. Y. Tagami: J. Chem. Phys. **54**, 4990 (1971).
45. P. Horn, H. Benoit, and G. Oster: J. Chim. Phys. **48**, 530 (1951).
46. K. Nagai: Polym. J. **3**, 67 (1972).
47. M. Arpin, C. Strazielle, G. Weill, and H. Benoit: Polymer **18**, 262 (1977).
48. P. J. Flory, P. R. Sundararajan, and L. C. DeBold: J. Am. Chem. Soc. **96**, 5015 (1974).
49. G. D. Patterson and P. J. Flory: J. Chem. Soc. Faraday Trans. 2 **68**, 1098 (1972).
50. G. D. Patterson and P. J. Flory: J. Chem. Soc. Faraday Trans. 2 **68**, 1111 (1972).
51. U. W. Suter and P. J. Flory: J. Chem. Soc. Faraday Trans. 2 **73**,1521 (1977).
52. P. J. Flory, E. Saiz, B. Erman, P. A. Irvine, and J. P. Hummel: J. Phys. Chem. **85**, 3215 (1981).
53. C. W. Carlson and P. J. Flory: J. Chem. Soc. Faraday Trans. 2 **73**, 1505 (1977).
54. D. Y. Yoon and P. J. Flory: Polymer **16**, 645 (1975).
55. T. Konishi, T. Yoshizaki, J. Shimada, and H. Yamakawa: Macromolecules **22**, 1921 (1989).
56. Y. Takaeda, T. Yoshizaki, and H. Yamakawa: Macromolecules **26**, 3742 (1993).
57. Y. Takaeda, T. Yoshizaki, and H. Yamakawa: Macromolecules **28**, 4167 (1995).
58. M. Fujii and H. Yamakawa: J. Chem. Phys. **72**, 6005 (1980).
59. H. Yamakawa, J. Shimada, and K. Nagasaka: J. Chem. Phys. **71**, 3573 (1979).
60. A. R. Edmonds: *Angular Momentum in Quantum Mechanics* (Princeton Univ., Princeton, 1974).
61. W. H. Stockmayer: Pure Appl. Chem. **15**, 539 (1967).
62. T. Yamada, T. Yoshizaki, and H. Yamakawa: Macromolecules **25**, 1487 (1992).
63. A. J. Bur and D. E. Roberts: J. Chem. Phys. **51**, 406 (1969).
64. S. Takada, T. Itou, H. Chikiri, Y. Einaga, and A. Teramoto: Macromolecules **22**, 973 (1989).
65. K. Nagai and T. Ishikawa: J. Chem. Phys. **43**, 4508 (1965).
66. P. J. Flory: *Statistical Mechanics of Chain Molecules* (Interscience, New York, 1969).
67. H. Benoit: Ann. Phys. (Paris) **6**, 561 (1951).
68. A. Peterlin and H. A. Stuart: J. Polym. Sci. **5**, 551 (1950).
69. W. H. Stockmayer and M. E. Baur: J. Am. Chem. Soc. **86**, 3485 (1964).
70. J. A. Schellman: Chem. Rev. **75**, 323 (1975).
71. H. Yamakawa: *Modern Theory of Polymer Solutions* (Harper & Row, New York, 1971).

6 Transport Properties

This chapter deals with the classical hydrodynamic theory of steady-state transport properties, such as the translational friction and diffusion coefficients and intrinsic viscosity, of the unperturbed HW chain, including the KP wormlike chain as a special case, on the basis of the cylinder and touched-bead models. An analysis of experimental data is made from various points of view, which are based on the present theory, especially for flexible polymers. In the same spirit as that in Chap. 5, use is then made of experimental data obtained for several flexible polymers in the Θ state over a wide range of molecular weight, including the oligomer region, and also for typical semi-flexible polymers. As a result, it is pointed out that there still remain several unsolved problems for flexible polymers even in the unperturbed state. It is convenient to begin by giving a general consideration of some aspects of polymer hydrodynamics which leads to the adoption of the present hydrodynamic models.

6.1 General Consideration of Polymer Hydrodynamics

As is well known, the transport theory of dilute polymer solutions is based on the idea that polymer molecules as sources of excess energy dissipation exert frictional forces on the solvent medium which is regarded as a continuous viscous fluid. Within the framework of classical hydrodynamics, the motion of the fluid with (shear) viscosity coefficient η_0 in steady flow may be described by the linearized Navier–Stokes equation (Stokes equation)

$$\eta_0 \nabla^2 \mathbf{v}(\mathbf{r}) - \nabla p(\mathbf{r}) + \mathbf{f}(\mathbf{r}) = \mathbf{0} \qquad (6.1)$$

with

$$\nabla \cdot \mathbf{v}(\mathbf{r}) = 0 \qquad (6.2)$$

for incompressible fluids, where $\mathbf{v}(\mathbf{r})$ and $p(\mathbf{r})$ are the velocity and pressure of the fluid at the point $\mathbf{r}$ in a Cartesian coordinate system, respectively, and $\mathbf{f}(\mathbf{r})$ is the force density, that is, the frictional force exerted on the fluid per unit volume at the same point. The fundamental solution of Eq. (6.1) with Eq. (6.2) is given by [1, 2]

$$v(\mathbf{r}) = \int \mathbf{T}(\mathbf{r} - \mathbf{r}') \cdot \mathbf{f}(\mathbf{r}')d\mathbf{r}' \,, \tag{6.3}$$

where $\mathbf{T}(\mathbf{r})$ is the Green function usually called the *Oseen hydrodynamic interaction tensor* and given by

$$\mathbf{T}(\mathbf{r}) = \frac{1}{8\pi\eta_0 r}(\mathbf{I} + \mathbf{e}_r\mathbf{e}_r) \tag{6.4}$$

with $\mathbf{I}$ the unit tensor and $\mathbf{e}_r$ the unit vector in the direction of $\mathbf{r}$.

In the case for which a point force $\mathbf{F}$ is exerted at the origin of the coordinate system, $\mathbf{f}(\mathbf{r})$ is given by

$$\mathbf{f}(\mathbf{r}) = \mathbf{F}\delta(\mathbf{r}) \tag{6.5}$$

with $\delta(\mathbf{r})$ being a Dirac delta function, and therefore we have, from Eqs. (6.3) and (6.5),

$$v(\mathbf{r}) = \mathbf{T}(\mathbf{r}) \cdot \mathbf{F} \,. \tag{6.6}$$

This is the basic equation in the well-known Kirkwood procedure [1, 3, 4] of polymer transport theory for bead models, in which the segments (beads) constituting the polymer chain are treated as point sources of friction. The solutions of linear coupled equations determining the frictional forces from Eq. (6.6) possess the Zwanzig singularities [5, 6] which lead to unphysical behavior of the transport properties, for example, negative translational diffusion coefficients of a rigid rod [6]. Such mathematical singularities always occur irrespective of the preaveraging or nonpreaveraging of the Oseen tensor $\mathbf{T}$, but they can be removed from the physically possible range of hydrodynamic interaction strength except in the case of rigid rods if the Stokes diameter of the (spherical) bead is assumed [7]. Thus the occurrence of the physical singularities is related to a particular spatial distribution of beads.

For finite bead models Rotne and Prager [8] applied a variational method to derive a correction to the Oseen tensor which gives an upper bound to the true positive definite diffusion tensor. It reads

$$\mathbf{T}_{\mathrm{m}}(\mathbf{r}) = \mathbf{T}(\mathbf{r}) + \frac{1}{16\pi\eta_0 r}\left(\frac{d_{\mathrm{b}}}{r}\right)^2\left(\frac{1}{3}\mathbf{I} - \mathbf{e}_r\mathbf{e}_r\right), \tag{6.7}$$

where $\mathbf{T}(\mathbf{r})$ is the Oseen tensor given by Eq. (6.4) and d_{b} is the diameter of the bead. This *modified Oseen tensor* $\mathbf{T}_{\mathrm{m}}$ may be derived by distributing point forces uniformly on the surface of the spherical bead [9] and is precisely the first-order correction in the case of translational motion [10]. Indeed, the use of Eq. (6.6) with $\mathbf{T}_{\mathrm{m}}$ in place of $\mathbf{T}$ removes the physical singularities for the translational diffusion coefficient of rigid rods [11]. It must however be noted that the use of the Oseen tensor, when preaveraged, also removes them accidentally [11]. Further, the modified Oseen tensor becomes identical with the Oseen tensor if preaveraged, as seen from Eq. (6.7). Even under these circumstances, the use of the former without preaveraging must be much better than the use of the latter for rigid discrete models composed of a rather small number of beads. Indeed, there have been many investigations of this kind [12, 13], including those of complex, rigid, biological macromolecules.

Now it is well known that in the extreme the number of beads in the chain is equal to one, the translational friction coefficient evaluated by the Kirkwood procedure takes the Stokes law value correctly, while it cannot give the Einstein intrinsic viscosity of the single bead. This is also the case with the use of the modified Oseen tensor. This defect may be removed by treating the polymer chain as a body of finite volume whose surface exerts the frictional force **f** per unit area and satisfies the nonslip boundary condition, as done by Edwards and co-workers [14, 15]. [Note that the finite volume of the body may be, to some extent, taken into account by Eq. (6.7).] In this case the fluid velocity **v** produced is given by Eq. (6.3) instead of by Eq. (6.6), although the integral in Eq. (6.3) must be replaced by the surface integral. In this chapter we consider two types of such polymer hydrodynamic models: cylinder models and touched-bead models. Necessarily, the results may be expressed in terms of dimensional parameters defining the body and also the basic (HW or KP) model parameters and may also be applied to short chains or the oligomer region.

For earlier theories for the KP chain following the Kirkwood procedure, the reader is referred to MTPS [1].

6.2 Hydrodynamic Models

6.2.1 Cylinder Model

For cylinder models the exact application of the present procedure mentioned in the last section is limited to short rigid cylinders. It requires an introduction of several mathematical approximations even in the limit of long Gaussian cylinders [15]. For long cylinders we therefore adopt an alternative, approximate but equivalent method, that is, the *Oseen–Burgers* (OB) *procedure* [16], which is less familiar in the polymer field but which was applied long ago to rigid, cylindrical bodies [16–18].

Consider a cylinder of length L and diameter d ($L \gg d$) whose axis as the chain contour obeys HW (or KP) statistics, and suppose that it is immersed in a solvent having an unperturbed velocity field $\mathbf{v}^0$ [19–23]. We replace the cylinder by a distribution $\mathbf{f}(s)$ of the frictional force (exerted on the fluid) per unit length along the cylinder axis as a function of the contour distance s from one end ($0 \leq s \leq L$). As depicted in Fig. 6.1, let $\hat{\mathbf{r}}$ be the normal radius vector from the contour point s_1, whose radius vector is $\mathbf{r}(s_1)$ in an external Cartesian coordinate system, to an arbitrary point on the cylinder surface, so that

$$|\hat{\mathbf{r}}| = \hat{r} = \frac{1}{2}d, \tag{6.8}$$

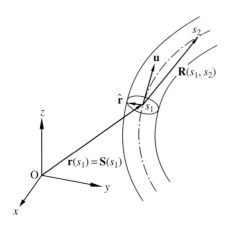

Fig. 6.1. Cylinder model for an evaluation of steady-state transport coefficients by the Oseen–Burgers procedure (see the text)

$$\hat{\mathbf{r}} \cdot \mathbf{u} = 0 \tag{6.9}$$

with $\mathbf{u}$ the unit vector tangential to the axis at the point s_1, and let $\mathbf{R}(s_1, s_2) = \mathbf{r}(s_2) - \mathbf{r}(s_1)$ be the vector distance between the contour points s_1 and s_2.

For an instantaneous configuration, the velocity $\mathbf{v}(\hat{\mathbf{r}})$ of the solvent at the point $\hat{\mathbf{r}}$ on the cylinder surface relative to the velocity $\mathbf{U}(\hat{\mathbf{r}})$ of the cylinder at $\hat{\mathbf{r}}$ may then be expressed, from Eq. (6.3), as

$$\mathbf{v}(\hat{\mathbf{r}}) = \mathbf{v}^0(\hat{\mathbf{r}}) - \mathbf{U}(\hat{\mathbf{r}}) + \int_0^L \mathbf{T}(\mathbf{R} - \hat{\mathbf{r}}) \cdot \mathbf{f}(s_2) ds_2 \,. \tag{6.10}$$

The OB procedure requires a nonslip boundary condition on the average, that is, that the values of $\mathbf{v}(\hat{\mathbf{r}})$ averaged over a normal cross section of the cylinder vanish for all values of s_1,

$$\left\langle \mathbf{v}(\hat{\mathbf{r}}) \right\rangle_{\hat{\mathbf{r}}} = \mathbf{0} \qquad \text{for } 0 \le s_1 \le L \,, \tag{6.11}$$

where $\langle \quad \rangle_{\hat{\mathbf{r}}}$ denotes the average over $\hat{\mathbf{r}}$, assuming its uniform distribution subject to the conditions given by Eqs. (6.8) and (6.9). Since the unperturbed flow field is assumed to be non-existent (translational diffusion) or linear in space (shear viscosity) and since the velocity $\mathbf{U}(\hat{\mathbf{r}})$ is derived from the translational or angular velocity of the cylinder, we have

$$\left\langle \mathbf{v}^0(\hat{\mathbf{r}}) \right\rangle_{\hat{\mathbf{r}}} = \mathbf{v}^0(s_1) \,, \tag{6.12}$$

$$\left\langle \mathbf{U}(\hat{\mathbf{r}}) \right\rangle_{\hat{\mathbf{r}}} = \mathbf{U}(s_1) \,, \tag{6.13}$$

so that Eq. (6.10) reduces to

$$\mathbf{U}(s_1) - \mathbf{v}^0(s_1) = \int_0^L \left\langle \mathbf{T}(\mathbf{R} - \hat{\mathbf{r}}) \right\rangle_{\hat{\mathbf{r}}} \cdot \mathbf{f}(s_2) ds_2 \,. \tag{6.14}$$

Now it is known that for spheroids (ellipsoids of revolution) exact expressions for the translational and rotatory diffusion coefficients D and D_r and intrinsic viscosity $[\eta]$ can be obtained from Eqs. (6.1) and (6.2) with the non-slip boundary condition. For a prolate spheroid of major axis L and minor axis d, the asymptotic factors, $\ln(L/d) + $ const., (along with the prefactors) involved in the exact D, D_r and $[\eta]$ for $L/d \gg 1$ are coincident with those obtained from Eq. (6.14), where in this case $\hat{\mathbf{r}}$ is not a constant but depends on s_1 [24]. This gives grounds for the application of the OB procedure to the long cylinder. In the present case, however, further developments require the preaveraging of the Oseen tensor in Eq. (6.14). Then it reduces to

$$\mathbf{U}(s_1) - \mathbf{v}^0(s_1) = \frac{1}{6\pi\eta_0} \int_0^L K(s_1, s_2)\mathbf{f}(s_2)ds_2 \qquad (6.15)$$

with

$$
\begin{aligned}
K(s_1, s_2) &= K(s; d) \\
&= \langle |\mathbf{R} - \hat{\mathbf{r}}|^{-1} \rangle,
\end{aligned}
\qquad (6.16)
$$

where $s = |s_1 - s_2|$ and $\langle \ \rangle$ denotes the averages over $\hat{\mathbf{r}}$ and chain configurations.

The problem is to evaluate the kernel K in the integral Eq. (6.15) determining the frictional force $\mathbf{f}$. In what follows, all lengths are measured as before in units of λ^{-1}. For convenience, we consider the kernel $K(L; d)$ with $s_1 = 0$ and $s_2 = L$. It may be evaluated from

$$K(L; d) = (2\pi)^{-1} \int d\mathbf{R} \int' d\hat{\mathbf{r}} |\mathbf{R} - \hat{\mathbf{r}}|^{-1} G(\mathbf{R} \,|\, \mathbf{u}_0 = \mathbf{e}_z; L), \qquad (6.17)$$

where the integration over $\hat{\mathbf{r}}$ is carried out under the conditions of Eqs. (6.8) and (6.9) with $\mathbf{u} = \mathbf{u}_0$ and G is the conditional distribution function given by Eq. (4.158). If we note that the Oseen tensor may be expressed as the inverse of its Fourier transform [15],

$$\mathbf{T}(\mathbf{r}) = (8\pi^3\eta_0)^{-1} \int k^{-2}(\mathbf{I} - \mathbf{e}_k\mathbf{e}_k)\exp(-i\mathbf{k} \cdot \mathbf{r})d\mathbf{k} \qquad (6.18)$$

with $\mathbf{e}_k$ the unit vector in the direction of $\mathbf{k}$, then $K(L; d)$ may also be written in the form [19]

$$
\begin{aligned}
K(L; d) &= 2\pi^{-1} \int_0^\infty \langle \exp[i\mathbf{k} \cdot (\mathbf{R} - \hat{\mathbf{r}})] \rangle dk \\
&= \pi^{-1} \int_0^\infty \int_0^\pi J_0(\hat{r}k \sin\chi) I(\mathbf{k} \,|\, \mathbf{u}_0 = \mathbf{e}_z; L) \sin\chi\, dk\, d\chi, \quad (6.19)
\end{aligned}
$$

where $\mathbf{k} = (k, \chi, \omega)$ in spherical polar coordinates, I is the characteristic function, and J_0 is the zeroth-order Bessel function of the first kind defined by

$$J_0(x) = \frac{2}{\pi} \int_0^1 \frac{\cos xt}{(1 - t^2)^{1/2}} dt \,. \tag{6.20}$$

It is convenient to evaluate $K(L; d)$ in different approximations in three ranges of L: $L \leq \sigma_1$, $\sigma_1 < L \leq \sigma_2$, and $L > \sigma_2$. We adopt an equation obtained in the second Daniels approximation from the second line of Eqs. (6.19) with the $\mathcal{I}_{0ll}^{00,00}(k; L)$ given by Eq. (4.172) for $L > \sigma_2$, an approximate expression, which can reproduce the values obtained by the weighting function method from Eq. (6.17) with $|\mathbf{R} - \hat{\mathbf{r}}|^{-1}$ being expressed in terms of the Legendre polynomials $P_l(\cos \alpha)$ given by Eq. (3.B.15) with α the angle between $\mathbf{R}$ and $\hat{\mathbf{r}}$, for $\sigma_1 < L \leq \sigma_2$, and an approximate $\epsilon3$ equation (for $d = 0$) from the ϵ method for $L \leq \sigma_1$. These three functions are joined at $L = \sigma_1$ and σ_2 following the procedure of Hearst and Stockmayer [25] as in Eq. (3.D.5).

The approximate interpolation formula for $K(L; d)$ so obtained is given by [26]

$$
\begin{aligned}
K(L; d) &= \left(\frac{6}{\pi c_\infty L}\right)^{1/2} \sum_{i=0}^{2} \sum_{j=0}^{i} B_{ij} d^{2j} (c_\infty L)^{-i} \\
&\quad + h(\sigma_2 - L)(c_\infty L)^{-1/2} \sum_{i=0}^{q} \sum_{j=0}^{2} C_{ij} d^{2j} (L - \sigma_2)^{i+3} \qquad \text{for } L > \sigma_1 \\
&= \left(L^2 + \frac{1}{4}d^2\right)^{-1/2} \left(1 + \sum_{i=1}^{5} f_{i0} L^i + \sum_{i=1}^{3} \sum_{j=1}^{2} f_{ij} d^{2j} L^i\right) \qquad \text{for } L \leq \sigma_1
\end{aligned}
\tag{6.21}
$$

with

$$B_{00} = 1\,, \quad B_{11} = -\frac{1}{8}\,, \quad B_{22} = \frac{63}{4480}\,, \quad f_{10} = \frac{1}{3}\,, \tag{6.22}$$

where c_∞ is given by Eq. (4.75); $h(x)$ is a unit step function defined by $h(x) = 1$ for $x \geq 0$ and $h(x) = 0$ for $x < 0$; f_{i0} ($i = 1$–3) are the coefficients of L^i in Eq. (4.232); and $B_{10}, B_{20}, B_{21}, C_{ij}, f_{i0}$ ($i = 4, 5$), f_{ij} ($i = 1$–3; $j = 1$, 2), σ_1, σ_2 and q are constants independent of L and d but dependent on the HW model parameters κ_0 and τ_0 and are to be determined numerically. From the practical point of view, however, we do not give the numerical results, since the cylinder model is mainly applied to typical semiflexible polymers, which may be represented by the KP chain in most cases.

In the particular case of the KP chain, for which $\kappa_0 = 0$ and $c_\infty = 1$, we have [19]

$$\sigma_1 = \sigma_2 = 2.278\,,$$

$$
\begin{aligned}
&B_{10} = -\frac{1}{40}\,, \quad B_{20} = -\frac{73}{4480}\,, \quad B_{21} = \frac{21}{320}\,, \quad C_{ij} = 0\,, \\
&f_{20} = 0.1130\,, \quad f_{30} = -0.02447\,, \quad f_{40} = f_{50} = 0\,, \\
&f_{11} = 0.04080\,, \quad f_{21} = -0.04736\,, \quad f_{31} = 0.009666\,, \\
&f_{12} = 0.004898\,, \quad f_{22} = -0.002270\,, \quad f_{32} = 0.0002060 \quad (\text{KP})\,.
\end{aligned}
\tag{6.23}
$$

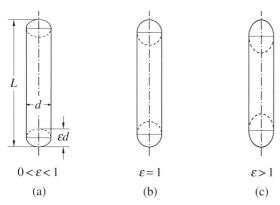

$$0 < \varepsilon < 1 \qquad\qquad \varepsilon = 1 \qquad\qquad \varepsilon > 1$$

$$\text{(a)} \qquad\qquad\quad \text{(b)} \qquad\qquad\quad \text{(c)}$$

Fig. 6.2a–c. Three types of spheroid-cylinders. The hemispheroids at the ends are: **a** oblate (pancake shaped); **b** spherical; **c** prolate (cigar shaped). The case (b) is a (prolate) spherocylinder

We note that the values of f_{20} and f_{30} in Eqs. (6.23) do not agree with those in Eq. (4.232) with $\kappa_0 = 0$.

Finally, we make a preliminary remark on the treatment in the range of small L, in which end effects must be taken into account. We assume that as L is decreased, the HW cylinder becomes a *spheroid-cylinder*, that is, a straight cylinder with oblate, spherical, or prolate hemispheroid caps at the ends such that its total length is L and the length of the intermediate cylinder part is $L - \epsilon d$, so that ϵ is the ratio of the principal diameters of the end spheroid, as depicted in Fig. 6.2 [27]. Its transport coefficients (for arbitrary L) are evaluated in Appendix 6.A. Extrapolation to them is made properly from the OB solutions obtained for the HW cylinder for $L/d \gg 1$.

6.2.2 Touched-Bead Model

Consider a chain composed of N identical spherical beads of diameter d_b whose centers are located on the HW (or KP) chain contour of total length L. The contour distance between the centers of two adjacent beads is set equal to the bead diameter, so that $Nd_b = L$. [Note that this relation between N and L is different from that in Sect. 5.2.3 (b).] Strictly speaking, two adjacent beads do not touch each other but slightly overlap since the contour distance between their centers is larger than the straight distance. However, the difference between these two distances is negligibly small, and therefore we call this model the touched-bead model.

Now suppose that the chain is immersed in the solvent having the unperturbed flow field $\mathbf{v}^0$ as in the case of the cylinder model. Let $\mathbf{r}_i$ be the vector position of the center of the ith bead ($i = 1, 2, \cdots, N$) and let $\hat{\mathbf{r}}_i$ be the radius vector from its center to an arbitrary point on its surface, so that $|\hat{\mathbf{r}}_i| = \hat{r}_i = d_b/2$, as depicted in Fig. 6.3. Under the nonslip boundary

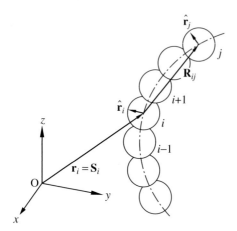

Fig. 6.3. Touched-bead model for an evaluation of steady-state transport coefficients (see the text)

condition on the surface of each bead, the velocity $\mathbf{U}_i(\hat{\mathbf{r}}_i)$ of the point $\hat{\mathbf{r}}_i$ of the ith bead may be expressed, from Eq. (6.3), as [28]

$$\mathbf{U}_i(\hat{\mathbf{r}}_i) - \mathbf{v}^0(\hat{\mathbf{r}}_i) = \int_{S_i} \mathbf{T}(\hat{\mathbf{r}}'_i - \hat{\mathbf{r}}_i) \cdot \mathbf{f}_i(\hat{\mathbf{r}}'_i) d\hat{\mathbf{r}}'_i$$

$$+ \sum_{\substack{j=1 \\ \neq i}}^{N} \int_{S_j} \mathbf{T}(\mathbf{R}_{ij} - \hat{\mathbf{r}}_i + \hat{\mathbf{r}}_j) \cdot \mathbf{f}_j(\hat{\mathbf{r}}_j) d\hat{\mathbf{r}}_j , \qquad (6.24)$$

where $\mathbf{R}_{ij} = \mathbf{r}_j - \mathbf{r}_i$ is the vector distance between the centers of the ith and jth beads, $\mathbf{f}_i(\hat{\mathbf{r}}_i)$ is the frictional force exerted on the fluid by the unit area at $\hat{\mathbf{r}}_i$ of the surface of the ith bead, and the integration is carried out over its surface (S_i).

Equation (6.24) is the coupled integral equation determining $\mathbf{f}_i$. In order to make its solution accessible, we expand the tensor $\mathbf{T}(\mathbf{R}_{ij} - \hat{\mathbf{r}}_i + \hat{\mathbf{r}}_j)$ in Eq. (6.24) in a Taylor series around $\hat{\mathbf{r}}_j - \hat{\mathbf{r}}_i = \mathbf{0}$ and neglect terms of $\mathcal{O}(R_{ij}^{-n})$ $(n \geq 2)$. This is equivalent to replacing this tensor by $\mathbf{T}(\mathbf{R}_{ij})$. Then, if the Oseen tensor $\mathbf{T}(\mathbf{R}_{ij})$ is preaveraged, Eq. (6.24) reduces to

$$\mathbf{U}_i(\hat{\mathbf{r}}_i) - \mathbf{v}^0(\hat{\mathbf{r}}_i) = \int_{S_i} \mathbf{T}(\hat{\mathbf{r}}'_i - \hat{\mathbf{r}}_i) \cdot \mathbf{f}_i(\hat{\mathbf{r}}'_i) d\hat{\mathbf{r}}'_i$$

$$+ \frac{1}{6\pi\eta_0} \sum_{\substack{j=1 \\ \neq i}}^{N} \langle R_{ij}^{-1} \rangle \mathbf{F}_j \qquad (6.25)$$

with

$$\langle R_{ij}^{-1} \rangle = \langle R^{-1}(|i - j|d_b) \rangle , \qquad (6.26)$$

$$\mathbf{F}_i = \int_{S_i} \mathbf{f}_i(\hat{\mathbf{r}}_i) d\hat{\mathbf{r}}_i , \qquad (6.27)$$

where $\mathbf{F}_i$ is the total frictional force exerted by the ith bead on the fluid.

We can construct an interpolation formula for the kernel $\langle R^{-1}(L)\rangle = K(L;0)$ in a manner similar to that in the case of $K(L;d)$ with finite d but as a function of L, κ_0, and τ_0 covering almost all important ranges of κ_0 and τ_0. If all lengths are measured in units of λ^{-1}, the result may be written in the form [29]

$$\langle R^{-1}(L)\rangle = \left[\frac{\langle R^2(L)\rangle_{\mathrm{KP}}}{\langle R^2(L)\rangle}\right]^{1/2} K_{\mathrm{KP}}(L;0)\left[1 + \kappa_0{}^2 \Gamma(L)\right], \qquad (6.28)$$

where $\langle R^2(L)\rangle$ is the mean-square end-to-end distance given by Eq. (4.83), $\langle R^2(L)\rangle_{\mathrm{KP}}$ is its KP value, and $K_{\mathrm{KP}}(L;0)$ is the KP kernel given by Eq. (6.21) with $c_\infty = 1$ and $d = 0$ and with Eqs. (6.22) and (6.23). In Eq. (6.28), $\Gamma(L)$ may be well approximated by

$$\Gamma(L) = \exp\left(-\frac{2}{L}\right)\sum_{k=1}^{2}\frac{A_k}{L^k} + \exp\left[-\left(2+\frac{3}{4}\nu\right)\right]\sum_{k=3}^{7}A_k L^k \qquad (6.29)$$

with

$$A_1 = \frac{3}{r^2(4+\tau_0{}^2)} - \frac{3}{10(9+\nu^2)(36+\nu^2)}$$
$$\times\left[1 + \frac{101+\kappa_0{}^2}{4+\tau_0{}^2} + \frac{3(160+7\kappa_0{}^2)}{(4+\tau_0{}^2)^2}\right], \qquad (6.30)$$

$$A_k = \left[1 + \delta_{k2}\left(\frac{1}{r^2}-1\right)\right]\sum_{i=0}^{7}\sum_{j=0}^{6}a_{ij}^k \nu^i \left(\frac{\tau_0}{\nu}\right)^{2j} \qquad (k = 2-7), \qquad (6.31)$$

where ν and r are given by Eqs. (4.76) and (4.77), respectively; and a_{ij}^k are numerical constants independent of L, κ_0, and τ_0 and their values are given in Appendix IV. The application of Eq. (6.28) is limited to the following ranges of κ_0 and τ_0 : $\nu \lesssim 6$ for $0 \leq \tau_0/\nu \leq 0.2$ and $\nu \leq 8$ for $0.2 \leq \tau_0/\nu \leq 1$. We note that Eq. (6.28) gives the exact linear term in Eq. (4.232) for $L \ll 1$ and the first Daniels approximation for $L \gg 1$.

6.3 Translational Friction Coefficient

6.3.1 Cylinder Model

Suppose that the center of mass of the HW cylinder possesses the translational velocity $\mathbf{U}$ in the vanishing unperturbed flow field,

$$\mathbf{v}^0 = \mathbf{0}. \qquad (6.32)$$

In what follows, all lengths are measured in units of λ^{-1}. If we take the configurational average of both sides of Eq. (6.15) and note that $\langle \mathbf{U}(s_1) \rangle = \mathbf{U}$ for all values of s_1, then we have

$$\int_0^L K(s_1, s_2) \langle \mathbf{f}(s_2) \rangle \, ds_2 = 6\pi\eta_0 \mathbf{U} \,. \tag{6.33}$$

The mean total frictional force $\langle \mathbf{F} \rangle$ is given by

$$\langle \mathbf{F} \rangle = \int_0^L \langle \mathbf{f}(s) \rangle \, ds = \Xi \mathbf{U} \,, \tag{6.34}$$

where Ξ is the translational friction coefficient of the cylinder.

Now, if we use the Kirkwood–Riseman (KR) approximation [1,3], $\langle \mathbf{f}(s) \rangle = L^{-1} \langle \mathbf{F} \rangle$, in the integral Eq. (6.33) to solve it analytically, we obtain [19, 22]

$$L^{-2}\Xi \int_0^L \int_0^L K(s_1, s_2) ds_1 ds_2 = 6\pi\eta_0 \,, \tag{6.35}$$

and therefore

$$\frac{3\pi\eta_0 L}{\Xi} \equiv f_{\mathrm{D}}(L; \kappa_0, \tau_0, d)$$

$$= L^{-1} \int_0^L (L - s) K(s; d) ds \,. \tag{6.36}$$

Then, if we assume the Einstein relation $D = k_{\mathrm{B}}T/\Xi$, the translational diffusion coefficient D (in the long-time limit) and sedimentation coefficient s may be expressed in terms of the function f_D defined by the first line of Eqs. (6.36) as

$$D = \left(\frac{k_{\mathrm{B}}T}{3\pi\eta_0 L} \right) f_D \,, \tag{6.37}$$

$$s = \frac{M(1 - \bar{v}\rho_0) D}{N_{\mathrm{A}} k_{\mathrm{B}} T} = \left[\frac{M(1 - \bar{v}\rho_0)}{3\pi\eta_0 N_{\mathrm{A}} L} \right] f_D \,, \tag{6.38}$$

where N_{A} is the Avogadro constant, M is the polymer molecular weight, $\bar{v}$ is its partial specific volume, and ρ_0 is the mass density of the solvent. Note that the Einstein relation does not hold for the exact D and Ξ for rigid, nonspherical molecules [11, 30] (see also Appendix 6.A), and therefore that Eq. (6.37) and the second of Eqs. (6.38) in general are not exactly valid, although the first of Eqs. (6.38) is correct.

The function f_D may be evaluated straightforwardly by substitution of Eqs. (6.21) with $L = s$ into the second line of Eqs. (6.36) and integration. However, the result is semianalytical, and moreover, not convenient for practical use because of its complexity. We therefore reconstruct an approximate but simpler and completely analytical interpolation formula for f_D on the basis of its calculated values. The result may be written in the form [22],

$$f_D = f_{D,\text{a-KP}} \Gamma_D(L; \kappa_0, \tau_0, d), \tag{6.39}$$

where $f_{D,\text{a-KP}}$ is the function f_D for the *associated KP chain* that is the KP chain whose Kuhn segment length is equal to c_∞, and is given by

$$f_{D,\text{a-KP}} = f_{D,\text{KP}}(c_\infty^{-1}L; c_\infty^{-1}d) \tag{6.40}$$

with $f_{D,\text{KP}}(L; d)$ the function f_D for the KP chain.

The function $f_{D,\text{KP}}$ evaluated directly by the use of the KP kernel $K(s; d)$ given by Eqs. (6.21) with Eqs. (6.22) and (6.23) in the second line of Eqs. (6.36) is not very complicated and is given by

$$f_{D,\text{KP}}(L; d) = F_1(L; \hat{L}, d) + h(L - \sigma_1)F_2(L; d) \tag{6.41}$$

with

$$
\begin{aligned}
F_1 = \sum_{i=0}^{3} f_{i0} \hat{L}^i \left[I_i\left(\frac{d}{\hat{L}}\right) - \frac{\hat{L}}{L} I_{i+1}\left(\frac{d}{\hat{L}}\right) \right] \\
+ \sum_{i=1}^{3} \sum_{j=1}^{2} f_{ij} d^{2j} \left[I_i\left(\frac{d}{\hat{L}}\right) - \frac{\hat{L}}{L} I_{i+1}\left(\frac{d}{\hat{L}}\right) \right],
\end{aligned} \tag{6.42}
$$

$$
\begin{aligned}
F_2 = \left(\frac{6}{\pi}\right)^{1/2} \sum_{i=0}^{2} \sum_{j=0}^{i} B_{ij} d^{2j} \left\{ \frac{L^{1/2-i}}{\left(i - \frac{1}{2}\right)\left(i - \frac{3}{2}\right)} \right. \\
\left. + \left[\frac{1}{i - \frac{1}{2}} - \frac{\sigma_1}{\left(i - \frac{3}{2}\right)L} \right] \sigma_1^{1/2-i} \right\},
\end{aligned} \tag{6.43}
$$

where $f_{00} = 1$, $\sigma_1 = 2.278$, and

$$
\begin{aligned}
\hat{L} = L \qquad & \text{for } L \le \sigma_1 \\
= \sigma_1 \qquad & \text{for } L > \sigma_1,
\end{aligned} \tag{6.44}
$$

$$
\begin{aligned}
I_0(x) &= -\ln x + \ln 2 + \ln\left[1 + \left(1 + \tfrac{1}{4}x^2\right)^{1/2}\right], \\
I_1(x) &= \left(1 + \tfrac{1}{4}x^2\right)^{1/2} - \tfrac{1}{2}x, \\
I_2(x) &= \tfrac{1}{2}\left(1 + \tfrac{1}{4}x^2\right)^{1/2} - \tfrac{1}{8}x^2 I_0(x), \\
I_3(x) &= \tfrac{1}{3}\left(1 + \tfrac{1}{4}x^2\right)^{1/2} - \tfrac{1}{6}x^2 I_1(x), \\
I_4(x) &= \tfrac{1}{4}\left(1 - \tfrac{3}{8}x^2\right)\left(1 + \tfrac{1}{4}x^2\right)^{1/2} + \tfrac{3}{128}x^4 I_0(x).
\end{aligned} \tag{6.45}
$$

A good approximation to the function Γ_D in Eq. (6.39), which must become unity in the limits of $L = 0$ and ∞, is of the form

$$\Gamma_D = 1 + \left(\frac{A_1}{L^{1/2}} + \frac{A_2}{L}\right)\left[1 - (1 + \xi L)e^{-\xi L}\right] + A_3 L e^{-\xi L} \tag{6.46}$$

with

$$\xi = 0.3 + 0.4\nu\,, \tag{6.47}$$

$$A_i = \sum_{k,l=0}^{3}\left(\sum_{j=0}^{2} a_{ij}^{kl} d^j + a_{i3}^{kl}\ln d\right)\nu^l\cos(k\pi\tau_0/\nu)\,, \tag{6.48}$$

where ν is given by Eq. (4.76) and a_{ij}^{kl} are numerical constants independent of L, κ_0, τ_0, and d with $a_{12}^{kl} \equiv 0$. As already noted, the cylinder model is mainly used for the KP chain, and we do not give the numerical results for a_{ij}^{kl}.

As shown in Appendix 6.A, the end effects on f_D are rather small; the $f_{D,\mathrm{KP}}$ given by Eq. (6.41) may be smoothly joined to that for the spheroid-cylinder in the range of small L ($L/d < 5$).

Now we consider two extreme cases: long rigid rods and Gaussian cylinders. The rod limit, which we indicate by the subscript (R), may be obtained by letting $L \to 0$ and $d \to 0$ at constant $L/d \equiv p$. Thus we have, from Eq. (6.39),

$$\lim_{p\to\infty}\lim_{\substack{L,d\to 0\\(\mathrm{const.}p)}} f_D = \lim_{p\to\infty} f_{D,(\mathrm{R})} \equiv f_{D,(\mathrm{R}^*)}$$

$$= \ln p + 2\ln 2 - 1 + \mathcal{O}(p^{-1})\,. \tag{6.49}$$

It is important to note that this asymptotic form is exactly correct; it happens to agree with the result derived by the OB procedure with the non-preaveraged Oseen tensor [24] (see Appendix 6.A).

The Gaussian cylinder with $\langle R^2\rangle = c_\infty L$, which we indicate by the subscript (G), may be obtained by letting $L \to \infty$ in the kernel given by Eq. (6.19). Then we have

$$I_{(\mathrm{G})}(\mathbf{k}\,|\,\mathbf{u}_0 = \mathbf{e}_z; L) = \exp\left(-\frac{1}{6}c_\infty L k^2\right)\,, \tag{6.50}$$

$$K_{(\mathrm{G})}(L;d) = \frac{2}{d}\,\mathrm{erf}\left[\left(\frac{3d^2}{8c_\infty L}\right)^{1/2}\right]\,, \tag{6.51}$$

where $\mathrm{erf}(x)$ is the error function defined by

$$\mathrm{erf}(x) = 2\pi^{-1/2}\int_0^x \exp(-t^2)dt\,. \tag{6.52}$$

Substitution of Eq. (6.51) with $L = s$ into the second line of Eqs. (6.36) and integration leads to

$$\begin{aligned}
f_{D,(\mathrm{G})} &= \left(\frac{6L}{c_\infty}\right)^{1/2}\left[\left(\frac{1}{4x}+x+\frac{1}{3}x^3\right)\mathrm{erf}(x)\right.\\
&\qquad\left.+\frac{1}{6\pi^{1/2}}(5+2x^2)\exp(-x^2)-x-\frac{1}{3}x^3\right]\\
&= \frac{4}{3}\left(\frac{6}{\pi c_\infty}\right)^{1/2}L^{1/2}\left(1-\frac{3\pi^{1/2}}{4}x+x^2-\frac{\pi^{1/2}}{4}x^3+\cdots\right) \tag{6.53}
\end{aligned}$$

with

$$x^2 = \frac{3d^2}{8c_\infty L} \, . \tag{6.54}$$

We note that although Edwards and Oliver [15] derived an expansion similar to the second of Eqs. (6.53), their numerical coefficients involve several mathematical errors.

The random-coil limit, which we indicate by the subscript (C), is obtained by letting further $L \to \infty$,

$$\lim_{L \to \infty} f_{D,(G)} \equiv f_{D,(C)} = \frac{4}{3} \left(\frac{6}{\pi c_\infty} \right)^{1/2} L^{1/2} \, , \tag{6.55}$$

so that we have

$$\Xi_{(C)} = \frac{9}{4} \left(\frac{\pi}{6} \right)^{1/2} \pi \eta_0 (c_\infty L)^{1/2} \, . \tag{6.56}$$

Equation (6.56) is equivalent to the KR equation in the nondraining limit [1,3]. Note that it may be directly obtained from the second line of Eqs. (6.36) with $K(s; d) = (6/\pi c_\infty s)^{1/2}$.

Finally, we examine numerically the behavior of f_D. Values of f_D calculated from Eq. (6.41) for the KP chain are plotted against $\log p$ in Fig. 6.4 for the indicated values of d. The dotted line R represents the values calculated from Eq. (6.A.39) for the spheroid-cylinder with $\epsilon = 1$ (prolate spherocylinder). It is seen that all the solid curves come in smooth contact with the dotted line at small p, as noted above. Figure 6.5 shows double-logarithmic plots of f_D against L. The solid curves represent the values for the KP chain for the indicated values of d, and the dashed curves represent the values calculated from the first of Eqs. (6.53) for the Gaussian cylinder. The KP chain is seen to be almost identical with the Gaussian chain at $d \simeq 1.0$. For the HW chain, we give numerical results on the basis of the touched-bead model in the next subsection.

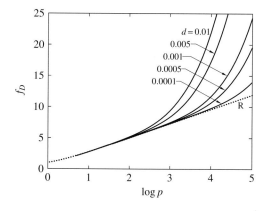

Fig. 6.4. Semi-logarithmic plots of f_D against p for the KP cylinder model for the indicated values of d. The *dotted line* R represents the values for the spheroid-cylinder with $\epsilon = 1$ (prolate spherocylinder)

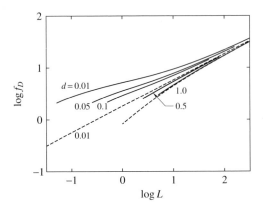

Fig. 6.5. Double-logarithmic plots of f_D against L for the KP cylinder model for the indicated values of d. The *dashed curves* represent the values for the Gaussian cylinder for $d = 0.01$ and 1.0

6.3.2 Touched-Bead Model

If we take the configurational average of both sides of Eq. (6.25), we have

$$8\pi\eta_0 \mathbf{U} = \int_{S_i} \mathbf{K}_i(\hat{\mathbf{r}}_i, \hat{\mathbf{r}}'_i) \cdot \langle \mathbf{f}_i(\hat{\mathbf{r}}'_i) \rangle d\hat{\mathbf{r}}'_i + \frac{4}{3} \sum_{\substack{j=1 \\ \neq i}}^{N} \langle R_{ij}^{-1} \rangle \langle \mathbf{F}_j \rangle , \qquad (6.57)$$

where we have put $\mathbf{v}^0 = \mathbf{0}$ and $\langle \mathbf{U}_i \rangle = \mathbf{U}$ as before, and the tensor $\mathbf{K}_i$ is defined by

$$\mathbf{K}_i(\hat{\mathbf{r}}_i, \hat{\mathbf{r}}'_i) = 8\pi\eta_0 \mathbf{T}(\hat{\mathbf{r}}'_i - \hat{\mathbf{r}}_i) . \qquad (6.58)$$

Now we define the inverse $\mathbf{K}_i^{-1}$ by

$$\delta^{(2)}(\hat{\mathbf{r}}_i - \hat{\mathbf{r}}'_i) = \int_{S_i} \mathbf{K}_i^{-1}(\hat{\mathbf{r}}_i, \hat{\mathbf{r}}''_i) \cdot \mathbf{K}_i(\hat{\mathbf{r}}''_i, \hat{\mathbf{r}}'_i) d\hat{\mathbf{r}}''_i$$

$$= \int_{S_i} \mathbf{K}_i(\hat{\mathbf{r}}_i, \hat{\mathbf{r}}''_i) \cdot \mathbf{K}_i^{-1}(\hat{\mathbf{r}}''_i, \hat{\mathbf{r}}'_i) d\hat{\mathbf{r}}''_i \qquad (6.59)$$

with $\delta^{(2)}(\mathbf{r})$ being a two-dimensional Dirac delta function. As in the first of Eqs. (6.A.8), the translational friction constant ζ $(= 3\pi\eta_0 d_b)$ [1] of the bead may then be expressed as

$$\zeta \mathbf{I} = 8\pi\eta_0 \int_{S_i} d\hat{\mathbf{r}}_i \int_{S_i} d\hat{\mathbf{r}}'_i \mathbf{K}_i^{-1}(\hat{\mathbf{r}}_i, \hat{\mathbf{r}}'_i) . \qquad (6.60)$$

Thus, multiplying both sides of Eq. (6.57) by $\mathbf{K}_i^{-1}(\hat{\mathbf{r}}''_i, \hat{\mathbf{r}}_i)$ from the left and integrating over $\hat{\mathbf{r}}''_i$ and $\hat{\mathbf{r}}_i$, we obtain

$$\langle \mathbf{F}_i \rangle + \frac{\zeta}{6\pi\eta_0} \sum_{\substack{j=1 \\ \neq i}}^{N} \langle R_{ij}^{-1} \rangle \langle \mathbf{F}_j \rangle = \zeta \mathbf{U} . \qquad (6.61)$$

This is just the KR equation determining the frictional forces $\langle \mathbf{F}_i \rangle$ in the case of translational friction [1,3]. The mean total frictional force $\langle \mathbf{F} \rangle$ is given by

$$\langle \mathbf{F} \rangle = \sum_{i=1}^{N} \langle \mathbf{F}_i \rangle = \Xi \mathbf{U} \, . \tag{6.62}$$

If we use the KR approximation [1,3], $\langle \mathbf{F}_i \rangle = N^{-1} \langle \mathbf{F} \rangle$, in Eq. (6.61), we readily have for the translational diffusion coefficient $D \, (= k_{\mathrm{B}} T / \Xi)$

$$D = \frac{k_{\mathrm{B}} T}{N \zeta} \left(1 + \frac{\zeta}{6 \pi \eta_0 N} \sum_{\substack{i=1 \\ i \neq j}}^{N} \sum_{j=1}^{N} \langle R_{ij}^{-1} \rangle \right) , \tag{6.63}$$

so that [31]

$$f_D(L; \kappa_0, \tau_0, d_{\mathrm{b}}) = 1 + \frac{d_{\mathrm{b}}}{L} \sum_{k=1}^{N-1} (L - k d_{\mathrm{b}}) \langle R^{-1}(k d_{\mathrm{b}}) \rangle , \tag{6.64}$$

where $N = L / d_{\mathrm{b}}$ and $\langle R^{-1}(L) \rangle$ is given by Eq. (6.28), all lengths being measured in units of λ^{-1}. Equation (6.63) is the well-known Kirkwood formula [1,4,32].

Figure 6.6 shows double-logarithmic plots of f_D against L for the indicated values of κ_0 and for $\tau_0 = 0$ and $d_{\mathrm{b}} = 0.15$ as typical cases corresponding to flexible polymers. The solid curves represent the values calculated from Eq. (6.64) for the touched-bead model. For comparison, in the case of $\kappa_0 = 0$ the values calculated from Eq. (6.41) with the relation $d = 0.891 d_{\mathrm{b}}$ ($d_{\mathrm{b}} = 0.15$) for the corresponding KP cylinder model are represented by the dashed curve. This relation between d and d_{b} has been obtained from a comparison between theoretical values of f_D for the touched-bead rod and the straight cylinder with hemisphere caps at the ends [24]. Even with this relation, the values of f_D for the two models are seen to differ appreciably from each

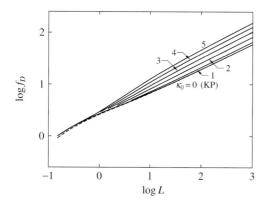

Fig. 6.6. Double-logarithmic plots of f_D against L for the HW touched-bead model for the indicated values of κ_0 and for $\tau_0 = 0$ and $d_{\mathrm{b}} = 0.15$. The *dashed curve* represents the values for the KP cylinder with $d = 0.891 d_{\mathrm{b}}$

other for small L for such flexible chains. For these, the touched-bead model is recommended. It is interesting to see that as κ_0 (helical nature) is increased, the plot changes from inverse S-shaped to S-shaped curves apart from the range of small L, all the slopes being, of course, equal to $1/2$ in the limit of $L \to \infty$.

6.4 Intrinsic Viscosity

6.4.1 Cylinder Model

Suppose that the HW cylinder is immersed in the solvent having the unperturbed flow field,

$$\mathbf{v}^0(\mathbf{r}) = \epsilon_0 \mathbf{e}_x \mathbf{e}_y \cdot \mathbf{r} , \tag{6.65}$$

where ϵ_0 is the velocity gradient and the molecular center of mass is fixed at the origin of the Cartesian coordinate system $(\mathbf{e}_x, \mathbf{e}_y, \mathbf{e}_z)$, so that the radius vector $\mathbf{r}(s)$ of the contour point s is identical with its vector distance $\mathbf{S}(s)$ from the center of mass, as depicted in Fig. 6.1. Then the cylinder rotates around the z axis with the angular velocity $\boldsymbol{\omega} = -(\epsilon_0/2)\mathbf{e}_z$, and the velocity $\mathbf{U}(s_1)$ of the contour point s_1 is given by

$$\mathbf{U}(s_1) - \mathbf{v}^0(s_1) = -\epsilon_0 \mathbf{m} \cdot \mathbf{S}(s_1) \tag{6.66}$$

with

$$\mathbf{m} = \frac{1}{2}(\mathbf{e}_x \mathbf{e}_y + \mathbf{e}_y \mathbf{e}_x) , \tag{6.67}$$

so that Eq. (6.15) becomes

$$6\pi\eta_0\epsilon_0 \mathbf{m} \cdot \mathbf{S}(s_1) = -\int_0^L K(s_1, s_2)\mathbf{f}(s_2)ds_2 . \tag{6.68}$$

The intrinsic viscosity $[\eta]$ may then be expressed in the form

$$[\eta] = -\frac{N_A}{M\eta_0\epsilon_0} \mathbf{m} : \int_0^L \langle \mathbf{f}(s)\mathbf{S}(s) \rangle ds . \tag{6.69}$$

In what follows, all lengths are measured in units of λ^{-1}. If we define a function $\varphi(s_1, s_2)$ by

$$\varphi(s_1, s_2) = -\frac{1}{\eta_0\epsilon_0} \mathbf{m} : \langle \mathbf{f}(s_1)\mathbf{S}(s_2) \rangle , \tag{6.70}$$

Eq. (6.69) may be rewritten as

$$[\eta] = \frac{N_A}{M} \int_0^L \varphi(s, s)ds . \tag{6.71}$$

If we multiply both sides of Eq. (6.68) by $\mathbf{S}(s_3)$ and take the configurational average, we obtain the integral equation for φ,

$$\int_0^L K(s_1, s_3)\, \varphi(s_3, s_2) ds_3 = \pi \langle \mathbf{S}(s_1) \cdot \mathbf{S}(s_2) \rangle, \qquad (6.72)$$

where we have exchanged s_2 for s_3.

Now it is convenient to change variables from s_1, s_2, and s_3 to x, y, and ξ, respectively, as follows,

$$x = \frac{2s_1}{L} - 1, \qquad (6.73)$$

and so on, and put

$$\psi(x, y) = \frac{1}{2}\varphi(s_1, s_2), \qquad (6.74)$$

$$g(x, y) = \frac{\pi}{L}\langle \mathbf{S}(s_1) \cdot \mathbf{S}(s_2) \rangle. \qquad (6.75)$$

Then Eqs. (6.71) and (6.72) reduce to

$$[\eta] = \frac{N_A L}{M}\int_{-1}^1 \psi(x, x) dx, \qquad (6.76)$$

$$\int_{-1}^1 K(x, \xi)\psi(\xi, y) d\xi = g(x, y), \qquad (6.77)$$

where $K(x, y) = K(s; d)$ with $s = |s_1 - s_2| = (L/2)|x - y|$.

The average in Eq. (6.75) may be evaluated from

$$\langle \mathbf{S}(s_1) \cdot \mathbf{S}(s_2) \rangle = \frac{1}{2L}\left[\int_0^L \langle R^2(s_1, s_2) \rangle ds_1 + \int_0^L \langle R^2(s_1, s_2) \rangle ds_2 \right]$$
$$- \frac{1}{2}\langle R^2(s_1, s_2) \rangle - \langle S^2 \rangle, \qquad (6.78)$$

where $\langle R^2(s_1, s_2) \rangle$ is given by Eq. (4.82) with $L = s$, and $\langle S^2 \rangle$ is the mean-square radius of gyration of the chain of contour length L and is given by Eq. (4.83). Thus we obtain, from Eqs. (6.75) and (6.78),

$$g(x, y) = \frac{{\tau_0}^2}{\nu^2} g_{KP}(x, y) + \frac{\pi {\kappa_0}^2}{2\nu^2 r^2}\left(x^2 + y^2 - 2|x - y| + \frac{2}{3} \right)$$
$$- \frac{\pi}{L^2}\mathrm{Re}\left\{ \frac{2c}{z} e^{-zL/2}\left[\cosh(zLx/2) + \cosh(zLy/2) \right] \right.$$
$$\left. + cLe^{-zL|x-y|/2} - \frac{2c}{z}\left[1 + \frac{1}{zL}(1 - e^{-zL}) \right] \right\} \qquad (6.79)$$

with

$$c = \frac{{\kappa_0}^2}{\nu^2 r^4}(4 - \nu^2 - 4i\nu), \qquad (6.80)$$

$$z = 2 + i\nu, \qquad (6.81)$$

where ν and r are given by Eqs. (4.76) and (4.77), respectively, z is the $z_{1,1}$ given by Eq. (4.45), Re indicates the real part, and g_{KP} is the g function for the KP chain and is given by

$$
g_{KP}(x,y) = \frac{\pi}{8L^2}\left\{ L^2\left(x^2 + y^2 - 2|x-y| + \frac{2}{3} \right) - 2Le^{-L|x-y|} \right.
$$
$$
\left. -2e^{-L}[\cosh(Lx) + \cosh(Ly)] + 2 + \frac{1}{L}(1 - e^{-2L}) \right\}. \quad (6.82)
$$

Now we consider the rod limit as in Eqs. (6.49) for f_D. We then have

$$
K_{(R)}(L;d) = \left(L^2 + \frac{1}{4}d^2 \right)^{-1/2}, \quad (6.83)
$$

$$
g_{(R)}(x,y) = \frac{\pi L}{4}xy. \quad (6.84)
$$

If we expand the solution $\psi(x,y)$ of the integral Eq. (6.77) with Eqs. (6.83) and (6.84) in terms of the Legendre polynomials $P_l(x)$, we can obtain its asymptotic solution in the limit of $p \to \infty$ [20]. Thus we have

$$
[\eta]_{(R^*)} = \frac{\pi N_A L^3}{24M}\left[\ln p + 2\ln 2 - \frac{7}{3} + \mathcal{O}(p^{-1}) \right]^{-1}. \quad (6.85)
$$

If we use the nonpreaveraged Oseen tensor, Eq. (6.85) is replaced by [33]

$$
[\eta]_{(R^*)} = \frac{2\pi N_A L^3}{45M}\left[\ln p + 2\ln 2 - \frac{25}{12} + \mathcal{O}(p^{-1}) \right]^{-1}, \quad (6.86)
$$

where the numerical prefactor $2/45$ is originally due to Kirkwood and Auer [34] and Ullman [35] (see also Appendix 6.A).

For the Gaussian cylinder, on the other hand, the kernel $K_{(G)}$ is given by Eq. (6.51) and the corresponding $g_{(G)}$ is given by

$$
g_{(G)}(x,y) = \frac{\pi c_\infty}{8}\left(x^2 + y^2 - 2|x-y| + \frac{2}{3} \right). \quad (6.87)
$$

Then we obtain, from Eqs. (6.76) and (6.77) with Eqs. (6.51) and (6.87), for the coil limit $[\eta]_{(C)}$

$$
\lim_{L\to\infty}[\eta]_{(G)} \equiv [\eta]_{(C)} = \Phi_\infty \frac{(c_\infty L)^{3/2}}{M} \quad (6.88)
$$

with $\Phi_\infty = 2.862 \times 10^{23}$ (mol^{-1}). The second of Eqs. (6.88) is equivalent to the KR equation in the nondraining limit [1,3]. This value of Φ_∞ is originally due to Auer and Gardner [36], who used a Gegenbauer polynomial expansion method to find the asymptotic solution, the value of their T_A being incorrect.

Note that the second of Eqs. (6.88) may be directly obtained by the use of the kernel $K(L; d) = (6/\pi c_\infty L)^{1/2}$ instead of Eq. (6.51).

For intermediate values of L, values of $[\eta]$ must be found by solving numerically the integral Eq. (6.77). On the basis of the numerical results, we may then construct approximate interpolation formulas. However, it must be noted that for rather large d (~ 1) corresponding to flexible polymers, the solution cannot be obtained for small L but is limited to the range of large L because of the nature of the kernel [20, 23].

We first consider the KP chain, for convenience. For $L \geq \sigma = 2.278$, the numerical results for $[\eta]_{\mathrm{KP}}$ may be represented by [20, 23]

$$[\eta]_{\mathrm{KP}} = \Phi_\infty \frac{L^{3/2}}{M} \left(1 - \sum_{i=1}^{4} C_i L^{-i/2}\right)^{-1} \qquad \text{for } L \geq \sigma \qquad (6.89)$$

with

$$C_i = \sum_{j=0}^{2} \alpha_{ij} d^j + \sum_{j=0}^{1} \beta_{ij} d^{2j} \ln d, \qquad (6.90)$$

where $\Phi_\infty = 2.870 \times 10^{23}$ (slightly different from the Auer–Gardner value); and α_{ij} and β_{ij} are numerical constants independent of L and d and their values are given in Table 6.1. We note that Eq. (6.89) with Eq. (6.90) is applicable for $L \geq 2.278$ when $d \leq 0.2$ and for $L^{1/2}/d \gtrsim 30$ when $0.2 < d < 1.0$.

For $L < \sigma$, we write $[\eta]_{\mathrm{KP}}$ in the form

$$[\eta]_{\mathrm{KP}} = [\eta]_{\mathrm{R}} f(L) \qquad \text{for } L < \sigma, \qquad (6.91)$$

where $[\eta]_{\mathrm{R}}$ is the intrinsic viscosity of the spheroid-cylinder such that its asymptotic form is given by Eq. (6.85) instead of by Eq. (6.86). Thus an interpolation formula for $[\eta]_{\mathrm{R}}$ is constructed to give the OB solution $[\eta]_{(\mathrm{R})}$ from Eq. (6.77) for $p \geq 100$ and to become Eq. (6.A.49) for the spheroid-cylinder for $p < 100$. The result reads [23]

Table 6.1. Values of α_{ij} and β_{ij} in Eq. (6.90)

d	i	α_{i0}	α_{i1}	α_{i2}	β_{i0}	β_{i1}
$[0, 0.1]^{\mathrm{a}}$	1	3.230981	−143.7458	−1906.263	2.463404	−1422.067
	2	−22.46149	1347.079	19387.400	−5.318869	13868.57
	3	54.81690	−3235.401	−49357.06	15.41744	−34447.63
	4	−32.91952	2306.793	36732.64	−8.516339	25198.11
$[0.1, 1.0]$	1	6.407860	−25.43785	23.33518	3.651970	−25.73698
	2	−115.0086	561.0286	−462.8501	−33.69143	523.8108
	3	318.0792	−1625.451	1451.374	92.13427	−1508.112
	4	−144.5268	661.6760	−1057.731	−42.41552	211.6622

^{a} $[a, b]$ means that $a \leq d \leq b$

$$[\eta]_\mathrm{R} = \frac{\pi N_A L^3}{24M} F(p; \epsilon) \tag{6.92}$$

with

$$\begin{aligned} F(p; \epsilon)^{-1} &= \ln p + 2\ln 2 - \frac{7}{3} + 0.548250(\ln p)^{-1} \\ &\quad -11.1231 p^{-1} \qquad \text{for } p \geq 100 \\ &= \frac{15}{16} F_\eta(p; \epsilon)^{-1} \qquad \text{for } \epsilon \leq p < 100, \end{aligned} \tag{6.93}$$

where F_η is given by Eq. (6.A.49) for $0.6 \leq \epsilon \leq 1.3$. From the numerical solutions for $[\eta]_\mathrm{KP}$ for $L < \sigma$ and $d \leq 0.1$ [20], the function f in Eq. (6.91) has been found to be almost independent of d and ϵ and may be approximated by

$$f(L) = 1 - \sum_{j=1}^{5} C_j L^j \tag{6.94}$$

with

$$\begin{aligned} C_1 &= 0.321593, \quad C_2 = 0.0466384, \quad C_3 = -0.106466, \\ C_4 &= 0.0379317, \quad C_5 = -0.00399576. \end{aligned} \tag{6.95}$$

We note that Eqs. (6.89) and (6.91) are joined smoothly at $L = \sigma$ for $d \leq 0.2$, although the latter has been constructed for $d \leq 0.1$.

Thus, as in the case of f_D, an interpolation formula for $[\eta]$ for the HW cylinder may be written in the form [23]

$$[\eta] = [\eta]_\text{a-KP} \Gamma_\eta(L; \kappa_0, \tau_0, d), \tag{6.96}$$

where $[\eta]_\text{a-KP}$ is the intrinsic viscosity of the associated KP chain and is given by

$$[\eta]_\text{a-KP} = c_\infty{}^3 [\eta]_\mathrm{KP}(c_\infty^{-1} L; c_\infty^{-1} d). \tag{6.97}$$

Note that for $[\eta]_\text{a-KP}$ the ranges of $L \gtrless \sigma$ for $[\eta]_\mathrm{KP}$ must be replaced by those of $L \gtrless \sigma c_\infty$. A good approximation to the function Γ_η, which must become unity in the limits of $L = 0$ and ∞, is of the form

$$\begin{aligned} \Gamma_\eta &= 1 + \left(\frac{A_1}{L^{1/2}} + \frac{A_2}{L}\right)\left[1 - (1 + \xi L)e^{-\xi L}\right] \\ &\quad + A_3 L^{-3/2}\left[1 - \left(1 + \xi L + \frac{1}{2}\xi^2 L^2\right)e^{-\xi L}\right] + A_4 L e^{-\xi L}, \end{aligned} \tag{6.98}$$

where ξ is given by Eq. (6.47), and A_i is of the same form as the A_i given by Eq. (6.48), the results for the numerical constants a_{ij}^{kl} involved not being given. We note that the range of application of Eq. (6.96) is limited to $d \lesssim 0.08$, although for larger d there are numerical solutions for large L.

Finally, we examine numerically the behavior of $[\eta]$. Figure 6.7 shows double-logarithmic plots of $M[\eta]/M_0[\eta]_\mathrm{E}p$ against p, where $[\eta]_\mathrm{E}$ is the Einstein intrinsic viscosity [1] of a rigid sphere of diameter d and is given by

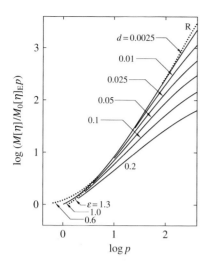

Fig. 6.7. Double-logarithmic plots of $M[\eta]/M_0[\eta]_E p$ against p for the KP cylinder model for the indicated values of d (see the text). The *dotted curves* R represent the values for the spheroid-cylinders with the indicated values of ϵ

$$[\eta]_E = \frac{5}{12}\pi N_A \left(\frac{d^3}{M_0}\right) \tag{6.99}$$

with M_0 its molecular weight. The solid curves represent the values calculated from Eqs. (6.89) and (6.91) for the KP chain with $\epsilon = 1$ for the indicated values of d. The dotted curves R represent the values calculated from Eq. (6.92) for the spheroid-cylinder with $\epsilon = 0.6$, 1.0, and 1.3, which differ appreciably from each other only for $p \lesssim 5$. Figure 6.8 shows double-logarithmic plots of $M[\eta]/\Phi_\infty L$ against L. The solid curves represent the values for the KP chain for the indicated values of d, and the dashed curves represent the values numerically obtained for the Gaussian cylinder. The KP chain is seen to be almost identical with the Gaussian chain at $d \simeq 1.0$ as in the case of f_D plotted in Fig. 6.5.

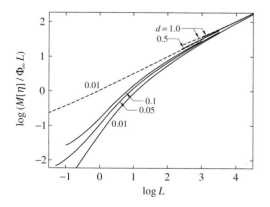

Fig. 6.8. Double-logarithmic plots of $M[\eta]/\Phi_\infty L$ against L for the KP cylinder model for the indicated values of d. The *dashed curves* represent the values for the Gaussian cylinder for $d = 0.01$ and 1.0

6.4.2 Touched-Bead Model

In the unperturbed flow field $\mathbf{v}^0$ given by Eq. (6.65), the velocity $\mathbf{U}_i(\hat{\mathbf{r}}_i)$ of the point $\hat{\mathbf{r}}_i$ of the ith bead is given by

$$\mathbf{U}_i(\hat{\mathbf{r}}_i) - \mathbf{v}^0(\hat{\mathbf{r}}_i) = -\epsilon_0 \mathbf{m} \cdot (\mathbf{S}_i + \hat{\mathbf{r}}_i), \qquad (6.100)$$

where the vector position $\mathbf{r}_i$ of the center of the ith bead is identical with its vector distance $\mathbf{S}_i$ from the center of mass, as depicted in Fig. 6.3. Thus Eq. (6.25) becomes

$$-8\pi\eta_0\epsilon_0 \mathbf{m} \cdot (\mathbf{S}_i + \hat{\mathbf{r}}_i) = \int_{S_i} \mathbf{K}_i(\hat{\mathbf{r}}_i, \hat{\mathbf{r}}_i') \cdot \mathbf{f}_i(\hat{\mathbf{r}}_i') d\hat{\mathbf{r}}_i' + \frac{4}{3}\sum_{\substack{j=1 \\ \neq i}}^N \langle R_{ij}^{-1}\rangle \mathbf{F}_j .$$

$$(6.101)$$

As shown in Appendix 6.B, $[\eta]$ may then be expressed in the form

$$[\eta] = -\frac{N_A}{M\eta_0\epsilon_0}\sum_{i=1}^N \left[\mathbf{m} : \langle \mathbf{F}_i\mathbf{S}_i\rangle + \mathbf{m} : \left\langle \int_{S_i} \mathbf{f}_i(\hat{\mathbf{r}}_i)\hat{\mathbf{r}}_i d\hat{\mathbf{r}}_i \right\rangle\right]. \quad (6.102)$$

Now, multiplying both sides of Eq. (6.101) by $\mathbf{K}_i^{-1}(\hat{\mathbf{r}}_i'', \hat{\mathbf{r}}_i)$ from the left and integrating over $\hat{\mathbf{r}}_i$, we find [28]

$$\mathbf{f}_i(\hat{\mathbf{r}}_i) = -8\pi\eta_0 \int_{S_i} \mathbf{K}_i^{-1}(\hat{\mathbf{r}}_i, \hat{\mathbf{r}}_i')\hat{\mathbf{r}}_i' d\hat{\mathbf{r}}_i' : \epsilon_0 \mathbf{m}$$

$$-8\pi\eta_0 \int_{S_i} \mathbf{K}_i^{-1}(\hat{\mathbf{r}}_i, \hat{\mathbf{r}}_i') d\hat{\mathbf{r}}_i'$$

$$\cdot \left(\epsilon_0 \mathbf{m} \cdot \mathbf{S}_i + \frac{1}{6\pi\eta_0}\sum_{\substack{j=1 \\ \neq i}}^N \langle R_{ij}^{-1}\rangle \mathbf{F}_j\right). \quad (6.103)$$

Integration of both sides of Eq. (6.103) over $\hat{\mathbf{r}}_i$ leads to

$$\mathbf{F}_i + \frac{\zeta}{6\pi\eta_0}\sum_{\substack{j=1 \\ \neq i}}^N \langle R_{ij}^{-1}\rangle \mathbf{F}_j = -\eta_0\phi_i : \epsilon_0 \mathbf{m} - \zeta\epsilon_0 \mathbf{m} \cdot \mathbf{S}_i , \quad (6.104)$$

where we have used Eq. (6.60) and ϕ_i is the shear force triadic [37], which is given in the present notation by

$$\phi_i = 8\pi \int_{S_i} d\hat{\mathbf{r}}_i \int_{S_i} d\hat{\mathbf{r}}_i' \mathbf{K}_i^{-1}(\hat{\mathbf{r}}_i, \hat{\mathbf{r}}_i')\hat{\mathbf{r}}_i' . \quad (6.105)$$

According to Brenner [37], ϕ_i vanishes for spherically isotropic bodies. Then, if we multiply both sides of Eq. (6.104) by $\mathbf{S}_j$ from the right and average them over chain configurations, we obtain

$$\langle \mathbf{F}_i \mathbf{S}_i \rangle + \frac{\zeta}{6\pi\eta_0} \sum_{\substack{k=1 \\ \neq i}}^{N} \langle R_{ik}^{-1} \rangle \langle \mathbf{F}_k \mathbf{S}_j \rangle = -\zeta\epsilon_0 \mathbf{m} \cdot \langle \mathbf{S}_i \mathbf{S}_j \rangle . \tag{6.106}$$

Thus the first term in the square brackets of Eq. (6.102) gives the KR intrinsic viscosity [1,3].

Next we evaluate the second term in the square brackets of Eq. (6.102). Multiplying both sides of Eq. (6.103) by $\hat{\mathbf{r}}_i$ from the left, integrating over $\hat{\mathbf{r}}_i$, and making a double-dot product of the result and the symmetric tensor $\mathbf{m}$, we obtain

$$-\mathbf{m} : \int_{S_i} \mathbf{f}_i(\hat{\mathbf{r}}_i)\hat{\mathbf{r}}_i d\hat{\mathbf{r}}_i = 8\pi\eta_0 \mathbf{m} : \int_{S_i} d\hat{\mathbf{r}}_i \int_{S_i} d\hat{\mathbf{r}}_i' \hat{\mathbf{r}}_i \mathbf{K}_i^{-1}(\hat{\mathbf{r}}_i, \hat{\mathbf{r}}_i')\hat{\mathbf{r}}_i' : \epsilon_0 \mathbf{m} , \tag{6.107}$$

where we have put $\mathbf{f}_i \hat{\mathbf{r}}_i = \hat{\mathbf{r}}_i \mathbf{f}_i$ since $\mathbf{f}_i \hat{\mathbf{r}}_i$ is a symmetric tensor, and used again the relation $\boldsymbol{\phi}_i = \mathbf{0}$. The right-hand side of Eq. (6.107), which we denote by σ, is given by

$$\sigma = -\mathbf{m} : \int_{S_i} \mathbf{f}_i^0(\hat{\mathbf{r}}_i)\hat{\mathbf{r}}_i d\hat{\mathbf{r}}_i , \tag{6.108}$$

where $\mathbf{f}_i^0$ is the solution of the integral equation [38],

$$\int_{S_i} \mathbf{K}_i(\hat{\mathbf{r}}_i, \hat{\mathbf{r}}_i') \cdot \mathbf{f}_i^0(\hat{\mathbf{r}}_i')d\hat{\mathbf{r}}_i' = -8\pi\eta_0\epsilon_0 \mathbf{m} \cdot \hat{\mathbf{r}}_i . \tag{6.109}$$

Therefore, $\mathbf{f}_i^0$ is given by the first term on the right-hand side of Eq. (6.103) and represents the frictional force distribution on the surface of the single isolated bead under the nonslip boundary condition when it rotates around its center with the angular velocity $-(\epsilon_0/2)\mathbf{e}_z$ in the flow field given by Eq. (6.65), so that σ represents the increment of the xy component of the stress tensor due to the single Einstein sphere,

$$\sigma = \frac{5}{12}\pi d_{\mathrm{b}}^3 \eta_0 \epsilon_0 . \tag{6.110}$$

Thus $[\eta]$ may be expressed in the form [28]

$$[\eta] = [\eta]^{(\mathrm{KR})} + [\eta]_{\mathrm{E}} , \tag{6.111}$$

where $[\eta]^{(\mathrm{KR})}$ and $[\eta]_{\mathrm{E}}$ are the KR and Einstein intrinsic viscosities [1] given by

$$[\eta]^{(\mathrm{KR})} = \frac{N_{\mathrm{A}}}{M} \sum_{i=1}^{N} \varphi_{ii} , \tag{6.112}$$

$$[\eta]_{\mathrm{E}} = \frac{5}{12}\pi N_{\mathrm{A}} \left(\frac{d_{\mathrm{b}}^3}{M_0}\right) , \tag{6.113}$$

respectively [compare Eq. (6.113) with Eq. (6.99)]. In Eq. (6.112), φ_{ij} is defined by

$$\varphi_{ij} = -\frac{1}{\eta_0 \epsilon_0} \mathbf{m} : \langle \mathbf{F}_i \mathbf{S}_j \rangle , \tag{6.114}$$

and is the solution of the linear coupled equations obtained from Eq. (6.106); that is,

$$\varphi_{ij} + \frac{1}{2} d_{\mathrm{b}} \sum_{\substack{k=1 \\ \neq i}}^{N} \langle R_{ik}^{-1} \rangle \varphi_{kj} = \frac{1}{2} \pi d_{\mathrm{b}} \langle \mathbf{S}_i \cdot \mathbf{S}_j \rangle . \tag{6.115}$$

In what follows, all lengths are measured in units of λ^{-1}. Then $\langle R_{ij}^{-1} \rangle$ is given by Eq. (6.26) and $\langle \mathbf{S}_i \cdot \mathbf{S}_j \rangle$ is given by

$$\langle \mathbf{S}_i \cdot \mathbf{S}_j \rangle = \frac{N d_{\mathrm{b}}}{\pi} g \left(\frac{2i-1}{N} - 1, \frac{2j-1}{N} - 1 \right) , \tag{6.116}$$

where $g(x, y)$ is given by Eq. (6.79).

Now we construct interpolation formulas for the KR intrinsic viscosities $[\eta]^{(\mathrm{KR})}$ of the KP and HW chains. The former may be written in the form

$$[\eta]_{\mathrm{KP}}^{(\mathrm{KR})} = 6^{3/2} \Phi_\infty \frac{\langle S^2 \rangle_{\mathrm{KP}}^{3/2}}{M} \Gamma_{\mathrm{KP}}(L; d_{\mathrm{b}}) , \tag{6.117}$$

where $\Phi_\infty = 2.870 \times 10^{23}$, and $\langle S^2 \rangle_{\mathrm{KP}}$ is the mean-square radius of gyration of the KP chain of total contour length $L = N d_{\mathrm{b}}$ and is given by Eq. (4.85). We evaluate the function Γ_{KP} by the use of values of $[\eta]^{(\mathrm{KR})}$ calculated from Eq. (6.112) with the numerical solutions of Eq. (6.115) for various values of d_{b} ranging from 0.01 to 0.8, where the number of beads N is limited to 2–1000, so that the contour length L is limited to small values for small d_{b}. For larger L, therefore, we adopt the values of $[\eta]_{\mathrm{KP}}$ of the KP cylinder model having the cylinder diameter d properly chosen as those of $[\eta]_{\mathrm{KP}}^{(\mathrm{KR})}$ of the KP touched-bead model. (Recall that for large L the solutions for the cylinder model can be obtained.) It has been found that the values of $[\eta]_{\mathrm{KP}}^{(\mathrm{KR})}$ obtained numerically from Eq. (6.112) are joined smoothly to those of $[\eta]_{\mathrm{KP}}$ of the KP cylinder model calculated from Eqs. (6.89) and (6.91) at $N(= L/d_{\mathrm{b}}) \simeq 1000$ for $0.01 \leq d_b \leq 0.8$ if we choose $d = 0.74 d_{\mathrm{b}}$ [28]. A good approximation to $\Gamma_{\mathrm{KP}}^{-1}$ so obtained is of the form

$$\Gamma_{\mathrm{KP}}^{-1} = 1 + e^{-5L} \sum_{i=0}^{3} C_i L^{i/2} + e^{-1/4L} \sum_{i=4}^{7} C_i L^{-(i-3)/2} \tag{6.118}$$

with

$$C_i = \sum_{j=0}^{2} \alpha_{ij} d_{\mathrm{b}}^{j} + \sum_{j=0}^{1} \beta_{ij} d_{\mathrm{b}}^{2j} \ln d_{\mathrm{b}} , \tag{6.119}$$

where α_{ij} and β_{ij} are numerical constants independent of L and d_{b} and their values are given in Table 6.2. We note that Eq. (6.117) with Eqs. (6.118) and

Table 6.2. Values of α_{ij} and β_{ij} in Eq. (6.119)

i	α_{i0}	α_{i1}	α_{i2}	β_{i0}	β_{i1}
0	-9.6291	$1.6198\ (2)$	$1.1316\ (2)$	-1.5358	$9.4913\ (2)$
1	-2.3491	$-1.4420\ (2)$	$-2.0502\ (3)$	-2.3605	$-3.4732\ (3)$
2	$5.4811\ (1)^{a}$	$-4.8402\ (2)$	$4.1942\ (3)$	$1.0550\ (1)$	$4.0771\ (3)$
3	$-6.2255\ (1)$	$7.8877\ (2)$	$-2.6846\ (3)$	$-1.1528\ (1)$	$-1.1290\ (3)$
4	$3.0814\ (-1)$	-4.5617	1.5182	-1.9421	-3.1301
5	-5.1619	$1.6758\ (1)$	-4.0308	$5.1951\ (-1)$	$1.2811\ (1)$
6	2.9298	$-1.3380\ (1)$	-2.6757	$1.1938\ (-1)$	-9.9978
7	$-6.2856\ (-1)$	1.6070	7.4332	$-8.2021\ (-2)$	-2.8832

^{a} $a(n)$ means $a \times 10^{n}$

(6.119) is applicable for $0.01 \le d_{\mathrm{b}} \le 0.8$, and for the integral values of L/d_{b} when $2 \le L/d_{\mathrm{b}} \le 1000$ and for all values of L/d_{b} when $L/d_{\mathrm{b}} \ge 1000$.

For the case of the HW touched-bead model, $[\eta]^{(\mathrm{KR})}$ may be written in the form

$$[\eta]^{(\mathrm{KR})} = \left(\frac{\langle S^2 \rangle}{\langle S^2 \rangle_{\mathrm{KP}}} \right)^{3/2} [\eta]^{(\mathrm{KR})}_{\mathrm{KP}} \Gamma^{(\mathrm{KR})}_{\eta} (L; \kappa_0, \tau_0, d_b). \qquad (6.120)$$

A good approximation to the function $\Gamma^{(\mathrm{KR})}_{\eta}$ constructed similarly but by proper extrapolations to $L = \infty$ in some cases is of the form

$$\Gamma^{(\mathrm{KR})}_{\eta} = 1 + \kappa_0^2 \left[e^{-\nu L/10} \sum_{i=1}^{2} A_i L^i + e^{-12/\nu L} \sum_{i=3}^{6} A_i L^{-(i-2)/2} \right] \qquad (6.121)$$

with

$$A_i = \sum_{k,l=0}^{3} \left[\sum_{j=0}^{2} a_{ij}^{kl} d_{\mathrm{b}}^{\,j} + \sum_{j=3}^{4} a_{ij}^{kl} d_{\mathrm{b}}^{\,2(j-2)} \ln d_{\mathrm{b}} \right] \nu^l \cos(k\pi\tau_0/\nu), \qquad (6.122)$$

where a_{ij}^{kl} are numerical constants independent of L, κ_0, τ_0, and d_{b} and their values are given in Appendix V. The ranges of κ_0 and τ_0 in which Eq. (6.120) with Eqs. (6.121) and (6.122) is applicable are such that κ_0 and τ_0 satisfy the conditions, $\nu < 8$, $\kappa_0 \le 7$, $\tau_0 \ge \kappa_0 - 5.5$, $\kappa_0 + \tau_0 > 0.5$, $\kappa_0 = 0.5i$, and $\tau_0 = j$ with i the positive integer and j the nonnegative integer; and the range of d_{b} is limited to $0.01 \le d_{\mathrm{b}} \le 0.8$ for $\tau_0 > 2\kappa_0$, to $0.01 \le d_{\mathrm{b}} \le 0.6$ for $\tau_0 < 2\kappa_0$ and $\kappa_0 \le 4$, to $0.01 \le d_{\mathrm{b}} \le 0.4$ for $4 < \kappa_0 \le 5$, and to $0.01 \le d_{\mathrm{b}} \le 0.2$ for $\kappa_0 > 5$.

Figure 6.9 shows double-logarithmic plots of $M[\eta]/\Phi_\infty L$ against L for the indicated values of κ_0 and for $\tau_0 = 0$ and $d_{\mathrm{b}} = 0.15$ as typical cases corresponding to flexible polymers. The solid curves represent the values calculated from Eq. (6.111) with Eqs. (6.113) and (6.120) for the touched-bead model. The dashed curves represent the corresponding values of $[\eta]^{(\mathrm{KR})}$ (without

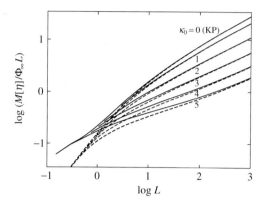

Fig. 6.9. Double-logarithmic plots of $M[\eta]/\Phi_\infty L$ against L for the HW touched-bead model for the indicated values of κ_0 and for $\tau_0 = 0$ and $d_b = 0.15$. The *solid and dashed curves* represent the values with and without the contribution of the Einstein intrinsic viscosity $[\eta]_E$, respectively

$[\eta]_E$). It is seen that the contribution of $[\eta]_E$ is very important in the oligomer region and that as κ_0 (helical nature) is increased, the plot changes from S-shaped to inverse S-shaped curves apart from the range of small L in contrast to the case of f_D plotted in Fig. 6.6, all the slopes becoming equal to $1/2$ in the limit of $L \to \infty$.

6.5 Analysis of Experimental Data

6.5.1 Basic Equations and Model Parameters

We begin by making a comparison of theory with experiment for several flexible and semiflexible polymers to determine their HW model parameters. For convenience, we first analyze data for the intrinsic viscosity $[\eta]$, using the molecular weight M instead of the total contour length L as usual. As already noted, we then adopt the HW (or KP) touched-bead model for flexible polymers and the KP cylinder model for semiflexible polymers, and write $[\eta]$ in the form

$$[\eta] = (\lambda^2 M_L)^{-2} f_\eta(\lambda L; \lambda^{-1}\kappa_0, \lambda^{-1}\tau_0, \lambda d_b) \quad \text{(HW)}$$
$$= (\lambda^2 M_L)^{-2} f_\eta(\lambda L; \lambda d) \quad \text{(KP)} \qquad (6.123)$$

with

$$\log M = \log(\lambda L) + \log(\lambda^{-1} M_L), \qquad (6.124)$$

where $L = N d_b$ for the touched-bead model. The function f_η is defined by

$$f_\eta(\lambda L) = \lambda^{-1} M_L[\bar\eta], \qquad (6.125)$$

where $[\bar\eta]$ is the intrinsic viscosity measured in units of $(\lambda^{-1})^3$ and is given by Eq. (6.111) for the HW model and by Eqs. (6.89) and (6.91) for the KP model. In the limit of $\lambda L \to \infty$, we have

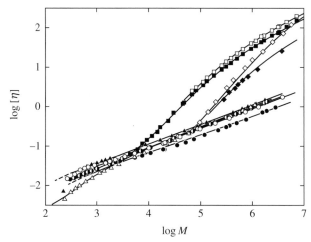

Fig. 6.10. Double-logarithmic plots of $[\eta]$ (in dL/g) against M for a-PS in cyclo-hexane at 34.5°C (○) [39–42], a-PMMA in acetonitrile at 44.0°C (●) [41,43,44], i-PMMA in acetonitrile at 28.0°C (◐) [45,46], PIB in IAIV at 25.0°C (▲) [41,47], PDMS in MEK at 20.0°C (△) [48], PHIC in n-butyl chloride at 25°C (■) [49], PHIC in n-hexane at 25°C (□) [50], DNA in 0.2 mol/l NaCl at 25°C (◆) [51–53], and schizophyllan in water at 25°C (◇) [54]. The *solid curves* represent the best-fit HW or KP theoretical values, each *dashed line* segment connecting the HW values for $N = 1$ and 2

$$\lim_{\lambda L \to \infty} \left[\frac{f_\eta(\lambda L)}{(\lambda L)^{1/2}} \right] = c_\infty^{3/2} \Phi_\infty \,, \tag{6.126}$$

where $c_\infty = 1$ for the KP chain and $\Phi_\infty = 2.870 \times 10^{23}$ (mol^{-1}).

Figure 6.10 shows double-logarithmic plots of $[\eta]$ (in dL/g) against M for a-PS ($f_r = 0.59$) in cyclohexane at 34.5°C (Θ) [39–42], a-PMMA ($f_r = 0.79$) in acetonitrile at 44.0°C (Θ) [41,43,44], i-PMMA ($f_r = 0.01$) in acetonitrile at 28.0°C (Θ) [45,46], polyisobutylene (PIB) in isoamyl isovalerate (IAIV) at 25.0°C (Θ) [41,47], PDMS in methyl ethyl ketone (MEK) at 20.0°C (Θ) [48], PHIC in n-butyl chloride at 25°C [49], PHIC in n-hexane at 25°C [50], DNA in 0.2 mol/l NaCl at 25°C [51–53], and schizophyllan in water at 25°C [54]. In the figure the solid curves represent the best-fit HW and KP theoretical values calculated from the first and second lines of Eqs. (6.123) for the flexible and semiflexible polymers, respectively, with the values of the model parameters listed in Table 6.3, where in the figure each dashed line segment connects the HW values for $N = 1$ and 2, and we have assumed $\epsilon = 1$ for the KP chain (cylinder). For the table we note that we have used the values of $\lambda^{-1}\kappa_0$ and $\lambda^{-1}\tau_0$ determined from equilibrium properties for a-PS and PDMS since all the parameters cannot be determined unambiguously, and that for the same reason, we have assumed that the PIB chain takes the 8_3 helix in dilute so-lution as well as in the crystalline state [55], so that it may be represented

Table 6.3. Values of the HW model parameters for typical flexible and semiflexible polymers from $[\eta]$ and D

Polymer (f_r)	Solvent	Temp. (°C)	$\lambda^{-1}\kappa_0$	$\lambda^{-1}\tau_0$	λ^{-1} (Å)	M_L (Å^{-1})	d_b (d) (Å)	Obs. (Ref.)
a-PS (0.59)	Cyclohexane	34.5	$(3.0)^a$	$(6.0)^a$	23.5	42.6	10.1	$[\eta]$ ([39–42])
			$(3.0)^a$	$(6.0)^a$	27.0	35.0	9.5	D ([31,41])
a-PMMA (0.79)	Acetonitrile	44.0	4.5	2.0	45.0	38.6	7.2	$[\eta]$ ([41,43,44])
			$(4.5)^b$	$(2.0)^b$	65.0	35.0	9.0	D ([41,56,57])
i-PMMA (0.01)	Acetonitrile	28.0	2.5	2.0	32.6	38.6	8.2	$[\eta]$ ([45,46])
			$(2.5)^b$	$(2.0)^b$	45.5	33.0	9.1	D ([45,57])
PIB	IAIV,Benzene	25.0	0	$\cdots$	12.7	24.1	6.4	$[\eta]$ ([41,47])
	IAIV		0	$\cdots$	18.7	24.1	6.9	D ([41,59])
PDMS	MEK	20.0	$(2.6)^c$	$(0)^c$	28.0	20.6	2.0	$[\eta]$ ([48])
	Bromo-cyclohexane	29.5	$(2.6)^c$	$(0)^c$	25.5	20.6	2.0	$[\eta]$ ([41,48])
			$(2.6)^c$	$(0)^c$	31.0	18.0	1.6	D ([41,48])
PHIC	n-Butyl chloride	25	0	$\cdots$	700	76.0	(15)	$[\eta]$ ([49])
	n-Hexane	25	0	$\cdots$	840	71.5	(16)	$[\eta]$ ([50])
			0	$\cdots$	840	71.5	(25)	s ([50])
DNA	0.2 mol/l NaCl	25	0	$\cdots$	1200	195	(15)	$[\eta]$ ([51–53])
			0	$\cdots$	1200	195	(25)	s ([51-53,58])
Schizo-phyllan	Water	25	0	$\cdots$	4000	215	(26)	$[\eta], s$ ([54])

[a] From $\langle \Gamma^2 \rangle$ (Table 5.1)

[b] From $[\eta]$

[c] From $\langle \mu^2 \rangle$ (see Sect. 5.4.1)

by the KP (touched-bead) chain with $M_L = 24.1$ Å^{-1}, taking the helix axis as its contour.

Before discussing the results for $[\eta]$, we analyze data for the translational diffusion coefficient D (or the sedimentation coefficient s). Corresponding to Eqs. (6.123), we write D in the form

$$\frac{\eta_0 M D}{k_B T} = \left(\frac{M_L}{3\pi}\right) f_D(\lambda L; \lambda^{-1}\kappa_0, \lambda^{-1}\tau_0, \lambda d_b) \quad \text{(HW)}$$

$$= \left(\frac{M_L}{3\pi}\right) f_D(\lambda L; \lambda d) \quad \text{(KP)}, \qquad (6.127)$$

where f_D is given by Eq. (6.64) for the HW model and by Eq. (6.41) for the KP model. In the limit of $\lambda L \to \infty$, we have

$$\lim_{\lambda L \to \infty} \left[\frac{f_D(\lambda L)}{(\lambda L)^{1/2}}\right] = \frac{\sqrt{6}}{2} c_\infty^{-1/2} \rho_\infty , \qquad (6.128)$$

where ρ_∞ is equal to the Kirkwood value 1.505 [3,4,32] (see the next subsection).

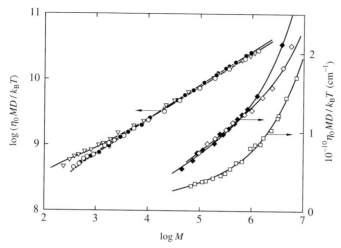

Fig. 6.11. Double-logarithmic plots of $\eta_0 MD/k_B T$ (in cm^{-1}) against M for a-PS in cyclohexane at 34.5°C (○) [31,41], a-PMMA in acetonitrile at 44.0°C (●) [41,56,57], and PDMS in bromocyclohexane at 29.5°C (▽) [41,48], and semi-logarithmic plots of $\eta_0 MD/k_B T$ (in cm^{-1}) against M for PHIC in n-hexane at 25°C (□) [50], DNA in 0.2 mol/l NaCl at 25°C (◆) [51–53,58], and schizophyllan in water at 25°C (◇) [54]. The *solid curves* represent the best-fit HW or KP theoretical values, each *dashed line* segment connecting the HW values for $N = 1$ and 2

Figure 6.11 shows double-logarithmic plots of $\eta_0 MD/k_B T$ (in cm^{-1}) against M for a-PS in cyclohexane at 34.5°C (Θ) [31, 41], a-PMMA in acetonitrile at 44.0°C (Θ) [41,56,57], and PDMS in bromocyclohexane at 29.5°C (Θ) [41, 48], and semi-logarithmic plots of $\eta_0 MD/k_B T$ against M for PHIC in n-hexane at 25°C [50], DNA in 0.2 mol/l NaCl at 25°C [51–53, 58], and schizophyllan in water at 25°C [54], where the values of D for the semiflexible polymers have been calculated from s using the first of Eqs. (6.38). In the figure the solid curves represent the best-fit HW and KP theoretical values calculated from the first and second lines of Eqs. (6.127) for the flexible and semiflexible polymers, respectively, with the values of the model parameters listed in Table 6.3, where in the figure each dashed line segment connects the HW values for $N = 1$ and 2. Note that we have used the values of $\lambda^{-1}\kappa_0$ and $\lambda^{-1}\tau_0$ determined from the equilibrium properties for a-PS and PDMS and from $[\eta]$ for a-PMMA. A similar analysis has also been made for i-PMMA in acetonitrile at 28.0°C (Θ) [45, 57] and PIB in IAIV at 25.0°C (Θ) [41, 59], the values of the model parameters determined being given in Table 6.3.

Now we are in a position to discuss the above results of analysis of $[\eta]$ and D. In general, there is seen to be rather good agreement between theory and experiment over a wide range of M. The most important fact that is observed in Fig. 6.10 is that for flexible polymers the exponent law for the relation between $[\eta]$ and M, that is, the Houwink–Mark–Sakurada relation holds only in a limited range of M, although the exponent becomes asymptotically equal to 1/2 for large M. In particular, it is interesting to see that for a-PMMA the double-logarithmic plot of $[\eta]$ against M follows an inverse S-shaped curve, exhibiting the asymptotic behavior only for $M \gtrsim 10^5$, as also predicted by the theory, that for PDMS the plot does not exhibit the asymptotic behavior in the range of M examined, and that for PIB and PDMS $[\eta]$ decreases sharply with decreasing M for small M, especially for the latter for $M \lesssim 10^4$. As seen from Fig. 6.11, on the other hand, the deviation of the double-logarithmic plot of MD against M from the asymptotic relation (with slope 1/2) for flexible polymers is rather small, but for a-PMMA the plot clearly follows an S-shaped curve corresponding to the plot of $[\eta]$. Such behavior of $[\eta]$ and D of a-PMMA is characteristic of the chain of strong helical nature. It is seen from Table 6.3 that for flexible polymers the values of λ^{-1} determined from $[\eta]$ are somewhat smaller than those from D, while the values of M_L determined from $[\eta]$ are somewhat larger than those from D. This is due to the disagreement between the theoretical and experimental values of Φ_∞ and ρ_∞, which may be regarded as arising from the preaveraging of the Oseen tensor (see the next subsection). On the other hand, the values of d_b obtained may be reasonable except for PDMS, considering the chemical structures of the chains. For PIB and PDMS, a further analysis of experimental data is made in later subsections.

For the semiflexible polymers, agreement between the values of both λ^{-1} and M_L from $[\eta]$ and s is not very bad. The reason for this is that the equations for the KP cylinder model for small λL (or M) have been obtained so as to be completely (for D) or almost (for $[\eta]$) free from the preaveraging approximation (see also the next subsection). It is interesting to note that the values of M_L obtained for DNA and schizophyllan are just those corresponding to their double and triple helices, respectively. For PHIC and DNA, however, the values of d obtained from $[\eta]$ are seen to be appreciably smaller than those from s. This may probably be due to the fact that the rough surface of the real semiflexible polymer chain has been replaced by the smooth cylinder surface [24]. For a similar analysis of $[\eta]$ and s for a wider variety of semiflexible polymers, the reader is referred to the review article by Norisuye [60].

6.5.2 Reduced Hydrodynamic Volume and Radius

The *reduced hydrodynamic volume* Φ and *radius* ρ^{-1} may be defined by

$$\Phi = \frac{V_H}{\langle S^2 \rangle^{3/2}}, \tag{6.129}$$

$$\rho^{-1} = \frac{R_H}{\langle S^2 \rangle^{1/2}}, \tag{6.130}$$

so that $\rho\Phi = V_H / R_H \langle S^2 \rangle$, where V_H and R_H are the hydrodynamic (molar) volume and radius defined by

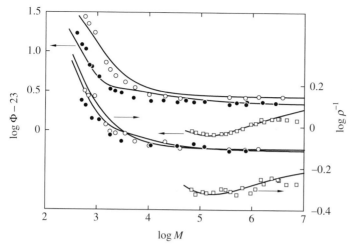

Fig. 6.12. Double-logarithmic plots of Φ and ρ^{-1} against M for a-PS in cyclohexane at 34.5°C ($\circ$) [31,39,41], a-PMMA in acetonitrile at 44.0°C ($\bullet$) [43,56,61], and PHIC in n-hexane at 25°C ($\square$) [50]. The *solid curves* represent the HW or KP theoretical values (see the text)

$$V_{\mathrm{H}} = 6^{-3/2} M [\eta] \,, \tag{6.131}$$

$$R_{\mathrm{H}} = \frac{k_{\mathrm{B}} T}{6\pi\eta_0 D} \,. \tag{6.132}$$

Note that Φ is just the Flory–Fox factor [1] and that Φ and ρ become Φ_∞ and ρ_∞, respectively, in the limit of $M \to \infty$. (It is unfortunate that the reduced radius is usually defined by the reciprocal of ρ instead of by ρ.)

Figure 6.12 shows double-logarithmic plots of Φ and ρ^{-1} against M for a-PS in cyclohexane at 34.5°C (Θ) [31, 39, 41], a-PMMA in acetonitrile at 44.0°C (Θ) [43, 56, 61], and PHIC in n-hexane at 25°C [50]. The solid curves represent the HW or KP theoretical values calculated from Eqs. (6.129) and (6.130) with Eqs. (6.131) and (6.132) with the values of the model parameters given in Table 6.3, where the theoretical values have been multiplied by the constant ratios of the experimental to theoretical Φ_∞ and ρ_∞^{-1}, respectively, for the flexible polymers because of the appreciable differences between their theoretical and experimental values (see below). It is interesting to see that both experimentally and theoretically, Φ and ρ^{-1} increase with decreasing M for small M, and in particular, they exhibit a minimum for PHIC. More important is the fact that even in the limit of $M \to \infty$, the values of Φ_∞ and ρ_∞^{-1} for a-PMMA are definitely different from those for a-PS, indicating that Φ and ρ are not necessarily universal constants in contradiction to the Flory view [1].

The experimental values of Φ_∞ and ρ_∞ (with polydispersity corrections) for the five flexible polymers in Fig 6.10 are summarized in Table 6.4 [41,45]. It is seen that the former can never be regarded as a universal constant, being

Table 6.4. Values of Φ_∞ and ρ_∞ for flexible polymers

Polymer (f_r)	Solvent	Temp. (°C)	$10^{-23}\Phi_\infty$ (mol^{-1})	ρ_∞	ρ_∞ (calc)
a-PS (0.59)	Cyclohexane	34.5	2.79±0.08	1.26±0.01	1.35
	trans-Decaline	21.0	2.75±0.09	1.27±0.01	1.35
a-PMMA (0.79)	Acetonitrile	44.0	2.34±0.06	1.29±0.02	1.34
	n-Butyl chloride	40.8	2.60±0.06	1.24±0.02	1.34
i-PMMA (0.01)	Acetonitrile	28.0	2.58±0.11	1.25±0.02	1.34
PIB	IAIV	25.0	2.71±0.06	1.27±0.01	1.36
PDMS	Bromocyclohexane	29.5	2.79±0.04	1.28±0.02	1.37

also clearly dependent on solvent for a-PMMA, and that they are apprecia-bly smaller than the Kirkwood values 2.87×10^{23} (exactly 2.862×10^{23}) of Φ_∞ [3,36] and 1.505 of ρ_∞ [3,4,32] (even the Zimm value 1.479 of ρ_∞ [62]), respectively, as has often been pointed out for flexible polymers [60]. If Φ_∞ depends on solvent as in the case of a-PMMA, some consideration is required in an analysis of experimental data since any existent theory of $[\eta]$ cannot explain this fact (see below). Thus we use a maneuver to remove the diffi-culty (for flexible polymers). It consists of introducing empirically a constant prefactor C_η into the right-hand side of the first line of Eqs. (6.123) as [43]

$$[\eta] = C_\eta (\lambda^2 M_L)^{-1} f_\eta \qquad (6.133)$$

in order to take into account the difference between observed values of Φ_∞ in two or more Θ solvents. For a-PMMA, for example, C_η is set equal to unity and 1.11 (= 2.60/2.34, the ratio of Φ_∞) in acetonitrile and n-butyl chloride, respectively. Then we obtain the value 7.9 Å (instead of 7.2 Å) for d_b for a-PMMA in n-butyl chloride at 40.8°C (Θ), the values of the other model parameters being the same as those in acetonitrile at 44.0°C (Θ) (in Table 6.3) [43].

Thus it is pertinent to give here a brief survey of theoretical investigations of Φ_∞ and ρ_∞ performed since the earlier theories. Fixman and Pyun [1,63, 64] evaluated long ago Φ_∞ for the Gaussian chain (spring-bead model) by perturbation theory with the use of the nonpreaveraged Oseen tensor or with fluctuating hydrodynamic interaction (HI) and showed its decrease below the Kirkwood value. In 1980, Zimm [65] carried out Monte Carlo evaluation of $[\eta]$ and D similarly for the Gaussian chain with fluctuating HI in the rigid-body ensemble approximation and found that the Kirkwood values of Φ_∞ and ρ_∞ are about 12 and 13% too high, respectively. Subsequently, Fixman [66,67] derived the decrease in ρ_∞ below the Kirkwood value, depending on the local structure and hence the stiffness of the chain, by introducing constraints on bond lengths and bond angles, or equivalently internal friction, in the chain with fluctuating HI. As shown in Chap. 9 (Appendix 9.A), a similar result can be obtained on the basis of the HW chain with partially fluctuating

				Fig. 6.13. Conformational
$10^{-23}\,\Phi_\infty$	0	1.5–2.8	9.23	change from a long rigid rod to a rigid sphere
ρ_∞	∞	1.5–1.2	0.775	through a random coil with the values of their Φ_∞, ρ_∞,
$N_A^{-1}\,\Phi_\infty\,\rho_\infty$	0.228	0.3–0.6	1.19	and $N_A^{-1}\Phi_\infty\rho_\infty$

(orientation-dependent) HI [68]. The values of ρ_∞ so calculated are given in the last column of Table 6.4. It is seen that they are smaller than the Kirkwood and Zimm values [3, 62], being consistent with the experimental values. However, the non-universality of ρ_∞ is rather small compared to that of Φ_∞ both experimentally and theoretically. For stiff polymers (with very large λ^{-1}), the above HW theory gives the Zimm value 1.479 of ρ_∞ and this is consistent with experimental results [68]. As for Φ_∞, on the other hand, even the HW model fails to explain its non-universality for flexible polymers and also its small values ($\sim 1.5 \times 10^{23}$) for semiflexible polymers [60]. Although it requires further theoretical investigations, the observed Φ_∞ for a-PMMA and semiflexible polymers may not be considered to have reached its true asymptotic limit [41].

Finally, it is interesting to examine the dependence of Φ_∞ and ρ_∞ on the global chain conformation. Figure 6.13 illustrates the conformational change from a long rigid rod to a rigid sphere through a random coil with the values of their Φ_∞, ρ_∞, and $N_A^{-1}\Phi_\infty\rho_\infty$. For the rigid rod, these values have been calculated from Eqs. (6.37), (6.49), and (6.86) with the relation $\langle S^2 \rangle = L^2/12$, and for the sphere, they have been calculated from the Einstein and Stokes equations with the relation $\langle S^2 \rangle = 3d_b^{\,2}/20$. As for the intermediate values, the lower bound of Φ_∞ and the upper bound of ρ_∞ correspond to semiflexible polymers, while the upper bound of Φ_∞ and the lower bound of ρ_∞ correspond to flexible polymers. It is seen that the changes in Φ_∞, ρ_∞, and $\Phi_\infty\rho_\infty$ are consistent with the conformational change above, the change in the product $\Phi_\infty\rho_\infty$ being rather insensitive.

6.5.3 Negative Intrinsic Viscosity

The values of $[\eta]$ of PIB and PDMS for very small M in the oligomer region have not been plotted in Fig. 6.10. The reason for this is that they become negative in that region. Figure 6.14 shows plots of $[\eta]$ against the number of repeat units x for PIB in IAIV at 25.0°C (Θ) and in benzene at 25.0°C (Θ) [47] and for PDMS in MEK at 20.0°C (Θ) and in bromocyclohexane

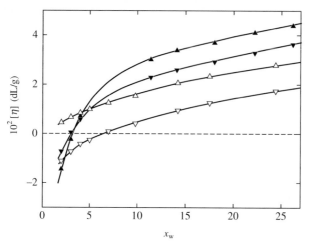

Fig. 6.14. Plots of $[\eta]$ against x for PIB in IAIV at 25.0°C ($\blacktriangle$) and in benzene at 25.0°C ($\blacktriangledown$) [47] and for PDMS in MEK at 20.0°C ($\triangle$) and in bromocyclohexane at 29.5°C ($\triangledown$) [48]. The *solid curves* connect the data points smoothly

at 29.5°C (Θ) [48]. The solid curves connect the data points smoothly. Such negative intrinsic viscosities have also been observed for certain oligomers, for example, n-alkane in benzene [69] and butadiene oligomers in Aroclor 1248 and/or 1254 [70]. In fact, the negative $[\eta]$ was discovered a long time ago for a variety of binary simple liquid mixtures, for example, for benzene in ethanol (at 25°C) [71] and CCl_4 in tetrachloroethylene (at 25°C) [72]. It means a decrease in the solution viscosity below that of the solvent by an addition of a solute, and this may be regarded as arising from specific interactions between solute and solvent molecules such that a liquid structure of some kind existing in the solvent is destroyed in the vicinity of a solute molecule.

Now all polymer transport theories have been developed so far within the framework of classical hydrodynamics and none of them can treat such effects of specific interactions on $[\eta]$. Necessarily, they give the positive $[\eta]_E$ of the polymer bead in Eq. (6.111). Thus we must remove the contributions of specific interactions from raw data for $[\eta]$ so that the corrected $[\eta]$ is at least positive for all possible values of M to be fit for an analysis by the use of the present theory of $[\eta]$. For this purpose, we rewrite Eq. (6.111) empirically in the form [47]

$$[\eta] = [\eta]^{(KR)} + [\eta]_E + \eta^* \,, \tag{6.134}$$

where η^* is an empirical additional nonnegative term. If we assume that this modification applies only to the term $[\eta]_E$ (not to $[\eta]^{(KR)}$), following Fixman [73], then η^* must be independent of M, so that

$$\lim_{M \to \infty} \left(\frac{\eta^*}{[\eta]} \right) = 0 \,. \tag{6.135}$$

As seen from Fig. 6.14, the difference between the values of $[\eta]$ of each polymer in the two Θ solvents is almost independent of x for $x \gtrsim 5$. (For smaller x, effects of chain ends are remarkable.) Let the subscripts (1) and (2) indicate the two solvents, where we assume that $[\eta]_{(1)} > [\eta]_{(2)}$ at a given M. In the present cases (PIB and PDMS) [47, 48], the difference $[\eta] - \eta^*$ may be regarded as independent of solvent, and therefore the difference $[\eta]_{(1)} - [\eta]_{(2)}$, which is nearly independent of x for $x \gtrsim 5$ and which we denote by $\Delta\eta$, may be given by

$$\Delta\eta \equiv [\eta]_{(1)} - [\eta]_{(2)} = \eta^*_{(1)} - \eta^*_{(2)} . \tag{6.136}$$

From the results in Fig. 6.14, we then have $\Delta\eta = 0.0078$ and 0.0115 dL/g for PIB and PDMS, respectively. In the analysis displayed in Fig. 6.10, we have assumed that $\eta^*_{(1)} = 0$. If this assumption is still adopted, we have $\eta^* = -0.0078$ and -0.0115 dL/g for PIB in benzene and PDMS in bromocyclohexane, respectively. For these two systems, the data for $[\eta] - \eta^*$ may then be analyzed by the use of Eq. (6.111) as before. Thus we obtain the results for them given in Table 6.3.

The negative intrinsic viscosity or the parameter η^* is discussed in relation to the high-frequency dynamic intrinsic viscosity in Chap. 10. However, a molecular-theoretical interpretation of η^* itself is one of the problems in the future.

6.5.4 Draining Effect

Figure 6.15 shows plots of $([\eta] - \eta^*)/M^{1/2}$ and $\eta_0 M^{1/2} D/k_B T$ against $M^{1/2}$ for a-PS in cyclohexane at $34.5°C$ (Θ) [31, 39–42] and PDMS in bromocyclohexane at $29.5°C$ (Θ) [41, 48]. The solid curves connect the data points smoothly, and the vertical line segments with shadows indicate the values of M above which the HW theoretical values of $\langle S^2 \rangle / M$ become almost independent of M for the respective polymers. It is seen that $([\eta] - \eta^*)/M^{1/2}$ decreases and $M^{1/2}D$ increases with decreasing M even in the range of M where the static properties such as $\langle S^2 \rangle$ exhibit the Gaussian chain behavior in the unperturbed state. This anomalous behavior should be regarded as the so-called "draining effect" [1]. By the term draining effect, we simply mean that for unperturbed flexible polymers the ratio $[\eta]/M^{1/2}$ decreases from its constant limiting value (for large M) with decreasing M.

The analysis of such data by the use of the HW transport theory leads inevitably to remarkably small values of d_b, as given in Table 6.3. The smallness of d_b suggests that the nonslip boundary condition on the bead surface may break down for PDMS. This indicates that its intermolecular interactions are rather small, thus possibly leading to its low glass transition temperature and bulk modulus. However, the individual PDMS chain is not so flexible as expected from these bulk properties, since the value of its λ^{-1} is almost the same as that for a-PS, as seen from Table 6.3.

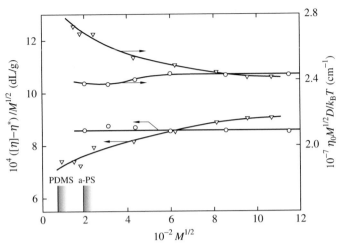

Fig. 6.15. Plots of $([\eta]-\eta^*)/M^{1/2}$ and $\eta_0 M^{1/2} D/k_\mathrm{B}T$ against $M^{1/2}$ for a-PS in cyclohexane at 34.5°C (with $\eta^* = 0$) (○) [31,39–42] and PDMS in bromocyclohexane at 29.5°C (with $\eta^* = -0.0115$ dL/g) (▽) [41,48]. The *solid curves* connect the data points smoothly. The HW theoretical values of $\langle S^2 \rangle/M$ are almost independent of M to the right of the respective vertical line segments

6.6 Ring Polymers

In this section we evaluate the translational friction coefficient Ξ and intrinsic viscosity $[\eta]$ for ring polymers by an application of the OB procedure to the KP cylinder model [74]. The results are applied to circular DNA.

6.6.1 Translational Friction Coefficient

For the KP cylinder ring the second line of Eqs. (6.36) for the function f_D becomes

$$f_D = \int_0^{L/2} K(s; L, d) ds \,, \tag{6.137}$$

where we have used the relation $K(s) = K(L - s)$ for rings. Following the same procedure as that used for the KP linear cylinder, we approximate $K(s)$ (with all lengths in units of λ^{-1}) by

$$K(s; L, d) = \left[\frac{6L}{\pi s(L - s)} \right]^{1/2} \left[1 - \frac{L(1 + 5d^2)}{40s(L - s)} - \frac{11}{120L} \right] \quad \text{for } \sigma < s \leq \frac{L}{2}$$

$$= \left(s^2 + \frac{1}{4}d^2 \right)^{-1/2} \left[1 + \sum_{i=1}^{3} f_i(L, d)s^i \right] \quad \text{for } 0 \leq s \leq \sigma$$

$$\tag{6.138}$$

with

Table 6.5. Values of σ_k and f_{ijk} in Eqs. (6.139) and (6.140)

	k			
	0	1	2	3
σ_k	2.18559	$-4.67985\,(-1)$	$4.91581\,(-1)$	$-1.50334\,(1)$
f_{10k}	$3.33333\,(-1)^{\text{a}}$	0	0	0
f_{11k}	$-4.50040\,(-2)$	$-2.75430\,(-1)$	1.19325	-5.59657
f_{12k}	$-2.20160\,(-2)$	$-8.22244\,(-2)$	$4.57470\,(-1)$	2.97966
f_{20k}	$1.19083\,(-1)$	$5.30304\,(-1)$	$9.99369\,(-1)$	-4.99560
f_{21k}	$5.18804\,(-3)$	$1.40740\,(-1)$	-2.39261	9.30255
f_{22k}	$1.58136\,(-2)$	$7.54396\,(-2)$	$-6.20245\,(-1)$	3.39914
f_{30k}	$-2.65957\,(-2)$	$-1.49946\,(-2)$	$-6.88179\,(-1)$	4.85298
f_{31k}	$9.15166\,(-4)$	$-5.33328\,(-2)$	1.03760	-4.61578
f_{32k}	$-2.97808\,(-3)$	$-1.87217\,(-2)$	$1.93592\,(-1)$	$-9.82380\,(-1)$

a $a(n)$ means $a \times 10^n$

$$\sigma = \sum_{k=0}^{3} \sigma_k L^{-k} \,, \tag{6.139}$$

$$f_i = \sum_{k=0}^{3}\sum_{j=0}^{2} f_{ijk} d^{2j} L^{-k} \,, \tag{6.140}$$

where σ_k and f_{ijk} are numerical constants independent of L and d and their values are given in Table 6.5. We note that Eqs. (6.138) are valid for $L \geq 3.480$ (in fact for relatively large L) and that Eqs. (6.138) with $d = 0$ for $K(s; L, 0) = \langle R^{-1} \rangle$ have been derived by the use of Eqs. (3.D.2) and (3.D.3).

Substitution of Eqs. (6.138) into Eq. (6.137) and integration leads to

$$
\begin{aligned}
f_D = \left(\frac{6L}{\pi}\right)^{1/2} &\left[\left(1 - \frac{11}{120L}\right) \sin^{-1}\left(\frac{L-2\sigma}{L}\right) - \frac{(L-2\sigma)(1+5d^2)}{20L\sigma^{1/2}(L-\sigma)^{1/2}} \right. \\
&+ \ln\left[\frac{2\sigma + (4\sigma^2 + d^2)^{1/2}}{d}\right] + f_1\left[\left(\sigma^2 + \frac{1}{4}d^2\right)^{1/2} - \frac{1}{2}d\right] \\
&+ \frac{1}{2}f_2\left\{\sigma\left(\sigma^2 + \frac{1}{4}d^2\right)^{1/2} - \frac{1}{4}d^2 \ln\left[\frac{2\sigma + (4\sigma^2 + d^2)^{1/2}}{d}\right]\right\} \\
&\left. + \frac{1}{3}f_3\left(\sigma^2 - \frac{1}{2}d^2\right)\left(\sigma^2 + \frac{1}{4}d^2\right)^{1/2} \right] \qquad \text{for } L \geq 3.480\,.
\end{aligned}
\tag{6.141}
$$

In the coil limit of $L \to \infty$, we have, from Eq. (6.141),

$$f_{D,(C)} = \frac{1}{2}(6\pi)^{1/2}L^{1/2} \,, \tag{6.142}$$

so that

$$\Xi_{(C)} = (6\pi)^{1/2}\eta_0 L^{1/2} \,. \tag{6.143}$$

This is identical with the KR value in the nondraining limit obtained by Bloomfield and Zimm [75] and by Fukatsu and Kurata [76]. We note that $f_{D,(C)}(\text{ring})/f_{D,(C)}(\text{linear}) = 3\pi/8$.

Next we consider the *rigid*-ring limit of f_D, which we indicate by the subscript (R). The kernel may then be given by

$$K_{(R)}(s; L; d) = \left(\frac{\pi}{L}\right)\left[\sin^2\left(\frac{\pi s}{L}\right) + \left(\frac{\pi}{2p}\right)^2\right]^{-1/2}\left[1 + \mathcal{O}(p^{-2})\right], \quad (6.144)$$

where $p = L/d$ as before, and we note that the contribution of neglected terms in Eq. (6.144) to $f_{D,(R)}$ does not exceed 1% for $p \geq 10$. Substitution of Eq. (6.144) into Eq. (6.137) and integration leads to

$$f_{D,(R)} = \left(\frac{4p^2}{4p^2 + \pi^2}\right)^{1/2} K\left[\left(\frac{4p^2}{4p^2 + \pi^2}\right)^{1/2}\right], \quad (6.145)$$

where $K(k)$ is the complete elliptic integral of the first kind defined by

$$K(k) = \int_0^{\pi/2} (1 - k^2 \sin^2\theta)^{-1/2} d\theta. \quad (6.146)$$

We then have, from Eq. (6.145),

$$f_{D,(R^*)} \equiv \lim_{p \to \infty} f_{D,(R)}$$

$$= \ln p + \ln\left(\frac{8}{\pi}\right) + \mathcal{O}(p^{-1}). \quad (6.147)$$

This is to be compared with the second line of Eqs. (6.49) for the rigid rod in the same limit. If we avoid the preaveraging of the Oseen tensor, the leading term of $f_{D,(R^*)}$ appearing in the translational diffusion coefficient D and sedimentation coefficient s for the rigid ring is replaced by $(11/12)\ln p$ [17, 30, 77, 78], the remaining terms also being altered.

6.6.2 Intrinsic Viscosity

In the integral Eq. (6.77) the kernel $K(x, \xi)$ is given by Eqs. (6.138) with $2s = L|x - \xi|$ and the g function is given by

$$g(x, y) = \frac{\pi}{L}\left[\langle S^2 \rangle - \frac{1}{2}\left\langle R^2\left(\frac{L}{2}|x - y|\right)\right\rangle\right], \quad (6.148)$$

where $\langle S^2 \rangle$ and $\langle R^2(s) \rangle$ are given by Eqs. (3.D.8) and (3.D.4), respectively. The numerical results obtained from Eq. (6.77) may then be expressed in the form

$$[\eta] = \Phi_\infty \frac{L^{3/2}}{M}\left(1 + \sum_{i=1}^4 C_i L^{-i/2}\right)^{-1} \quad \text{for } L \geq 3.480 \quad (6.149)$$

Table 6.6a. Values of α_{ij} in Eq. (6.150)

d	i	α_{i0}	α_{i1}	α_{i2}	α_{i3}	α_{i4}
[0.001, 0.1][a]	1	0.809231	−40.8202	−483.899	⋯	⋯
	2	−13.7690	380.429	5197.48	⋯	⋯
	3	35.0883	−1079.70	−14530.3	⋯	⋯
	4	−28.6643	927.876	12010.0	⋯	⋯
[0.1, 1.0]	1	−2.17381	−11.3578	249.523	−729.371	489.172
	2	112.769	−851.870	−21390.1	56909.8	−34787.5
	3	−1680.23	24753.1	498848	−1314310	792477
	4	7043.32	−142907	−2883470	7668650	−4648720

[a] $[a, b]$ means that $a \le d \le b$

Table 6.6b. Values of β_{ij} in Eq. (6.150)

d	i	β_{i0}	β_{i1}	β_{i2}
[0.001, 0.1][a]	1	−2.53944	−339.266	⋯
	2	0.818816	3517.90	⋯
	3	−1.44344	−9855.73	⋯
	4	0.571812	8221.82	⋯
[0.1, 1.0]	1	−3.58885	74.3257	−335.732
	2	41.8243	−9944.26	22067.0
	3	−526.628	244353	497280
	4	2177.01	−1407520	2937180

[a] $[a, b]$ means that $a \le d \le b$

with Φ_∞ being equal to the value 1.854×10^{23} obtained from the exact asymptotic solution [75, 76, 79] and with

$$C_i = \sum_{j=0}^{4} \alpha_{ij} d^j + \sum_{j=0}^{2} \beta_{ij} d^{2j} \ln d \,, \tag{6.150}$$

where α_{ij} and β_{ij} are numerical constants independent of L and d and their values are given in Tables 6.6a and 6.6b, respectively. Thus we have $[\eta]_{(C)}(\text{ring})/[\eta]_{(C)}(\text{linear}) = 0.648$ in the limit of $L \to \infty$.

Finally, we consider the rigid-ring limit of $[\eta]$. The kernel is given by Eq. (6.144) and the g function is given by

$$g_{(R)}(x, y) = \left(\frac{L}{4\pi}\right) \cos(\pi|x - y|) \,. \tag{6.151}$$

If we expand the solution $\psi(x, y)$ of the integral Eq. (6.77) with Eqs. (6.144) and (6.151) in a Fourier series, we can obtain its asymptotic solution in the limit of $p \to \infty$ [74]. Thus we have

$$[\eta]_{(R^*)} = \frac{N_A L^3}{8\pi M}\left[\ln p + \ln\left(\frac{8}{\pi}\right) - 2 + \mathcal{O}(p^{-1})\right]^{-1}. \qquad (6.152)$$

If we avoid the preaveraging of the Oseen tensor, Eq. (6.152) is replaced by [74]

$$[\eta]_{(R^*)} = \frac{17 N_A L^3}{120\pi M}\left[\ln p + \ln\left(\frac{8}{\pi}\right) - \frac{144}{85} + \mathcal{O}(p^{-1})\right]^{-1}. \qquad (6.153)$$

Thus the ratio $[\eta]_{(R^*)}(\text{ring})/[\eta]_{(R^*)}$ (linear) is equal to $3/\pi$ and $51/16\pi$ in the preaveraging and nonpreaveraging cases, respectively.

6.6.3 Application to DNA

In this subsection we make a comparison of theory with experiment using experimental data for the sedimentation coefficient s [80–87] and intrinsic viscosity $[\eta]$ [85,86] obtained for (untwisted) circular DNA. Figure 6.16 shows double-logarithmic plots of f_D (from s) and $[\eta]$ (in dL/g) against M. The solid curves represent the theoretical values calculated from Eqs. (6.141) and (6.149) with the values of the model parameters given in Table 6.3 for linear DNA. For comparison, the corresponding theoretical values for linear DNA are represented by the dashed curves. There is seen to be rather good agreement between theory and experiment.

The problems of twisted circular DNA (DNA topoisomers) are considered in the next chapter.

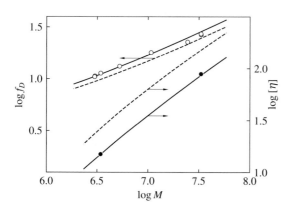

Fig. 6.16. Double-logarithmic plots of f_D and $[\eta]$ (in dL/g) against M for circular DNA [80–87]. The *solid and dashed curves* represent the KP theoretical values for ring and linear chains, respectively (see the text)

Appendix 6.A Transport Coefficients of Spheroid-Cylinders

In this appendix we evaluate the translational and rotatory friction (or diffusion) coefficients and intrinsic viscosity of the spheroid-cylinder defined in

Sect. 6.2.1 and depicted in Fig. 6.2 [27]. We introduce external $(\mathbf{e}_x, \mathbf{e}_y, \mathbf{e}_z)$ and molecular $(\mathbf{e}_1, \mathbf{e}_2, \mathbf{e}_3)$ Cartesian coordinate systems, choosing the center of mass of the body as the origin of the latter. The superscript (e) is used to indicate vectors and tensors expressed in the external system, and no superscript is used for those in the molecular system. The spatial configuration of the body may be determined by the vector position $\mathbf{R}_c^{(e)} \equiv \mathbf{R}_c = (x_c, y_c, z_c)$ of the center of mass in the external system and the Euler angles $\Theta = (\alpha, \beta, \gamma)$ defining the orientation of the molecular system with respect to the external system. The matrix transforming the external coordinates to the molecular coordinates, which we denote by $\mathbf{A}$, is then identical with the matrix $\mathbf{Q}$ given by Eq. (4.96) with (α, β, γ) in place of $(\tilde{\theta}, \tilde{\phi}, \tilde{\psi})$.

Now let $\mathbf{U}_c$ be the instantaneous velocity of the center of mass of the body, let $\boldsymbol{\Omega}$ be its instantaneous angular velocity, and let $\mathbf{v}^0$ be the unperturbed flow field of a solvent. Under the nonslip boundary condition, the frictional force $\mathbf{f}(\mathbf{r})$ exerted by the unit area at $\mathbf{r}$ of the surface of the body satisfies the integral equation

$$8\pi\eta_0\left[\mathbf{U}_c + \boldsymbol{\Omega} \times \mathbf{r}_1 - \mathbf{v}^0(\mathbf{r}_1)\right] = \int_S \mathbf{K}(\mathbf{r}_1, \mathbf{r}_2) \cdot \mathbf{f}(\mathbf{r}_2) d\mathbf{r}_2, \qquad (6.A.1)$$

where $\mathbf{r}_1$ and $\mathbf{r}_2$ are the vector positions of two arbitrary points on the surface of the body, $\mathbf{K}(\mathbf{r}_1, \mathbf{r}_2)$ is defined by

$$\mathbf{K}(\mathbf{r}_1, \mathbf{r}_2) = 8\pi\eta_0 \mathbf{T}(\mathbf{r}_1 - \mathbf{r}_2) \qquad (6.A.2)$$

as in Eq. (6.58), and the integration in Eq. (6.A.1) is carried out over the surface of the body. If we define the inverse $\mathbf{K}^{-1}(\mathbf{r}_1, \mathbf{r}_2)$ by

$$\delta^{(2)}(\mathbf{r}_1 - \mathbf{r}_2)\mathbf{I} = \int_S \mathbf{K}^{-1}(\mathbf{r}_1, \mathbf{r}_3) \cdot \mathbf{K}(\mathbf{r}_3, \mathbf{r}_2) d\mathbf{r}_3$$

$$= \int_S \mathbf{K}(\mathbf{r}_1, \mathbf{r}_3) \cdot \mathbf{K}^{-1}(\mathbf{r}_3, \mathbf{r}_2) d\mathbf{r}_3 \qquad (6.A.3)$$

with $\delta^{(2)}(\mathbf{r})$ a two-dimensional Dirac delta function and with $\mathbf{I}$ the unit tensor as in Eqs. (6.59), then the formal solution of Eq. (6.A.1) is obtained as

$$\mathbf{f}(\mathbf{r}_1) = 8\pi\eta_0 \int_S \mathbf{K}^{-1}(\mathbf{r}_1, \mathbf{r}_2) \cdot \left[\mathbf{U}_c + \boldsymbol{\Omega} \times \mathbf{r}_2 - \mathbf{v}^0(\mathbf{r}_2)\right] d\mathbf{r}_2. \qquad (6.A.4)$$

We first put $\mathbf{v}^0 = \mathbf{0}$ to consider the translational and rotatory friction tensors of the body, which we denote by $\boldsymbol{\Xi}$ and $\boldsymbol{\Xi}_{c,r}$, respectively. The total frictional force $\mathbf{F}$ and torque $\mathbf{T}_c$ (about the center of mass) exerted by the body on the solvent are then given by

$$\mathbf{F} = \boldsymbol{\Xi} \cdot \mathbf{U}_c = \int_S \mathbf{f}(\mathbf{r}) d\mathbf{r}, \qquad (6.A.5)$$

$$\mathbf{T}_c = \boldsymbol{\Xi}_{c,r} \cdot \boldsymbol{\Omega} = \int_S \mathbf{r} \times \mathbf{f}(\mathbf{r}) d\mathbf{r} = \int_S \mathbf{B}_c(\mathbf{r})^T \cdot \mathbf{f}(\mathbf{r}) d\mathbf{r} , \qquad (6.A.6)$$

where the superscript T indicates the transpose and the tensor $\mathbf{B}_c$ is given by

$$\mathbf{B}_c(\mathbf{r}) = \begin{pmatrix} 0 & r_3 & -r_2 \\ -r_3 & 0 & r_1 \\ r_2 & -r_1 & 0 \end{pmatrix} \qquad (6.A.7)$$

with $\mathbf{r} = r_1\mathbf{e}_1 + r_2\mathbf{e}_2 + r_3\mathbf{e}_3$. If we substitute Eq. (6.A.4) into Eqs. (6.A.5) and (6.A.6), we find

$$\boldsymbol{\Xi} = 8\pi\eta_0 \int_S \int_S \mathbf{K}^{-1}(\mathbf{r}_1, \mathbf{r}_2) d\mathbf{r}_1 d\mathbf{r}_2 = 8\pi\eta_0 \int_S \boldsymbol{\Psi}_1(\mathbf{r}) d\mathbf{r} , \qquad (6.A.8)$$

$$\boldsymbol{\Xi}_{c,r} = 8\pi\eta_0 \int_S \int_S \mathbf{B}_c(\mathbf{r}_1)^T \cdot \mathbf{K}^{-1}(\mathbf{r}_1, \mathbf{r}_2) \cdot \mathbf{B}_c(\mathbf{r}_2) d\mathbf{r}_1 d\mathbf{r}_2$$

$$= 8\pi\eta_0 \int_S \mathbf{B}_c(\mathbf{r})^T \cdot \boldsymbol{\Psi}_2(\mathbf{r}) d\mathbf{r} , \qquad (6.A.9)$$

where the tensors $\boldsymbol{\Psi}_1$ and $\boldsymbol{\Psi}_2$ are the solutions of the integral equations

$$\int_S \mathbf{K}(\mathbf{r}_1, \mathbf{r}_2) \cdot \boldsymbol{\Psi}_1(\mathbf{r}_2) d\mathbf{r}_2 = \mathbf{I} , \qquad (6.A.10)$$

$$\int_S \mathbf{K}(\mathbf{r}_1, \mathbf{r}_2) \cdot \boldsymbol{\Psi}_2(\mathbf{r}_2) d\mathbf{r}_2 = \mathbf{B}_c(\mathbf{r}_1) . \qquad (6.A.11)$$

Now, if we take $\mathbf{e}_3$ along the axis of revolution of the body, then $\boldsymbol{\Xi}$ and $\boldsymbol{\Xi}_{c,r}$ and hence the translational diffusion tensor $\mathbf{D}_c = k_B T \boldsymbol{\Xi}^{-1}$ (of the center of mass) and rotatory diffusion tensor $\mathbf{D}_r = k_B T \boldsymbol{\Xi}_{c,r}^{-1}$ are diagonalized. We denote their principal values by Ξ_j, $\Xi_{r,j}$, D_j, and $D_{r,j}$, respectively, so that

$$D_j = \frac{k_B T}{\Xi_j} , \qquad (6.A.12)$$

$$D_{r,j} = \frac{k_B T}{\Xi_{r,j}} \qquad (6.A.13)$$

with $D_1 = D_2$ and $D_{r,1} = D_{r,2}$ (and $\Xi_1 = \Xi_2$ and $\Xi_{r,1} = \Xi_{r,2}$). The mean translational diffusion coefficient D averaged over the orientation of the body is given by

$$D = \frac{1}{3} \text{Tr} \, \mathbf{D}_c^{(e)} = \frac{1}{3} \text{Tr} \, \mathbf{D}_c = \frac{1}{3} k_B T (2\Xi_1^{-1} + \Xi_3^{-1}) . \qquad (6.A.14)$$

Next we consider the intrinsic viscosity $[\eta]$. The unperturbed flow field $\mathbf{v}^0$ given by Eq. (6.65) may be expressed in the molecular coordinate system as follows,

$$\mathbf{v}^0(\mathbf{r}) = \epsilon_0 \mathbf{A} \cdot \mathbf{e}_x \mathbf{e}_y \cdot \mathbf{A}^T \cdot \mathbf{r} . \qquad (6.A.15)$$

We may then put $\mathbf{U}_c = \mathbf{0}$, so that

$$\mathbf{\Omega} \times \mathbf{r} - \mathbf{v}^0(\mathbf{r}) = -\epsilon_0 \mathbf{m} \cdot \mathbf{r} \qquad (6.A.16)$$

with

$$\mathbf{m} = \frac{1}{2}\mathbf{A} \cdot (\mathbf{e}_x\mathbf{e}_y + \mathbf{e}_y\mathbf{e}_x) \cdot \mathbf{A}^T , \qquad (6.A.17)$$

since the body rotates about the z axis with the angular velocity $-\epsilon_0/2$ in the limit of $\epsilon_0 = 0$ [27]. Thus Eq. (6.A.4) becomes

$$\mathbf{f}(\mathbf{r}_1) = -8\pi\eta_0\epsilon_0 \int_S \mathbf{K}^{-1}(\mathbf{r}_1, \mathbf{r}_2) \cdot \mathbf{m} \cdot \mathbf{r}_2 d\mathbf{r}_2 . \qquad (6.A.18)$$

As shown in Appendix 6.B, $[\eta]$ of the body may be expressed in terms of the surface integral as

$$[\eta] = -\frac{N_A}{M\eta_0\epsilon_0} \int_S \mathbf{e}_x \cdot \left\langle \mathbf{A}^T \cdot \mathbf{f}(\mathbf{r})\mathbf{r} \cdot \mathbf{A} \right\rangle \cdot \mathbf{e}_y d\mathbf{r} = \frac{8\pi N_A}{M} \int_S \Psi(\mathbf{r}) d\mathbf{r} \quad (6.A.19)$$

with

$$\Psi(\mathbf{r}_1) = \int_S \mathbf{e}_x \cdot \left\langle \mathbf{A}^T \cdot \mathbf{K}^{-1}(\mathbf{r}_1, \mathbf{r}_2) \cdot \mathbf{m} \cdot \mathbf{r}_2\mathbf{r}_1 \cdot \mathbf{A} \right\rangle \cdot \mathbf{e}_y d\mathbf{r}_2 . \quad (6.A.20)$$

The orientational average in Eq. (6.A.20) may be evaluated by expanding the matrices $\mathbf{A}$ and $\mathbf{m}$ in terms of the Wigner functions $\mathcal{D}_l^{mj}(\alpha, \beta, \gamma)$ [88]. The function Ψ may then be expressed as

$$\Psi(\mathbf{r}) = \frac{1}{3}(-\psi_{11}r_1 - \psi_{12}r_2 + 2\psi_{13}r_3) + (\psi_{22}r_3 + \psi_{23}r_2) + (\psi_{31}r_3 + \psi_{33}r_1)$$
$$+(\psi_{41}r_2 + \psi_{42}r_1) + (\psi_{51}r_1 - \psi_{52}r_2) , \qquad (6.A.21)$$

where $\mathbf{r} = (r_1, r_2, r_3)$, and the vectors $\boldsymbol{\psi}_j = (\psi_{j1}, \psi_{j2}, \psi_{j3})$ $(j = 1\text{–}5)$ are the solutions of the integral equations

$$\int_S \mathbf{K}(\mathbf{r}_1, \mathbf{r}_2) \cdot \boldsymbol{\psi}_j(\mathbf{r}_2) d\mathbf{r}_2 = \frac{1}{5}\mathbf{m}_j \cdot \mathbf{r}_1 \qquad (6.A.22)$$

with

$$\mathbf{m}_1 = -\mathbf{e}_1\mathbf{e}_1 - \mathbf{e}_2\mathbf{e}_2 + 2\mathbf{e}_3\mathbf{e}_3 ,$$
$$\mathbf{m}_2 = \mathbf{e}_2\mathbf{e}_3 + \mathbf{e}_3\mathbf{e}_2 , \qquad \mathbf{m}_3 = \mathbf{e}_3\mathbf{e}_1 + \mathbf{e}_1\mathbf{e}_3 , \qquad (6.A.23)$$
$$\mathbf{m}_4 = \mathbf{e}_1\mathbf{e}_2 + \mathbf{e}_2\mathbf{e}_1 , \qquad \mathbf{m}_5 = \mathbf{e}_1\mathbf{e}_1 - \mathbf{e}_2\mathbf{e}_2 .$$

We find exact numerical solutions of all the integral equations above for small $p = L/d$ and also solutions in the OB approximation (with the non-preaveraged Oseen tensor) for large p. We first consider the latter. General expressions for the transport coefficients of the spheroid-cylinder in the OB approximation may be derived by replacing the integrals over the surface by those over the contour distance s $(-L/2 \leq s \leq L/2)$, where $\boldsymbol{\Psi}_1$, $\boldsymbol{\Psi}_2$, and ψ_j $(j = 1\text{–}5)$ are then functions of s with $r_1 = r_2 = 0$ and $r_3 = s$.

Then the function f_D defined by Eq. (6.37) may be expressed as [27]

$$f_D = \frac{1}{4}\left(2F_1^{-1} + F_2^{-1}\right), \tag{6.A.24}$$

where

$$F_j = \int_{-1}^{1} \Psi_{1j}(x)dx \tag{6.A.25}$$

with $x = 2s/L$. In Eq. (6.A.25), Ψ_{1j} $(j = 1, 2)$ are the solutions of the integral equations

$$\int_{-1}^{1} K_j(x_1, x_2)\Psi_{1j}(x_2)dx_2 = 1, \tag{6.A.26}$$

where $K_j(x_1, x_2)$ $(j = 1, 2)$ are given by

$$K_1 = \frac{2p^2(x_1 - x_2)^2 + 3[1 - h(x_1)]}{d[p^2(x_1 - x_2)^2 + 1 - h(x_1)]^{3/2}}, \tag{6.A.27}$$

$$K_2 = \frac{4p^2(x_1 - x_2)^2 + 2[1 - h(x_1)]}{d[p^2(x_1 - x_2)^2 + 1 - h(x_1)]^{3/2}}, \tag{6.A.28}$$

with $p = L/d$ and with

$$
\begin{aligned}
h(x) &= 0 & \text{for } 0 \leq |x| < 1 - \frac{\epsilon}{p} \\
&= \left[1 - \frac{p(1 - |x|)}{\epsilon}\right]^2 & \text{for } 1 - \frac{\epsilon}{p} < |x| \leq 1. \tag{6.A.29}
\end{aligned}
$$

In the limit of $p \to \infty$ (with $h = \epsilon = 0$), Eq. (6.A.24) becomes the second line of Eqs. (6.49) [24] if we find the asymptotic solution of Eq. (6.A.26) by a Legendre polynomial expansion method [20].

The rotatory diffusion coefficient $D_{r,1}$ and intrinsic viscosity $[\eta]$ may be expressed as [33]

$$D_{r,1} = \frac{3k_B T}{\pi \eta_0 L^3 F_r}, \tag{6.A.30}$$

$$[\eta] = \frac{\pi N_A L^3}{90M}(3F_r + F_{\eta\infty}), \tag{6.A.31}$$

where

$$F_r = 3\int_{-1}^{1} x\Psi_{21}(x)dx, \tag{6.A.32}$$

$$F_{\eta\infty} = 6\int_{-1}^{1} x\Psi_{22}(x)dx \tag{6.A.33}$$

with Ψ_{2j} $(j = 1, 2)$ the solutions of the integral equations

$$\int_{-1}^{1} K_j(x_1, x_2)\Psi_{2j}(x_2)dx_2 = x_1 . \tag{6.A.34}$$

In the OB approximation we have $\Xi_{r,3} = 0$ from Eqs. (6.A.9) since $r_1 = r_2 = 0$, so that the rotation of the cylinder about its axis cannot be considered. In the limit of $p \to \infty$, we find [33] by the Legendre polynomial expansion method

$$F_r^{-1} = \ln p + 2\ln 2 - \frac{11}{6} + \mathcal{O}(p^{-1}), \tag{6.A.35}$$

$$F_{\eta\infty}^{-1} = \ln p + 2\ln 2 - \frac{17}{6} + \mathcal{O}(p^{-1}), \tag{6.A.36}$$

so that Eq. (6.A.31) becomes Eq. (6.86).

In the following, we complete expressions for f_D, $D_{r,1}$, and $[\eta]$.

(a) Translational Diffusion Coefficient

As mentioned above and in Sect. 6.3.1, the OB approximation along with the preaveraging of the Oseen tensor and with the KR approximation [3] as in Eq. (6.35) can give the correct asymptotic result for f_D as given by Eq. (6.49). We therefore evaluate f_D in this way from

$$f_D = \frac{1}{2}L \int_{-1}^{1}\int_{-1}^{1} K(x_1, x_2)dx_1dx_2 \tag{6.A.37}$$

with

$$K(x_1, x_2) = \left\{ (x_1 - x_2)^2 + \left[1 - h(x_1)\right]p^{-2} \right\}^{-1/2}. \tag{6.A.38}$$

The result reads

$$f_D = \sinh^{-1}(2p - \epsilon) - \frac{\epsilon}{p}\sinh^{-1}\epsilon - \frac{1}{2p}\left\{ \left[(2p - \epsilon)^2 + 1\right]^{1/2} - (\epsilon^2 + 1)^{1/2} \right\}$$

$$- \frac{\epsilon}{2p}\ln\left(2(p - \epsilon)\{2p - \epsilon + \left[(2p - \epsilon)^2 + 1\right]^{1/2}\} + 1 \right)$$

$$+ \frac{\epsilon}{2p}\ln\left[4p(p - \epsilon) + 1\right] + f_D' \tag{6.A.39}$$

with

$$f_D' = \frac{\epsilon}{2p(\epsilon^2 - 1)^{1/2}}\left\{ \epsilon\ln\left[\epsilon^2 + (\epsilon^4 - 1)^{1/2}\right] + (2p - \epsilon) \right.$$

$$\left. \times \ln\left| \frac{\epsilon(2p - \epsilon) - (\epsilon^2 - 1)^{1/2}[(2p - \epsilon)^2 + 1]^{1/2}}{2p[(\epsilon^2 - 1)^{1/2} - \epsilon] + 1} \right| \right\} \qquad \text{for } \epsilon > 1$$

$$= \frac{1}{2p}\left\{ 2p - \left[(2p - 1)^2 + 1\right]^{1/2} + \sqrt{2} \right\} \qquad \text{for } \epsilon = 1$$

$$= \frac{\epsilon}{2p(1 - \epsilon^2)^{1/2}}\left[2\epsilon\tan^{-1}\left(\frac{1 - \epsilon^2}{1 + \epsilon^2}\right)^{1/2} + (2p - \epsilon) \right]$$

$$\times \left\{ \sin^{-1} \frac{\epsilon(2p-\epsilon)}{[4p(p-\epsilon)+1]^{1/2}} + \sin^{-1} \frac{1-2p\epsilon}{[4p(p-\epsilon)+1]^{1/2}} \right\} \right] \quad \text{for } \epsilon < 1.$$

$$(6.A.40)$$

Equation (6.A.39) with Eqs. (6.A.40) is the desired result and reduces to the result of Norisuye et al. [89] when $\epsilon = 1$ (prolate spherocylinder), and to the well-known result [90, 91] for the spheroid when $\epsilon = p$. The latter, which we denote by $f_{D,(\text{SD})}$, is given by

$$f_{D,(\text{SD})} = \epsilon F(\epsilon) \qquad (6.A.41)$$

with

$$\begin{aligned} F(\epsilon) &= \frac{1}{(\epsilon^2 - 1)^{1/2}} \cosh^{-1} \epsilon & \text{for } \epsilon > 1 \\ &= 1 & \text{for } \epsilon = 1 \\ &= \frac{1}{(1 - \epsilon^2)^{1/2}} \cos^{-1} \epsilon & \text{for } \epsilon < 1. \end{aligned} \qquad (6.A.42)$$

Figure 6.A.1 shows double-logarithmic plots of f_D against p. The solid and dot-dashed curves represent the values calculated from Eq. (6.A.39) for the spheroid-cylinders with $\epsilon = 0.5$ and 1.0 and from Eq. (6.A.41) for the spheroid ($\epsilon = p$), respectively, and the unfilled ($\epsilon \neq p$) and filled ($\epsilon = p$) circles represent the exact numerical solutions. The former values are seen to agree well with the latter. Thus we may adopt Eq. (6.A.39) with Eqs. (6.A.40) as a useful interpolation formula for f_D for the spheroid-cylinder. When $\epsilon = 0$, Eq. (6.A.39) gives the $f_{D,(\text{R})}$ from Eq. (6.39), and the values for this limiting case are represented by the dotted curve. It is seen that the end effects may be ignored for $p \gtrsim 5$. For comparison, the values calculated from the Broersma equation [18], which are not very different from those from his new version [92], are represented by the dashed curve B. Indeed, his solution of the integral equation is not exact, although asymptotically correct in the limit of $p \to \infty$.

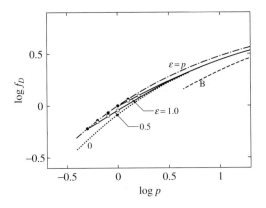

Fig. 6.A.1. Double-logarithmic plots of f_D against p. The *solid, dot-dashed, and dotted curves* represent the values calculated from Eq. (6.A.39) for the spheroid-cylinders, spheroid ($\epsilon = p$), and cylinder ($\epsilon = 0$), respectively, and the *unfilled and filled circles* represent the exact numerical solutions. The *dashed curve* B represents the values by Broersma [18] for cylinders

(b) *Rotatory Diffusion Coefficient*

We construct an interpolation formula for $D_{r,1}$ on the basis of the exact numerical solutions and the OB asymptotic solution above. The result for $F_r = F_r(p, \epsilon)$ in Eq. (6.A.30) reads

$$F_r(p, \epsilon)^{-1} = \ln p + 2\ln 2 - \frac{11}{6} + \frac{a_{r0}(\epsilon)}{\ln(1+p)} + \sum_{i=1}^{6} a_{ri}(\epsilon) p^{-i/4} \quad (6.A.43)$$

with

$$a_{r0}(\epsilon) = \left[\ln(1+\epsilon)\right]\left[f_r(\epsilon)^{-1} - \ln \epsilon - 2\ln 2 + \frac{11}{6} - \sum_{i=1}^{6} a_{ri}(\epsilon) \epsilon^{-i/4}\right],$$
$$(6.A.44)$$

$$a_{ri}(\epsilon) = \sum_{j=0}^{2} a_{rij}\, \epsilon^j, \quad (6.A.45)$$

$$f_r(\epsilon)^{-1} = F_r(\epsilon, \epsilon)^{-1} = \frac{\pi \eta_0 d^3 \epsilon^3}{3 k_B T} D_{r,1,(SD)}, \quad (6.A.46)$$

where a_{rij} are numerical constants and their values are given in Table 6.A.1; and $D_{r,1,(SD)}$ is the rotatory diffusion coefficient $D_{r,1}$ of the spheroid and is given by [90, 91]

$$\frac{\pi \eta_0 d^3 D_{r,1,(SD)}}{k_B T} = \frac{3}{2(\epsilon^4 - 1)}\left[(2\epsilon^2 - 1)F - \epsilon\right] \quad \text{for } \epsilon \neq 1$$
$$= 1 \quad \text{for } \epsilon = 1 \quad (6.A.47)$$

with F being given by Eqs. (6.A.42). Note that at $\epsilon = p$, Eq. (6.A.30) with Eq. (6.A.43) gives the exact solution given by Eqs. (6.A.47) for the spheroid. The range of application of Eq. (6.A.43) is limited to $0.6 \lesssim \epsilon \lesssim 1.3$. We note that $D_{r,3,(SD)}$ is given by [90, 91]

Table 6.A.1. Values of a_{rij} in Eq. (6.A.45)

i	a_{ri0}	a_{ri1}	a_{ri2}
1	2.23068	20.8613	−10.0473
2	−13.8396	−96.9314	48.1626
3	33.9241	288.840	−148.672
4	−29.0266	−411.528	221.719
5	8.13792	299.915	−167.783
6	1.26984	−82.2022	47.6616

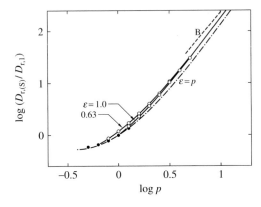

Fig. 6.A.2. Double-logarithmic plots of $D_{r,(S)}/D_{r,1}$ against p (see the text). The *solid and dot-dashed curves* represent the values calculated from Eq. (6.A.30) with Eq. (6.A.43) for the spheroid-cylinders and spheroid ($\epsilon = p$), respectively, and the *unfilled and filled circles* represent the exact numerical solutions. The *dashed curve* B represents the values by Broersma [18] for cylinders

$$\frac{\pi \eta_0 d^3 D_{r,3,(SD)}}{k_B T} = \frac{3}{2(\epsilon^2 - 1)}(\epsilon - F) \qquad \text{for } \epsilon \neq 1$$
$$= 1 \qquad \text{for } \epsilon = 1, \qquad (6.A.48)$$

Figure 6.A.2 shows double-logarithmic plots of $D_{r,(S)}/D_{r,1}$ against p, where $D_{r,(S)}$ is the rotatory diffusion coefficient of the Stokes sphere and is equal to $D_{r,1,(SD)}$ with $\epsilon = 1$. The solid and dot-dashed curves represent the values calculated from Eq. (6.A.30) with Eq. (6.A.43) for the spheroid-cylinders with $\epsilon = 0.63$ and 1.0 and from Eqs. (6.A.47) for the spheroid ($\epsilon = p$), respectively, and the unfilled ($\epsilon \neq p$) and filled ($\epsilon = p$) circles represent the exact numerical solutions. It is seen that the former values agree well with the latter, and that the end effects are rather small even for small p. In the figure are also shown the values calculated from the Broersma equation [18], which is also correct only for large p. (Note that they are not very different from the values from his new version [92].)

(c) Intrinsic Viscosity

As in the case of $D_{r,1}$, we construct an interpolation formula for $[\eta]$ on the basis of the exact numerical solutions and the OB asymptotic solution above (with the nonpreaveraged Oseen tensor). The result reads

$$[\eta] = \frac{2\pi N_A L^3}{45M} F_\eta(p, \epsilon) \qquad (6.A.49)$$

with

$$F_\eta(p, \epsilon)^{-1} = \ln p + 2\ln 2 - \frac{15}{12} + \frac{a_{\eta 0}(\epsilon)}{\ln(1 + p)} + \sum_{i=1}^{5} a_{\eta i}(\epsilon) p^{-i/4},$$
$$(6.A.50)$$

Table 6.A.2. Values of $a_{\eta ij}$ in Eq. (6.A.52)

i	$a_{\eta i0}$	$a_{\eta i1}$	$a_{\eta i2}$
1	5.94814	−4.90678	2.56381
2	−22.7705	14.6631	−7.24894
3	42.5200	−25.8741	11.4158
4	−25.8372	15.2681	−4.32430
5	7.48088	−4.22595	0.298512

$$a_{\eta 0}(\epsilon) = \left[\ln(1+\epsilon)\right]\left[f_\eta(\epsilon)^{-1} - \ln\epsilon - 2\ln 2 + \frac{15}{12} - \sum_{i=1}^{5} a_{\eta i}(\epsilon)\,e^{-i/4}\right],$$

$$(6.A.51)$$

$$a_{\eta i}(\epsilon) = \sum_{j=0}^{2} a_{\eta ij}\,e^{j}, \tag{6.A.52}$$

$$f_\eta(\epsilon) = F_\eta(\epsilon,\epsilon) = \frac{45M}{2\pi N_A d^3 \epsilon^3}[\eta]_{(\mathrm{SD})}, \tag{6.A.53}$$

where $a_{\eta ij}$ are numerical constants and their values are given in Table 6.A.2; and $[\eta]_{(\mathrm{SD})}$ is the intrinsic viscosity of the spheroid and is given by [93,94]

$$
\begin{aligned}
[\eta]_{(\mathrm{SD})} = {} & \frac{\pi N_A d^3}{30M}\epsilon(\epsilon^2 - 1)^2\left\{\frac{2[-(4\epsilon^2 - 1)F + 2\epsilon^3 + \epsilon]}{3\epsilon(3F + 2\epsilon^3 - 5\epsilon)[(2\epsilon^2 + 1)F - 3\epsilon]}\right. \\
& + \frac{28}{3\epsilon(3F + 2\epsilon^3 - 5\epsilon)} + \frac{4}{(\epsilon^2 + 1)(-3\epsilon F + \epsilon^2 + 2)} \\
& \left. + \frac{2(\epsilon^2 - 1)}{\epsilon(\epsilon^2 + 1)[(2\epsilon^2 - 1)F - \epsilon]}\right\} \quad \text{for } \epsilon \neq 1 \\
= {} & \frac{5\pi N_A d^3}{12M} \quad \text{for } \epsilon = 1
\end{aligned}
\tag{6.A.54}
$$

with F being given by Eqs. (6.A.42). Note that at $\epsilon = p$, Eq. (6.A.49) gives the exact solution given by Eqs. (6.A.54) for the spheroid. We also note that the derivation of the first of Eqs. (6.A.54) by Simha [93] is not correct, although his result happens to be correct, as shown by Saito [94], the second being originally due to Einstein [1,95]. The range of application of Eq. (6.A.49) is limited to $0.6 \lesssim \epsilon \lesssim 1.3$.

Figure 6.A.3 shows double-logarithmic plots of $M[\eta]/M_0[\eta]_{\mathrm{E}}p$ against p, where $[\eta]_{\mathrm{E}}$ is given by Eq. (6.99). The solid and dot-dashed curves represent the values calculated from Eq. (6.A.49) for the spheroid-cylinders with the indicated values of ϵ and from Eqs. (6.A.54) for the spheroid ($\epsilon = p$), respectively, and the unfilled ($\epsilon \neq p$) and filled ($\epsilon = p$) circles represent the exact

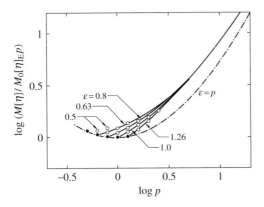

Fig. 6.A.3. Double-logarithmic plots of $M[\eta]/M_0[\eta]_{\mathrm{E}}p$ against p (see the text). The *solid and dot-dashed curves* represent the values calculated from Eq. (6.A.49) for the spheroid-cylinders and spheroid ($\epsilon = p$), respectively, and the *unfilled and filled circles* represent the exact numerical solutions

numerical solutions. It is seen that the end effects on $[\eta]$ are more remarkable than those on D and $D_{\mathrm{r},1}$.

Finally, we note that in the case of cylinders we have found that the values of $[\eta]$ with the nonpreaveraged and preaveraged Oseen tensors agree with each other to within 1% for $60 \lesssim p \lesssim 150$ [27]. This fact has been used in the construction of the interpolation formula for $[\eta]_{\mathrm{R}}$ given by Eq. (6.92).

Appendix 6.B Excess Stress Tensor for the Touched-Bead Model

In this appendix we derive an expression for the excess stress tensor due to an addition of a single touched-bead (or generally subbody) model chain to an incompressible fluid with viscosity coefficient η_0 [38]. In the unperturbed flow field $\mathbf{v}^0$ given by Eq. (6.65) with ϵ_0 the velocity gradient, the intrinsic viscosity $[\eta]$ may be written in the form

$$[\eta] = \frac{N_{\mathrm{A}}}{M \eta_0 \epsilon_0} \langle \sigma'_{xy} \rangle = \frac{N_{\mathrm{A}}}{M \eta_0 \epsilon_0} \mathbf{m} : \langle \boldsymbol{\sigma}' \rangle \,, \qquad (6.B.1)$$

where $\mathbf{m}$ is given by Eq. (6.67), and σ'_{xy} is the xy component of the excess stress tensor $\boldsymbol{\sigma}'$ for the single chain.

Now the equation of motion for the (incompressible) fluid in steady flow may be written in the form [1]

$$\nabla \cdot \boldsymbol{\sigma}(\mathbf{r}) + \mathbf{f}(\mathbf{r}) = \mathbf{0} \,, \qquad (6.B.2)$$

where $\mathbf{f}(\mathbf{r})$ is the force density due to the external (frictional) force exerted on the fluid (per unit volume) at a point $\mathbf{r}$, and $\boldsymbol{\sigma}$ is the stress tensor given by

$$\boldsymbol{\sigma}(\mathbf{r}) = -p(\mathbf{r})\mathbf{I} + \eta_0 \big\{ \nabla \mathbf{v}(\mathbf{r}) + \big[\nabla \mathbf{v}(\mathbf{r}) \big]^{T} \big\} \qquad (6.B.3)$$

with $\mathbf{v}(\mathbf{r})$ the fluid velocity, p the pressure, $\mathbf{I}$ the unit tensor, and the superscript T indicating the transpose. Note that substitution of Eq. (6.B.3) with Eq. (6.2) into Eq. (6.B.2) leads to Eq. (6.1). In the present case of the single chain composed of N beads (subbodies), $\mathbf{f}(\mathbf{r})$ is given by

$$\mathbf{f}(\mathbf{r}) = \sum_{j=1}^{N} \int_{S_j} \delta(\mathbf{r} - \mathbf{r}_j - \hat{\mathbf{r}}_j) \mathbf{f}_j(\hat{\mathbf{r}}_j) d\hat{\mathbf{r}}_j \,, \tag{6.B.4}$$

where $\mathbf{r}_j$ is the vector position of the center of the jth bead, $\hat{\mathbf{r}}_j$ is the radius vector from its center to an arbitrary point on its surface, $\mathbf{f}_j(\hat{\mathbf{r}}_j)$ is the frictional force exerted by the unit area at $\hat{\mathbf{r}}_j$ on the fluid, and the integration is carried out over its surface (S_j) (see Fig. 6.3).

The stress tensor $\boldsymbol{\sigma}$ may be written as a sum of the stress tensor $\boldsymbol{\sigma}_0$ of the pure fluid and the excess stress tensor $\boldsymbol{\sigma}'$ due to the force density $\mathbf{f}$; that is, $\boldsymbol{\sigma} = \boldsymbol{\sigma}_0 + \boldsymbol{\sigma}'$. Equation (6.B.2) may therefore be rewritten as

$$\nabla \cdot \boldsymbol{\sigma}_0 = \mathbf{0} \,, \tag{6.B.5}$$

$$\nabla \cdot \boldsymbol{\sigma}' + \mathbf{f} = \mathbf{0} \,. \tag{6.B.6}$$

We take the Fourier transform of both sides of Eq. (6.B.6),

$$i\mathbf{k} \cdot \tilde{\boldsymbol{\sigma}}'(\mathbf{k}) + \tilde{\mathbf{f}}(\mathbf{k}) = \mathbf{0} \,, \tag{6.B.7}$$

where

$$\tilde{\boldsymbol{\sigma}}'(\mathbf{k}) = \int \boldsymbol{\sigma}'(\mathbf{r}) \exp(i\mathbf{k} \cdot \mathbf{r}) d\mathbf{r} \,, \tag{6.B.8}$$

$$\tilde{\mathbf{f}}(\mathbf{k}) = \int \mathbf{f}(\mathbf{r}) \exp(i\mathbf{k} \cdot \mathbf{r}) d\mathbf{r}$$

$$= \sum_{j=1}^{N} \int_{S_j} \mathbf{f}_j(\hat{\mathbf{r}}_j) \exp\left[i\mathbf{k} \cdot (\mathbf{r}_j + \hat{\mathbf{r}}_j)\right] d\hat{\mathbf{r}}_j \,. \tag{6.B.9}$$

Let $\mathbf{R}_c$ be the vector position of the center of mass of the chain, and let $\mathbf{S}_j$ be the vector distance from it to the center of the jth bead. We have $\mathbf{r}_j = \mathbf{R}_c + \mathbf{S}_j$, and the second line of Eqs. (6.B.9) may be rewritten as

$$\tilde{\mathbf{f}}(\mathbf{k}) = \exp(i\mathbf{k} \cdot \mathbf{R}_c) \sum_{j=1}^{N} \mathbf{F}_j + i\mathbf{k} \cdot \exp(i\mathbf{k} \cdot \mathbf{R}_c)$$

$$\times \sum_{j=1}^{N} \int_{S_j} \left\{ \int_0^1 \exp\left[i\xi \mathbf{k} \cdot (\mathbf{S}_j + \hat{\mathbf{r}}_j)\right] d\xi \right\} (\mathbf{S}_j + \hat{\mathbf{r}}_j) \mathbf{f}_j(\hat{\mathbf{r}}_j) d\hat{\mathbf{r}}_j \,, \tag{6.B.10}$$

where $\mathbf{F}_j$ is the total frictional force exerted by the jth bead and is given by Eq. (6.27). Under the condition of ordinary viscosity measurements, there is not any external force other than shear flow field, so that the total frictional force (sum of $\mathbf{F}_j$) must vanish. We then obtain from Eqs. (6.B.7) and (6.B.10)

$$\tilde{\boldsymbol{\sigma}}'(\mathbf{k}) = -\exp(i\mathbf{k}\cdot\mathbf{R}_c)\sum_{j=1}^{N}\int_{S_j}\left\{\int_0^1 \exp\left[i\xi\mathbf{k}\cdot(\mathbf{S}_j+\hat{\mathbf{r}}_j)\right]d\xi\right\}$$
$$\times(\mathbf{S}_j+\hat{\mathbf{r}}_j)\mathbf{f}_j(\hat{\mathbf{r}}_j)d\hat{\mathbf{r}}_j\,. \tag{6.B.11}$$

Finally, we take the configurational average of both sides of Eq. (6.B.11), noting that $\mathbf{R}_c$ is distributed uniformly in the fluid and that the average over $\mathbf{R}_c$ may be taken independently of the other variables. We then obtain

$$\langle\tilde{\boldsymbol{\sigma}}'(\mathbf{k})\rangle = -(2\pi)^3\delta(\mathbf{k})\sum_{j=1}^{N}\left[\langle\mathbf{S}_j\mathbf{F}_j\rangle + \left\langle\int_{S_j}\hat{\mathbf{r}}_j\mathbf{f}_j(\hat{\mathbf{r}}_j)d\hat{\mathbf{r}}_j\right\rangle\right]. \tag{6.B.12}$$

Thus, by Fourier inversion of Eq. (6.B.12), we obtain

$$\langle\boldsymbol{\sigma}'\rangle = -\sum_{j=1}^{N}\left[\langle\mathbf{F}_j\mathbf{S}_j\rangle + \left\langle\int_{S_i}\mathbf{f}_j(\hat{\mathbf{r}}_j)\hat{\mathbf{r}}_j d\hat{\mathbf{r}}_j\right\rangle\right]. \tag{6.B.13}$$

Substitution of Eq. (6.B.13) into the second of Eqs. (6.B.1) leads to Eq. (6.102). In the case of a single rigid body, it also reduces to the first of Eqs. (6.A.19).

References

1. H. Yamakawa: *Modern Theory of Polymer Solutions* (Harper & Row, New York, 1971).
2. M. Doi and S. F. Edwards: *The Theory of Polymer Dynamics* (Clarendon Press, Oxford, 1986).
3. J. G. Kirkwood and J. Riseman: J. Chem. Phys. **16**, 565 (1948).
4. J. G. Kirkwood: J. Polym. Sci. **12**, 1 (1954).
5. R. E. DeWames, W. F. Hall, and M. C. Shen: J. Chem. Phys. **46**, 2782 (1967).
6. R. Zwanzig, J. Kiefer, and G. H. Weiss: Proc. Natl. Acad. Sci. USA **60**, 381 (1968).
7. H. Yamakawa: Ann. Rev. Phys. Chem. **25**, 179 (1974).
8. J. Rotne and S. Prager: J. Chem. Phys. **50**, 4831 (1969).
9. H. Yamakawa: J. Chem. Phys. **53**, 436 (1970).
10. T. Yoshizaki and H. Yamakawa: J. Chem. Phys. **73**, 578 (1980).
11. H. Yamakawa and G. Tanaka: J. Chem. Phys. **57**, 1537 (1972).
12. J. García de la Torre and V. A. Bloomfield: Q. Rev. Biophys. **14**, 81 (1981).
13. H. Yamakawa: Ann. Rev. Phys. Chem. **35**, 23 (1984).
14. S. F. Edwards and G. J. Papadopoulos: J. Phys. A **1**, 173 (1968).
15. S. F. Edwards and M. A. Oliver: J. Phys. A **4**, 1 (1971).
16. J. M. Burgers: *Second Report on Viscosity and Plasticity of the Amsterdam Academy of Sciences* (North-Holland, Amsterdam), Chap. 3.
17. C. -M. Tchen: J. Appl. Phys. **25**, 463 (1954).
18. S. Broersma: J. Chem. Phys. **32**, 1632 (1960).

19. H. Yamakawa and M. Fujii: Macromolecules **6**, 407 (1973).
20. H. Yamakawa and M. Fujii: Macromolecules **7**, 128 (1974).
21. J. Shimada and H. Yamakawa: Macromolecules **9**, 583 (1976).
22. H. Yamakawa and T. Yoshizaki: Macromolecules **12**, 32 (1979).
23. H. Yamakawa and T. Yoshizaki: Macromolecules **13**, 633 (1980).
24. H. Yamakawa: Macromolecules **16**, 1928 (1983).
25. J. E. Hearst and W. H. Stockmayer: J. Chem. Phys. **37**, 1425 (1962).
26. H. Yamakawa, J. Shimada, and M. Fujii: J. Chem. Phys. **68**, 2140 (1978).
27. T. Yoshizaki and H. Yamakawa: J. Chem. Phys. **72**, 57 (1980).
28. T. Yoshizaki, I. Nitta, and H. Yamakawa: Macromolecules **21**, 165 (1988).
29. H. Yamakawa and T. Yoshizaki: J. Chem. Phys. **78**, 572 (1983).
30. H. Yamakawa and J. Yamaki: J. Chem. Phys. **57**, 1542 (1972).
31. T. Yamada, T. Yoshizaki, and H. Yamakawa: Macromolecules **25**, 377 (1992).
32. J. G. Kirkwood: Rec. Trav. Chim. **68**, 649 (1949).
33. H. Yamakawa: Macromolecules **8**, 339 (1975).
34. J. G. Kirkwood and P. L. Auer: J. Chem. Phys. **19**, 281 (1951).
35. R. Ullman: J. Chem. Phys. **40**, 2422 (1964).
36. P. L. Auer and C. S. Gardner: J. Chem. Phys. **23**, 1546 (1955).
37. H. Brenner: Chem. Eng. Sci. **19**, 631 (1964).
38. T. Yoshizaki and H. Yamakawa: J. Chem. Phys. **88**, 1313 (1988).
39. Y. Einaga, H. Koyama, T. Konishi, and H. Yamakawa: Macromolecules **22**, 3419 (1989).
40. T. Konishi, T. Yoshizaki, T. Saito, Y. Einaga, and H. Yamakawa: Macromolecules **23**, 290 (1990).
41. T. Konishi, T. Yoshizaki, and H. Yamakawa: Macromolecules **24**, 5614 (1991).
42. F. Abe, Y. Einaga, and H. Yamakawa: Macromolecules **26**, 1891 (1993).
43. Y. Fujii, Y. Tamai, T. Konishi, and H. Yamakawa: Macromolecules **24**, 1608 (1991).
44. F. Abe, K. Horita, Y. Einaga, and H. Yamakawa: Macromolecules **27**, 725 (1994).
45. N. Sawatari, T. Konishi, T. Yoshizaki, and H. Yamakawa: Macromolecules **28**, 1089 (1995).
46. M. Kamijo, F. Abe, Y. Einaga, and H. Yamakawa: Macromolecules **28**, 1095 (1995).
47. F. Abe, Y. Einaga, and H. Yamakawa: Macromolecules **24**, 4423 (1991).
48. T. Yamada, H. Koyama, T. Yoshizaki, Y. Einaga, and H. Yamakawa: Macromolecules **26**, 2566 (1993).
49. M. Kuwata, H. Murakami, T. Norisuye, and H. Fujita: Macromolecules **17**, 2731 (1984).
50. H. Murakami, T. Norisuye, and H. Fujita: Macromolecules **13**, 345 (1980).
51. J. E. Godfrey: Biophys. Chem. **5**, 285 (1976).
52. J. E. Godfrey and H. Eisenberg: Biophys. Chem. **5**, 301 (1976).
53. D. Jolly and H. Eisenberg: Biopolymers **15**, 61 (1976).
54. T. Yanaki, T. Norisuye, and H. Fujita: Macromolecules **13**, 1462 (1980).
55. H. Kusanagi, H. Tadokoro, and Y. Chatani: Polym. J. **9**, 181 (1977).
56. K. Dehara, T. Yoshizaki, and H. Yamakawa: Macromolecules **26**, 5137 (1993).
57. T. Arai, N. Sawatari, T. Yoshizaki, Y. Einaga, and H. Yamakawa: Macromolecules **29**, 2309 (1996).
58. M. T. Record, Jr., C. P. Woodbury, and R. B. Inman: Biopolymers **14**, 393 (1975).
59. M. Osa, F. Abe, T. Yoshizaki, Y. Einaga, and H. Yamakawa: Macromolecules **29**, 2302 (1996).
60. T. Norisuye: Prog. Polym. Sci. **18**, 543 (1993).

61. Y. Tamai, T. Konishi, Y. Einaga, M. Fujii, and H. Yamakawa: Macromolecules **23**, 4067 (1990).
62. B. H. Zimm: J. Chem. Phys. **24**, 269 (1956).
63. M. Fixman: J. Chem. Phys. **42**, 3831 (1965).
64. C. W. Pyun and M. Fixman: J. Chem. Phys. **42**, 3838 (1965).
65. B. H. Zimm: Macromolecules **13**, 592 (1980).
66. M. Fixman: J. Chem. Phys. **80**, 6324 (1984); **84**, 4085 (1986).
67. M. Fixman: Faraday Discuss. No. 83, 199 (1987); J. Chem. Phys. **89**, 2442 (1988).
68. H. Yamakawa and T. Yoshizaki: J. Chem. Phys. **91**, 7900 (1989).
69. P. Rempp: J. Polym. Sci. **23**, 83 (1957).
70. E. D. von Meerwall, S. Amelar, M. A. Smeltzly, and T. P. Lodge: Macromolecules **22**, 295 (1989).
71. A. E. Dunstan: J. Chem. Soc. London **85**, 817 (1904).
72. von W. Herz and W. Rathmann: Z. Elektrochem. **19**, 589 (1913).
73. M. Fixman: J. Chem. Phys. **92**, 6858 (1990).
74. M. Fujii and H. Yamakawa: Macromolecules **8**, 792 (1975).
75. V. A. Bloomfield and B. H. Zimm: J. Chem. Phys. **44**, 315 (1966).
76. M. Fukatsu and M. Kurata: J. Chem. Phys. **44**, 4539 (1966).
77. R. Zwanzig: J. Chem. Phys. **45**, 1858 (1966).
78. E. Paul and R. M. Mazo: J. Chem. Phys. **48**, 2378 (1968); **51**, 1102 (1969).
79. G. Tanaka and H. Yamakawa: Polym. J. **4**, 446 (1973).
80. L. V. Crawford and P. H. Bleck: Virology **24**, 388 (1964).
81. J. Vinograd, J. Lebowitz, R. Radloff, R. Watson, and P. Laipis: Proc. Natl. Acad. Sci. USA **53**, 1104 (1965).
82. L. V. Crawford: J. Mol. Biol. **13**, 362 (1965).
83. C. Bode and A. D. Kaiser: J. Mol. Biol. **14**, 399 (1965).
84. I. B. David and D. R. Wolstenholme: J. Mol. Biol. **28**, 233 (1967).
85. A. Opschoor, P. H. Pouwels, C. M. Knijnenburg, and J. B. T. Aten: J. Mol. Biol. **37**, 13 (1968).
86. J. R. Dawson and J. A. Harpst: Biopolymers **10**, 2499 (1971).
87. D. A. Ostrander and H. B. Gray, Jr.: Biopolymers **12**, 1387 (1973).
88. H. Yamakawa, T. Yoshizaki, and M. Fujii: Macromolecules **10**, 934 (1977).
89. T. Norisuye, M. Motowoka, and H. Fujita: Macromolecules **12**, 320 (1979).
90. G. B. Jeffery: Proc. Roy. Soc. (London) Ser. A **102**, 161 (1922).
91. F. Perrin: J. Phys. Rad. **7**, 1 (1936).
92. S. Broersma: J. Chem. Phys. **74**, 6989 (1981).
93. R. Simha: J. Phys. Chem. **44**, 25 (1940).
94. N. Saito: J. Phys. Soc. Japan **6**, 297 (1951).
95. A. Einstein: Ann. Phys. (Leipzig) **19**, 289 (1906); **34**, 591 (1911).

7 Applications to Circular DNA

In this chapter the statistical-mechanical and transport theories of the HW chain developed so far are applied to some interesting problems of circular DNA such as cyclization of linear DNA and analysis of circular DNA topoisomers (topological isomers) or supercoiled forms. These problems may be treated theoretically by modeling duplex DNA as the KP1 chain (or sometimes the original KP chain), which is a special case of the HW chain. From the statistical-mechanical standpoint, all kinds of ring-closure probabilities for these KP chains, which do not necessarily concern DNA problems, are also considered in this chapter. Relevant experimental data are analyzed by the use of the present theories in order to determine the stiffness parameter (Kuhn segment length) and torsional force constant of duplex DNA. However, all aspects of the problem of the supercoiling of DNA are not discussed since it is beyond the scope of this book.

7.1 Ring-Closure Probabilities

7.1.1 Definitions

Closed circular duplex DNA molecules are formed by covalently joining the (cohesive) ends of the linear molecules. The efficiency of this cyclization reaction may be described by the *ring-closure probability with the end orientations specified*, or the *Jacobson–Stockmayer* (*J*) *factor* [1], as defined as the ratio of equilibrium constants for cyclization and bimolecular association. For the evaluation of the J factor and also all related DNA problems, we may represent duplex DNA by a special case of the HW chain with $\kappa_0 = 0$ and $\tau_0 \neq 0$, that is, the KP1 chain, affixing a localized Cartesian coordinate system $[\mathbf{e}_\xi(s), \mathbf{e}_\eta(s), \mathbf{e}_\zeta(s)]$ to the chain at the contour (helix axis) point s with $\mathbf{e}_\xi$ pointing to one of the sugar phosphate backbones [see Fig. 4.4b]. Then the parameter τ_0 is equal to the twist rate of the linear DNA in its undeformed state (at the minimum of energy) with the pitch of its strand helix equal to $2\pi/\tau_0$. We adopt the stiffness parameter λ^{-1} (equal to the Kuhn segment length and twice the persistence length for $\kappa_0 = 0$) and Poisson's ratio σ (or the torsional force constant β) as the two other parameters that are required to define the model (KP1 chain) completely. It must be noted that similar elastic models were adopted by Fuller [2], Benham [3], Le Bret [4], and Tanaka and Takahashi [5] in the study of the supercoiling of DNA and by Barkley and Zimm [6] in the study of its dynamics.

Now we consider the Green function $G(\mathbf{R}, \Omega \,|\, \Omega_0; L)$ for the chain of total contour length L defined in Sect. 4.2.1, where $\mathbf{R}$ is the end-to-end vector distance, $\Omega = \Omega(L)$, and $\Omega_0 = \Omega(0)$ with $\Omega(s) = [\theta(s), \phi(s), \psi(s)]$ ($0 \leq \theta \leq \pi$, $0 \leq \phi \leq 2\pi$, $0 \leq \psi \leq 2\pi$) being the Euler angles defining the orientation of the localized coordinate system at s ($0 \leq s \leq L$) with respect to an external coordinate system. Following the Jacobson–Stockmayer theory [1] and its extension [7, 8], the J factor (in molecules per unit volume) may be related to the ring-closure probability with the end orientations specified $G(\mathbf{0}, \Omega_0 \,|\, \Omega_0; L)$ as

$$J = 8\pi^2 G(\mathbf{0}, \Omega_0 \,|\, \Omega_0; L). \tag{7.1}$$

However, the reaction product is not a homogeneous species but rather a mixture of topological isomers, that is, *topoisomers* of closed circular DNA with different linking numbers [9], as illustrated in Fig. 7.1. The *linking number*, which is an integer and which we denote by N, is defined as the number of complete revolutions made by one strand about the (unknotted) DNA axis when the axis is constrained to lie in a plane [2, 9, 10], or the number of rotations the localized coordinate system at s completes about the contour (of the closed DNA in a plane) as s is changed from 0 to L. The number N is a topological parameter and is independent of chain configuration (or deformation).

We can then consider the *N-dependent ring-closure probability* $G(\mathbf{0}, \Omega_0 \,|\, \Omega_0; N, L)$, so that

$$G(\mathbf{0}, \Omega_0 \,|\, \Omega_0; L) = \sum_{N=-\infty}^{\infty} G(\mathbf{0}, \Omega_0 \,|\, \Omega_0; N, L). \tag{7.2}$$

It is clear that this G depends on N as $|\Delta N|$, where

$$\Delta N = N - \frac{\tau_0 L}{2\pi} \equiv N - \overline{N}. \tag{7.3}$$

Note that $\overline{N} = \tau_0 L / 2\pi$ is equal to the number of helix turns in the linear

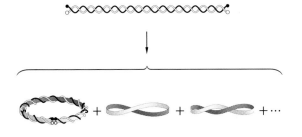

Fig. 7.1. Illustration of the formation of closed circular DNA topoisomers with different linking numbers

DNA fragment of length L in its undeformed state, so that ΔN is not necessarily an integer. Its meaning is the following: if the linear chain, which is initially in the undeformed state, is deformed so that its contour is always confined in a plane, we must twist one end by ΔN turns with respect to the other in order to join them to obtain the closed DNA with the linking number N. We also note that $\overline{N}$ is equal to $n_{\rm bp}/n_0$ with $n_{\rm bp} = L/l_{\rm bp}$, where $n_{\rm bp}$ is the number of base pairs in the DNA fragment, $l_{\rm bp}$ is the distance between them, and n_0 is the *helix repeat*, that is, the number of base pairs per helix turn. In this book we assume the following values: $l_{\rm bp} = 3.4$Å and $n_0 = 10.46$.

7.1.2 Linking-Number-Dependent Ring-Closure Probability

We begin by considering the N-dependent ring-closure probability $G(\mathbf{0}, \Omega_0 \,|\, \Omega_0; N, L)$ for small L. It may be effectively evaluated by replacing the continuous chain by an equivalent discrete chain composed of $n + 1$ segments, extrapolation to $n = \infty$ being made at the final stage [11]. For the continuous KP1 chain, its total potential energy E is given, from Eq. (4.4) with $\kappa_0 = 0$ or Eq. (4.15), by

$$E = \frac{1}{2} \int_0^L \left[\alpha(\omega_\xi{}^2 + \omega_\eta{}^2) + \beta(\omega_\zeta - \tau_0)^2\right] ds\,, \qquad (7.4)$$

where α and β are related to the parameters λ^{-1} and σ by Eqs. (3.37) and (4.5), respectively. In what follows, all lengths are measured in units of λ^{-1} and $k_{\rm B}T$ is chosen to be unity unless otherwise noted.

Now we consider the discrete chain. Its $n+1$ segments are numbered 0, 1, $\cdots$, n, each having length L/n except for the end ones of length $L/2n$. We can affix a localized coordinate system $(\mathbf{e}_{\xi_p}, \mathbf{e}_{\eta_p}, \mathbf{e}_{\zeta_p})$ to the pth segment ($p = 0$, 1, $\cdots$, n) corresponding to the system $[\mathbf{e}_\xi(s), \mathbf{e}_\eta(s), \mathbf{e}_\zeta(s)]$ at $s = pL/n$ of the continuous chain and denote the associated Euler angles by $\Omega_p = (\theta_p, \phi_p, \psi_p)$. The total potential energy $E(\Omega_1, \cdots, \Omega_n)$ (in units of $k_{\rm B}T$) of the discrete chain with Ω_0 fixed may be written, from Eqs. (7.4) and (4.10), as

$$E\big(\{\Omega_n\}\big) = \sum_{p=1}^{n} u(\Omega_{p-1}, \Omega_p)\,, \qquad (7.5)$$

where $\{\Omega_n\} = \Omega_1, \cdots, \Omega_n$, and

$$u(\Omega_{p-1}, \Omega_p) = u^{(0)}(\Omega_{p-1}, \Omega_p) - \frac{1}{2\sin^2 \theta_{p-1}}\left(\frac{L}{n}\right) \qquad (7.6)$$

with

$$u^{(0)}(\Omega_{p-1}, \Omega_p) = \frac{n}{4L}\left\{(\Delta\theta_p)^2 + (\Delta\phi_p)^2 \sin^2\left[\tfrac{1}{2}(\theta_p + \theta_{p-1})\right]\right\}$$

$$+ \frac{n}{4(1+\sigma)L}\left\{\Delta\phi_p \cos\left[\tfrac{1}{2}(\theta_p + \theta_{p-1})\right] + \Delta\psi_p - \frac{\tau_0 L}{n}\right\}^2. \qquad (7.7)$$

We note here that Ω_p should rather be determined successively from Ω_0 with given $\Delta\Omega_p = \Omega_p - \Omega_{p-1} = (\Delta\theta_p, \Delta\phi_p, \Delta\psi_p) = (\theta_p - \theta_{p-1}, \phi_p - \phi_{p-1},$ $\psi_p - \psi_{p-1})$ $(p = 1, \cdots, n)$, so that $-\infty < \theta_p, \phi_p, \psi_p < \infty$, and also that we have added to $u^{(0)}$ the infinitesimally small potential given by the second term on the right-hand side of Eq. (7.6) in order to make it possible to evaluate the configuration integral over $\{\Omega_n\}$.

The partition function Z is then given by

$$Z = \int \exp\left[-E(\{\Omega_n\})\right]d\{\Omega_n\} \tag{7.8}$$

with $d\{\Omega_n\} = d\Omega_1 \cdots d\Omega_n$ and $d\Omega_p = |\sin\theta_p|d\theta_p d\phi_p d\psi_p$. If we change variables from Ω_p to $\Delta\Omega'_p = (\Delta\theta'_p, \Delta\phi'_p, \Delta\psi'_p) = [\Delta\theta_p, \Delta\phi_p \sin(\theta_{p-1} + \frac{1}{2}\Delta\theta_p),$ $\Delta\phi_p \cos(\theta_{p-1} + \frac{1}{2}\Delta\theta_p) + \Delta\psi_p - \tau_0 L/n]$, we can carry out successively the integrations over $\Omega_n, \Omega_{n-1}, \cdots, \Omega_1$ in this order to find

$$Z = \left(\frac{4\pi L}{n}\right)^{3n/2} (1+\sigma)^{n/2}\left[1 - \frac{1}{4}L + \mathcal{O}(L^2)\right]. \tag{7.9}$$

Now we proceed to evaluate the ring-closure probability $G(\mathbf{0}, \Omega_0 \,|\, \Omega_0; N, n)$ for the discrete chain, which tends to $G(\mathbf{0}, \Omega_0 \,|\, \Omega_0; N, L)$ for the continuous chain in the limit of $n \to \infty$ at constant L. It is evident that G is symmetric about the initial unit tangent vector $\mathbf{u}_0$, and therefore we may remove the degree of freedom of rotation about it. Suppose that the joint of the closed chain is fixed at the origin of the external coordinate system $(\mathbf{e}_x, \mathbf{e}_y, \mathbf{e}_z)$ so that $\mathbf{e}_{\zeta_0} = \mathbf{u}_0$ coincides with $\mathbf{e}_x$, as depicted in Fig. 7.2. Let Lh be the distance of the center M of the $(n/2)$th segment from the x axis, assuming that n is even, let Φ_M $(0 \le \Phi_M \le 2\pi)$ be the rotation angle of M about the x axis, where $\Phi_M = 0$ when M lies in the xy plane, and let $G(\mathbf{0}, \Omega_0 \,|\, \Omega_0; h, \Phi_M, N, n)$ be the ring-closure probability with h and Φ_M also specified. We change Φ_M in such a way that E and hence this G remain unchanged. Then $\mathbf{e}_{\xi_0}$ and $\mathbf{e}_{\eta_0}$ must change with Φ_M, but we have

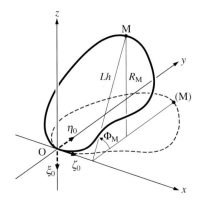

Fig. 7.2. Removal of the degree of freedom of rotation of the closed ring about $\mathbf{u}_0 = \mathbf{e}_{\zeta_0}$, which is placed to coincide with $\mathbf{e}_x$ (see the text)

$$G(\mathbf{0}, \Omega_0 \mid \Omega_0; N, n) = \int_0^\infty dh \int_0^{2\pi} d\Phi_{\mathrm{M}} L h\, G(\mathbf{0}, \Omega_0 \mid \Omega_0; h, \Phi_{\mathrm{M}}, N, n)$$

$$= 2\pi L \int_0^\infty dh\, h\, G(\mathbf{0}, \Omega_0 \mid \Omega_0; h, 0, N, n)\,. \qquad (7.10)$$

Thus we may consider only the configurations such that M lies in the xy plane with $\mathbf{e}_{\zeta_0} = \mathbf{e}_x$. At this stage, we reaffix all localized coordinate systems so that $\mathbf{e}_{\xi_0} = -\mathbf{e}_z$ and $\mathbf{e}_{\eta_0} = \mathbf{e}_y$. This does not alter E and G. The new situation is indicated by the dashed lines in Fig. 7.2. Thus we reinterpret the second line of Eqs. (7.10) in this fashion. Further, we note that if R_{M} is the z component of the radial vector of M, then $\Phi_{\mathrm{M}} = 0$ is equivalent to $R_{\mathrm{M}} = 0$.

It is easy to see that h, R_{M}, and the components R_x, R_y, and R_z of $\mathbf{R}$ in the external coordinate system are functions of $\{\Theta_n\} = \Theta_1, \cdots, \Theta_n$ with $\Theta_p = (\theta_p, \phi_p)$. When the fluctuation around the most probable configuration at the minimum of energy is small for small L, $G(\mathbf{0}, \Omega_0 \mid \Omega_0; h, 0, N, n)$ may be expressed as

$$G(\mathbf{0}, \Omega_0 \mid \Omega_0; h, 0, N, n) = Z^{-1} \int e^{-E} d\{\Omega_{n-1}\}/d\mathbf{R} dR_{\mathrm{M}} dh\,, \qquad (7.11)$$

where $d\{\Omega_{n-1}\} = d\Omega_1 \cdots d\Omega_{n-1}$ and the integration is carried out over $\{\Omega_{n-1}\}$ with the boundary conditions $\Omega_0 = (\pi/2,\, 0,\, 0)$ and $\Omega_n = (\pi/2, 2\pi, 2\pi N)$ [see also Eq. (7.16) below] and subject to the constraints

$$R_\alpha(\{\Theta_n\}) = 0 \qquad (\alpha = x, y, z, \mathrm{M})\,, \qquad (7.12)$$

$$h(\{\Theta_n\}) = h\,. \qquad (7.13)$$

If we remove the constraints of Eq. (7.12) by introducing Fourier representations of Dirac delta functions [12], we obtain, from Eqs. (7.10) and (7.11),

$$G(\mathbf{0}, \Omega_0 \mid \Omega_0; N, n) = (2\pi)^{-3} L Z^{-1} \int h(\{\Theta_n\}) \exp\left[-E(\{\Omega_n\})\right.$$

$$\left. +i \sum_\alpha k_\alpha R_\alpha(\{\Theta_n\})\right] d\mathbf{k} d\{\Omega_{n-1}\} \qquad (7.14)$$

with i the imaginary unit and with $d\mathbf{k} = dk_x dk_y dk_z dk_{\mathrm{M}}$. The evaluation of this integral consists of three steps: (1) determination of the most probable (closed) configuration $\{\Omega_n^*\}$, (2) expansion of E, R_α and h in terms of fluctuations in $\{\Omega_n\}$ around $\{\Omega_n^*\}$, and (3) integration over $\mathbf{k}$ and these fluctuations. Note that N may then be specified only for the most probable configuration.

First, the configuration $\{\Omega_n^*\}$ may be determined from the necessary condition for the extremum that the energy E becomes a minimum with the boundary conditions above and subject to the constraints of Eq. (7.12); that is,

$$\nabla_{\Omega_p}\left(E + L^{-2} \sum_\alpha \gamma_\alpha R_\alpha\right) = 0 \qquad (p = 1, \cdots, n-1) \qquad (7.15)$$

at $\{\Omega_n\} = \{\Omega_n^*\}$, where $\nabla_{\Omega_p} = (\partial/\partial\theta_p, \partial/\partial\phi_p, \partial/\partial\psi_p)$ and γ_α are (reduced) Lagrange multipliers. It is evident that one of the possible configurations $\{\Omega_n^*\}$ is the one for which the contour is a circle of radius $L/2\pi$ and ω_ζ is a constant independent of s. In his study of the supercoiling of closed circular DNA, Le Bret [4] treated the mechanical problem equivalent to the above variational principle and showed that this configuration is stable or metastable as far as $|\Delta N|/(1+\sigma) < 3^{1/2}$. For $|\Delta N|/(1+\sigma) > 3^{1/2}$, the circular configuration is never stable but will spontaneously assume superhelical forms such as the figure-of-eight-shaped (8-shaped) configuration, as illustrated in Fig. 7.1. Since G must be much smaller for these configurations, we confine ourselves to the case of circular configurations with $|\Delta N|/(1+\sigma) < 3^{1/2}$, for which

$$\Omega_p^* = (\theta_p^*, \phi_p^*, \psi_p^*) = \left(\frac{\pi}{2}, \frac{2\pi p}{n}, \frac{2\pi Np}{n}\right) \tag{7.16}$$

with

$$\gamma_\alpha = 0 \quad \text{for all } \alpha. \tag{7.17}$$

At $\{\Omega_n\} = \{\Omega_n^*\}$, we then have

$$E^* = \frac{\pi^2}{L}\left[1 + \frac{(\Delta N)^2}{1+\sigma}\right], \tag{7.18}$$

$$h^* = \pi^{-1}, \tag{7.19}$$

where h^* has been extrapolated to $n = \infty$, for simplicity.

Next we consider the fluctuations by setting $\{\Omega_n\} = \{\Omega_n^* + \delta\Omega_n\} = \Omega_1^* + \delta\Omega_1, \cdots, \Omega_n^* + \delta\Omega_n$ with $\Omega_p^* + \delta\Omega_p = (\theta_p^* + \delta\theta_p, \phi_p^* + \delta\phi_p, \psi_p^* + \delta\psi_p)$. However, in order to carry out the integration, it is convenient to change variables further from $\{\delta\Omega_n\}$ to $\{\delta\Omega_n'\}$ by $\delta\Omega_p' = (\delta\theta_p', \delta\phi_p', \delta\psi_p') = (\delta\theta_p, \delta\phi_p \cos\delta\theta_p, \delta\psi_p)$. Correspondingly, $\{\Theta_n\}$ is transformed to $\{\delta\Theta_n'\}$. Then $E(\{\Omega_n\})$, $R_\alpha(\{\Theta_n\})$, and $h(\{\Theta_n\})$ may be expanded as power series in $\{\delta\Omega_n'\}$ or $\{\delta\Theta_n'\}$. We write the results in the form

$$E = E^* + E_0(\{\delta\Omega_n'\}) + \tilde{E}(\{\delta\Omega_n'\}), \tag{7.20}$$

$$R_\alpha = R_{\alpha,0}(\{\delta\Theta_n'\}) + \tilde{R}_\alpha(\{\delta\Theta_n'\}), \tag{7.21}$$

$$h = h^* + \tilde{h}(\{\delta\Theta_n'\}), \tag{7.22}$$

where E_0 and $R_{\alpha,0}$ are the main parts of the fluctuations and are given by

$$E_0 = \frac{n}{4L}\sum_{p=1}^{n}\Big\{(\delta\theta_p - \delta\theta_{p-1})^2 + (\delta\phi_p' - \delta\phi_{p-1}')^2 - \tfrac{1}{4}(\Delta\phi^*)^2(\delta\theta_p + \delta\theta_{p-1})^2$$

$$+ \frac{1}{1+\sigma}\Big[(\delta\psi_p - \delta\psi_{p-1})^2 + \tfrac{1}{4}(\Delta\phi^*)^2(\delta\theta_p + \delta\theta_{p-1})^2$$

$$- \Delta\phi^*(\delta\theta_p + \delta\theta_{p-1})(\delta\psi_p - \delta\psi_{p-1}) - \Delta\psi^*(\delta\theta_p + \delta\theta_{p-1})(\delta\phi_p' - \delta\phi_{p-1}')\Big]\Big\}$$

$$\tag{7.23}$$

with $\Delta\phi^* = \phi_p^* - \phi_{p-1}^* = 2\pi/n$ and $\Delta\psi^* = \psi_p^* - \psi_{p-1}^* - \tau_0 L/n = 2\pi\Delta N/n$, and by

$$R_{x,0} = -\frac{L}{n}\sum_{p=1}^{n}(1 - \tfrac{1}{2}\delta_{pn})\delta\phi_p' \sin\phi_p^*, \quad R_{y,0} = \frac{L}{n}\sum_{p=1}^{n}(1 - \tfrac{1}{2}\delta_{pn})\delta\phi_p' \cos\phi_p^*,$$

$$R_{z,0} = -\frac{L}{n}\sum_{p=1}^{n}(1 - \tfrac{1}{2}\delta_{pn})\delta\theta_p, \quad R_{M,0} = -\frac{L}{n}\sum_{p=1}^{n/2}(1 - \tfrac{1}{2}\delta_{p,n/2})\delta\theta_p$$

$$(7.24)$$

with δ_{pn} the Kronecker delta. In Eqs. (7.20)–(7.22), $\tilde{E}$, $\tilde{R}_\alpha$, and $\tilde{h}$ are higher-order terms, for which we omit explicit expressions.

As seen from Eq. (7.23), E_0 is a quadratic form in $\{\delta\Omega'_{n-1}\}$. If we transform it to a diagonal form by an orthogonal transformation Q and $\{\delta\Omega'_{n-1}\}$ to new variables by Q, the required integral over $\mathbf{k}$ and $\{\delta\Omega'_{n-1}\}$ becomes a form similar to that often encountered in random-flight statistics [12] and can readily be evaluated, although with some devices, the details being omitted. Thus the final result (with $n = \infty$) may be written in the form

$$G(\mathbf{0}, \Omega_0 \,|\, \Omega_0; N, L) = C_0 L^{-13/2} \exp\left\{-\frac{\pi^2}{L}\left[1 + \frac{(\Delta N)^2}{1+\sigma}\right] + (C_1 + \tfrac{1}{4})L\right\}, \quad (7.25)$$

where C_0 and C_1 are functions of ΔN and σ but must be determined numerically by solving numerically the eigenvalue problem for the matrix associated with the above quadratic form. Good interpolation formulas for C_0 and C_1 so found for $0 \le |\Delta N|/(1+\sigma) \le 1.45$ are

$$C_0 = \frac{1}{(1+\sigma)^{1/2}}\sum_{j=0}^{7} a_{0j}\left(\frac{\Delta N}{1+\sigma}\right)^{2j}, \quad (7.26)$$

$$C_1 = \sum_{j=0}^{7}\left(a_{1j}^{(0)} + \frac{a_{1j}^{(1)}}{1+\sigma}\right)\left(\frac{\Delta N}{1+\sigma}\right)^{2j}, \quad (7.27)$$

where a_{0j}, $a_{1j}^{(0)}$, and $a_{1j}^{(1)}$ are numerical constants and their values are given in Table 7.1. For $|\Delta N|/(1+\sigma) > 1.45$, $G(\mathbf{0}, \Omega_0 \,|\, \Omega_0; N, L)$ almost vanishes for $L \lesssim 3$ (see below), so that we may then put $C_0 = 0$. The range of application of Eq. (7.25) is limited to $L \lesssim 2.5$.

In order to examine numerically the behavior of the above G, we introduce the N-dependent J factor $J_N(L)$ defined by

$$J_N(L) = 8\pi^2 G(\mathbf{0}, \Omega_0 \,|\, \Omega_0; N, L), \quad (7.28)$$

corresponding to Eq. (7.1), for later convenience. Values of $J_N(L)$ calculated from Eq. (7.28) with Eqs. (7.25)–(7.27) for $\sigma = 0$ are plotted against L in Fig. 7.3 for the indicated values of ΔN. It is seen that $J_N(L)$ exhibits a maximum for $0 \le |\Delta N| \lesssim 1$ and that at constant $L \lesssim 3$, it decreases with increasing

Table 7.1. Values of a_{0j}, $a_{1j}^{(0)}$, and $a_{1j}^{(1)}$ in Eqs. (7.26) and (7.27)

i	a_{0j}	$a_{1j}^{(0)}$	$a_{1j}^{(1)}$
0	2.784	0.2639	−0.0383
1	2.113	0.1399	−0.0827
2	0.6558	−0.1131	0.0125
3	1.719	0.6500	−0.2170
4	−2.478	−1.1223	0.3961
5	2.588	1.0320	−0.3991
6	−1.210	−0.4601	0.1899
7	0.2437	0.0829	−0.0367

$|\Delta N|$ and becomes negligibly small for $|\Delta N| \gtrsim 1.4$. In this connection, recall that the circular configuration is never stable for $|\Delta N|/(1 + \sigma) > 3^{1/2}$. We note that as σ is increased from 0 to 0.5, J_N with $\Delta N = 0$ and also its dependence on ΔN become small. In any case, it may be concluded that J_N only with $N = N^*$ and $N^* \pm 1$ make significant contribution, where N^* is an integer closest to $\overline{N}$.

Finally, we must make some remarks on the specification of the linking number. It is related to the imposition of nonperiodic boundary conditions on the distribution functions. Evaluation of them with such boundary conditions is possible near the rigid-rod limit [6] and also near the rigid-ring limit as above. However, it is difficult when large fluctuations are allowed, as seen from the above developments. Indeed, in Chap. 4, the differential equations satisfied by the Green functions with periodic boundary conditions have been derived, so that they may be expanded in terms of the Wigner functions $\mathcal{D}_l^{mj}$ with the nonnegative integers l (see also the next subsection). Of course, nonperiodic boundary conditions can be imposed in mechanical (not statistical) problems such as the determination of the most stable configuration under constraints [3, 4].

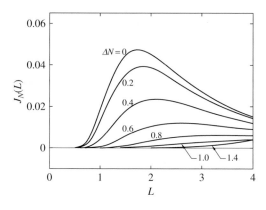

Fig. 7.3. Plots of J_N against L for $\sigma = 0$ and for the indicated values of ΔN

7.1.3 Ring-Closure Probability
with the End Orientations Specified

We proceed to evaluate the ring-closure probability $G(\mathbf{0}, \Omega_0 \,|\, \Omega_0; L)$ [11]. It is then convenient to introduce a parameter (auxillary variable) r $(0 \leq r \leq 1/2)$ defined by

$$r = |\overline{N} - N^*| \,. \tag{7.29}$$

Its meaning is the following: if the linear chain, which is initially in the undeformed state, is deformed so that its contour is always confined in a plane, we must twist or untwist one end by *at least* r turns with respect to the other in order to join them so that $\Omega = \Omega_0$. Since N^* is a step function of L, r is a periodic function of L; that is,

$$r(L) = |\overline{N}| - k \qquad \text{for } k \leq |\overline{N}| \leq k + \tfrac{1}{2}$$

$$= 1 - |\overline{N}| + k \qquad \text{for } k + \tfrac{1}{2} < |\overline{N}| < k + 1 \tag{7.30}$$

with k being nonnegative integers. Then the three values N^* and $N^* \pm 1$ of N mentioned above correspond to $\Delta N = r$ and $r \pm 1$, so that the J factor defined by Eq. (7.1) may be expressed as

$$J(L) = \sum_{\Delta N - r = -1}^{1} J_N(L) \tag{7.31}$$

provided that L is small $(L \lesssim 2.5)$.

Next we derive expressions for $J(L)$ for large L. If we put $\mathbf{R} = \mathbf{0}$ and $\Omega = \Omega_0 = (0, 0, 0)$ in Eq. (4.155) for $G(\mathbf{R}, \Omega \,|\, \Omega_0; L)$ and use the relations $\mathcal{G}^{\cdots}_{\cdots}(0, L) = 0$ for $l_3 \neq 0$, $\mathcal{D}^{mj}_l(0, 0, 0) = \left[(2l+1)/8\pi^2\right]^{1/2} \delta_{mj}$ [Eq. (4.C.6)], and $Y_0^0 = (4\pi)^{-1/2}$ with Eqs. (4.156), (5.92), and (5.95), we obtain, from Eq. (7.1),

$$J(L) = (4\pi)^{-1/2} \sum_{j=0}^{\infty} \sum_{l=j}^{\infty} (2 - \delta_{j0})(2l+1)\, \bar{\mathcal{G}}^{00,jj}_{ll0}(0, L) \,, \tag{7.32}$$

where $\bar{\mathcal{G}}^{\cdots}_{\cdots}$ is the real part of $\mathcal{G}^{\cdots}_{\cdots}$. Then, for the KP1 chain, if we follow the developments in Sect. 4.6.1, we obtain the interesting relation

$$\mathcal{G}^{00,jj}_{ll0}(R; L; \sigma, \tau_0) = \exp\left[-(\sigma j^2 + ij\tau_0)L\right] \mathcal{G}^{00,jj}_{ll0}(R; L; \sigma = \tau_0 = 0) \,, \tag{7.33}$$

where we have used Eq. (4.88) for $g^{jj'}_l(L)$ and note that $\mathcal{G}^{\cdots}_{\cdots}$ is real for $\sigma = \tau_0 = 0$. Substitution of Eq. (7.33) into Eq. (7.32) leads to

$$J(L) = \sum_{j=0}^{\infty} F_j(L) \cos(j\tau_0 L) \,, \tag{7.34}$$

where we note that $\cos(j\tau_0 L) = \cos(2\pi jr)$ from Eq. (7.29), and F_j is given by

$$F_j(L) = (2 - \delta_{j0})(4\pi)^{-1/2} \exp(-\sigma j^2 L)$$
$$\times \sum_{l=j}^{\infty} (2l + 1)\mathcal{G}_{ll0}^{00,jj}(0; L; \sigma = \tau_0 = 0). \qquad (7.35)$$

We note that if we integrate both sides of Eq. (4.155) over ψ and ψ_0, divide them by 2π, and put $\mathbf{R} = \mathbf{0}$ and $\mathbf{u} = \mathbf{u}_0 = \mathbf{e}_z$, we obtain for the ring-closure probability $G(\mathbf{0}, \mathbf{u}_0 \,|\, \mathbf{u}_0; L)$

$$G(\mathbf{0}, \mathbf{u}_0 \,|\, \mathbf{u}_0; L) = (4\pi)^{-1} F_0(L). \qquad (7.36)$$

Now, in the Daniels approximation, $\mathcal{G}_{\cdots}^{\cdots}$ is expanded in inverse powers of L, suppressing all exponential terms of order $\exp(-\text{const.} \, L)$, so that for the KP1 chain $\mathcal{G}_{ll0}^{00,jj}(R; L)$ may be set equal to zero for $j \neq 0$, as seen from Eq. (7.33). We then have

$$J(L) = F_0(L). \qquad (7.37)$$

For the KP chain, we have the Daniels expansion of $G(\mathbf{0}, \mathbf{u} \,|\, \mathbf{u}_0; L)$ from Eq. (3.83) with $\mathbf{R} = \mathbf{0}$. In the second Daniels approximation, it is given by [13, 14]

$$G(\mathbf{0}, \mathbf{u} \,|\, \mathbf{u}_0 = \mathbf{e}_z; L) = (4\pi)^{-1} \left(\frac{3}{2\pi L}\right)^{3/2} \left[1 - \frac{5}{8L} - \frac{79}{640L^2}\right.$$
$$\left. - \frac{3}{4}\left(\frac{1}{L} - \frac{1}{8L^2}\right) P_1(\cos\theta) + \frac{1}{12L^2} P_2(\cos\theta) + \mathcal{O}(L^{-3})\right], \quad (7.38)$$

where P_n is the Legendre polynomial and $\mathbf{u} = (1, \theta, \phi)$ in spherical polar coordinates. From Eq. (7.37) with Eqs. (7.36) and (7.38) with $\mathbf{u} = \mathbf{u}_0$, we then find

$$J(L) = \left(\frac{3}{2\pi L}\right)^{3/2} \left[1 - \frac{11}{8L} + \frac{103}{1920L^2} + \mathcal{O}(L^{-3})\right]. \qquad (7.39)$$

Thus, in this approximation, $J(L)$ is independent of σ and τ_0, as is also seen from Eqs. (7.35) and (7.37).

On the other hand, the sth approximation to $\mathcal{G}_{ll0}^{00,jj}(R; L)$ by the weighting function method may be written, from Eq. (4.203), in the form

$$\mathcal{G}_{ll0}^{00,jj}(R; L) = \left(\frac{3}{2\langle R^2\rangle}\right)^{3/2} w(\rho) \sum_{n=0}^{s} M_{l,n}^j(L) \rho^{2n}, \qquad (7.40)$$

where ρ is given by Eq. (4.180) and we choose as the weighting function $w(\rho)$ the function $w_{\mathrm{II}}(\rho)$ given by Eq. (4.208). With values of $\mathcal{G}_{\cdots}^{\cdots}(0; L; \sigma = \tau_0 = 0)$ so evaluated with $s = 6$ and for $0 \leq j \leq l \leq 5$, interpolation formulas for $F_j(L)$ are constructed. The results are

$$F_0(L) = \sum_{k=0}^{3} f_{0k} L^{-k-3/2} ,$$

$$F_1(L) = \exp\left[-(2+\sigma)L\right] \sum_{k=0}^{4} f_{1k} L^{-k} , \tag{7.41}$$

$$F_j(L) = 0 \quad \text{for } j \geq 2$$

with

$$f_{00} = 0.3346 , \quad f_{01} = -0.4810 , \quad f_{02} = -0.04212 , \quad f_{03} = 0.1495 ,$$

$$f_{10} = -0.1856 , \quad f_{11} = 2.353 , \quad f_{12} = 2.344 , \quad f_{13} = -18.47 , \quad f_{14} = 16.37 . \tag{7.42}$$

The range of application of Eqs. (7.41) is limited to $2 \lesssim L \lesssim 4$ (strictly $2.8 \lesssim L \lesssim 4$). For $L \gtrsim 4$, we may use the Daniels approximation, that is, Eq. (7.39), since there the values by the weighting function method agree well with those from Eq. (7.39).

Finally, we construct an empirical interpolation formula for $J(L)$ for intermediate L. Let L_0 be the value of L at which the value of J given by Eq. (7.31) agrees with that of J given by Eq. (7.34) with Eqs. (7.41), both for a given value of r. (It may be determined graphically.) Let J_1 and J_1' be the values of the former J and its first derivative with respect to L at $L = L_0 - 0.4 \equiv L_1$, respectively, and let J_2 and J_2' be those of the latter J and its first derivative at $L = L_0 + 0.4 \equiv L_2$, respectively. J_1' and J_2' may be calculated from

$$J_1' = J_N(L_1) \sum_{\Delta N - r = -1}^{1} \left\{ -\frac{13}{2L_1} + \frac{\pi^2}{L_1^2} \left[1 + \frac{(\Delta N)^2}{1+\sigma} \right] + C_1 + \frac{1}{4} \right\}, \tag{7.43}$$

$$J_2' = -\sum_{k=0}^{3} \left(\frac{3}{2} + k \right) f_{0k} L_2^{-5/2-k} - \cos(2\pi r) \left\{ (2+\sigma) F_1(L_2) \right.$$

$$\left. + \exp\left[-(2+\sigma)L_2\right] \sum_{k=1}^{4} k f_{1k} L_2^{-1-k} \right\}, \tag{7.44}$$

where in Eq. (7.43), $J_N(L)$ and C_1 are given by Eq. (7.28) with Eq. (7.25) and Eq. (7.27), respectively, and in Eq. (7.44), F_1 and f_{jk} are given by the second of Eqs. (7.41) and Eqs. (7.42), respectively. Then a good interpolation formula for $J(L)$ at constant r, which we denote by $J(L, r)$, is

$$J(L, r) = J_1 + J_1'(L - L_1) - 1.5625 \left[3(J_1 - J_2) + 0.8(2J_1' + J_2') \right] (L - L_1)^2$$

$$+ 1.9531 \left[2(J_1 - J_2) + 0.8(J_1' + J_2') \right] (L - L_1)^3 \quad (L_1 < L < L_2). \tag{7.45}$$

$J(L)$ as an explicit function of L in this range may be calculated from Eq. (7.45) with Eqs. (7.30).

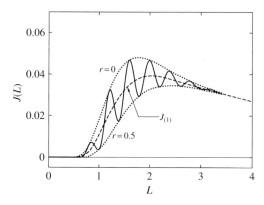

Fig. 7.4. Plots of J against L for $\sigma = 0$ and $\tau_0 = 5\pi$ (*solid curve*). The *dashed curve* represents the values of $J_{(1)}$ and the *dotted curves* indicate the upper ($r = 0$) and lower ($r = 0.5$) bounds of J

Values of $J(L)$ so calculated as an explicit function of L with Eqs. (7.30) for $\sigma = 0$ and $\tau_0 = 5\pi$ are represented by the solid curve in Fig. 7.4. The dotted curves with $r = 0$ and 0.5 indicate the upper and lower bounds, respectively. It is interesting to see that $J(L)$ stays at zero for very small L, then increases oscillating between the bounds, and finally decreases monotonically. This is consistent with experimental results, as shown later. The oscillation is due to the fact that if the number of base pairs in the DNA fragment is not an integral muliple of the helix repeat, the need to twist the DNA helix in order to make the strand ends meet decreases the J factor significantly in this range.

7.1.4 Other Ring-Closure Probabilities

In this subsection we apply the method developed in Sect. 7.1.2 to the evaluation of the ring-closure probabilities $G(\mathbf{0}, \mathbf{u}_0 \,|\, \mathbf{u}_0; L)$ and $G(\mathbf{0}; L)$ (for the KP1 chain) for small L, which do not necessarily concern DNA problems. (For large L, they have already been evaluated.) Since both are related only to the behavior of the chain contour, the final results may be obtained correctly even if we do not consider the torsional energy from the start as in the case of the original KP chain. Therefore, we drop the term proportional to $(1 + \sigma)^{-1}$ from the potential energy E of the discrete chain and denote the rest by $E_B(\{\Theta_n\})$. We give only the results with a brief description of the derivation [11].

(*a*) $G(\mathbf{0}, \mathbf{u}_0 \,|\, \mathbf{u}_0; L)$

The corresponding $G(\mathbf{0}, \mathbf{u}_0 \,|\, \mathbf{u}_0; n)$ for the discrete chain may be written in a form similar to Eq. (7.14) as follows,

$$G(\mathbf{0}, \mathbf{u}_0 \,|\, \mathbf{u}_0; n) = (2\pi)^{-3} L Z_B^{-1} \int h(\{\Theta_n\}) \exp\Big[-E_B(\{\Theta_n\})$$

$$+ i \sum_\alpha k_\alpha R_\alpha(\{\Theta_n\})\Big] d\mathbf{k} d\{\Theta_{n-1}\}, \qquad (7.46)$$

where Z_B is the partition function given by

$$Z_B = \int \exp\left[-E_B(\{\Theta_n\})\right]d\{\Theta_n\} = \left(\frac{4\pi L}{n}\right)^n \left[1 - \frac{1}{4}L + \mathcal{O}(L^2)\right] \quad (7.47)$$

with $d\{\Theta_n\} = d\Theta_1 \cdots d\Theta_n$ and $d\Theta_p = |\sin\theta_p|d\theta_p d\phi_p$. The most probable (closed) configuration is a circle of radius $L/2\pi$; that is,

$$\Theta_p^* = \left(\frac{\pi}{2}, \frac{2\pi p}{n}\right) \quad (7.48)$$

with

$$\gamma_\alpha = 0 \qquad \text{for all } \alpha, \quad (7.49)$$

so that we have

$$E_B^* = \frac{\pi^2}{L}, \quad (7.50)$$

$$h^* = \pi^{-1}. \quad (7.51)$$

In this case, the prefactor C_0 can be evaluated analytically, and we obtain the final result

$$G(\mathbf{0}, \mathbf{u}_0 \mid \mathbf{u}_0; L) = \pi^2 L^{-6} \exp\left(-\frac{\pi^2}{L} + 0.514\,L\right), \quad (7.52)$$

which is valid for $L < 1.9$.

We define the J factor $J_{(1)}(L)$ by

$$J_{(1)}(L) = 4\pi G(\mathbf{0}, \mathbf{u}_0 \mid \mathbf{u}_0; L). \quad (7.53)$$

For large L, $J_{(1)}(L)$ is seen to be given, from Eqs. (7.36), (7.37), and (7.53), by

$$J_{(1)}(L) = F_0(L). \quad (7.54)$$

Equation (7.54) with the first of Eqs. (7.41) is valid for $2.8 \lesssim L \lesssim 4$. For $L \gtrsim 4$, $J_{(1)}(L)$ may be equated to $J(L)$ given by Eq. (7.39). A good interpolation formula for $J_{(1)}(L)$ for intermediate L is

$$J_{(1)}(L) = 0.03882 + 0.003494\,(L - 1.9) - 0.01618\,(L - 1.9)^2$$
$$+0.008601\,(L - 1.9)^3 \qquad (1.9 < L < 2.7). \quad (7.55)$$

Values of $J_{(1)}(L)$ calculated from Eq. (7.53) with Eq. (7.52) and Eqs. (7.54) and (7.55) are represented by the dashed curve in Fig. 7.4. It is seen that $J(L)$ oscillates around $J_{(1)}(L)$.

(b) $G(\mathbf{0}; L)$

The (angle-independent) ring-closure probability $G(\mathbf{0}; L)$ may be given by

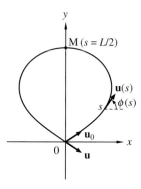

Fig. 7.5. Determination of the most probable configuration of the closed KP chain without the end orientations specified (see the text)

$$G(\mathbf{0}; L) = \int G(\mathbf{0}, \mathbf{u} \,|\, \mathbf{u}_0; L)d\mathbf{u} \equiv G(\mathbf{0}, \,|\, \mathbf{u}_0; L) \,. \qquad (7.56)$$

The corresponding $G(\mathbf{0}, \,|\, \mathbf{u}_0; n)$ for the discrete chain is given by Eq. (7.46) with $d\{\Theta_n\}$ in place of $d\{\Theta_{n-1}\}$. We solve the associated variational problem for the continuous chain [14]. For this purpose, we choose $\mathbf{u}_0$ and $\mathbf{u}$ to be in the xy plane so that the x axis bisects the angle between $\mathbf{u}_0$ and $\mathbf{u}$, as depicted in Fig. 7.5. The most probable (closed) configuration without the end orientations specified is then symmetric about the y axis, so that we need only to consider the range of $0 \leq s \leq L/2$.

It is clear that

$$\theta_p^* = \frac{\pi}{2} \qquad (7.57)$$

for the most probable (closed) configuration with $s = pL/n$. If we define the angle $x(s)$ by

$$x = \frac{1}{2}(\pi - \phi) \qquad (7.58)$$

for $0 \leq s \leq L/2$, then the solution of the Euler equation for x [Eq. (7.15) in the continuous limit] may be obtained as [14]

$$\int_0^{x^*} (1 - k_0^2 \sin^2 x)^{-1/2} dx = \frac{2}{L} k_0^{-1} K(k_0^{-1}) \left(\frac{L}{2} - s \right) , \qquad (7.59)$$

where k_0 is the solution of the equation

$$K(k_0^{-1}) = 2E(k_0^{-1}) \,, \qquad (7.60)$$

and $K(k)$ and $E(k)$ are the complete elliptic integrals of the first and second kinds, respectively, the former being given by Eq. (6.146). The solution $\phi^*(s) = \phi_p^*$ may be written, from Eqs. (7.58) and (7.59), in the form

$$\cos \phi_p^* = 1 - 2\mathrm{cn}^2(v \,|\, k_0) \qquad (7.61)$$

with

$$v = 2k_0^{-1}K(k_0^{-1})\left(\frac{1}{2} - \frac{p}{n}\right), \tag{7.62}$$

where $\mathrm{cn}\,(v\,|\,k_0)$ is the Jacobian elliptic function (whose parameter is k_0) defined by

$$\mathrm{cn}\,v = \cos\varphi, \tag{7.63}$$

$$v = \int_0^\varphi (1 - k_0^2\sin^2 x)^{-1/2}dx = F(\varphi\,|\,k_0) \tag{7.64}$$

with $F(\varphi\,|\,k)$ the incomplete elliptic integral of the first kind. The Lagrange multipliers in Eq. (7.15) are found to be

$$\gamma_x = -2\left[K(k_0^{-1})\right]^2, \tag{7.65}$$
$$\gamma_y = \gamma_z = \gamma_{\mathrm{M}} = 0.$$

Since the solution of Eq. (7.60) is $k_0 = 1.100$ [14], we have $K(k_0^{-1}) = 2.321$, $\gamma_x = -10.77$, and

$$E_{\mathrm{B}}^* = \frac{7.027}{L}, \tag{7.66}$$

$$h^* = 0.2554. \tag{7.67}$$

We note that $\phi_0^* = 0.860$ ($= 49°18'$), that Lh is the distance of M from the initial tangent $\mathbf{u}_0$ (not from the x axis), and that the most probable configuration is characterized by the vanishing curvature ($d\phi/ds = 0$) at the chain ends. In this case, γ_x does not vanish, and therefore $E_{\mathrm{B},0}$ and terms involving γ_x form a quadratic form. Thus we obtain the final result

$$G(\mathbf{0}; L) = 28.01\,L^{-5}\exp\left(-\frac{7.027}{L} + 0.492\,L\right). \tag{7.68}$$

We note that in the earlier evaluation [14], the factor L^{-1} appears in place of L^{-5} in Eq. (7.68) because of the approximate treatment of the fluctuation.

For large L, we obtain, from the first of Eqs. (3.90) with Eq. (3.85),

$$G(\mathbf{0}; L) = \left(\frac{3}{2\pi L}\right)^{3/2}\left[1 - \frac{5}{8L} - \frac{79}{640L^2} + \mathcal{O}(L^{-3})\right] \tag{7.69}$$

in the second Daniels approximation, corresponding to Eq. (7.39). The values of $G(\mathbf{0}; L)$ by the weighting function method agree with those from Eq. (7.68) for $2 \lesssim L \lesssim 4$ and those from Eq. (7.69) for $L \gtrsim 4$ to within 1%.

Finally, we consider the J factor $J_{(0)}(L)$ defined by

$$J_{(0)}(L) = G(\mathbf{0}; L). \tag{7.70}$$

Its values calculated from Eq. (7.68) for $L < 4$ are represented by the solid curve in Fig. 7.6. As was expected, $J_{(0)}(L)$ is larger than $J_{(1)}(L)$ and $J(L)$ in the range displayed. For comparison, the corresponding values in the earlier evaluation (YS) [14] are represented by the dot-dashed curve, and the Monte Carlo values of Hagerman [15] are represented by the unfilled circles. The former values are somewhat overestimated near the peak, while the latter values are rather in good agreement with those calculated from Eq. (7.68).

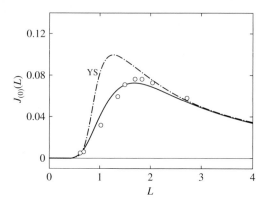

Fig. 7.6. Plots of $J_{(0)}$ against L. The *solid curve* represents the values calculated from Eq. (7.68), the *dot-dashed curve* represents the earlier (YS) values [14], and the *unfilled circles* represent the Monte Carlo values of Hagerman [15]

7.1.5 Comparison with Experiment

In this subsection we make a comparison of theory with experiment using experimental data obtained by Baldwin and co-workers [16–18] for DNA. Their earlier data [16] for the J factor as a function of L or n_{bp} ($= L/l_{bp}$) are difficult to analyze precisely, the data points being distributed around the curve of $J_{(1)}(L)$ against L [11]. Figure 7.7 shows double-logarithmic plots of J (in mol/L) against n_{bp} with more accurate data subsequently obtained by Shore and Baldwin [17], although in the narrow range of n_{bp}. The solid curve represents the best-fit theoretical values calculated with $\lambda^{-1} = 900$ Å and $\sigma = -0.4$, while the dashed curve represents those with $\lambda^{-1} = 950$ Å and $\sigma = -0.2$, ignoring the largest three observed values. These two sets of estimates of λ^{-1} and σ lead to the values 3.0×10^{-19} and 2.4×10^{-19} erg cm of the torsional force constant β, respectively. The above values of λ^{-1} are somewhat smaller than those from the transport properties (see Table 6.3). Further, recall that σ may be assumed to be zero for flexible chains (although $\sigma \simeq 0.5$ for most of polymeric materials in the bulk). Since we are considering

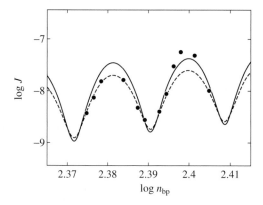

Fig. 7.7. Double-logarithmic plots of J (in mol/L) against n_{bp} for DNA. The *filled circles* represent the experimental values of Shore and Baldwin [17]. The *solid curve* represents the best-fit theoretical values calculated with $\lambda^{-1} = 900$ Å and $\beta = 3.0 \times 10^{-19}$ erg cm, and the *dashed curve* represents those with $\lambda^{-1} = 950$ Å and $\beta = 2.4 \times 10^{-19}$ erg cm

local elasticity on the atomic or molecular level, the assumption of $\sigma = 0$ or < 0 is not necessarily surprising. The same data were analyzed by a Monte Carlo method by Levene and Crothers [19], who obtained $\lambda^{-1} = 950$ Å and $\beta = 3.8 \times 10^{-19}$ erg cm. For a wide variety of methods of determination and estimates of λ^{-1} and β for DNA, the reader is referred to the review article by Hagerman [20].

Next we consider the distribution of topoisomers. The fraction f_N of the topoisomers with the linking number N is given by

$$f_N = J_N/J,\tag{7.71}$$

where J_N is given by Eq. (7.28) with Eq. (7.25) for small L, and N may be assumed to take the three values N^* and $N^* \pm 1$. These equations are adapted to an analysis [21] of data obtained by Shore and Baldwin [18] for the topoisomer distribution as a function of the amount of ethidium bromide (Et) bound, which unwinds the double helix. If ϕ_{Et} is the angle (in degrees) by which the binding of an Et molecule unwinds the helix, the binding of ν Et molecules per base pair will change the number of helix turns by $\delta = \nu\phi_{Et}n_{bp}/360$. (Note that $\phi_{Et} \simeq 26°$ [9].) We assume that the DNA double helix with Et bound may then still be regarded as the KP1 chain with λ^{-1} and σ remaining unchanged but with the constant torsion $\tilde{\tau}_0$ given by

$$\frac{\tilde{\tau}_0 L}{2\pi} = \frac{\tau_0 L}{2\pi} - \delta.\tag{7.72}$$

Therefore, the fraction $\tilde{f}_N$ of the topoisomers with the linking number N in the presence of bound Et is given by Eq. (7.71) with $\tilde{\tau}_0$ and $\tilde{N}^*$ in place of τ_0 and N^*, respectively, where $\tilde{N}^*$ is an integer closest to $\tilde{\tau}_0 L/2\pi$. Under the experimental conditions of Shore and Baldwin [18], in the absence of bound Et there exists only one topoisomer with $N = N^*$, while in its presence there exist only two topoisomers formed by minimal undertwisting and overtwisting. Let $\langle N \rangle_0$ and $\langle N \rangle$ be the average linking numbers in the absence and presence of bound Et, respectively. We then have

$$\langle N \rangle_0 = N^*,\tag{7.73}$$

$$\langle N \rangle = \tilde{N}^* \tilde{f}_{\tilde{N}^*} + (\tilde{N}^* - 1)\tilde{f}_{\tilde{N}^*-1} \quad \text{for } \tilde{\tau}_0 L/2\pi \leq \tilde{N}^*$$
$$= \tilde{N}^* \tilde{f}_{\tilde{N}^*} + (\tilde{N}^* + 1)\tilde{f}_{\tilde{N}^*+1} \quad \text{for } \tilde{\tau}_0 L/2\pi > \tilde{N}^*.\tag{7.74}$$

Figure 7.8 shows plots of $\langle N \rangle_0 - \langle N \rangle$ against δ with the data of Shore and Baldwin [18] for $n_{bp} = 247$. The values of δ have been calculated from the equation $\delta = 7.7 \times 10^{-3} n_{bp} c_{Et}$, where c_{Et} is the concentration of Et (in μg/mL). The solid curve represents the best-fit theoretical values calculated from Eqs. (7.73) and (7.74) with $\lambda^{-1} = 900$ Å and $\sigma = -0.5$ ($\beta = 3.6 \times 10^{-19}$erg cm). It is interesting to see that $\langle N \rangle_0 - \langle N \rangle$ changes in steps. Theoretically, it changes one-by-one with every unit step of δ. This is due to the fact that $\tilde{J}_N$ at δ is equal to $\tilde{J}_{N-1}$ at $\delta + 1$, as seen from

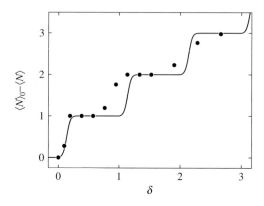

Fig. 7.8. Plots of $\langle N \rangle_0 - \langle N \rangle$ against δ for DNA with $n_{\mathrm{bp}} =$ 247. The *filled circles* represent the experimental values of Shore and Baldwin [18], and the curve represents the best-fit theoretical values calculated with $\lambda^{-1} = 900$ Å and $\beta = 3.6 \times 10^{-19}$ erg cm

Eqs. (7.25) and (7.28) with Eqs. (7.3) and (7.72). However, each observed step becomes progressively broader. This may be regarded as arising mainly from the broadening of the topoisomer distribution by a superimposed distribution of the number of Et molecules bound [18]. The topoisomer distribution for larger L is considered in the next section.

7.2 Topoisomer Statistics

7.2.1 Basic Concepts and Equations

In the present and following subsections we treat the distribution of topoisomers or the N-dependent ring-closure probability $G(\mathbf{0}, \Omega_0 \,|\, \Omega_0; N, L)$, which we simply denote here by $P(N; L)$, and also other related quantities for larger L [22, 23]. For this purpose, we must introduce two new quantities: the *twist* Tw and the *writhe* Wr [9]. The linking number N is then given by the sum of them [2],

$$N = Tw + Wr \,. \qquad (7.75)$$

Now Tw is the contribution from the twisting of the strands about the helix axis and Wr is that from the bending of the helix axis. Both are dependent on chain configuration (or deformation) and can vary continuously. In the present notation, Tw is defined by [2]

$$Tw = (2\pi)^{-1} \int_0^L \omega_\zeta(s) ds \,. \qquad (7.76)$$

Then Eq. (7.75) is rather the defining equation for Wr. The expression for it suitable for the present purpose is the one derived by Le Bret [4]; that is,

$$Wr = -(2\pi)^{-1} \int_0^L \left(\frac{d\phi}{ds} \right) \cos\theta ds + Wr(z) \,, \qquad (7.77)$$

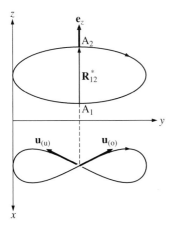

Fig. 7.9. Symmetrical projections of the most probable configuration of the closed curve with $Wr = 0.37$ and $Wr(z) = 1$ onto the xy and yz planes (see the text)

where $Wr(z)$ is the *directional writhing number* in the direction of $\mathbf{e}_z$ [2]. $Wr(z)$ takes only integral values and is defined as follows [2]. Suppose that the closed chain is projected onto the xy plane and traced in a fixed direction. Let $\mathbf{u}_{(o)}$ and $\mathbf{u}_{(u)}$ be the unit tangent vectors of the overcrossing and undercrossing contours at an intersection, respectively. Let n_r be the number of intersections for which the triple $(\mathbf{u}_{(o)}, \mathbf{u}_{(u)}, \mathbf{e}_z)$ is right-handed, and let n_l be the number of those for which the triple is left-handed. Then $Wr(z)$ is given by

$$Wr(z) = n_r - n_l. \tag{7.78}$$

For example, we have $Wr(z) = 0$ for a circle confined in the xy plane and $Wr(z) = 1$ ($n_r = 1$ and $n_l = 0$) for the contour depicted in Fig. 7.9, where the overcrossing contour is continuous and the undercrossing contour is broken. When the overcrossing and undercrossing contours in Fig. 7.9 are interchanged, $Wr(z)$ jumps from 1 to -1. Note that as long as the fluctuation in the configuration is small around a given fixed configuration, $Wr(z)$ may be regarded as remaining constant and independent of configuration, and then the expression for Wr becomes very simple.

We then derive a useful general expression for the distribution $P(N; L)$ written in terms of Wr. In what follows, all lengths are measured in units of λ^{-1} and $k_B T$ is chosen to be unity as before. For the discrete chain, $P(N; L)$ may be written in the form

$$P(N; L) = Z^{-1} \int' \exp\left[-E(\{\Omega_n\})\right] d\{\Omega_{n-1}\}, \tag{7.79}$$

where E and Z are given by Eqs. (7.5) and (7.8), respectively, and the prime on the integration sign indicates that the integration is carried out under the restrictions that the chain is closed and that the linking number is equal to N (along with the boundary condition $\Omega_0 = \Omega_n$).

We can carry out analytically the integration in Eq. (7.79) over $\{\psi_{n-1}\}$ associated with the torsional part, as follows. We first fix the chain contour, that is, $\{\Theta_n\}$ and determine the most probable configuration $\{\psi_{n-1}^*\}$ under this restriction. The condition $\partial E/\partial \psi_p = 0$ ($p = 1, \cdots, n-1$) for the extremum may be written, from Eqs, (7.5)–(7.7), as

$$(\phi_p - \phi_{p-1})\cos\left[\tfrac{1}{2}(\theta_p + \theta_{p-1})\right] + \psi_p^* - \psi_{p-1}^* = \frac{\tau_0 L}{n} + c$$

$$(p = 1, \cdots, n),\qquad (7.80)$$

where c is a constant independent of p. It is seen from the third of Eqs. (4.10) that the left-hand side of Eq. (7.80) corresponds to $(L/n)\omega_\zeta(s)$ at $s = pL/n$ for the continuous chain. In the most probable configuration, therefore, we have $\omega_\zeta(s) = \tau_0 + cn/L$, so that $\omega_\zeta(s)$ is independent of s. From Eqs. (7.75) and (7.76), c is then found to be $2\pi(\Delta N - Wr)/n$. This completes the determination of $\{\psi_{n-1}^*\}$ from Eq. (7.80). From Eqs. (7.5)–(7.7) and (7.80), E may then be expressed *exactly* (without the infinitesimally small terms added) in terms of the fluctuations $\{\delta\psi_{n-1}\}$ in $\{\psi_{n-1}\}$ around $\{\psi_{n-1}^*\}$ as

$$E = E_B\left(\{\Theta_n\}\right) + \frac{\pi^2(\Delta N - Wr)^2}{(1+\sigma)L} + \frac{n}{4(1+\sigma)L}\sum_{p=1}^{n}(\delta\psi_p - \delta\psi_{p-1})^2 \quad (7.81)$$

with $\delta\psi_0 = \delta\psi_n = 0$, where the second and third terms on the right-hand side of Eq. (7.81) are the minimum of the torsional energy and its fluctuating term, respectively. By the use of Eq. (7.81), the integration over $\{\psi_{n-1}\}$ (or $\{\delta\psi_{n-1}\}$) in Eq. (7.79) can be carried out analytically, and $P(N; L)$ may be expressed in the form of a convolution integral

$$P(N; L) = \frac{1}{[4\pi(1+\sigma)L]^{1/2}}\int P(Wr; L)\exp\left[-\frac{\pi^2(\Delta N - Wr)^2}{(1+\sigma)L}\right]dWr$$

$$(7.82)$$

with $P(Wr; L)$ being the Wr-dependent ring-closure probability (irrespective of the value of N) given by

$$P(Wr; L) = Z_B^{-1}\int' \exp(-E_B)d\{\Theta_{n-1}\}/dWr,\qquad (7.83)$$

where Z_B is given by Eqs. (7.47) and the prime on the integration sign indicates that the integration is carried out under the restriction that the chain is closed (and its writhe is equal to Wr).

The convolution form of Eq. (7.82) indicates that the variance $\langle(\Delta N)^2\rangle$ of the linking number is equal to the sum of the variances of Tw and Wr,

$$\langle(\Delta N)^2\rangle = \langle(\Delta Tw)^2\rangle + \langle Wr^2\rangle,\qquad (7.84)$$

where $\Delta Tw = Tw - \overline{N} = \Delta N - Wr$, and we have, from Eq. (7.82),

$$\langle(\Delta Tw)^2\rangle = (2\pi^2)^{-1}(1+\sigma)L.\qquad (7.85)$$

Finally, we note that for the discrete chain Eq. (7.77) is replaced by

$$Wr = -(2\pi)^{-1} \sum_{p=1}^{n} (\phi_p - \phi_{p-1}) \cos\left[-\tfrac{1}{2}(\theta_p + \theta_{p-1})\right] + Wr(z). \quad (7.86)$$

7.2.2 Distribution of the Writhe

As seen from Eq. (7.82), the evaluation of $P(N; L)$ is reduced to that of the distribution of the writhe $P(Wr; L)$. The latter may be written in a form similar to Eq. (7.14) as follows,

$$P(Wr; L) = (2\pi)^{-4} L Z_{\mathrm{B}}^{-1} \int h(\{\Theta_n\}) \exp\Big\{ -E_{\mathrm{B}}(\{\Theta_n\})$$

$$+ i \sum_\alpha k_\alpha R_\alpha(\{\Theta_n\}) + i k_w [Wr(\{\Theta_n\}) - Wr] \Big\} d\mathbf{k} d\{\Theta_{n-1}\}, \quad (7.87)$$

where the sum over α is taken over x, y, z, and M, and $d\mathbf{k} = dk_x dk_y dk_z dk_{\mathrm{M}} \times dk_w$. We may evaluate $P(Wr; L)$ only for $Wr \geq 0$, since $P(Wr; L) = P(-Wr; L)$. However, the evaluation is limited to the following three ranges: (1) $Wr \simeq 0$, (2) $0 \ll Wr \leq 1$, and (3) $1 < Wr \lesssim 2$. In each case, evaluation is carried out by taking proper account of the small fluctuations in $\{\Theta_{n-1}\}$ around $\{\Theta_{n-1}^*\}$. Thus we finally construct an empirical interpolation formula valid for all values of Wr. It is then convenient to write $P(Wr; L)$ in the form

$$P(Wr; L) = \pi^2 L^{-7} \exp\left(-\frac{\pi^2}{L}\right) \tilde{P}(Wr; L) \quad (7.88)$$

and derive expressions for $\tilde{P}(Wr; L)$. Note that $LP/\tilde{P}$ is just equal to the leading term of $G(\mathbf{0}, \mathbf{u}_0 \mid \mathbf{u}_0; L)$.

We first consider the range (1). The most probable configuration for $Wr = 0$ is clearly a circle of radius $L/2\pi$ with the bending energy $E_{\mathrm{B}}^* = \pi^2/L$. If only the fluctuations of first order are retained, the integrations over k_α ($\alpha = x$, y, z, M) and $\{\Theta_{n-1}\}$ in Eq. (7.87) can be carried out analytically, but that over k_w must be numerically treated. It can then be analytically shown that $\tilde{P}(Wr; L)$ depends on Wr and L as Wr/L for $Wr \simeq 0$. A good interpolation formula so found is

$$\ln \tilde{P}_0(Wr; L) = \sum_{j=0}^{6} 10^j a_j \left(\frac{Wr}{L}\right)^{2j} \qquad \text{for} \quad \frac{Wr}{L} \leq 0.32$$

$$= 4.22414 - 2\sqrt{3}\pi^2 \left(\frac{Wr}{L}\right) \qquad \text{for} \quad \frac{Wr}{L} > 0.32, \quad (7.89)$$

where the subscript 0 on $\tilde{P}$ indicates that it is valid for $Wr \simeq 0$; and a_j are numerical constants and their values are given in Table 7.2.

Next we consider the range (2). The most probable configuration for a given Wr ranging from 0 to 1 was determined by Le Bret [4]; it changes

Table 7.2. Values of a_j, b_j, c_j, and d_j in Eqs. (7.89), (7.92) and (7.93)

j	a_j	b_j	c_j	d_j
0	1.9379	68.381	$\cdots$	-0.197997609403
1	-17.412	63.638	8.7456	-0.059664102410
2	26.565	30.812	-0.42137	-0.03384594250
3	-46.347	-47.432	3.7180	0.06596504601
4	55.487	2.6680	-4.0179	-0.0154304201201
5	-37.040	$\cdots$	2.5937	$\cdots$
6	10.218	$\cdots$	$\cdots$	$\cdots$

from the circle with $Wr = 0$ to the 8-shaped configuration with $Wr = 1$. Figure 7.9 shows as an example the most probable configuration with $Wr = 0.37$ and $Wr(z) = 1$ (determined following Le Bret). In this section the most probable configuration with $0 \ll Wr \leq 1$ is also referred to as the 8-shaped configuration, for convenience. Further, the contour points A_1 and A_2 corresponding to the crossing in the xy plane are called the "nodes" of the 8-shaped configuration.

Before proceeding to make further developments, we must make two remarks. The first concerns a kind of asymmetry of the shape of the 8-shaped configuration. When the most probable configuration is an 8-shaped configuration, it is necessary to constrain h in addition to $\mathbf{u}_0$ and $R_M (= 0)$ in contrast to the case of the circle. The imposition of the constraint on h is equivalent to specifying the segment number $\hat{p}$, or the contour distance $\hat{s}$, of one of the nodes. Thus the integration over $\{\Theta_{n-1}\}$ in Eq. (7.87) may be carried out first over $\{\Theta_{n-1}\}$ with h fixed, and then over h, where in the latter integration we may change variables from h to $\hat{s}$ ($0 \leq \hat{s} \leq L/2$). The second remark concerns the value of $Wr(z)$. Consider the most probable configuration for $Wr = Wr^*$ ($0 \ll Wr^* \leq 1$) and $Wr(z) = 1$, and allow the fluctuations around it under the restriction that the first term on the right-hand side of Eq. (7.86) is equal to $Wr^* - 1$. If we consider formally the mathematical fluctuations in $\{\Theta_{n-1}\}$, the chain, which is then *phantom*, is allowed to cross itself in the course of the deformation from the most probable configuration. The configurations that result may then be classified into two types: one with $Wr(z) = 1$ and the other with $Wr(z) = -1$ (with the configurations with $|Wr(z)| \geq 2$ being ignored). From Eq. (7.86), we have $Wr = Wr^*$ and $Wr = Wr^* - 2$ for the configurations with $Wr(z) = 1$ and $Wr(z) = -1$, respectively. In order to evaluate $P(Wr = Wr^*; L)$, we must therefore inhibit the fluctuations leading to $Wr(z) = -1$. We note that even the small fluctuations may actually lead to the latter case for the phantom chain with $Wr^* \simeq 1$, and that the inhibited configurations make contribution to $P(Wr = Wr^* - 2; L)$ (see below). This requirement may be taken into account, although only approximately, by imposing the constraint that only the fluctuations that satisfy $\mathbf{e}_{12}^* \cdot \mathbf{R}_{12} > 0$ are allowed, where $\mathbf{R}_{12}$ is the vector distance between the contour points $\hat{s}$ (corresponding to A_1) and $\hat{s} + L/2$ (corresponding to A_2), and $\mathbf{e}_{12}^*$ is the unit vector in the direction of $\mathbf{R}_{12} = \mathbf{R}_{12}^*$ in the most probable configuration. This constraint can be imposed on the configuration integral by the use of a Fourier representation of a unit step function.

Thus, considering the above two remarks, we may evaluate $P(Wr; L)$ from Eq. (7.87) by introducing in the integrand a factor Δ,

$$\Delta = (2\pi)^{-2} \int_0^{L/2} d\hat{s} \left| \frac{dh}{d\hat{s}} \right| \int_0^\infty da \int dk_h dk_a$$

$$\times \exp\left\{ ik_h \left[h(\{\Theta_n\}) - h(\hat{s}) \right] + ik_a \left[\mathbf{e}_{12}^* \cdot \mathbf{R}_{12}(\{\Theta_n\}) - a \right] \right\}. \quad (7.90)$$

If only the fluctuations of first order are retained, we can carry out analytically all the integrations except over $\hat{s}$, the result being almost independent of $\hat{s}$. Thus, multiplying it by $L/2$, we obtain

$$\tilde{P}_1(Wr; L) = C_0(Wr) \left\{ 1 - \frac{1}{2} \mathrm{erfc} \left[\frac{C_1(Wr)}{L^{1/2}} \right] \right\} \exp\left[-\frac{\pi^2}{L} g(Wr) \right] \quad (7.91)$$

with good interpolation formulas for C_0 and C_1,

$$C_0(Wr) = \sum_{j=0}^4 b_j Wr^j, \qquad C_1(Wr) = \sum_{j=1}^5 c_j (1 - Wr)^j, \qquad (7.92)$$

where the subscript 1 on $\tilde{P}$ indicates that it is valid for $0 \ll Wr \leq 1$; b_j and c_j are numerical constants and their values are given in Table 7.2; $\mathrm{erfc}(x) = 1 - \mathrm{erf}(x)$ is the complementary error function, $\mathrm{erf}(x)$ being given by Eq. (6.52); and $g(Wr)$ is the function defined by Le Bret [4] so that $\pi^2 [1 + g(Wr)]/L$ is the bending energy in the most probable configuration for a given Wr, and is given by

$$g(Wr) = 2\sqrt{3}\, Wr - \frac{11}{8} Wr^2 + \sum_{j=0}^4 d_j Wr^{3+j} \qquad (7.93)$$

with d_j being numerical constants whose values are given in Table 7.2. In Eq. (7.91), the factor $(1 - \frac{1}{2}\mathrm{erfc})$ represents the effect of the constraint $\mathbf{e}_{12}^* \cdot \mathbf{R}_{12} > 0$. We note that if this constraint is removed, this factor reduces to 1, and that when $Wr = 1$, it is equal to $\frac{1}{2}$ [since $\mathrm{erfc}(0) = 1$], reflecting the fact that the fluctuations leading to $Wr(z) = 1$ and -1 are equally probable.

In the range (3), we may derive an expression for $\tilde{P}(Wr; L)$ indirectly without recourse to the most probable configuration. As mentioned above, the configurations with $Wr(z) = -1$ make contribution to $P(Wr = Wr^* - 2; L)$. This is just what we desire here. The expression for it can readily be derived from the fact that the sum of $P(Wr = Wr^*; L)$ and $P(Wr = Wr^* - 2; L)$ should be equal to the probability $P(Wr = Wr^*; L)$ without any restriction on $Wr(z)$, which is given by Eq. (7.88) with $\tilde{P} = \tilde{P}_1$ given by Eq. (7.91) without the factor $(1 - \frac{1}{2}\mathrm{erfc})$. Thus, if we use the relations $P(Wr; L) = P(-Wr; L)$ and $\mathrm{erfc}(-x) = 2 - \mathrm{erfc}(x)$ and if we put $2 - Wr^* = Wr\ (1 < Wr < 2)$, then $\tilde{P}(Wr; L)$ valid for $1 < Wr \lesssim 2$, which we denote by $\tilde{P}_1'(Wr; L)$, is given by

$$\tilde{P}_1'(Wr; L) = \tilde{P}_1(Wr; L) \tag{7.94}$$

with

$$C_0(Wr) = C_0(2 - Wr), \quad C_1(Wr) = -C_1(2 - Wr),$$
$$g(Wr) = g(2 - Wr) \qquad (1 < Wr \lesssim 2). \tag{7.95}$$

Finally, we construct an empirical interpolation formula for $\tilde{P}$ such that it gives $\tilde{P}_0$ for $Wr \simeq 0$, $\tilde{P}_1$ for $0 \ll Wr \leq 1$, and $\tilde{P}_1'$ for $1 < Wr \lesssim 2$. The result reads

$$\begin{aligned}\tilde{P}(Wr; L) &= \tilde{P}_0(Wr; L)\,\tilde{P}_1(Wr; L)\,Q(Wr; L) & \text{for } 0 \leq Wr \leq 1 \\ &= \tilde{P}_0(Wr; L)\,\tilde{P}_1'(Wr; L)\,Q(Wr; L) & \text{for } 1 < Wr < 2 \\ &= 0 & \text{for } Wr \geq 2 \end{aligned} \tag{7.96}$$

with

$$Q(Wr; L) = \frac{1}{C_0(0)}\exp\left(\frac{2\sqrt{3}\pi^2 Wr}{L}\right). \tag{7.97}$$

The range of application of Eqs. (7.96) is limited to $L \lesssim 5$. Values of $\ln \tilde{P}$ calculated from Eqs. (7.96) with Eq. (7.97) are plotted against Wr^2 in Fig. 7.10 for the indicated values of L. It is seen that the plots are not linear, indicating that the distribution of Wr is not Gaussian at least in the range of $L \lesssim 4$. We note that Le Bret [24] and Chen [25] evaluated the distribution of Wr for (unknotted) cyclic, freely jointed chains of 10–150 bonds by Monte Carlo methods and found that the results are almost Gaussian.

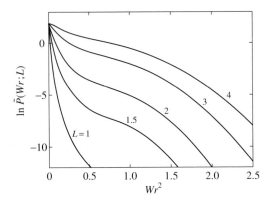

Fig. 7.10. Plots of $\ln \tilde{P}(Wr; L)$ against Wr^2 for the indicated values of L. The values are those from the interpolation formula, Eqs. (7.96)

7.2.3 Moments of the Writhe

The moments $\langle Wr^{2m}\rangle$ of Wr may be evaluated from

$$\langle Wr^{2m}\rangle = \frac{\displaystyle\int_{-\infty}^{\infty} Wr^{2m}\tilde{P}(Wr;L)dWr}{\displaystyle\int_{-\infty}^{\infty}\tilde{P}(Wr;L)dWr} \tag{7.98}$$

with the unnormalized distribution function of Wr. This is a defining equation for them. We can obtain numerical results, in particular, for $\langle Wr^2\rangle$ from Eq. (7.98), which is valid for $L \lesssim 4$.

For very small L, however, we can derive the expansion of $\langle Wr^{2m}\rangle$ around $L = 0$. If we expand the exponential term in the integrand of Eq. (7.82), then $P(N;L)$ may be expressed as an expansion in terms of $\Delta N/(1+\sigma)$ and $(1+\sigma)^{-1}$, the expansion coefficients being expressed in terms of $\langle Wr^{2m}\rangle$. On the other hand, $P(N;L)$ given by Eq. (7.25) may also be written as an expansion in terms of $\Delta N/(1+\sigma)$ and $(1+\sigma)^{-1}$. From a comparison of these two expansions, $\langle Wr^{2m}\rangle$ may be obtained. The results so obtained are summarized as follows. The second moment is given by

$$\langle Wr^2\rangle = 0.00384974\,L^2\left[1 + 0.164\,L + \mathcal{O}(L^2)\right], \tag{7.99}$$

and the leading terms of the ratios $\langle Wr^{2m}\rangle/\langle Wr^2\rangle^m$ for $m = 2-5$ are 3.7198, 28.599, 377.87, and 7744.6, respectively. These ratios are appreciably larger than the corresponding values 3, 15, 105, and 945 expected for the (one-dimensional) Gaussian distribution, indicating that the distribution of Wr for small L is much broader than the Gaussian distribution having the same $\langle Wr^2\rangle$ (see also Fig. 7.10).

For large L, we evaluate $\langle Wr^2\rangle$ by a Monte Carlo method on the basis of the circular original KP (KP,o) chain [23]. For this purpose, we consider the equivalent discrete chain composed of n segments (instead of $n+1$ segments) numbered 1, 2, $\cdots$, n, each having length $\Delta s = L/n$. Then $\langle Wr^2\rangle$ may be written in the form

$$\langle Wr^2\rangle = \int Wr^2 P(\{\Theta_n\};\,L)d\{\Theta_n\} \tag{7.100}$$

with

$$P(\{\Theta_n\};L) = \text{const. } \exp\left[-E_{\mathrm{B}}(\{\Theta_n\})\right]. \tag{7.101}$$

Thus we generate (unknotted) circular KP,o Monte Carlo chains that obey the Boltzmann distribution $P(\{\Theta_n\};L)$. The procedure is essentially the same as that used by Frank-Kamenetskii et al. [26]. Then $\langle Wr^2\rangle$ may be computed as a simple arithmetic mean of Wr^2 over the generated configurations $\{\Theta_n\}$; that is,

$$\langle Wr^2\rangle = \mathcal{N}^{-1}\sum_{\{\Theta_n\}} Wr^2\,, \tag{7.102}$$

where $\mathcal{N}$ is the number of generated chains in the ensemble. In Eq. (7.102), however, we do not use Eq. (7.86) for Wr but the equation generalized by Le Bret [4],

$$Wr = (2\pi)^{-1} \sum_{p=1}^{n} \{ \sin^{-1} [\sin \eta_p \sin(\phi_{p+1} - \chi_p)]$$
$$- \sin^{-1} [\sin \eta_p \sin(\phi_p - \chi_p)] \} + Wr(z), \qquad (7.103)$$

where $\mathbf{n}_p = (1, \eta_p, \chi_p)$ $(0 < \eta_p < \pi/2)$ in spherical polar coordinates is the unit vector perpendicular to $\mathbf{u}_p$ and $\mathbf{u}_{p+1}$. Note that Eq. (7.103) reduces to Eq. (7.86) when the difference between $\mathbf{u}_p$ and $\mathbf{u}_{p+1}$ is small.

Figure 7.11 shows plots of $\langle Wr^2 \rangle / L$ against $\log L$. The unfilled circles represent the present Monte Carlo values [23] and the vertical bars without circles represent those of Frank-Kamenetskii et al. [26], both for the circular KP,o chain. For comparison, the figure also includes the Monte Carlo values of Vologodskii et al. [27], Le Bret [24], and Chen [25] for the circular, freely jointed chain, where we have assumed that the number of Kuhn segments in the chain is equal to L. We note that all these values were obtained for the unknotted chains without excluded volume. It is seen that the coil-limiting value of $\langle Wr^2 \rangle / L$ for the KP,o chain is appreciably larger than that for the freely jointed chain. In Fig. 7.11, the solid curve represents the theoretical values numerically calculated from Eq. (7.98), and it is seen that there is good agreement between the theoretical and Monte Carlo values for the KP,o chain for $L \lesssim 3$. Thus a good interpolation formula for $\langle Wr^2 \rangle$ constructed on the basis of these results is

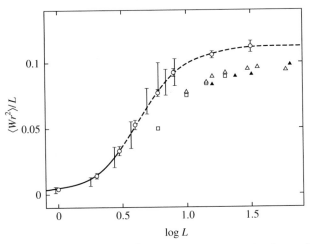

Fig. 7.11. Plots of $\langle Wr^2 \rangle / L$ against $\log L$ with Monte Carlo data; $\circ$ [23], I [26] for KP,o chain; $\square$ [27], $\triangle$ [24], $\blacktriangle$ [25] for freely jointed chain. The *solid and dashed curves* represent the values calculated from Eqs. (7.98) and (7.104), respectively

$$\langle Wr^2 \rangle = \frac{0.112L^3}{10.71 + L^2} \exp(-9.882L^{-2.5}) + 0.00385L^2 \exp(-2L)(1 + 1.491L$$

$$+8.423L^2 - 14.09L^3 + 17.55L^4 - 5.552L^5 + 0.6018L^6). \qquad (7.104)$$

In Fig. 7.11, the dashed curve represents the values calculated from Eq. (7.104).

7.2.4 Distribution of the Linking Number

Values of $\ln\left[P(N;L)/P(\Delta N = 0;L)\right]$ calculated as a function of $(\Delta N)^2$ from Eq. (7.82) with Eqs. (7.88) and (7.96) for $\sigma = -0.3$ are represented by the solid curves in Fig. 7.12 for the indicated values of L. [Recall that $P(N;L)$ depends on N as $|\Delta N|$.] For comparison, the corresponding values calculated from Eq. (7.25) are represented by the dashed curves. Recall that this $P(N;L)$ diverges at $|\Delta N|/(1 + \sigma) \gtrsim 3^{1/2}$. It is seen that the deviation (of the solid curves) from linearity is small for $L \geq 3$ and so even for $L < 3$ provided that $|\Delta N| \lesssim 0.5$. In other words, $P(N;L)$ is Gaussian in such a range of $|\Delta N|$ (under ordinary experimental conditions) at least for $L \lesssim 5$. For large L, the Monte Carlo results of Le Bret [24] and Chen [25] show that $P(Wr;L)$ is Gaussian, as already mentioned, so that it follows from Eq. (7.82) that $P(N;L)$ must also be Gaussian there. Thus it may be concluded that $P(N;L)$ is almost Gaussian for all values of L. This is consistent with the experimental finding [18, 28–31] for the distribution f_N of topoisomers,

$$f_N = \text{const.}\left[-\frac{K(\Delta N)^2}{RT}\right], \qquad (7.105)$$

where K is a constant called the *apparent twisting coefficient* and R is the molar gas constant.

Now we consider the dependence of K on L (or n_{bp}). For the Gaussian distribution f_N, K is related to its variance by the equation

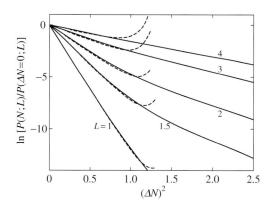

Fig. 7.12. Plots of $\ln[P(N;L)/P(\Delta N = 0;L)]$ against $(\Delta N)^2$ for $\sigma = -0.3$ and for the indicated values of L. The *solid and dashed curves* represent the values calculated from Eqs. (7.82) and (7.25), respectively

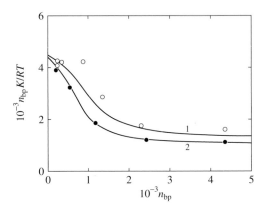

Fig. 7.13. Plots of $n_{\mathrm{bp}}K/RT$ against n_{bp} for DNA; ○, in 50 mmol/l NaCl and 10 mmol/l MgCl$_2$ at 20°C [18]; ●, in 10 mmol/l MgCl$_2$ at 37°C [31]. *Curve 1* represents the best-fit theoretical values calculated for the former data with $\lambda^{-1} = 1350$ Å and $\beta = 3.1 \times 10^{-19}$ erg cm, and *curve 2* represents those for the latter with $\lambda^{-1} = 1050$ Å and $\beta = 3.2 \times 10^{-19}$ erg cm

$$\frac{K}{RT} = \frac{1}{2\langle(\Delta N)^2\rangle}, \tag{7.106}$$

so that by the use of Eqs. (7.84) and (7.85), we have

$$\frac{n_{\mathrm{bp}}K}{RT} = \Gamma_0\left[1 + \frac{2\pi^2\langle Wr^2\rangle}{(1+\sigma)L}\right]^{-1}, \tag{7.107}$$

where Γ_0 (unreduced) is given by

$$\Gamma_0 = \frac{2\pi^2\beta}{l_{\mathrm{bp}}k_{\mathrm{B}}T}. \tag{7.108}$$

Figure 7.13 shows plots of $n_{\mathrm{bp}}K/RT$ against n_{bp}. The unfilled and filled circles represent the experimental values of Shore and Baldwin [18] in 50 mmol/l NaCl and 10 mmol/l MgCl$_2$ at 20°C and those of Horowitz and Wang [31] in 10 mmol/l MgCl$_2$ at 37°C, respectively. Curve 1 represents the best-fit theoretical values calculated from Eq. (7.107) with Eq. (7.104) for the former data with $\lambda^{-1} = 1350$ Å and $\beta = 3.1 \times 10^{-19}$ erg cm, and curve 2 represents those for the latter with $\lambda^{-1} = 1050$ Å and $\beta = 3.2 \times 10^{-19}$ erg cm. It is seen that the theory may explain well the behavior of K, although the data of the two groups are somewhat different from each other. In any case, it is important to see that as n_{bp} is increased, $n_{\mathrm{bp}}K$ first decreases and then becomes a constant. This decrease arises from the fluctuation in the writhe Wr.

7.2.5 Mean-Square Radii of Gyration

In this subsection we evaluate the mean-square radius of gyration $\langle S^2\rangle_N$ of the circular KP1 chain with the linking number N fixed and also the mean-square radii of gyration $\langle S^2\rangle_{Wr}$ and $\langle S^2\rangle$ of the circular KP,o chain with the writhe Wr fixed and without any restriction, respectively, by the Monte Carlo method as used for the evaluation of $\langle Wr^2 \rangle$ [23].

For this purpose, we consider the same discrete chain as that in Sect. 7.2.3. The distribution function $P(\{\Omega_n\} \,|\, N; L)$ of $\{\Omega_n\}$ for the circular KP1 chain with the linking number N is given by

$$P(\{\Omega_n\} \,|\, N; L) = \text{const.} \, \exp\left[-E(\{\Omega_n\})\right], \tag{7.109}$$

where E is given by Eq. (7.5) with $\Omega_0 \equiv \Omega_N$ and $\Delta s = L/n$ but without the infinitesimally small terms added. Then $\langle S^2 \rangle_N$ may be evaluated from

$$\langle S^2 \rangle_N = \int S^2(\{\Theta_n\}) P(\{\Omega_n\} | N; L) d\{\Omega_n\}, \tag{7.110}$$

where $S^2(\{\Theta_n\})$ is the squared radius of gyration for the configuration $\{\Theta_n\}$. We can carry out analytically the integration in Eq. (7.110) over $\{\psi_n\}$ as in the derivation of Eq. (7.82) and obtain

$$\langle S^2 \rangle_N = C_N^{-1} \int S^2(\{\Theta_n\}) P(\{\Theta_n\}; L)$$
$$\times \exp\left\{ -\frac{\pi^2 \left[\Delta N - Wr(\{\Theta_n\})\right]^2}{(1+\sigma)L} \right\} d\{\Theta_n\} \tag{7.111}$$

with

$$C_N = \int P(\{\Theta_n\}; L) \exp\left\{ -\frac{\pi^2 \left[\Delta N - Wr(\{\Theta_n\})\right]^2}{(1+\sigma)L} \right\} d\{\Theta_n\}, \tag{7.112}$$

where $P(\{\Theta_n\}; L)$ is given by Eq. (7.101).

$\langle S^2 \rangle_{Wr}$ is closely related to $\langle S^2 \rangle_N$. It can be shown from Eq. (7.111) that

$$\langle S^2 \rangle_N = \int \langle S^2 \rangle_{Wr} P(Wr \,|\, N; L) \, dWr, \tag{7.113}$$

where $P(Wr|N; L)$ is the conditional distribution function of Wr for the circular KP1 chain with N fixed and is given, from Eq. (7.82) for $P(N; L)$, by

$$P(Wr|N; L) = \frac{1}{\left[4\pi(1+\sigma)L\right]^{1/2}} \frac{P(Wr; L)}{P(N; L)}$$
$$\times \exp\left[-\frac{\pi^2(\Delta N - Wr)^2}{(1+\sigma)L} \right]. \tag{7.114}$$

We note that $P(Wr; L)$ is defined by Eq. (7.83) for the KP,o chain. On the other hand, $\langle S^2 \rangle$ may be expressed in terms of $\langle S^2 \rangle_{Wr}$ as

$$\langle S^2 \rangle = \int \langle S^2 \rangle_{Wr} P(Wr; L)\, dWr\,. \tag{7.115}$$

We note that $\langle S^2 \rangle$ is just equal to $\langle S^2 \rangle_N$ with $(1+\sigma) \to \infty$, that is, with the vanishing torsional energy, as seen from Eqs. (7.113)–(7.115).

Before presenting numerical results, we consider the limiting cases. For small L, we can evaluate analytically $\langle S^2 \rangle_N$ and $\langle S^2 \rangle$ for the discrete chain, followed by extrapolation to $n = \infty$, as in Sect. 7.1.2. The result for $\langle S^2 \rangle_N$ so derived from Eq. (7.110) reads

$$\langle S^2 \rangle_N = \frac{L^2}{4\pi^2}\left[1 - C_S L + \mathcal{O}(L^2)\right], \tag{7.116}$$

where C_S is a function of $\Delta N/(1+\sigma) \equiv a$ and is given by

$$C_S(a) = 0.1140 \exp(0.21687\, a^2 - 0.063708\, a^4 + 0.075371\, a^6)$$
$$(|a| \le 1.45)\,. \tag{7.117}$$

As noted above, $\langle S^2 \rangle$ is obtained, from Eq. (7.116) with $(1+\sigma) \to \infty$ $(a \to 0)$, as

$$\langle S^2 \rangle = \frac{L^2}{4\pi^2}\left[1 - 0.1140\, L + \mathcal{O}(L^2)\right]. \tag{7.118}$$

On the other hand, note that for large L, $\langle S^2 \rangle$ is given by Eq. (3.D.9).

Now we proceed to the Monte Carlo evaluation. As in Eq. (7.102), values of $\langle S^2 \rangle_{Wr}$ for $Wr = m\Delta$ may be computed as an arithmetic mean of S^2 over the generated configurations $\{\Theta_n\}$ whose Wr lies between $\left(m - \frac{1}{2}\right)\Delta$ and $\left(m + \frac{1}{2}\right)\Delta$, where m is an integer and Δ is a constant in the range of 0.05–0.25 properly chosen for each value of L. The averages $\langle S^2 \rangle$ may be taken over all the generated configurations. Figure 7.14 shows plots of the ratio $\langle S^2 \rangle_{Wr}/\langle S^2 \rangle$ against Wr for the indicated values of L [23]. The dot-dashed curve ($L = 0$) represents the theoretical values calculated for the most probable configuration [4] for a given Wr (with $\langle S^2 \rangle = \langle S^2 \rangle_{Wr=0}$ for $L = 0$). Recall that this configuration changes from the circle with $Wr = 0$ to the 8-shaped configuration with $Wr = 1$. It is seen that as Wr is increased from 0, $\langle S^2 \rangle_{Wr}$ decreases rather rapidly for small L and gradually for large L, indicating that the chain takes an open form for small Wr and a more compact form for large Wr.

Figure 7.15 shows plots of the ratio $12\langle S^2 \rangle/L$ (of $\langle S^2 \rangle$ to its coil-limiting value $L/12$) against $\log L$. The filled circles represent the above Monte Carlo values (for the unknotted chain) [23]. The dotted curve represents the values of $12\langle S^2 \rangle/L = 3L/\pi^2$ for the rigid ring, and the dot-dashed curve and the dashed curve represent the values calculated from Eqs. (7.118) and (3.D.9), respectively. It is seen that the ratio increases monotonically from 0 to about unity with increasing L, and that the Monte Carlo values agree well with the theoretical values calculated from Eq. (7.118) for $L \lesssim 3$ and those from Eq. (3.D.9) for $L \gtrsim 3$. However, the Monte Carlo value for $L = 32$ is somewhat

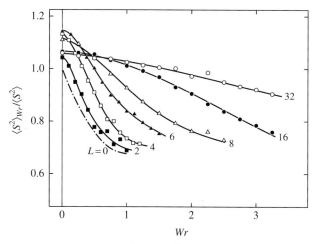

Fig. 7.14. Plots of $\langle S^2 \rangle_{Wr}/\langle S^2 \rangle$ against Wr with Monte Carlo data for the indicated values of L [23]. The *dot-dashed curve* ($L = 0$) represents the theoretical values calculated for the most probable configuration [4] for a given Wr

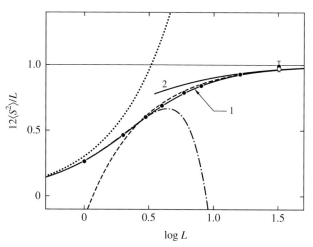

Fig. 7.15. Plots of $12\langle S^2 \rangle/L$ against $\log L$ with Monte Carlo data [23]. The *dotted curve* represents the values for the rigid ring, and the *dot-dashed and dashed curves* represent the values calculated from Eqs. (7.118) and (3.D.9), respectively. The *solid curves* 1 and 2 represent the values calculated from Eqs. (7.119) and (3.D.8), respectively

larger than that calculated from Eq. (3.D.9), although within the indicated error bounds. This may be regarded as arising from the fact that the contribution from the knotted configurations is included in Eq. (3.D.9). In fact, it has been found that the difference between the Monte Carlo values of $\langle S^2 \rangle$ above and those computed without discarding the knotted configurations is negligibly small for $L \leq 16$ but is somewhat large for $L = 32$ [23]. (A similar difference was found for the freely jointed chain [24, 32].) The latter value of $\langle S^2 \rangle$ at $L = 32$ is represented by the unfilled circle in Fig. 7.15. Indeed, this value is closer to that calculated from Eq. (3.D.9) than is the former.

An empirical interpolation formula for $\langle S^2 \rangle$ constructed for the unknotted chain for $L \leq 16$ on the basis of the Monte Carlo values (except for $L = 32$) along with Eqs. (7.118) and (3.D.9) is

$$\langle S^2 \rangle = \frac{L^2}{4\pi^2} (1 - 0.1140\, L - 0.0055258\, L^2$$
$$+ 0.0022471\, L^3 - 0.00013155\, L^4) \qquad \text{for } L \leq 6$$

$$= \frac{L}{12} \left[1 - \frac{7}{6L} - 0.025 \exp(-0.01\, L^2) \right] \qquad \text{for } L > 6. \quad (7.119)$$

We note that if the contribution from the knotted configurations is included, $\langle S^2 \rangle$ for any L may be calculated from Eqs. (7.119). In Fig. 7.15, the solid curves 1 and 2 represent the values calculated from Eqs. (7.119) and (3.D.8), respectively, which are seen almost to agree with each other for $L \gtrsim 15$. Indeed, Eq. (3.D.8) is applicable only for such relatively large L, as already noted.

It is seen from Eq. (7.111) that the Monte Carlo values of $\langle S^2 \rangle_N = \langle S^2 \rangle_{\Delta N}$ may be computed as a *continuous* function of ΔN from

$$\langle S^2 \rangle_{\Delta N} = \frac{\displaystyle\sum_{\{\Theta_n\}} \tilde{f}_{\Delta N}(\{\Theta_n\}) S^2(\{\Theta_n\})}{\displaystyle\sum_{\{\Theta_n\}} \tilde{f}_{\Delta N}(\{\Theta_n\})} \qquad (7.120)$$

with

$$\tilde{f}_{\Delta N} = \exp\left[-\frac{\pi^2 (\Delta N - Wr)^2}{(1+\sigma)L} \right]. \qquad (7.121)$$

Values of the ratio $\langle S^2 \rangle_{\Delta N} / \langle S^2 \rangle_{\Delta N=0}$ so obtained for $\sigma = -0.3$ are plotted against $\Delta N/L$ in Fig. 7.16 for the indicated values of L [23]. The values represented by the dashed curves (for large ΔN) are not very accurate. It is seen that $\langle S^2 \rangle_{\Delta N}$ decreases with increasing ΔN, and that the dependence of $\langle S^2 \rangle_{\Delta N} / \langle S^2 \rangle_{\Delta N=0}$ on $\Delta N/L$ is rather insensitive to change in L for $L \gtrsim 8$, while it is sensitive for smaller L. We note that the dependence of $\langle S^2 \rangle_{\Delta N=0}$ on L for $\sigma = -0.3$ is similar to that of $\langle S^2 \rangle$; the ratio $\langle S^2 \rangle_{\Delta N=0} / \langle S^2 \rangle$ exceeds unity only slightly, having the maximum value 1.07 at $L \simeq 6$.

Finally, it is important to note that the values of $|Wr|$ and $|\Delta N|$ considered above are not so large that the chain cannot take the typical interwound form [9].

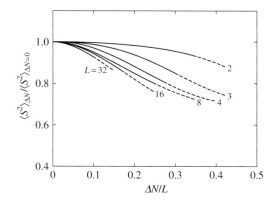

Fig. 7.16. Plots of $\langle S^2 \rangle_{\Delta N}$ $/\langle S^2 \rangle_{\Delta N=0}$ against $\Delta N/L$ with Monte Carlo data for $\sigma = -0.3$ and for the indicated values of L [23]

7.3 Translational Friction Coefficient of Topoisomers

In this section we evaluate the translational friction and sedimentation co-efficients, or the function f_D defined by the first line of Eqs. (6.36), for the DNA topoisomer with the linking number N by an application of the OB procedure to the KP1 cylinder ring [33]. For convenience, the f_D function for the circular KP1 chain with the linking number N is denoted by $f_{D,N}$. As in the case of f_D (for the circular KP,o chain), $f_{D,N}$ may be given formally by Eq. (6.137), that is,

$$f_{D,N} = \int_0^{L/2} K(s; N, L, d)\, ds\,, \qquad (7.122)$$

where the kernel K depends also on N. It may be expressed as

$$K(s; N, L, d) = \big\langle A(R, \Theta, L, d) \big\rangle_N \qquad (7.123)$$

with

$$A(R, \Theta, L, d) = \langle |\mathbf{R} - \hat{\mathbf{r}}|^{-1} \rangle_{\hat{\mathbf{r}}}$$

$$= \frac{1}{\left(R^2 + \frac{1}{4}d^2\right)^{1/2}} \sum_{m=0}^{\infty} \frac{(2m-1)!!(4m-1)!!}{2^m m!(2m)!}$$

$$\times \left[\frac{Rd\sin\Theta}{2(R^2 + \frac{1}{4}d^2)} \right]^{2m}, \qquad (7.124)$$

where $\mathbf{R}$ and $\hat{\mathbf{r}}$ have the same meaning as those in Eq. (6.14) with $R = |\mathbf{R}|$ and $|\hat{\mathbf{r}}| = d/2$, Θ is the angle between $\mathbf{R}$ and $\mathbf{u}(s_1)$, and $\langle \cdots \rangle_N$ denotes the configurational average for the circular KP1 chain with the linking number N.

Now we consider the same discrete circular KP1 chain as that in the last section with $\Delta s = L/n$ and $s = p\Delta s = (j-i)\Delta s$ $(p = 1, \cdots, n)$. For this discrete chain, the kernel may be calculated from

$$K(p\Delta s; N, L, d) = \langle A_{ij} \rangle_N \qquad (7.125)$$

with

$$A_{ij} = A(R_{ij}, \Theta_{ij}, L, d), \qquad (7.126)$$

where $\mathbf{R}_{ij}$ is the vector distance between the centers of the ith and jth segments with $R_{ij} = |\mathbf{R}_{ij}|$, and Θ_{ij} is the angle between $\mathbf{R}_{ij}$ and the direction $\mathbf{u}_i$ of the ith segment. As in Eq. (7.120), $\langle A_{ij} \rangle_N$ may then be computed by the Monte Carlo method from

$$\langle A_{ij} \rangle_N = \frac{\displaystyle\sum_{\{\Theta_n\}} \tilde{f}_{\Delta N}(\{\Theta_n\}) A_{ij}(\{\Theta_n\})}{\displaystyle\sum_{\{\Theta_n\}} \tilde{f}_{\Delta N}(\{\Theta_n\})} . \qquad (7.127)$$

Thus Eq. (7.122) for $f_{D,N} = f_{D,\Delta N}$ may be reduced to [33]

$$f_{D,\Delta N} = C(\Delta s, d) + \Delta s \sum_{p=1}^{n/2} \left(1 - \tfrac{1}{2}\delta_{p1} - \tfrac{1}{2}\delta_{p,n/2}\right) K(p\Delta s; N, L, d), \qquad (7.128)$$

where C is the contribution from $K(s)$ in the range of s from 0 to Δs (at $s \simeq 0$), the explicit expression being omitted, and n is assumed to be even. Values of the ratio $f_{D,\Delta N}/f_{D,\Delta N=0}$ calculated from Eq. (7.128) for $\sigma = -0.3$ and $d = 0.025$ are plotted against $\Delta N/L$ in Fig. 7.17 for the indicated values of L, corresponding to Fig. 7.16. It is seen that $f_{D,\Delta N}$ increases with increasing ΔN, and that the dependence of $f_{D,\Delta N}/f_{D,\Delta N=0}$ on $\Delta N/L$ is rather insensitive to change in L for $L \gtrsim 8$, while it is sensitive for smaller L. We note that the ratio $f_{D,\Delta N}/f_{D,\Delta N=0}$ is almost independent of d in the range of $0.015 \leq d \leq 0.03$, although $f_{D,\Delta N}$ itself depends on d. Although the present results are limited to the range of small ΔN, they are consistent with the experimental finding [34–36] that as $|\Delta N|$ is increased from 0, $f_{D,\Delta N}$ (or $s_{\Delta N}$) first increases rather rapidly, then exhibits a broad maximum, and finally increases steadily.

Experimentally, however, the ratio $f_{D,\Delta N}/f_D^*$ has been determined, where f_D^* is the f_D function for the corresponding nicked DNA that contains one single-chain scission per molecule. Therefore, the comparison of theory with experiment requires some comments. The DNA helix has been considered to be essentially continuous in the nicked DNA; a single-strand break in the DNA has little effect on the bending of the chain axis [37], and the ends of the strands are aligned across the nick [17]. We therefore assume that the linking number N of the nicked DNA fluctuates, taking only integral values. Then the fraction of the ones with the linking number N may be equated to the fraction f_N of the topoisomers with the linking number N. Thus f_D^* may be given by

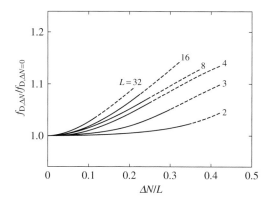

Fig. 7.17. Plots of $f_{D,\Delta N}$ $/f_{D,\Delta N=0}$ against $\Delta N/L$ for $\sigma = -0.3$ and $d = 0.025$ and for the indicated values of L. The values are those from Eq. (7.128)

$$f_D^* = \sum_{N=-\infty}^{\infty} f_{D,\Delta N} f_N \qquad (7.129)$$

with

$$f_N = \frac{\displaystyle\sum_{\{\Theta_n\}} \tilde{f}_{\Delta N}(\{\Theta_n\})}{\displaystyle\sum_{N=-\infty}^{\infty} \sum_{\{\Theta_n\}} \tilde{f}_{\Delta N}(\{\Theta_n\})}. \qquad (7.130)$$

It has been numerically found from Eq. (7.129) that f_D^* agrees with f_D for the circular KP,o chain to within 0.01% for $L \gtrsim 6$ [33]. This also justifies the analysis in Sect. 6.6.3, where the experimental data for f_D^* (for $L \gtrsim 12$) have been analyzed regarding the nicked DNA as the circular KP,o chain.

Finally, brief mention is made of earlier theories. Fukatsu and Kurata [38] and Bloomfield [39] considered a multiple-ring molecule composed of flexible rings of equal size linearly connected. However, this is not a realistic model; it has been pointed out [34, 35] that if the number of rings is assumed to be nearly equal to $|\Delta N|$, the theoretical values of $f_{D,N}$ are considerably larger than the experimental values. On the other hand, Gray [40] considered a rigid (rodlike) interwound superhelix to evaluate f_D but did not make an analysis of its dependence on ΔN. Camerini-Otero and Felsenfeld [41] evaluated $f_{D,N}$ for the same superhelix and showed that it changes only slowly with ΔN for relatively large ΔN, in agreement with experiment [35]. However, the rigid molecule is not realistic at least for small ΔN.

For various aspects of the problem of the supercoiling of DNA, the reader is referred to the review article by Vologodskii and Cozzarelli [42].

References

1. H. Jacobson and W. H. Stockmayer: J. Chem. Phys. **18**, 1600 (1950).
2. F. B. Fuller: Proc. Natl. Acad. Sci. USA **68**, 815 (1971).

3. C. J. Benham: Proc. Natl. Acad. Sci. USA **74**, 2397 (1977); Biopolymers **18**, 609 (1979).
4. M. Le Bret: Biopolymers **18**, 1709 (1979); **23**, 1835 (1984).
5. F. Tanaka and H. Takahashi: J. Chem. Phys. **83**, 6017 (1985).
6. M. D. Barkley and B. H. Zimm: J. Chem. Phys. **70**, 2991 (1979).
7. P. J. Flory: *Statistical Mechanics of Chain Molecules* (Interscience, New York, 1969).
8. P. J. Flory, U. W. Suter, and M. Mutter: J. Am. Chem. Soc. **98**, 5733 (1976).
9. C. R. Cantor and P. R. Schimmel: *Biophysical Chemistry* (Freeman, San Francisco, 1980), Part III.
10. F. H. C. Crick: Proc. Natl. Acad. Sci. USA **73**, 2639 (1976).
11. J. Shimada and H. Yamakawa: Macromolecules **17**, 689 (1984).
12. H. Yamakawa: *Modern Theory of Polymer Solutions* (Harper & Row, New York, 1971).
13. W. Gobush, H. Yamakawa, W. H. Stockmayer, and W. S. Magee: J. Chem. Phys. **57**, 2839 (1972).
14. H. Yamakawa and W. H. Stockmayer: J. Chem. Phys. **57**, 2843 (1972).
15. P. J. Hagerman: Biopolymers **24**, 1881 (1985).
16. D. Shore, J. Langowski, and R. L. Baldwin: Proc. Natl. Acad. Sci. USA **78**, 4833 (1981).
17. D. Shore and R. L. Baldwin: J. Mol. Biol. **170**, 957 (1983).
18. D. Shore and R. L. Baldwin: J. Mol. Biol. **170**, 983 (1983).
19. S. D. Levene and D. M. Crothers: J. Mol. Biol. **189**, 73 (1986).
20. P. J. Hagerman: Ann. Rev. Biophys. Biophys. Chem. **17**, 265 (1988).
21. J. Shimada and H. Yamakawa: Biopolymers **23**, 853 (1984).
22. J. Shimada and H. Yamakawa: J. Mol. Biol. **184**, 319 (1985)
23. J. Shimada and H. Yamakawa: Biopolymers **27**, 657 (1988).
24. M. Le Bret: Biopolymers **19**, 619 (1980).
25. Y. Chen: J. Chem. Phys. **75**, 2447 (1981).
26. M. D. Frank-Kamenetskii, A. V. Lukashin, V. V. Anshelevich, and A. V. Vologodskii: J. Biomol. Struct. Dynam. **2**, 1005 (1985).
27. A. V. Vologodskii, V. V. Anshelevich, A. V. Lukashin, and M. D. Frank-Kamenetskii: Nature **280**, 294 (1979).
28. R. E. Depew and J. C. Wang: Proc. Natl. Acad. Sci. USA **72**, 4275 (1975).
29. D. E. Pulleyblank, M. Shure, D. Tang, J. Vinograd, and H. -P. Vosberg: Proc. Natl. Acad. Sci. USA **72**, 4280 (1975).
30. W. R. Bauer: Ann. Rev. Biophys. Bioeng. **7**, 287 (1978).
31. D. S. Horowitz and J. C. Wang: J. Mol. Biol. **173**, 75 (1984).
32. G. ten Brinke and G. Hadziioannou: Macromolecules **20**, 480 (1987).
33. J. Shimada and H. Yamakawa: Biopolymers **27**, 675 (1988).
34. J. C. Wang: J. Mol. Biol. **43**, 25 (1969).
35. W. B. Upholt, H. B. Gray, and J. Vinograd: J. Mol. Biol. **61**, 21 (1971).
36. J. C. Wang: J. Mol. Biol. **87**, 797 (1974); **89**, 783 (1974).
37. J. B. Hays and B. H. Zimm: J. Mol. Biol. **48**, 297 (1970).
38. M. Fukatsu and M. Kurata: J. Chem. Phys. **44**, 4539 (1966).
39. V. A. Bloomfield: Proc. Natl. Acad. Sci. USA **55**, 717 (1966).
40. H. B. Gray: Biopolymers **5**, 1009 (1967).
41. R. D. Camerini-Otero and G. Felsenfeld: Proc. Natl. Acad. Sci. USA **74**, 1708 (1978).
42. A. V. Vologodskii and N. R. Cozzarelli: Ann. Rev. Biophys. Biomol. Struct. **23**, 609 (1994).

8 Excluded-Volume Effects

This chapter deals with the theory of the excluded-volume effects in dilute solution, such as various kinds of expansion factors and the second and third virial coefficients, developed on the basis of the perturbed HW chain which enables us to take account of both effects of excluded volume and chain stiffness. Necessarily, the derived theory is no longer the two-parameter (TP) theory [1], but it may give an explanation of experimental results [2] obtained in this field since the late 1970s, which all indicate that the TP theory breaks down. There are also some causes other than chain stiffness that lead to its breakdown. On the experimental side, it has for long been a difficult task to determine accurately the expansion factors since it is impossible to determine directly unperturbed chain dimensions in good solvents. However, this has proved possible by extending the measurement range to the oligomer region where the excluded-volume effect disappears. Thus an extensive comparison of the new non-TP theory with experiment is made mainly using experimental data recently obtained for several flexible polymers. As for semiflexible polymers with small excluded volume, some remarks are made without a detailed analysis.

8.1 End-Distance and Gyration-Radius Expansion Factors

8.1.1 Perturbation Theory

Consider the HW chain of total contour length L on which $n + 1$ beads (segments) are arrayed with spacing a between them along the contour, so that $L = na$ [3], and suppose that there exist excluded-volume interactions between them expressed in terms of the usual binary-cluster integral β [1]. By an application of the formulation for the random-flight chain [1], the mean-square end-to-end distance $\langle R^2 \rangle$ for this (perturbed) HW chain may then be written in the form

$$\langle R^2 \rangle = \langle R^2 \rangle_0 + \beta \sum_{i<j} \int R^2 \big[P_0(\mathbf{R}; L) P_0(\mathbf{0}_{ij}; L)$$

$$- P_0(\mathbf{R}, \mathbf{0}_{ij}; L) \big] d\mathbf{R} + \cdots , \qquad (8.1)$$

where the subscript 0 indicates the unperturbed value (without excluded volume), $\mathbf{0}_{ij}$ means that $\mathbf{R}_{ij} = \mathbf{0}$, $P_0(\mathbf{R}, \mathbf{R}_{ij}; L)$ is the (unperturbed) distribution function of $\mathbf{R}$ ($= \mathbf{R}_{0n}$) and $\mathbf{R}_{ij}$, and so on, with $\mathbf{R}_{ij}$ being the vector distance between the ith and jth beads ($i, j = 0, 1, 2, \cdots, n$), so that $P_0(\mathbf{R}; L)$ is identical with the Green function $G(\mathbf{R}; L)$ introduced in Chap. 4. In what follows, all lengths are measured in units of λ^{-1} and $k_{\mathrm{B}}T$ is chosen to be unity unless otherwise noted, and we assume as before that Poisson's ratio σ is zero for flexible chains. Then $\langle R^2 \rangle_0$ is given by Eq. (4.82) and the conventional excluded-volume parameter z is defined in the limit of $n \to \infty$ by [1]

$$z = \left(\frac{3}{2\pi \langle R^2 \rangle_0} \right)^{3/2} n^2 \beta \quad (n \to \infty), \tag{8.2}$$

where $\langle R^2 \rangle_0 = c_\infty L$ in Eq. (8.2) with c_∞ being given by Eq. (4.75) or (5.10).

Now, if the end-distance expansion factor α_R is defined as usual by

$$\langle R^2 \rangle = \langle R^2 \rangle_0 \alpha_R^2, \tag{8.3}$$

the first-order perturbation theory of α_R^2 for the HW chain may be written, from Eqs. (8.1) and (8.3), in the form

$$\alpha_R^2 = 1 + K(L; \kappa_0, \tau_0)z + \cdots, \tag{8.4}$$

where $K(L; \kappa_0, \tau_0)$ must become equal to $4/3$ in the limit of $L \to 0$ [1], and z is redefined, from Eq. (8.2), by [3,4]

$$z = \left(\frac{3}{2\pi} \right)^{3/2} BL^{1/2} \tag{8.5}$$

with B the *excluded-volume strength* defined by

$$B = \frac{\beta}{a^2 c_\infty^{3/2}}. \tag{8.6}$$

Similarly, the gyration-radius expansion factor α_S is defined by

$$\langle S^2 \rangle = \langle S^2 \rangle_0 \alpha_S^2, \tag{8.7}$$

where $\langle S^2 \rangle_0$ is given by Eq. (4.83). Since the first-order perturbation coefficient of α_S^2 must be equal to $134/105$ in the limit of $L \to \infty$ [1], we simply assume that

$$\alpha_S^2 = 1 + \frac{67}{70} K(L; \kappa_0, \tau_0)z + \cdots. \tag{8.8}$$

If the sums in Eq. (8.1) are replaced by integrals, the coefficient $K(L; \kappa_0, \tau_0)$ as a function of L (and also κ_0 and τ_0), which we simply denote by $K(L)$, may then be evaluated from

$$K(L) = \frac{F(L)}{L^{1/2} \langle R^2 \rangle_0}, \tag{8.9}$$

where

$$F(L) = \left(\frac{2\pi c_\infty}{3}\right)^{3/2} \int_0^L ds_1 \int_{s_1}^L ds_2 \left[G(0;s)\langle R^2\rangle_0 \right. $$
$$\left. - \int R^2 P_0(\mathbf{R}, \mathbf{0}_{s_1 s_2}; L)d\mathbf{R}\right] \tag{8.10}$$

with $s_1 = ia$, $s_2 = ja$, and $s = s_2 - s_1$.

In order to evaluate the integral over $\mathbf{R}$ in Eq. (8.10), it is convenient to introduce the trivariate distribution function $P_0(\mathbf{R}_1, \mathbf{R}_2, \mathbf{R}_{12} \,|\, \Omega_1 = 0; L)$ of $\mathbf{R}_1 = \mathbf{r}(s_1) - \mathbf{r}(0)$, $\mathbf{R}_2 = \mathbf{r}(L) - \mathbf{r}(s_2)$, and $\mathbf{R}_{12} = \mathbf{r}(s_2) - \mathbf{r}(s_1)$ with $\Omega_1 = \Omega(s_1) = 0$ [5], where $\mathbf{r}(s)$ is the radius vector of the contour point s ($0 \le s \le L$), and $\Omega(s) = [\theta(s), \phi(s), \psi(s)]$ is the Euler angles defining the orientation of a localized Cartesian coordinate system $[\mathbf{e}_\xi(s), \mathbf{e}_\eta(s), \mathbf{e}_\zeta(s)]$ affixed to the chain at s with respect to an external coordinate system $(\mathbf{e}_x, \mathbf{e}_y, \mathbf{e}_z)$ as before. This P_0 with $\mathbf{R}_{12} = \mathbf{0}$ may be evaluated from

$$P_0(\mathbf{R}_1, \mathbf{R}_2, \mathbf{0}_{s_1 s_2} \,|\, \Omega_1; L) = \int G(\mathbf{R}_1, \Omega_1 \,|\, \Omega_0; s_1)\, G(\mathbf{0}, \Omega_2 \,|\, \Omega_1; s)$$
$$\times G(\mathbf{R}_2, \Omega \,|\, \Omega_2; L - s_2)d\Omega_0 d\Omega_2 d\Omega \tag{8.11}$$

with $\Omega_2 = \Omega(s_2)$, where $G(\mathbf{R}, \Omega \,|\, \Omega_0; L)$ is given by Eq. (4.155), and $G(\mathbf{0}, \Omega \,|\, \Omega_0; L)$ with $\Omega_0 = 0$, which we simply call the *angle-dependent ring-closure probability*, is obtained from the former as

$$G(\mathbf{0}, \Omega \,|\, 0; L) = \sum_{l,j,j'} c_l h_l^{jj'}(L)\, \mathcal{D}_l^{j'j}(\Omega) \tag{8.12}$$

with c_l being given by Eq. (4.54) and with $h_l^{jj'}(L)$ (not to be confused with the angular correlation functions in the Flory system in Sect. 4.4.2) being given by

$$h_l^{jj'}(L) = (4\pi)^{-1/2} \mathcal{G}_{ll0}^{00,jj'}(0;L). \tag{8.13}$$

We note that $h_0^{00}(L)$ is equal to the angle-independent ring-closure probability $G(\mathbf{0}; L)$,

$$G(\mathbf{0}; L) = h_0^{00}(L). \tag{8.14}$$

Carrying out the integrations over Ω_0, Ω_2, and Ω, we obtain

$$P_0(\mathbf{R}_1, \mathbf{R}_2, \mathbf{0}_{s_1 s_2} \,|\, \Omega_1; L) = G(\mathbf{R}_1; s_1)\, G(\mathbf{0}; s)\, G(\mathbf{R}_2; L - s_2)$$
$$+ (4\pi)^{-1} \sum_{l \ge 1} \sum_{j,j'} (-1)^{j'} (2l + 1) \mathcal{G}_{0ll}^{00,0j'}(R_1; s_1)$$
$$\times h_l^{jj'}(s)\, \mathcal{G}_{0ll}^{00,0j}(R_2; L - s_2) P_l(\cos\gamma), \tag{8.15}$$

where γ is the angle between $\mathbf{R}_1$ and $\mathbf{R}_2$, and P_l are the Legendre polynomials. In deriving Eq. (8.15), we have used Eq. (4.C.8) with $\mathcal{D}_0^{00} = (8\pi^2)^{-1/2}$, Eq. (5.105), the relation,

$$G_{0ll}^{0m,0j} = (-1)^{(m+|m|)/2} G_{0ll}^{00,0j} \,, \tag{8.16}$$

[which can be derived from Eqs. (4.152) and (4.156)], and Eqs. (4.159) and (3.B.15).

By the use of the relation $\mathbf{R} = \mathbf{R}_1 + \mathbf{R}_2$, we then have

$$\int R^2 P_0(\mathbf{R}, \mathbf{0}_{s_1 s_2}; L) d\mathbf{R} = \int R^2 P_0(\mathbf{R}_1, \mathbf{R}_2, \mathbf{0}_{s_1 s_2} \,|\, \Omega_1; L) d\mathbf{R}_1 d\mathbf{R}_2$$

$$= \left(\langle R_1{}^2 \rangle_0 + \langle R_2{}^2 \rangle_0 \right) h_0^{00}(s) + \frac{8\pi}{3} \sum_{j,j'} (-1)^{j'}$$

$$\times A_{j'}(s_1) A_j(L - s_2) h_1^{jj'}(s) \tag{8.17}$$

with

$$A_j(L) = 3^{1/2} \int_0^\infty R^3 G_{011}^{00,0j}(R; L) dR$$

$$= (-1)^{(j+|j|)/2} \langle R Y_1^{j*}(\Theta, \Phi) \rangle_{\Omega_0 = 0} \,, \tag{8.18}$$

where the second line of Eqs. (8.18) has been derived from Eq. (4.157) by the use of Eqs. (3.B.5) and (8.16), and $\mathbf{R} = (R, \Theta, \Phi)$ in spherical polar coordinates. Thus A_j may be expressed in terms of the components of the persistence vector. In the particular case of $\tau_0 = 0$, the result reads

$$A_j(L) = \left(\frac{3}{4\pi} \right)^{1/2} \sum_{k=-1}^{1} d_{jk} \left\{ \frac{1 - \exp[-(2 + ik\kappa_0)L]}{2 + ik\kappa_0} \right\}, \tag{8.19}$$

where $\mathbf{d}_j = (d_{jk})$ with $-1 \le k \le 1$ are the vectors defined by $\mathbf{d}_0 = \left(\frac{1}{2}, 0, \frac{1}{2} \right)$ and $\mathbf{d}_{\pm 1} = \pm 2^{-3/2} i(1, 0, -1)$.

As seen from the second of Eqs. (8.17), we need the components $h_l^{jj'}$ with $l = 0$ and 1. It can be shown that the symmetry relations for $h_l^{jj'}$ are the same as those for $g_l^{jj'}$ given by Eqs. (4.128), and then the only required components $h_1^{jj'}$ are h_1^{00}, $h_1^{0(-1)}$, $h_1^{1(-1)}$, and $h_1^{(-1)(-1)}$. Thus the problem reduces to an evaluation of the ring-closure probabilities given by Eqs. (8.12) and (8.14) [5].

8.1.2 Ring-Closure Probabilities and the First-Order Coefficient

We first evaluate the ring-closure probabilities for the (unperturbed) HW chain for small L for the special case of $\kappa_0 \ne 0$ and $\tau_0 = 0$ by modifying the procedure developed in Chap. 7 for the KP chain. At the final stage, the results for the first-order perturbation coefficient $K(L)$ for other cases, for which the direct evaluation is difficult, are inferred. We replace the continuous chain by the equivalent discrete chain composed of $n + 1$ segments as before. For the present case, the total potential energy E of the former is given by

Eq. (7.4) with $\omega_\eta - \kappa_0$ in place of ω_η and with $\tau_0 = 0$, and that of the latter $E(\{\Omega_n\})$ is given by Eq. (7.5) with Eq. (7.6) and with

$$u^{(0)}(\Omega_{p-1}, \Omega_p) = \frac{1}{4Ln}(\kappa_0 L)^2 + \frac{n}{4L}\left((\Delta\theta_p)^2 + (\Delta\phi_p)^2 + (\Delta\psi_p)^2\right.$$
$$+ 2\Delta\phi_p\Delta\psi_p\cos\left[\tfrac{1}{2}(\theta_p + \theta_{p-1})\right]$$
$$-\frac{2\kappa_0 L}{n}\left\{\Delta\theta_p\cos\left[\tfrac{1}{2}(\psi_p + \psi_{p-1})\right]\right.$$
$$\left.\left.+ \Delta\phi_p\sin\left[\tfrac{1}{2}(\theta_p + \theta_{p-1})\right]\sin\left[\tfrac{1}{2}(\psi_p + \psi_{p-1})\right]\right\}\right).(8.20)$$

The partition function Z is then given by Eq. (7.9) with $\sigma = 0$.

As in Eq. (7.56) for the KP chain, the angle-independent ring-closure probability $G(\mathbf{0}; L) = G(\mathbf{0}, |\Omega_0; L)$ is the integral of $G(\mathbf{0}, \Omega|\Omega_0; L)$ over Ω, and therefore may be written in the form

$$G(\mathbf{0}; L) = Z^{-1}\int' \exp[-E(\Omega_n)]d\{\Omega_n\}, \qquad (8.21)$$

where the prime on the integration sign indicates that the integration over $\{\Omega_n\}$ is carried out under the restriction that $\mathbf{R}(\{\Theta_n\}) = (R_x, R_y, R_z) = \mathbf{0}$ with Ω_0 being fixed. If we remove the constraint $\mathbf{R} = \mathbf{0}$ by introducing a Fourier representation of a three-dimensional Dirac delta function, we obtain

$$G(\mathbf{0}; L) = (2\pi)^{-3}Z^{-1}\int \exp\left[-E(\{\Omega_n\})\right.$$
$$\left. + i\sum_\alpha k_\alpha R_\alpha(\{\Theta_n\})\right]d\mathbf{k}d\{\Omega_n\}, \qquad (8.22)$$

where the sum over α is taken over x, y, and z with $d\mathbf{k} = dk_x dk_y dk_z$. In the present case, Ω_0 is fixed in such a way that $\mathbf{e}_{\zeta_0}(= \mathbf{u}_0)$ and $\mathbf{e}_{\xi_0}$ lie in the xy plane (with Ω_n being unfixed), where the joint of the closed chain is fixed at the origin of the external coordinate system, and then the point M [the center of the $(n/2)$th segment] is no longer symmetrically distributed about $\mathbf{u}_0$ for the HW chain with $\kappa_0 \neq 0$, so that we cannot impose the constraint R_M (the z component of the radius vector of M) $= 0$ in Eq. (8.21), or in other words, we cannot reaffix the localized coordinate systems, as done for the KP chain (see Fig. 7.2).

The integration over $\{\Omega_n\}$ in Eq. (8.22) may be carried out as before over the fluctuations in $\{\Omega_n\}$ around the most probable configuration $\{\Omega_n^*\}$ at the minimum of energy with $\mathbf{R} = \mathbf{0}$ fixed. When $\tau_0 = 0$, this configuration must be planar and can be determined by a slight modification of the previous formulation for the KP chain with $\kappa_0 = 0$. If we choose $\mathbf{u}_0$ and $\mathbf{u}^*$ to be in the xy plane so that the x axis bisects the angle between them, as depicted in Fig. 8.1a, that is, if we choose $\Omega_0 = \left(\frac{\pi}{2}, \phi_0^*, \frac{\pi}{2}\right)$, then we have $\Omega_p^* = \left(\frac{\pi}{2}, \phi_p^*, \frac{\pi}{2}\right)$,

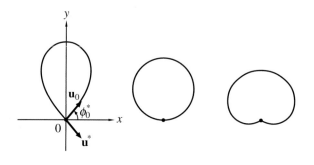

(a) $\kappa_0 L = 0$ (b) $\kappa_0 L = 2\pi$ (c) $\kappa_0 L = 9.1$

Fig. 8.1a–c. Typical, most probable closed configurations of the HW chain with $\tau_0 = 0$. The x axis bisects the angle between the unit tangent vectors $\mathbf{u}_0$ and $\mathbf{u}^*$ at the ends

where we note that ϕ_0^* depends on L and κ_0. In this case, the total potential energy E of the continuous chain may simply be given by

$$E = \frac{1}{2} \int_0^{L/2} \left(\frac{d\phi}{ds} - \kappa_0 \right)^2 ds \,, \tag{8.23}$$

so that the Euler equation is the same as that for $\kappa_0 = 0$.

We first obtain its solution for a given value of ϕ_0^*, and then determine the value of ϕ_0^* at which the configuration is most stable for a given value of $\kappa_0 L$. There are two possible cases: $\phi_0^* \geq 0$ and $\phi_0^* < 0$, although $\phi_0^* \geq 0$ in the previous case of $\kappa_0 = 0$. Thus Eq. (7.61) is replaced by

$$\cos \phi_p^* = (1 - 2\epsilon)\left[1 - 2\operatorname{cn}^2 (v \,|\, k_0)\right], \tag{8.24}$$

where $\epsilon = 0$ for $\phi_0^* \geq 0$ and $\epsilon = 1$ for $\phi_0^* < 0$, and

$$v = 2d \left(\frac{1}{2} - \frac{p}{n} \right) - 2\epsilon K(k_0) \left(\frac{p}{n} \right) \tag{8.25}$$

with

$$d = F(\delta \,|\, k_0) \,, \tag{8.26}$$

$$\delta = \frac{\pi}{2}(1 - \epsilon) - \frac{1}{2}\phi_0^* \,. \tag{8.27}$$

The parameter k_0 may be determined from

$$\tfrac{1}{2}\left[d + \epsilon K(k_0)\right](2 - k_0{}^2) = E(d \,|\, k_0) + \epsilon E(k_0) \,, \tag{8.28}$$

where $E(k_0)$ and $E(d \,|\, k_0)$ are the complete and incomplete elliptic integrals of the second kind, respectively. For the configuration that satisfies the Euler equation for a given ϕ_0^*, we have

$$E = \frac{1}{4L} \left\{ 8\left[d + \epsilon K(k_0)\right]^2 (2 - k_0)^2 - 4\kappa_0 L(\pi - \phi_0^*) + (\kappa_0 L)^2 \right\}. \tag{8.29}$$

Table 8.1. Values of $a_j^{(k)}$ in Eqs. (8.30), (8.31), and (8.33)

j	$a_j^{(1)}$	$a_j^{(2)}$	$a_j^{(3)}$	$a_j^{(4)}$
0	7.027	1.720	−0.04963	0.8250
1	−22.83	−2.467	0.04240	1.756
2	19.26	0.4013	−0.1809	−0.8090
3	−1.799	−3.082	0.2189	−0.1076
4	2.154	4.745	−0.1450	0.1881
5	−1.748	−3.091	0.04580	−0.04086
6	· · ·	· · ·	−0.005382	· · ·

The value of ϕ_0^* at which the configuration is most stable for a given $\kappa_0 L$ may be determined from $\partial E/\partial \phi_0^* = 0$. Then ϕ_p^* and E^* for the most probable configuration are given by Eqs. (8.24) and (8.29), respectively, with this value of ϕ_0^*. We note that ϕ_p^* and LE^* depend on κ_0 and L as $\kappa_0 L$. In Fig. 8.1 are depicted as examples the most probable configurations for the indicated three values of $\kappa_0 L$, where we note that $\phi_0^* \simeq 0$ for the case (b).

Good interpolation formulas constructed for the product $A = LE^*$ and ϕ_0^* on the basis of the numerical solutions so obtained are given by

$$A = LE^* = \sum_{j=0}^{5} a_j^{(1)} \left(\frac{\kappa_0 L}{10} \right)^j , \qquad (8.30)$$

$$\phi_0^* = \sum_{j=0}^{5} a_j^{(2)} \left(\frac{\kappa_0 L}{10} \right)^j , \qquad (8.31)$$

where $a_j^{(1)}$ and $a_j^{(2)}$ are numerical constants and their values are given in Table 8.1. The integral in Eq. (8.22) is then convergent only for $0 < \kappa_0 L \lesssim 9.3$, and we obtain the final result for $G(\mathbf{0}; L)$,

$$G(\mathbf{0}; L) = C_0 L^{-9/2} \exp \left(-\frac{A}{L} + C_1 L \right) \qquad (8.32)$$

with

$$C_0^{-1} = (\kappa_0 L)^{1/2} \left[(\kappa_0 L)^{1/2} - 3.05 \right] \sum_{j=0}^{6} a_j^{(3)} (\kappa_0 L)^{j/2} ,$$

$$C_1^{-1} = \sum_{j=0}^{5} a_j^{(4)} (\kappa_0 L)^{j/2} , \qquad (8.33)$$

where $a_j^{(3)}$ and $a_j^{(4)}$ are numerical constants and their values are also given in Table 8.1. The range of application of Eq. (8.32) with Eqs. (8.33) is limited to $0.3 \leq \kappa_0 L \leq 8.5$. We note that interpolation formulas for C_0 and C_1 for $\kappa_0 L \simeq 0$ or 0.93 cannot be constructed since their values become very large there, and that Eq. (7.68) for $G(0; L)$ for the KP chain cannot be obtained from Eq. (8.32) by taking the limit $\kappa_0 L \to 0$.

For the evaluation of the angle-dependent ring-closure probability $G(0, \Omega \,|\, \Omega_0; L)$ or the components $h_1^{jj'}(L)$, it is convenient to introduce the conditional distribution function $G(\Omega_n \,|\, 0, \Omega_0; n)$ of Ω_n for the closed discrete chain with Ω_0 fixed, which is related to $G(0, \Omega_n \,|\, \Omega_0; n)$ by

$$G(0, \Omega_n \,|\, \Omega_0; n) = G(0; L)G(\Omega_n \,|\, 0, \Omega_0; n). \tag{8.34}$$

This conditional distribution function with $\Omega_0 = \Omega_0^*$ may be approximated by the Gaussian distribution having the moments of the fluctuation $\delta\Omega_n = \Omega_n - \Omega_n^*$ with $\mathbf{R} = 0$ fixed. Then $h_1^{jj'}(L)$ may be evaluated from

$$h_1^{jj'}(L) = c_l^{-1} \int \mathcal{D}_l^{j'j*}(\Omega_n)G(0, \Omega_n \,|\, 0; n)d\Omega_n \tag{8.35}$$

with Eq. (8.34), where Eq. (8.35) has been obtained from Eq. (8.12). Thus the results are given by

$$h_1^{jj'}(L) = H^{jj'}(L)G(0; L) \tag{8.36}$$

with

$$H^{00}(L) = \cos(2\phi_0^* - C_\phi L) \exp\left[-(f_1 + f_2)L\right],$$

$$H^{0(-1)}(L) = -\frac{\sqrt{2}}{2} \sin(2\phi_0^* - C_\phi L) \exp\left[-(f_1 + f_2)L\right],$$

$$H^{(\pm 1)(-1)}(L) = \pm\Bigg(\tfrac{1}{4}\sin(2\phi_0^* - C_\phi L)\exp\left[-(f_2 + f_3)L\right]$$

$$\times\Big\{\exp\left[-f_1(1+g)^2 L\right] - \exp\left[-f_1(1-g)^2 L\right]\Big\} \tag{8.37}$$

$$-\tfrac{1}{2}\cos(2\phi_0^* - C_\phi L)\exp\left[-(f_1 g^2 + f_2 + f_3)L\right]\Bigg)$$

$$+\tfrac{1}{4}\exp(-f_3 L)\Big\{\exp\left[-f_1(1+g)^2 L\right] + \exp\left[-f_1(1-g)^2 L\right]\Big\},$$

where ϕ_0^* is given by Eq. (8.31), and

$$f_1^{-1} = -g^2 f_3^{-1} + \sum_{j=0}^{5} b_j^{(1)}(\kappa_0 L)^{j/2},$$

$$f_2^{-1} = \sum_{j=0}^{5} b_j^{(2)}\left(\frac{\kappa_0 L}{10}\right)^j,$$

$$f_3^{-1} = \sum_{j=0}^{5} b_j^{(3)}(\kappa_0 L)^{j/2}, \tag{8.38}$$

$$g = f_3 \sum_{j=0}^{5} b_j^{(4)}(\kappa_0 L)^{j/2},$$

Table 8.2. Values of $b_j^{(k)}$ in Eqs. (8.38) and (8.39)

j	$b_j^{(1)}$	$b_j^{(2)}$	$b_j^{(3)}$	$b_j^{(4)}$	$b_j^{(5)}$	$b_j^{(6)}$
0	0.9909	4.129	0.8997	1.056	-0.9176	-9.618
1	0.7265	-0.9494	-0.3629	-0.3106	2.625	9.333
2	-0.3687	0.1634	2.649	-0.5967	-3.575	-3.627
3	0.2974	-5.633	-2.297	0.6325	2.497	0.7032
4	-0.1776	8.465	0.8518	-0.2217	-0.8695	-0.06779
5	0.02857	-5.299	-0.1220	0.02285	0.1222	0.002622

$$C_\phi = \sum_{j=0}^{5} b_j^{(5)} (\kappa_0 L)^{j/2} \qquad \text{for } 0.3 \leq \kappa_0 L \leq 4$$

$$= \sum_{j=0}^{5} b_j^{(6)} (\kappa_0 L)^{j} \qquad \text{for } 4 < \kappa_0 L \leq 8.5 . \tag{8.39}$$

In Eqs. (8.38) and (8.39), $b_j^{(k)}$ are numerical constants and their values are given in Table 8.2.

We note that for the KP chain only the component h_1^{00} is required, the other components $h_1^{jj'}$ being unnecessary in the evaluation of the expansion factors. It is given by Eq. (8.36) with Eq. (7.68) and with

$$H^{00}(L) = \cos(1.720 + 0.06104\,L) \exp(-0.5077\,L) \qquad \text{(KP)} . \tag{8.40}$$

Next we derive analytical expressions for the ring-closure probabilities valid for large L and for arbitrary κ_0 and τ_0 from the Daniels-type distribution as given by Eq. (4.177). The results read

$$G(0; L) = \left(\frac{3}{2\pi c_\infty L} \right)^{3/2} \left(1 - \left\{ \frac{5}{8} - \frac{9\kappa_0^2}{r^2(4 + \tau_0^2)} + \frac{3\kappa_0^2}{2(9 + \nu^2)(36 + \nu^2)} \right. \right.$$
$$\left. \left. \times \left[1 + \frac{101 + \kappa_0^2}{4 + \tau_0^2} + \frac{3(160 + 7\kappa_0^2)}{(4 + \tau_0^2)^2} \right] \right\} \frac{1}{L} + \mathcal{O}(L^{-2}) \right), \tag{8.41}$$

$$h_1^{jj'}(L) = \left(\frac{3}{2\pi c_\infty L} \right)^{3/2} \left[\frac{3(-1)^{j'+1} b_j b_{j'}}{c_\infty L} + \mathcal{O}(L^{-2}) \right] \tag{8.42}$$

with

$$b_0 = \frac{c_\infty}{2\sqrt{3}} ,$$
$$b_1 = -b_{-1}^* = \frac{\kappa_0(2 - i\tau_0)}{2\sqrt{6}\,r^2} , \tag{8.43}$$

where ν and r are given by Eqs. (4.76) and (4.77), respectively.

Interpolation formulas for $G(\mathbf{0}; L)$ and $h_1^{jj'}(L)$ for intermediate L may be constructed as in Eq. (7.45) [5], but the results are omitted.

Now, in order to carry out the integrations over s_1 and s_2 in Eq. (8.10), we change variables from s_1 and s_2 to s and s_1. Then the integration over s_1, which is of a convolution type, is straightforward, and the remaining integration over s is carried out partly numerically. Thus the final results for $F(L)$ and hence $K(L)$ are obtained numerically. We first examine the behavior of $K(L)$ for the special case of moderately large κ_0 and $\tau_0 = 0$. Figure 8.2 shows plots of $L^{1/2}K(L)$ against L. The solid curve HW represents the values for the HW chain with $\kappa_0 = 4$ and $\tau_0 = 0$, and the dot-dashed curve KP represents the values for the KP chain, which we denote by $K_{\mathrm{KP}}(L)$. For comparison, the corresponding values for the latter in the earlier approximate evaluation (YS) [3] and the coil-limiting values $4L^{1/2}/3$ are also represented by the dashed (YS) and dotted (C) curves, respectively. It is seen that the present and YS values for the KP chain and those for the HW chain are rather close to each other. This is also the case with the HW chains with $2.5 \le \kappa_0 \le 6$ and $\tau_0 = 0$. It is also seen that all curves are almost straight lines for large L, so that the primary effect of chain stiffness is just to reduce the values of $L^{1/2}K(L)$ for large L from the coil-limiting values by a constant which is insensitive to change of the helical nature.

We then consider the behavior of $K(L)$ for other cases. This is in general very difficult, but the behavior can be inferred for two special cases: (1) $\kappa_0 \simeq 0$ (and any τ_0) and (2) $1 \ll \kappa_0 \lesssim \tau_0$, both of which lie in domain III of a (κ_0, τ_0)-plane of Fig. 4.13. For the first case, we have $c_\infty \simeq 1$ and $\kappa_0^2\nu^{-2} \simeq 0$, so that $\langle R^2 \rangle_0$ given by Eq. (4.82) for the HW chain may be approximated by $\langle R^2 \rangle_{0,\mathrm{KP}}$ for the KP chain. We may therefore treat the HW chain approximately as the KP chain as far as the statistics of the chain contour is concerned, and then we have $K(L) \simeq K_{\mathrm{KP}}(L)$.

For the second case, ν is very large, so that in Eq. (4.82) we may neglect terms of $\mathcal{O}(\nu^{-2})$, the factors $\kappa_0^2\nu^{-2}$ and $\tau_0^2\nu^{-2}$ being of order unity. Then Eq. (4.82) reduces to

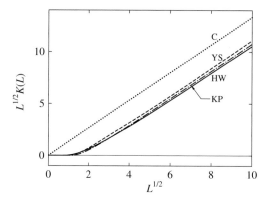

Fig. 8.2. Plots of $L^{1/2}K(L)$ against L. The *solid curve* HW represents the values for the HW chain with $\kappa_0 = 4$ and $\tau_0 = 0$. The *dot-dashed and dashed curves* represent the present [5] and earlier (YS) [3] values for the KP chain, respectively, and the *dotted curve* C the coil-limiting values

$$\bar{\lambda}^2 \langle R^2 \rangle_0 = \bar{\lambda}\bar{L} - \tfrac{1}{2}\left[1 - \exp(-2\bar{\lambda}\bar{L})\right] \qquad (8.44)$$

with $\bar{L} = (\tau_0/\nu)L$ and $\bar{\lambda} = \nu/\tau_0$. This means that $\langle R^2 \rangle_0$ for the HW chain of contour length L is approximately equal to that for the KP chain of contour length $\bar{L}$ and with stiffness parameter $\bar{\lambda}^{-1}$. Thus we may approximate the HW chain by this KP chain as far as the statistics of the (coarse-grained) chain contour is concerned. We introduce the excluded-volume effect into this KP chain. Let $\bar{\beta}$ and $\bar{a}$ be the binary-cluster integral and the spacing, respectively, between its beads, all lengths being measured in units of $\bar{\lambda}^{-1}$ (not λ^{-1}). Since its reduced total contour length is equal to $\bar{\lambda}\bar{L}$, we obtain, from Eq. (8.4) with Eqs. (8.5) and (8.6), for α_R^2 for this KP chain

$$\alpha_R^2 = 1 + \left(\frac{3}{2\pi}\right)^{3/2}\left(\frac{\bar{\beta}}{\bar{a}^2}\right)L^{1/2}K_{\mathrm{KP}}(L) + \cdots , \qquad (8.45)$$

where we have put $\bar{\lambda}\bar{L} = L$ with L being the contour length of the original HW chain. If we compare Eq. (8.45) with Eq. (8.4) for the original HW chain, we obtain the relations $K(L) = K_{\mathrm{KP}}(L)$ and $\beta/a^2 c_\infty^{3/2} = \bar{\beta}/\bar{a}^2$.

Thus it has been shown above that the relation $K(L) = K_{\mathrm{KP}}(L)$ holds for both the HW chain of strong helical nature and the KP-like chains. It may then be expected that this relation holds for any κ_0 and τ_0. A good interpolation formula thus constructed for $K(L)$ for the HW chain is given by [5]

$$
\begin{aligned}
K(L) &= \frac{4}{3} - \frac{2.711}{L^{1/2}} + \frac{7}{6L} && \text{for } L > 6 \\
&= \frac{1}{L^{1/2}}\exp\left(-\frac{6.611}{L} + 0.9198 + 0.03516\,L\right) && \text{for } L \le 6 . \quad (8.46)
\end{aligned}
$$

It is important to see that the first-order deviation of $K(L)$ from its coil-limiting value $4/3$ is of order $L^{-1/2}$, and therefore that as L is increased, $K(L)$ approaches more slowly its coil-limiting value than do the ratios $\langle R^2 \rangle_0/L$ and $\langle S^2 \rangle_0/L$. This suggests that the effects of chain stiffness on α_R and α_S remain rather large even for such large L where $\langle R^2 \rangle_0/L$ and $\langle S^2 \rangle_0/L$ reach almost their respective coil-limiting values c_∞ and $c_\infty/6$. Further, note that $K(L)$ becomes zero extremely rapidly at small L.

Finally, some comments must be briefly made on other theories. Chen and Noolandi [6] have evaluated $\langle R^2 \rangle$ and $\langle S^2 \rangle$ for the KP chain with excluded volume by an application of the renormalization scaling. However, it can be shown that their $K(L)$ is proportional to $L^{3/2}$ in the limit of $L \to 0$, so that it approaches zero gradually compared to the $K(L)$ given by Eqs. (8.46) [7]. This deficiency is due to the fact that they have not treated the ring-closure probability. Weill and des Cloizeaux [8] also considered the fact that the excluded-volume effect vanishes for very small L, but their theory is still essentially a TP theory [2].

8.1.3 Effects of Chain Stiffness – Quasi-Two-Parameter Scheme

As is well known, the TP theory claims that the expansion factor α (α_R or α_S) is a function only of the excluded-volume parameter z, that is,

$$\alpha = \alpha(z) \qquad \text{(TP)}. \tag{8.47}$$

As shown in the last subsection, however, the consideration of the chain stiffness on the basis of the HW chain leads to the breakdown of this scheme. Thus we introduce a parameter $\tilde{z}$ defined by

$$\tilde{z} = \frac{3}{4} K(L) z, \tag{8.48}$$

and assume that α is a function only of $\tilde{z}$, that is,

$$\alpha = \alpha(\tilde{z}) \qquad \text{(QTP)}, \tag{8.49}$$

where the function $\alpha(\tilde{z})$ may be obtained by replacing z by $\tilde{z}$ in a TP expression for $\alpha(z)$. This is referred to as the *quasi-two-parameter* (QTP) *scheme*; it is sometimes called the YSS (Yamakawa–Stockmayer–Shimada) scheme, based on their treatments [3–5]. For convenience, the parameter $\tilde{z}$ is referred to as the *intramolecular scaled excluded-volume parameter*. As seen from Eqs. (8.46) and (8.48), $\tilde{z}$ also slowly approaches its coil-limiting value z as L is increased.

As for TP expressions for α, it is reasonable and convenient to adopt the Domb–Barrett equations [9]. We then have

$$\alpha_R^2 = \left[1 + 10\tilde{z} + \left(\frac{70\pi}{9} + \frac{10}{3}\right)\tilde{z}^2 + 8\pi^{3/2}\tilde{z}^3\right]^{2/15}, \tag{8.50}$$

$$\alpha_S^2 = \left[0.933 + 0.067\exp(-0.85\,\tilde{z} - 1.39\,\tilde{z}^2)\right]\alpha_R^2, \tag{8.51}$$

where in Eq. (8.51) α_R^2 is given by Eq. (8.50). We note that these equations give the expansions

$$\alpha_R^2 = 1 + 1.333\,\tilde{z} - 2.075\,\tilde{z}^2 + \cdots, \tag{8.52}$$

$$\alpha_S^2 = 1 + 1.276\,\tilde{z} - 2.220\,\tilde{z}^2 + \cdots, \tag{8.53}$$

so that Eq. (8.50) gives the exact second-order perturbation theory of α_R^2, while Eq. (8.51) gives the second-order coefficient somewhat larger than the exact value 2.082 [1].

Now we evaluate numerically $\langle S^2 \rangle$ for RIS chains with excluded volume by a Monte Carlo method in order to examine the validity of Eq. (8.51) [4]. For simplicity, we consider the three-state (0°, 120°, and −120°) RIS chain with the tetrahedral bond angles and with the statistical weight matrices for polymethylene [10]. The skeletal atoms in the chain of n bonds are numbered 0, 1, 2, $\cdots$, n, where n is assumed to be even. Suppose that there are excluded-volume interactions between even-numbered atoms, for convenience. A positive excluded-volume interaction energy ϵ (in units of $k_\mathrm{B}T$) is assigned to each of pairs of the $2i$th and $2j$th atoms only when $|2i - 2j| \geq 5$ and only when they are located either at the same lattice site or at nearest-neighbor sites. We assume the bond length equal to 1.53 Å and the molecular weight 14 of the repeat unit. The generated chain may be represented by the KP chain with or without excluded volume. Thus, from a comparison of the two chains without excluded volume ($\epsilon = 0$) with respect to $\langle S^2 \rangle_0$, the KP model parameters λ^{-1} and M_L for the former are determined to be 11.5 Å and 11.0 Å^{-1}, respectively.

Figure 8.3 shows double-logarithmic plots of $6\langle S^2 \rangle/L$ against L with the Monte Carlo data for $\epsilon = 0.2$ (unfilled circles) and 0.07 (filled circles). The dashed curve ($B = 0$) represents the unperturbed KP values calculated from Eq. (4.85). The solid curves represent the best-fit theoretical values calculated from Eq. (8.7) with Eqs. (4.85), (8.5), (8.46), (8.48), and (8.51) with the indicated values of B (reduced). There is good agreement between the theoretical and Monte Carlo values. The dotted curve C represents the theoretical values similarly calculated with $B = 0.23$ but with the coil-limiting value $4/3$ for $K(L)$ for all values of L. This curve coincides with the corresponding solid curve for $L \gtrsim 10^3$, indicating that the effect of chain stiffness on α_S remains appreciable up to such large L. Further, it is interesting to see that the excluded-volume effect appears (the solid curves deviate from the dashed curve) at $L = 3$–5.

Finally, we note that corresponding to Eqs. (8.3) and (8.7), the mean-square electric dipole moment $\langle \mu^2 \rangle$ may be written in the form

$$\langle \mu^2 \rangle = \langle \mu^2 \rangle_0 \alpha_\mu^2, \tag{8.54}$$

where α_μ is the dipole-moment expansion factor. The perturbation theory of

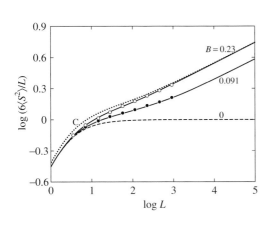

Fig. 8.3. Double-logarithmic plots of $6\langle S^2 \rangle/L$ against L with Monte Carlo data for $\epsilon = 0.2$ (○) and 0.07 (●) [4]. The *dashed curve* ($B = 0$) represents the unperturbed KP values calculated from Eq. (4.85). The *solid curves* represent the best-fit theoretical values calculated from Eq. (8.7) with Eqs. (8.45) and (8.51) with the indicated values of B (reduced), and the *dotted curve* C the theoretical values similarly calculated with $B = 0.23$ but with the coil-limiting value $4/3$ for $K(L)$

$\alpha_\mu{}^2$ for the HW chain is given in Appendix 8.A. In particular, the case of type-B (perpendicular) dipoles is discussed rather in detail. Note that $\alpha_\mu = \alpha_R$ for type-A (parallel) dipoles.

8.1.4 Comparison with Experiment

There is a difficulty in determining experimentally the expansion factor α_S from $\langle S^2 \rangle$ in a given good solvent since it is impossible to determine directly the unperturbed dimension $\langle S^2 \rangle_0$ in that good solvent. (Recall that $\langle S^2 \rangle_0$ may in general depend on solvent and temperature.) However, the (intramolecular) excluded-volume effect must disappear in the oligomer region (as also seen from Fig. 8.3), so that we have $\langle S^2 \rangle = \langle S^2 \rangle_0$ there. Therefore, if we choose the solvent and temperature so that in the oligomer region $\langle S^2 \rangle$ coincides with the unperturbed mean-square radius of gyration in a proper Θ solvent (at $T = \Theta$), which we denote by $\langle S^2 \rangle_\Theta$, then the latter may be regarded as equal to the unperturbed dimension $\langle S^2 \rangle_0$ in that good solvent for *all* values of the molecular weight M; that is,

$$\langle S^2 \rangle_0 = \langle S^2 \rangle_\Theta . \tag{8.55}$$

Taking this Θ solvent as a reference standard, we may then determine α_S from

$$\alpha_S{}^2 = \frac{\langle S^2 \rangle}{\langle S^2 \rangle_\Theta} . \tag{8.56}$$

In this subsection we make a comparison of theory with experiment using those experimental data for which the experimental requirement of Eq. (8.55) is fulfilled.

We first summarize necessary basic equations, in which lengths are not reduced by λ^{-1}. We adopt as before the Domb–Barrett equation for α_S,

$$\alpha_S{}^2 = \left[1 + 10\tilde{z} + \left(\frac{70\pi}{9} + \frac{10}{3} \right) \tilde{z}^2 + 8\pi^{3/2}\tilde{z}^3 \right]^{2/15}$$
$$\times \left[0.933 + 0.067 \exp(-0.85\,\tilde{z} - 1.39\,\tilde{z}^2) \right] \tag{8.57}$$

with

$$\tilde{z} = \frac{3}{4} K(\lambda L) z , \tag{8.58}$$

where $K(L)$ is given by Eqs. (8.46) and

$$z = \left(\frac{3}{2\pi} \right)^{3/2} (\lambda B)(\lambda L)^{1/2} . \tag{8.59}$$

We note that B is given by Eq. (8.6) with β and a unreduced, and L is related to the number of repeat units (degree of polymerization) x by the equation

$$L = \left(\frac{M_0}{M_L} \right) x = ax , \tag{8.60}$$

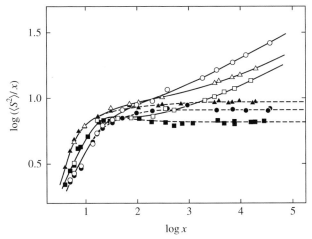

Fig. 8.4. Double-logarithmic plots of $\langle S^2 \rangle/x$ (in Å^2) against x for a-PS in cyclo-hexane at 34.5°C (Θ) ($\bullet$) and in toluene at 15.0°C ($\circ$) [7], a-PMMA in acetonitrile at 44.0°C (Θ) ($\blacksquare$) and in acetone at 25.0°C ($\square$) [11], and i-PMMA in acetonitrile at 28.0°C (Θ) ($\blacktriangle$) and in acetone at 25.0°C ($\triangle$) [12], where most of the data in the Θ solvents have been reproduced from Fig. 5.1. The *dashed and solid curves* connect smoothly the data points in the Θ and good solvents, respectively

where M_0 is the molecular weight of the repeat unit, and in the second of Eqs. (8.60) it has been taken as a single bead (with $a = M_0/M_L$).

Figure 8.4 shows double-logarithmic plots of $\langle S^2 \rangle/x$ (in Å^2) against x for a-PS in cyclohexane at 34.5°C (Θ) and in toluene at 15.0°C [7], a-PMMA in acetonitrile at 44.0°C (Θ) and in acetone at 25.0°C [11], and i-PMMA in acetonitrile at 28.0°C (Θ) and in acetone at 25.0°C [12], where most of the data in the Θ solvents have been reproduced from Fig. 5.1. The dashed and solid curves connect smoothly the data points in the Θ and good solvents, respectively. It is seen that for each polymer the values of $\langle S^2 \rangle$ in the good solvent agree well with those of $\langle S^2 \rangle_\Theta$ in the Θ solvent in the oligomer region. This indicates that the relation of Eq. (8.55) holds, so that α_S^2 may be calculated from Eq. (8.56) with the experimental values of $\langle S^2 \rangle$ and $\langle S^2 \rangle_\Theta$. We note that for these flexible polymers the critical value of λL for the onset of the excluded-volume effect is 2.0–2.5.

The values of α_S^2 thus determined are double-logarithmically plotted against x in Fig. 8.5 for a-PS in toluene at 15.0°C [7] and in benzene at 25.0°C [13, 14], a-PMMA in acetone at 25.0°C [11] and in chloroform at 25.0°C [11], and i-PMMA in acetone at 25.0°C [12] and in chloroform at 25.0°C [12]. Here, the values for a-PS in benzene at 25.0°C have been calculated using *trans*-decalin at 21.0°C (Θ) as a reference standard, in which $\langle S^2 \rangle_\Theta/x$ at large x is 6% smaller than that in cyclohexane at 34.5°C (Θ) [13]. In the figure the solid curves represent the best-fit QTP theoretical values calculated from Eq. (8.57) with Eqs. (8.58)–(8.60) with the values of λ^{-1} and M_L

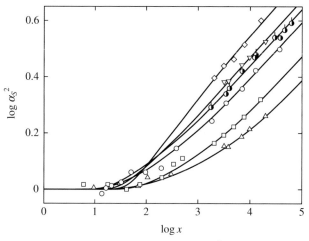

Fig. 8.5. Double-logarithmic plots of α_S^2 against x for a-PS in toluene at 15.0°C ($\bigcirc$) [7] and in benzene at 25.0°C ($\circlefthalfblack$,$\circtophalfblack$) [13,14], a-PMMA in acetone at 25.0°C ($\square$) [11] and in chloroform at 25.0°C ($\lozenge$) [11], and i-PMMA in acetone at 25.0°C ($\triangle$) [12] and in chloroform at 25.0°C ($\triangledown$) [12]. The *solid curves* represent the best-fit QTP theory values calculated from Eq. (8.57) (see the text)

in Table 5.1 and proper values of the reduced excluded-volume strength λB, and its values so determined are given in Table 8.3. The values of λB similarly determined for a-PS in MEK at 35.0°C [13] and in 4-*tert*-butyltoluene at 50.0°C [15], a-PMMA in nitroethane at 30.0°C [11], PIB in n-heptane at 25.0°C [16], and PDMS in toluene at 25.0°C [17] are also given in Table 8.3 along with the respective reference standards (Θ solvents). It also includes the values of β, per repeat unit, calculated from Eq. (8.6) with the values of the HW model parameters. It is interesting to see that the values of β for a-

Table 8.3. Values of the excluded-volume strength for typical flexible polymers from $\langle S^2 \rangle$

Polymer (f_r)	Solvent	Temp. (°C)	Θ-Solvent (Θ°C)	λB	β (Å³)	Ref.
a-PS (0.59)	Toluene	15.0	Cyclohexane (34.5)	0.26	33	[7]
	4-*tert*-Butyltoluene	50.0	Cyclohexane (34.5)	0.10	12	[15]
	Benzene	25.0	*trans*-Decalin (21.0)	0.33	40	[13]
	MEK	35.0	*trans*-Decalin (21.0)	0.060	7	[13]
a-PMMA (0.79)	Acetone	25.0	Acetonitrile (44.0)	0.22	12	[11]
	Chloroform	25.0	Acetonitrile (44.0)	1.15	62	[11]
	Nitroethane	30.0	Acetonitrile (44.0)	0.52	28	[11]
i-PMMA (0.01)	Acetone	25.0	Acetonitrile (28.0)	0.10	12	[12]
	Chloroform	25.0	Acetonitrile (28.0)	0.55	65	[12]
PIB	n-Heptane	25.0	IAIV (25.0)	0.083	6	[16]
PDMS	Toluene	25.0	Bromo-cyclohexane (29.5)	0.14	10	[17]

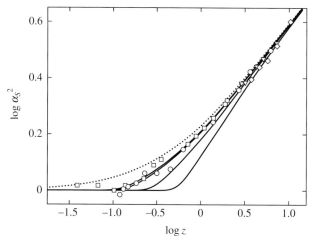

Fig. 8.6. Double-logarithmic plots of α_S^2 against z for a-PS in toluene at 15.0°C (○) [7], a-PMMA in acetone at 25.0°C (□) [11] and in chloroform at 25.0°C (◇) [11], and i-PMMA in chloroform at 25.0°C (▽) [12]. The *solid curves* represent the QTP theory values calculated from Eq. (8.57), and the *dotted curve* the TP theory values calculated with $\tilde{z} = z$ (see the text)

and i-PMMAs in the same solvent are almost the same, indicating that β is independent of the stereochemical structure of the polymer chain.

The values of α_S^2 in Fig. 8.5 for a-PS in toluene, a-PMMA in acetone and in chloroform, and i-PMMA in chloroform are double-logarithmically plotted against z in Fig. 8.6, where values of z have been calculated from Eq. (8.59) with Eqs. (8.60) with the above values of the parameters. The solid curves represent the QTP theory values calculated from Eq. (8.57) with Eqs. (8.58) and (8.59) with the values of λB, and the dotted curve represents the TP theory values calculated from Eq. (8.57) with $\tilde{z} = z$. There is good agreement between the QTP theoretical and experimental values. The solid curves (or data points) do not form a single-composite curve but deviate downward progressively from the dotted curve with decreasing z (or M) because of the effect of chain stiffness. This effect becomes more significant as λB is increased, or in other words, as the solvent quality becomes better. It is surprising to see that the effect on α_S remains rather large even at $z \simeq 10$ or at very large $M \simeq 10^6$.

Figure 8.7 shows double-logarithmic plots of α_S^2 against $\tilde{z}$ with the same data as those in Fig. 8.5 along with those for a-PS in MEK at 35.0°C [13], a-PMMA in nitroethane at 30.0°C [11], PIB in n-heptane at 25.0°C [16], and PDMS in toluene at 25.0°C [17], where values of $\tilde{z}$ have been calculated from Eq. (8.58) with the above values of z. The solid curve represents the QTP theory values calculated from Eq. (8.57). Although it is natural from the procedure of determining λB that all the data points form a single-composite curve and are fitted by the solid curve, there is excellent agreement

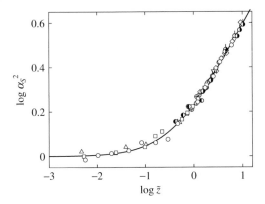

Fig. 8.7. Double-logarithmic plots of α_S^2 against $\tilde{z}$ with the same data as those in Fig. 8.5 along with those for a-PS in MEK at 35.0°C (●) [13], a-PMMA in nitroethane at 30.0°C (◇) [11], PIB in n-heptane at 25.0°C (⊕) [16], and PDMS in toluene at 25.0°C (⊖) [17]. The *solid curve* represents the QTP theory values calculated from Eq. (8.57)

between theory and experiment over the whole range of $\tilde{z}$ or M studied irrespective of the differences in polymer species (chain stiffness and local chain conformation) and solvent condition. The results imply that α_S is a function only of $\tilde{z}$, or in other words, the QTP scheme is valid for α_S.

Finally, it is pertinent to make some remarks on the excluded-volume effect in semiflexible polymers. Norisuye and co-workers [18, 19] have found that the critical value of λL for the onset of the excluded-volume effect for them is 20–50, being one order of magnitude larger than that for flexible polymers. This is due to the fact that λB is very small for these semiflexible polymers; for semiflexible polymers with large λB the critical value of λL is close to that for flexible polymers [19]. In any case, the QTP scheme for α_S seems valid also for semiflexible polymers, although $\langle S^2 \rangle_\Theta$ for them cannot be determined since there is no proper Θ solvent. It may rather be considered that this scheme enables us to determine the unperturbed dimension $\langle S^2 \rangle_0$ for semiflexible polymers in good solvents.

8.2 Viscosity- and Hydrodynamic-Radius Expansion Factors

8.2.1 Effects of Chain Stiffness and Fluctuating Hydrodynamic Interaction

The viscosity-radius expansion factor α_η for the intrinsic viscosity $[\eta]$ is defined as usual by [1]

$$[\eta] = [\eta]_0 \alpha_\eta^3 , \tag{8.61}$$

or

$$V_H = V_{H,0} \alpha_\eta^3 , \tag{8.62}$$

where V_H is the hydrodynamic (molar) volume defined by Eq. (6.131), and the subscript 0 indicates the unperturbed value as before. Similarly, the expansion factor α_H for the hydrodynamic radius R_H defined by Eq. (6.132) is defined by

$$R_H = R_{H,0}\alpha_H . \tag{8.63}$$

Note that α_H is identical with α_f in the earlier notation [1].

Now α_η and α_H may be written in the form

$$\alpha_\eta = \alpha_\eta^{(0)} h_\eta , \tag{8.64}$$

$$\alpha_H = \alpha_H^{(0)} h_H , \tag{8.65}$$

where h_η and h_H represent possible effects of fluctuating hydrodynamic interaction (HI), and $\alpha_\eta^{(0)}$ and $\alpha_H^{(0)}$ are the respective parts without these effects. In the conventional TP theory with $h_\eta = h_H = 1$, $\alpha^{(0)}$ ($\alpha_\eta^{(0)}$ or $\alpha_H^{(0)}$) is a function only of z, that is,

$$\alpha^{(0)} = \alpha^{(0)}(z) \qquad \text{(TP)} , \tag{8.66}$$

while in the QTP scheme $\alpha^{(0)}$ and h (h_η or h_H) must also be functions only of $\tilde{z}$, that is,

$$\alpha^{(0)} = \alpha^{(0)}(\tilde{z}) \qquad \text{(QTP)} , \tag{8.67}$$

$$h = h(\tilde{z}) \qquad \text{(QTP)} . \tag{8.68}$$

It is then convenient to adopt the Barrett equations [20, 21] for $\alpha_\eta^{(0)}$ and $\alpha_H^{(0)}$,

$$\alpha_\eta^{(0)} = (1 + 3.8\,\tilde{z} + 1.9\,\tilde{z}^2)^{0.1} , \tag{8.69}$$

$$\alpha_H^{(0)} = (1 + 5.93\,\tilde{z} + 3.59\,\tilde{z}^2)^{0.1} , \tag{8.70}$$

which give the respective, exact first-order perturbation theories in the Kirkwood–Riseman scheme [22],

$$(\alpha_\eta^{(0)})^3 = 1 + 1.14\,\tilde{z} - \cdots , \tag{8.71}$$

or

$$\alpha_\eta^{(0)} = 1 + 0.38\,\tilde{z} - \cdots , \tag{8.72}$$

$$\alpha_H^{(0)} = 1 + 0.593\,\tilde{z} - \cdots . \tag{8.73}$$

We note that the original Barrett equation for $\alpha_H^{(0)}$, in which the coefficient 5.93 of $\tilde{z}$ in Eq. (8.70) is replaced by 6.09, gives the Stockmayer–Albrecht value 0.609 [23] (from the Kirkwood formula) instead of 0.593 for the first-order perturbation coefficient, and that the Fixman–Pyun scheme [24, 25] gives the value 1.06 [26] instead of 1.14 for the first-order perturbation coefficient of $(\alpha_\eta^{(0)})^3$.

As mentioned in Sect. 6.5.2, the unperturbed reduced hydrodynamic radius $\rho_{\infty,0}{}^{-1}$ in the coil limit may be evaluated on the basis of the HW chain with partially fluctuating HI [27], the results being given in Table 6.4. Similarly, $h_H = h_H(z)$ may easily be evaluated for Gaussian chains in the uniform-expansion approximation [1], and then in the QTP scheme we have [28]

$$h_H = \frac{0.88}{1 - 0.12\,\alpha_S{}^{-0.43}}, \tag{8.74}$$

where $\alpha_S = \alpha_S(\tilde{z})$ is given by Eq. (8.57). Equation (8.74) predicts that h_H decreases from unity to 0.88 as $\tilde{z}$ (or z) is increased from 0 to ∞. As also mentioned in Sect. 6.5.2, the corresponding theory of the Flory–Fox factor $\Phi_{\infty,0}$ cannot be developed; and therefore there is no available theory of h_η either.

Before making a comparison of theory with experiment, we evaluate the intrinsic viscosity $[\eta]$ and translational diffusion coefficient D for the polymethylene-like RIS chain with excluded volume as considered in Sect. 8.1.3 by Monte Carlo methods to examine numerically the behavior of α_η and α_H [29]. In the present case, suppose that the chain is composed of $n + 1$ beads with fluctuating HI between them, each of Stokes diameter d_b, as well as with excluded-volume interactions. The evaluation is carried out in the Zimm rigid-body ensemble approximation [30]. This approximation may be considered to cause no significant errors in the ratios $\alpha_\eta{}^3 = [\eta]/[\eta]_0$ and $\alpha_H = R_H/R_{H,0}$. Figure 8.8 shows plots of α (α_η or α_H) against $\tilde{z}$ with the Monte Carlo data

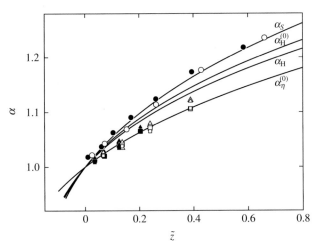

Fig. 8.8. Plots of α (α_η or α_H) against $\tilde{z}$ with Monte Carlo data; α_η ($\triangle,\blacktriangle$) and α_H ($\square,\blacksquare$) for $\epsilon = 0.2$ (*unfilled symbols*) and 0.05 (*filled ones*) and for $d_b/l = 1.0$ (*large symbols*) and 0.5 (*small ones*) [29]. The data for α_S (from Fig. 8.3) are also plotted. The *solid curves* represent the theoretical values of the indicated expansion factors (see the text)

thus obtained for $\epsilon = 0.2$ (unfilled symbols) and 0.05 (filled symbols) and for $d_b/l = 1.0$ (large symbols) and 0.5 (small symbols). For comparison, the data for α_S are also plotted. Here, we note that B has been determined to be 0.21 and 0.11 for $\epsilon = 0.2$ and 0.05, respectively, as in Fig. 8.3, and that values of $\tilde{z}$ have been calculated from Eq. (8.48). The solid curves represent the theoretical values of the indicated expansion factors calculated from Eqs. (8.57), (8.65), (8.69), (8.70), and (8.74). It is seen that the data points for each expansion factor form a single-composite curve. It is more important to see that the Monte Carlo values of α_η agree with the theoretical values from Eq. (8.69) (with $h_\eta = 1$), while the Monte Carlo values of α_H are much smaller than the theoretical values from Eq. (8.70) and even from Eq. (8.65) with Eqs. (8.70) and (8.74) and are rather close to the Monte Carlo values of α_η. The results, although in the range of small excluded volume, are consistent with those from the experimental data presented in the next subsection.

8.2.2 Comparison with Experiment

There are at least three cases to be considered in determining experimentally the expansion factors α_η and α_H. We begin by discussing them in order.

A first case is the easiest case for which it can be confirmed that $\langle S^2 \rangle = \langle S^2 \rangle_\Theta$ and also $[\eta] = [\eta]_\Theta$ and $R_H = R_{H,\Theta}$ in the oligomer region, so that Eq. (8.55) and also the relations

$$[\eta]_0 = [\eta]_\Theta , \tag{8.75}$$

$$R_{H,0} = R_{H,\Theta} , \tag{8.76}$$

hold for all values of M, the Flory–Fox factor Φ_Θ being independent of solvent. In this case we may determine α_η^3 and α_H from the equations $\alpha_\eta^3 = [\eta]/[\eta]_\Theta$ and $\alpha_H = R_H/R_{H,\Theta}$, respectively, and then α_η and α_H must be universal functions of α_S in the QTP scheme (see below). Examples of this case are a-PS [13, 16] and (perhaps) PIB [16] in some solvents.

A second is the case for which the relations of Eqs. (8.55) and (8.76) approximately hold but Eq. (8.75) is invalid, that is, $[\eta]_0 \neq [\eta]_\Theta$, so that $\Phi_0 \neq \Phi_\Theta$, because of the dependence on solvent of Φ_Θ (and also Φ_0). [Note that Φ_0 (Φ_Θ) is defined by Eq. (6.129) with Eq. (6.131) with $[\eta]_0$ ($[\eta]_\Theta$) and $\langle S^2 \rangle_0$ ($\langle S^2 \rangle_\Theta$).] In this case, if we define an apparent viscosity-radius expansion factor $\bar{\alpha}_\eta$ by the equation

$$[\eta] = [\eta]_\Theta \bar{\alpha}_\eta^3 , \tag{8.77}$$

we have

$$\bar{\alpha}_\eta^3 = C_\eta \alpha_\eta^3 = C_\eta \alpha_\Phi \alpha_S^3 , \tag{8.78}$$

where C_η and α_Φ are defined by

$$C_\eta = \Phi_0/\Phi_\Theta , \tag{8.79}$$

$$\alpha_\Phi = \Phi/\Phi_0 . \tag{8.80}$$

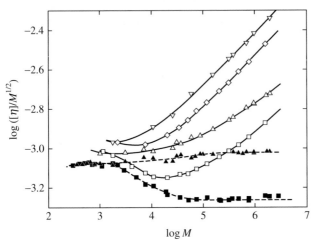

Fig. 8.9. Double-logarithmic plots of $[\eta]/M^{1/2}$ ($[\eta]$ in dL/g) against M for a- and i-PMMAs [11,12]. The symbols have the same meaning as those in Figs. 8.4 and 8.5. The *solid and dashed curves* connect the data points smoothly

In the QTP scheme, both α_η and α_Φ must be functions of α_S, and $\bar{\alpha}_\eta^{\,3}$ must be of the form

$$\bar{\alpha}_\eta^{\,3} = C_\eta f(\alpha_S) \qquad \text{(QTP)}, \tag{8.81}$$

where f is a function only of α_S. Note that the coefficient C_η is essentially identical with the constant prefactor in Eq. (6.133) and is to be determined experimentally. Examples of this case are a- and i-PMMAs [11, 12] (see Sect. 6.5.2).

Figure 8.9 shows double-logarithmic plots of $[\eta]/M^{1/2}$ ($[\eta]$ in dL/g) against M for a- and i-PMMAs [11, 12]. The symbols have the same meaning as those in Figs. 8.4 and 8.5. The solid and dashed curves connect the data points smoothly. It is interesting to see that in the oligomer region ($M \lesssim 2 \times 10^3$) the values of $[\eta]$ for the two PMMAs coincide with each other in each solvent, indicating that the average chain dimension in that region and also the hydrodynamic bead diameter d_b are independent of the stereochemical composition f_r. However, the dependence of $[\eta]/M^{1/2}$ on M for each PMMA varies depending on solvent in the oligomer region. This implies that for PMMA d_b as well as Φ_Θ depends on solvent. For simplicity, therefore, in the following analysis of α_η for PMMA we confine ourselves to the range of large M ($\gtrsim 10^4$) in which the possible effect of the solvent dependence of d_b may be regarded as negligibly small. (The solvent dependence of d_b is considered in a third case.)

Figure 8.10 shows double-logarithmic plots of $\bar{\alpha}_\eta^{\,3}$ determined from Eq. (8.77) against α_S^3 for a-PS in cyclohexane at 36.0–55.0°C [14], in toluene at 15.0°C [16], in benzene at 25.0°C [13,14], and in MEK at 35.0°C [13], and a-PMMA in acetone at 25.0°C, in nitroethane at 30.0°C, and in chloroform at 25.0°C [11]. The solid and dashed curves connect smoothly the data points

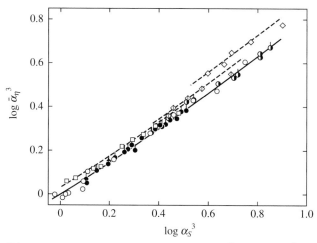

Fig. 8.10. Double-logarithmic plots of $\bar{\alpha}_\eta^3$ against α_S^3 for a-PS in cyclohexane at 36.0–55.0°C (●) [14], in toluene at 15.0°C (○) [16], in benzene at 25.0°C (◐,◑) [13,14], and in MEK at 35.0°C (◖) [13], and a-PMMA in acetone at 25.0°C (□), in nitroethane at 30.0°C (◇), and in chloroform at 25.0°C (◇) [11]. The *solid and dashed curves* connect smoothly the data points for a-PS and a-PMMA, respectively

for a-PS and a-PMMA, respectively. It is seen that the data points (dashed curve) for a-PMMA in each solvent deviate upward from those (solid curve) for a-PS by a certain constant independent of α_S^3. This constant may be equated to $\log C_\eta$, as seen from Eq. (8.78) or (8.81). The values of C_η so estimated for a-PMMA from the separations between the solid and dashed curves are 1.08, 1.11, and 1.25 for the acetone, nitroethane, and chloroform solutions, respectively. Thus we may determine α_η^3 from the equation $\alpha_\eta^3 = C_\eta^{-1}[\eta]/[\eta]_\Theta$ with these values of C_η. We note that $C_\eta = 1$ for a-PS since its Φ_Θ (or Φ_0) is independent of solvent (see Table 6.4), and that for a-PS the plots of α_η^3 against α_S^3 form a single-composite curve, as shown in Fig. 8.10, if both α_η and α_S are correctly determined [13] (see Table 8.3).

A third case is such that the relations of Eqs. (8.75) and (8.76) hold for large M along with Eq. (8.55) (for all values of M) but they do not hold in the oligomer region because of the specific interaction η^* between polymer and solvent molecules (Sect. 6.5.3) and/or the solvent dependence of d_b as above. (The validity of the above relations may be verified by the formation of a single-composite curve of double-logarithmic plots of α_η^3 against α_S^3.) An example of this case is PDMS [17].

Figure 8.11 shows double-logarithmic plots of $R_H/M^{1/2}$ (R_H in Å) against M for PDMS in toluene at 25.0°C [17] and in bromocyclohexane at 29.5°C (Θ) [31], where the raw data for D_Θ are the same as those in Fig. 6.11. The solid curve and the dashed curve (1) connect the respective data points smoothly. Clearly the difference between the values of R_H and $R_{H,\Theta}$ in the oligomer region arises from the solvent dependence of d_b, as in the case of

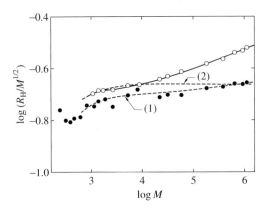

Fig. 8.11. Double-logarithmic plots of $R_H/M^{1/2}$ (R_H in Å) against M for PDMS in toluene at 25.0°C ($\circ$) [17] and in bromocyclohexane at 29.5°C ($\ominus$) ($\bullet$) [31]. The *solid curve* and the *dashed curve* (1) connect the respective data points smoothly. The *dashed curve* (2) represents the values of $R_{H,0}$ in toluene used as reference standards to calculate α_H (see the text)

PMMA. The problem is then to determine $R_{H,0}$ for PDMS in toluene. This may be done by means of the theory as follows. It may be calculated by multiplying the interpolated (smoothed) value of $R_{H,\Theta}$ [the dashed curve (1)] by the ratio of the unperturbed HW theoretical value of R_H in toluene to that in bromocyclohexane. These theoretical values may be calculated with the values of the HW model parameter determined from D_Θ and given in Table 6.3 but with the values of d_b which give good agreement between the theoretical and experimental values of R_H in the oligomer region in the respective solvents. The values of d_b so determined are 2.2 and 1.4 Å in toluene and bromocyclohexane, respectively. (Note that the latter value of d_b is somewhat smaller than that in Table 6.3.) The dashed curve (2) represents the values of $R_{H,0}$ so evaluated. We may then determine α_H in toluene from the equation $\alpha_H = R_H/R_{H,0}$ with these values of $R_{H,0}$.

As discussed in Sect. 6.5.3, $[\eta]$ for PDMS in bromocyclohexane becomes negative in the oligomer region because of the above specific interaction η^*. Thus we must also consider this effect in the determination of α_η. Figure 8.12

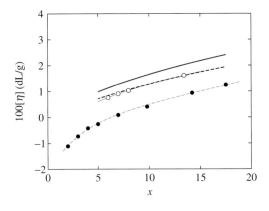

Fig. 8.12. Plots of $[\eta]$ against x for PDMS oligomers [17, 31]. The symbols have the same meaning as those in Fig. 8.11. The *light solid and dashed curves* connect the respective data points smoothly. The *heavy solid and dashed curves* represent the HW theoretical values (see the text)

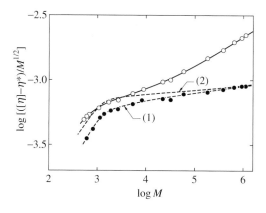

Fig. 8.13. Double-logarithmic plots of $([\eta] - \eta^*)/M^{1/2}$ ($[\eta]$ in dL/g) for PDMS [17,31]. The symbols have the same meaning as those in Fig. 8.11. The *solid curve* and the *dashed curve* (1) connect the respective data points smoothly. The *dashed curve* (2) represents the values of $([\eta] - \eta^*)_0$ in toluene used as reference standards to calculate $\alpha_\eta{}^3$ (see the text)

shows plots of $[\eta]$ against x for PDMS in toluene at 25.0°C [17] and in bromo-cyclohexane at 29.5°C (Θ) [31], where the latter data have been reproduced from Fig. 6.14. The light solid and dashed curves connect the respective data points smoothly. The heavy solid and dashed curves represent the HW theoretical values calculated with the values of the HW model parameters determined from $[\eta]_\Theta$ and given in Table 6.3 and with the above values of d_b, considering the physical requirement that the values of d_b from D_Θ and $[\eta]_\Theta$ in the same solvent must be the same. For PDMS in toluene, η^* is then found to be -0.0038 dL/g as an average of the differences between the four values of $[\eta]$ in toluene (unfilled circles) and the corresponding theoretical values (the corresponding points on the heavy solid curve). As for PDMS in bromocyclohexane, it is reestimated to be -0.0078 dL/g, this value being somewhat different from that in Sect. 6.5.3. (This is due to the difference between the present and previous values of d_b.)

Values of $([\eta] - \eta^*)/M^{1/2}$ ($[\eta]$ in dL/g) calculated with the experimental values of $[\eta]$ [17, 31] and the values of η^* thus determined for PDMS in the two solvents are double-logarithmically plotted against M in Fig. 8.13, where the symbols have the same meaning as those in Fig. 8.12. The solid curve and the dashed curve (1) connect the respective data points smoothly. The dashed curve (2) represents the values of $([\eta] - \eta^*)_0$ in toluene evaluated by adopting the same maneuver as that in the evaluation of $R_{H,0}$. We may then determine $\alpha_\eta{}^3$ in toluene from the equation $\alpha_\eta{}^3 = ([\eta] - \eta^*)/([\eta] - \eta^*)_0$ with these values of $([\eta] - \eta^*)_0$.

Now we proceed to make a comparison of theory with experiment for α_η and α_H. We first examine the behavior of $\alpha_\eta{}^3$ as a function of z. Figure 8.14 shows double-logarithmic plots of $\alpha_\eta{}^3$ against z for a-PS in toluene at 15.0°C [16], in benzene at 25.0°C [13, 14], and in MEK at 35.0°C [13, 14], and a-PMMA in acetone at 25.0°C, in nitroethane at 30.0°C, and in chloroform at 25.0°C [11], where values of z have been calculated as in Fig. 8.6 (with the values of the HW model parameters and λB determined from $\langle S^2 \rangle$). The solid curves represent the QTP theory values calculated from Eq. (8.64) with

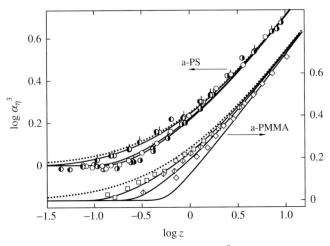

Fig. 8.14. Double-logarithmic plots of α_η^3 against z for a-PS in toluene at 15.0°C ($\bigcirc$) [16], in benzene at 25.0°C ($\bullet$,$\circ)$ [13,14], and in MEK at 35.0°C ($\bullet$,$\bullet$) [13,14], and a-PMMA in acetone at 25.0°C ($\square$), in nitroethane at 30.0°C ($\Diamond$), and in chloroform at 25.0°C ($\Diamond$) [11]. The *solid curves* represent the QTP theory values calculated from Eq. (8.64) with $h_\eta = 1$, and the *dotted curves* the TP theory values with $\tilde{z} = z$ (see the text)

Eq. (8.69) and $h_\eta = 1$ with the values of λB, and the dotted curves represent the TP theory values with $\tilde{z} = z$. There is good agreement between the QTP theoretical and experimental values except for a-PMMA in chloroform and in nitroethane, for which the theoretical values deviate downward from the experimental values for $z \lesssim 2.5$. This discrepancy may probably be due to an overestimate of experimental α_η^3 resulting from the solvent dependence of d_b mentioned above. However, the behavior of α_η^3 as a function of z is similar to that of α_S^2 in Fig. 8.6.

Figure 8.15 shows double-logarithmic plots of α_η^3 against $\tilde{z}$ with the

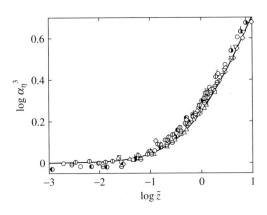

Fig. 8.15. Double-logarithmic plots of α_η^3 against $\tilde{z}$ with the same data as those in Fig. 8.14 along with those for i-PMMA in acetone at 25.0°C ($\triangle$) and in chloroform at 25.0°C ($\triangledown$) [12], PIB in *n*-heptane at 25.0°C (Φ) [16], and PDMS in toluene at 25.0°C ($\ominus$) [17]. The *solid curve* represents the QTP theory values calculated from Eq. (8.64) with $h_\eta = 1$ (see the text)

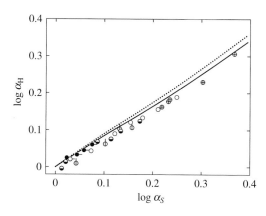

Fig. 8.16. Double-logarithmic plots of α_H against α_S for a-PS in toluene at 15.0°C (○) [15], in 4-*tert*-butyltoluene at 50.0°C (◖) [15], and in cyclohexane at 44.5°C (●) [32], PIB in *n*-heptane at 25.0°C (◔) [33], and PIP in cyclohexane at 35.0°C (⊕) [34]. The *solid curve* represents the theoretical values calculated from Eq. (8.65) with Eq. (8.57), and the *dotted curve* those of $\alpha_H^{(0)}$ calculated from Eq. (8.70) with Eq. (8.57) (see the text)

same data as those in Fig. 8.14 along with those for i-PMMA in acetone at 25.0°C and in chloroform at 25.0°C [12], PIB in *n*-heptane at 25.0°C [16], and PDMS in toluene at 25.0°C [17], where values of $\tilde{z}$ have been calculated as in Fig. 8.7. (We have adopted the values of the HW model parameters: $\lambda^{-1} = 14.0$ Å and $M_L = 24.1$ Å^{-1} for PIB from $\langle S^2 \rangle_\Theta$ and $\lambda^{-1} = 25.5$ Å and $M_L = 20.6$ Å^{-1} for PDMS from $[\eta]_\Theta$.) The solid curve represents the QTP theory values calculated from Eq. (8.64) with Eq. (8.69) and $h_\eta = 1$. It is seen that all the data points nearly form a single-composite curve and are fitted by the solid curve. Thus it may be concluded that α_η is a function only of $\tilde{z}$, or in other words, the QTP scheme is valid for α_η as well as for α_S, indicating also that there is no draining effect in α_η.

Next we examine the behavior of α_H. Figure 8.16 shows double-logarithmic plots of α_H against α_S for a-PS in toluene at 15.0°C [15], in 4-*tert*-butyltoluene at 50.0°C [15], and in cyclohexane at 44.5°C [32], PIB in *n*-heptane at 25.0°C [33], and *cis*-polyisoprene (PIP) in cyclohexane at 35.0°C [34]. The solid curve represents the theoretical values calculated from Eq. (8.65) with Eqs. (8.57), (8.70), and (8.74), and the dotted curve represents those of $\alpha_H^{(0)}$ calculated from Eq. (8.70) with Eq. (8.57). It is seen that the data points deviate downward from the dotted curve and even from the solid curve, indicating that Eq. (8.74) for h_H underestimates the effect of fluctuating HI.

Figure 8.17 shows double-logarithmic plots of α_H against $\tilde{z}$ for a-PS in toluene at 15.0°C and in 4-*tert*-butyltoluene at 50.0°C [15], a-PMMA in acetone at 25.0°C [35], i-PMMA in acetone at 25.0°C [35], PIB in *n*-heptane at 25.0°C [36], and PDMS in toluene at 25.0°C [17], where we have used the same values of $\tilde{z}$ as those in Fig. 8.15. The solid curve represents the theoretical values calculated from Eq. (8.65) with Eqs. (8.57), (8.70), and (8.74), and the dotted curve represents those of $\alpha_H^{(0)}$ calculated from Eq. (8.70). It is seen that all the data points are located even below the solid curve but nearly form a single-composite curve (except for PDMS), indicating that the QTP scheme is valid for α_H as well as for α_S and α_η. However, it is again

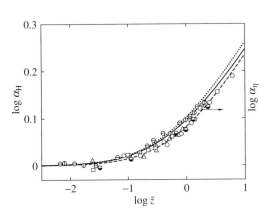

Fig. 8.17. Double-logarithmic plots of α_H against $\tilde{z}$ for a-PS in toluene at 15.0°C (○) and in 4-*tert*-butyltoluene at 50.0°C (●) [15], a-PMMA in acetone at 25.0°C (□) [35], i-PMMA in acetone at 25.0°C (△) [35], PIB in *n*-heptane at 25.0°C (◐) [36], and PDMS in toluene at 25.0°C (⊖) [17]. The *solid curve* represents the theoretical values calculated from Eq. (8.65), the *dotted curve* those of $\alpha_H^{(0)}$ calculated from Eq. (8.70), and the *dashed curve* those of $\alpha_\eta = \alpha_\eta^{(0)}$ calculated from Eq. (8.69) (see the text)

surprising to see that the data points closely follow the dashed curve which represents the theoretical values of $\alpha_\eta = \alpha_\eta^{(0)}$ calculated from Eq. (8.69), and therefore coincide with the data points for α_η within experimental error; and thus the results are consistent with those in Fig. 8.8 for the Monte Carlo data.

In order to confirm this near the Θ temperature, values of α_H for a-PS in cyclohexane [32, 37, 38] are plotted against $\tilde{z}$ in Fig. 8.18, where values of $\tilde{z}$ have been calculated with the values of λB determined from the second virial coefficient A_2 (see Sect. 8.5.1). The solid and dashed straight lines represent the theoretical values of $\alpha_H = \alpha_H^{(0)}$ and $\alpha_\eta = \alpha_\eta^{(0)}$ calculated from Eqs. (8.73) and (8.72), respectively. Clearly the first-order perturbation theory of $\alpha_H^{(0)}$ does not fit the experimental data even for small $|\tilde{z}|$, while the corresponding theory of $\alpha_\eta^{(0)}$ is seen to be valid apparently for α_H as well as for α_η over a rather wide range of $\tilde{z}$. The problem that remains is to develop a complete analytical theory of h_H.

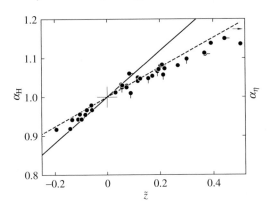

Fig. 8.18. Plots of α_H against $\tilde{z}$ for a-PS in cyclohexane near the Θ temperature (●,●,●) [32,37,38]. The *solid and dashed straight lines* represent the theoretical values of $\alpha_H = \alpha_H^{(0)}$ and $\alpha_\eta = \alpha_\eta^{(0)}$ calculated from Eqs. (8.73) and (8.72), respectively

8.3 Second Virial Coefficient

8.3.1 Perturbation Theory

We begin by developing the perturbation theory of the second virial coefficient A_2 [3], adopting the same model as that used to evaluate the expansion factors α_R and α_S in Sect. 8.1.1. It may be written in the form [1,3]

$$A_2 = (N_A n^2 \beta / 2M^2) h \,, \tag{8.82}$$

where we have assumed that $n \gg 1$, and h is the so-called h function that represents the correction to the single-contact term $(N_A n^2 \beta / 2M^2)$. In what follows, all lengths are measured in units of λ^{-1} as usual unless otherwise noted. Since we have the relation $n^2 \beta = c_\infty^{3/2} L^2 B$ from Eq. (8.6) with $L = na$, Eq. (8.82) may then be rewritten in the form

$$A_2 = (N_A c_\infty^{3/2} L^2 B / 2M^2) h \tag{8.83}$$

with

$$h = 1 - Q(L)z + \cdots \,, \tag{8.84}$$

where the coefficient $Q(L)$ of the double-contact term must approach its coil-limiting value 2.865 as L is increased. Even for the HW chain, it may be evaluated on the basis of the KP chain as in the case of the coefficient $K(L)$ in α_R and α_S.

Now, following the formulation for the random-flight chain [1], $Q(L)$ may be evaluated from

$$Q(L) = 2L^{-5/2} H(L) \tag{8.85}$$

with

$$H(L) = \int_0^L \int_0^L (L - s_1)(L - s_2) J(s_1, s_2) ds_1 ds_2 \,, \tag{8.86}$$

$$J(s_1, s_2) = (3/2\pi)^{-3/2} P_0(\mathbf{0}_{y_1 y_2})_{x_1 x_2}$$
$$= (3/2\pi)^{-3/2} \int G(\mathbf{R}; s_1) G(\mathbf{R}; s_2) d\mathbf{R} \,, \tag{8.87}$$

where $s_i = y_i - x_i$ $(i = 1, 2)$, and $P_0(\mathbf{0}_{y_1 y_2})_{x_1 x_2}$ is the (unperturbed) conditional probability density that, given an initial contact between the (x_1/a)th bead of chain 1 and the (x_2/a)th bead of chain 2, there is an additional contact between the (y_1/a)th and (y_2/a)th beads. Note that $J(s_1, s_2) = J(s_2, s_1)$.

By the use of Eq. (3.88) with Eq. (3.85) and also Eq. (3.103), we have, from the second line of Eqs. (8.87),

$$J(s_1, s_2) = \frac{1}{(s_1 + s_2)^{3/2}} \left[1 + \frac{1}{8(s_1 + s_2)} + \frac{1223}{1920(s_1 + s_2)^2} + \cdots \right]$$
$$\text{for } s_1, s_2 \gg 1 \,, \tag{8.88}$$

$$\lim_{s_2 \to 0} J(s_1, s_2) = \frac{1}{s_1^{3/2}} \left(1 - \frac{5}{8s_1} - \frac{79}{640s_1^2} + \cdots \right) \qquad \text{for } s_1 \gg 1, \quad (8.89)$$

$$\lim_{s_1, s_2 \to 0} J(s_1, s_2) = \left(\frac{3}{2\pi} \right)^{-3/2} \frac{1}{4\pi s_2^2} \delta(s_1 - s_2) . \qquad (8.90)$$

We then construct an interpolation formula for $J(s_1, s_2)$, which is valid for s_1 larger than some small positive value σ, from Eqs. (8.88) and (8.89). When $\sigma = 0.931$, which value has no great significance, the result is

$$J(s_1, s_2) = \sum_{j=0}^{2} f_i(\xi) s_1^{-3/2-i} \qquad \text{for } s_1 \geq \sigma , \qquad (8.91)$$

where

$$\xi = s_2/s_1 , \qquad (8.92)$$

$$f_i = c_i(1 + \xi)^{-3/2-i} \qquad \text{for } \xi \geq \sigma/s_1$$

$$= \sum_{j=0}^{3} a_{ij} \xi^j \qquad \text{for } \xi < \sigma/s_1 \qquad (8.93)$$

with

$$c_0 = 1 , \qquad c_1 = \frac{1}{8} , \qquad c_2 = \frac{1223}{1920} ,$$

$$a_{00} = 1 , \qquad a_{10} = -\frac{5}{8} , \qquad a_{20} = -\frac{79}{640} . \qquad (8.94)$$

The remaining coefficients a_{ij} are determined as functions of σ/s_1 so that the two f_is given by the first and second lines of Eqs. (8.93) have the same first and second derivatives at $\xi = \sigma/s_1$, but the results [3] are omitted. We note that the double-contact approximation does not suffice for the complete evaluation of A_2 for very small L [3].

Thus we evaluate $H(L)$ only for $L \geq \sigma$. It may then be split into three parts,

$$H = H_0 + H_1 + 2H_2 \qquad \text{for } L \geq \sigma , \qquad (8.95)$$

where H_0 is the part of the double integral of Eq. (8.86) for $0 \leq s_1, s_2 \leq \sigma$, H_1 for $\sigma \leq s_1, s_2 \leq L$, and H_2 for $\sigma \leq s_1 \leq L$ and $0 \leq s_2 \leq \sigma$. If we use Eq. (8.91), H_1 and H_2 can be evaluated straightforwardly, but the results are omitted. For the evaluation of H_0, we consider a function $H(t_1, t_2; L)$ defined by

$$H(t_1, t_2; L) = \int_0^{t_1} (L - s_1) I(s_1, t_2; L) ds_1 \qquad (8.96)$$

with

$$I(s_1, t_2; L) = \int_0^{t_2} (L - s_2) J(s_1, s_2) ds_2 , \qquad (8.97)$$

so that

$$H_0 = H(\sigma, \sigma; L). \tag{8.98}$$

We can construct an interpolation formula for $I(s_1, t_2 \to 0; L)$ from Eqs. (8.89), (8.90), and (8.97), and then have the limiting form

$$\lim_{t_2 \to 0} H(\sigma, t_2; L) = \int_d^\sigma (L - s_1) I(s_1, t_2 \to 0; L) ds_1, \tag{8.99}$$

where d is a cutoff parameter. Recalling the symmetry property of J (and H), we can then have the limiting form, $\lim_{t_1 \to 0} H(t_1, \sigma; L)$, and therefore $\lim_{t_1 \to 0} \partial H(t_1, \sigma; L)/\partial t_1$. On the other hand, we have, from Eq. (8.96),

$$\frac{\partial H(t_1, \sigma; L)}{\partial t_1} = (L - t_1) I(t_1, \sigma; L), \tag{8.100}$$

where $I(t_1, \sigma; L)$ for $t_1 \geq \sigma$ may be evaluated from Eq. (8.97) with Eq. (8.91). We can then construct an interpolation formula for $\partial H(t_1, \sigma; L)/\partial t_1$. Thus we obtain

$$H_0 = H(0, \sigma; L) + \int_0^\sigma \frac{\partial H(t_1, \sigma; L)}{\partial t_1} dt_1. \tag{8.101}$$

Summing up all terms in Eq. (8.95) thus evaluated, we obtain $H(L; d)$ and hence $Q(L; d)$. For flexible chains, we may take $d = 0.2$–0.5. Fortunately, however, $Q(L; d)$ is found to be insensitive to change in d in that range. We therefore choose $d = 0.3$ for all flexible polymers, for simplicity. Then $Q(L)$ (for $L \gtrsim 1$) is given in a very good approximation by [39]

$$\begin{aligned}
Q(L) = &-\frac{128\sqrt{2}}{15} - \frac{2.531}{L^{1/2}} - \frac{2.586}{L} + \frac{1.985}{L^{3/2}} - \frac{1.984}{L^2} - \frac{0.9292}{L^{5/2}} + \frac{0.1223}{L^3} \\
&+ \frac{8}{5} x^{5/2} + \frac{2}{3} x^{3/2} \left(8 + \frac{1}{6L} \right) + x^{1/2} \left(8 - \frac{13.53}{L} + \frac{0.2804}{L^2} \right) \\
&- \frac{1}{x^{1/2} L} \left(0.3333 - \frac{5.724}{L} + \frac{0.7974}{L^2} \right) - \frac{1}{x^{3/2} L^2} \left(0.3398 - \frac{0.7146}{L} \right)
\end{aligned} \tag{8.102}$$

with

$$x = 1 + \frac{0.961}{L}. \tag{8.103}$$

It is seen that $Q(L)$ also approaches slowly its coil-limiting value as L is increased, the first-order deviation from the latter being of order $L^{-1/2}$ as in the case of $K(L)$.

8.3.2 Effects of Chain Stiffness and Local Chain Conformations

As in the TP theory, the *interpenetration function* Ψ may be defined by [1]

$$A_2 = 4\pi^{3/2} N_A \frac{\langle S^2 \rangle^{3/2}}{M^2} \Psi , \qquad (8.104)$$

but it is now given, from Eqs. (8.83) and (8.104), by

$$\Psi = \left(\frac{6 \langle S^2 \rangle_0}{c_\infty L} \right)^{-3/2} \bar{z} h \qquad (8.105)$$

with

$$\bar{z} = z/\alpha_S^3 . \qquad (8.106)$$

We assume that h is a function only of a parameter $\hat{z}$ defined by

$$\hat{z} = \tilde{z}/\alpha_S^3 \qquad (8.107)$$

with

$$\tilde{z} = \left[\frac{Q(L)}{2.865} \right] z . \qquad (8.108)$$

Corresponding to the intramolecular scaled excluded-volume parameter $\tilde{z}$ defined by Eq. (8.48), the parameter $\tilde{z}$ is referred to as the *intermolecular scaled excluded-volume parameter*. It is seen that $\tilde{z}$ and $\hat{z}$ slowly approach their respective coil-limiting values z and $\bar{z}$ as L is increased.

Now we determine the functional form of $h(\hat{z})$ that may be combined with Eq. (8.51) or (8.57) for α_S [39]. This can be done in such a way that in the coil limit ($L \to \infty$) the values of $h(\bar{z})$ as a function of z with the α_S given by Eq. (8.57) (with $\tilde{z} = z$) are as close as possible to those of the Barrett function $h(z)$ [40] (with the intramolecular excluded-volume effect) at any z; that is,

$$h(\bar{z}) \simeq (1 + 14.3\, z + 57.3\, z^2)^{-0.2} . \qquad (8.109)$$

Replacing $\bar{z}$ by $\hat{z}$ in the function $h(\bar{z})$ so found, we obtain the desired function

$$h(\hat{z}) = (1 + 7.74\, \hat{z} + 52.3\, \hat{z}^{27/10})^{-10/27} . \qquad (8.110)$$

We note that in the coil limit the difference between the values of Ψ calculated from Eq. (8.105) with Eq. (8.110) and with the Barrett equation for h does not exceed 3%. As seen from Eqs. (8.105)–(8.108) and (8.110), h is a function of $\hat{z}$ and $\tilde{z}$, and $\bar{z} h$ is a function of z, $\tilde{z}$, and $\tilde{z}$, so that neither the TP nor the QTP scheme is valid for Ψ even apart from its prefactor; and moreover, the latter depends on L and also the HW model parameters κ_0 and τ_0. In the coil limit (TP theory) Eq. (8.105) reduces to

$$\Psi = \bar{z} h(z) \qquad \text{(TP)} , \qquad (8.111)$$

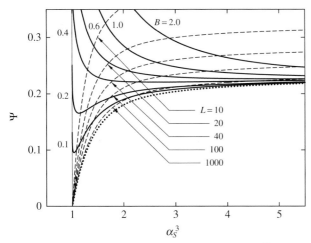

Fig. 8.19. Plots of the theoretical Ψ against α_S^3 for a-PS. The *solid and dashed curves* represent the values at constant B and L, respectively. The *dotted curve* represents the TP theory values

where $h(z)$ is given by Eq. (8.110) with $\hat{z} = \bar{z}$ (and with $\tilde{\bar{z}} = \tilde{z} = z$). We note that Nickel [41] and Chen and Noolandi [42] have also developed non-TP theories of Ψ, but the derived equations cannot explain all experimental results for flexible polymers.

We examine the behavior of Ψ taking as examples a-PS and a-PMMA, for which we assume the values of the HW model parameters given in Table 5.1. Values of Ψ as a function of α_S^3 calculated from Eq. (8.105) with Eqs. (8.57),

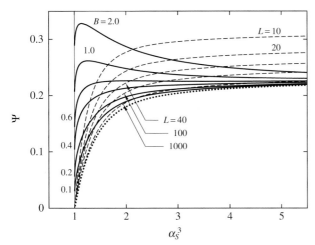

Fig. 8.20. Plots of the theoretical Ψ against α_S^3 for a-PMMA. The curves have the same meaning as those in Fig. 8.19

(8.106)–(8.108), and (8.110) are plotted in Figs. 8.19 and 8.20 for a-PS and a-PMMA, respectively. The dotted curves represent the TP theory values calculated from Eq. (8.111). The solid curves represent the values for the case in which L (or M) is changed at constant B, while the dashed curves represent the values for the case in which B is changed at constant L (or M). It is seen that the TP theory prediction is obtained as the asymptotic limit of $L \to \infty$ or $B \to 0$, that for finite L and B, Ψ always deviate upward from the TP theory prediction, and that the behavior of Ψ depends remarkably on chain stiffness and local chain conformation.

8.3.3 Effects of Chain Ends

In this subsection we consider possible effects of chain ends on A_2, which must become appreciable as L (or M) is decreased [39]. However, note that the effects on the expansion factors α must be vanishingly small since the probability densities for intramolecular contacts between beads (or the excluded-volume effect itself) are very small for small L because of chain stiffness.

For the present purpose, we consider a chain composed of $n + 1$ beads numbered 0, 1, 2, $\cdots$, j, $\cdots$, n from one end to the other and attach the label "0" to the $n - 1$ intermediate beads with $j = 1, 2, \cdots, n - 1$, the label "1" to the end bead with $j = 0$, and the label "2" to the other end bead with $j = n$, where the two end beads are different from the $n - 1$ intermediate ones and also from each other in species. For simplicity, we take into account the effects only on the single-contact term $A_2^{(1)}$ for small n. From the general formulation of A_2 [1], $A_2^{(1)}$ may then be written in the form

$$A_2^{(1)} = (N_A/2M^2) \sum \beta_{2,kl} , \tag{8.112}$$

where $\beta_{2,kl}$ is the binary-cluster integral for two beads with the labels k and l $(= 0, 1, 2)$, and the sum is taken over all possible sets of such two beads. The latter may be expressed as

$$\sum \beta_{2,kl} = -\sum_{i_1} \sum_{i_2} \int \chi_{i_1 i_2} d\mathbf{R}_{i_1 i_2} , \tag{8.113}$$

where $\chi_{i_1 i_2}$ is the χ function of the distance $\mathbf{R}_{i_1 i_2}$ between the i_1th bead of chain 1 and the i_2th bead of chain 2 [1]. Then there are six kinds of binary-cluster integrals, as schematically depicted in Fig. 8.21, where the numerical prefactor of each $\beta_{2,kl}$ represents its symmetry factor.

Now we define *excess* binary-cluster integrals β_{kl} by

$$\beta_{2,kl} = \beta_2 + \beta_{kl} \tag{8.114}$$

with

$$\beta \equiv \beta_2 \equiv \beta_{2,00} . \tag{8.115}$$

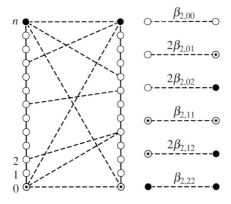

Fig. 8.21. Six kinds of intermolecular contacts and binary-cluster integrals $\beta_{2,kl}$ (see the text)

We then have

$$\sum \beta_{2,kl} = (n+1)^2\beta + 4(n+1)\beta_{2,1} + 4(\beta_{2,2} - 2\beta_{2,1})\,, \qquad (8.116)$$

where $\beta_{2,1}$ and $\beta_{2,2}$ are the *effective* excess binary-cluster integrals associated with the chain end beads and defined by

$$2\beta_{2,1} = \beta_{01} + \beta_{02}\,,$$
$$4\beta_{2,2} = \beta_{11} + 2\beta_{12} + \beta_{22}\,. \qquad (8.117)$$

Thus A_2 in general may be written in the form

$$A_2 = A_2^{(\mathrm{HW})} + A_2^{(\mathrm{E})}\,, \qquad (8.118)$$

where $A_2^{(\mathrm{HW})}$ is the part of A_2 without the effects of chain ends, or A_2 for the (fictitious) chain composed of $n+1$ identical beads, and $A_2^{(\mathrm{E})}$ represents the contribution of the effects of chain ends to A_2. The first term $A_2^{(\mathrm{HW})}$ is therefore given by Eq. (8.83) or (8.104), and the second term $A_2^{(\mathrm{E})}$ is given, from Eq. (8.112) with Eq. (8.116), by

$$A_2^{(\mathrm{E})} = a_{2,1}M^{-1} + a_{2,2}M^{-2}\,, \qquad (8.119)$$

where

$$a_{2,1} = 2N_{\mathrm{A}}\beta_{2,1}/M_0\,,$$
$$a_{2,2} = 2N_{\mathrm{A}}\Delta\beta_{2,2} \qquad (8.120)$$

with M_0 the molecular weight of the bead and with

$$\Delta\beta_{2,2} = \beta_{2,2} - 2\beta_{2,1}\,. \qquad (8.121)$$

At the Θ temperature, which is now defined as the temperature at which A_2 vanishes for large M, $A_2^{(\mathrm{HW})}$ and β must vanish, so that A_2 at the Θ temperature, which we denote by $A_{2,\Theta}$, is given by

$$A_{2,\Theta} = A_2^{(E)}. \tag{8.122}$$

This indicates that $A_{2,\Theta}$ does not vanish except at large M, depending on M.

8.3.4 Comparison with Experiment

Huber and Stockmayer [43] found experimentally that $A_{2,\Theta}$ does not vanish but increases with decreasing M for small M for a-PS in cyclohexane, and then this finding was confirmed by others [44, 45] for a-PS and also for a-PMMA. This may be regarded as arising from the effects of chain ends. Thus we first make a comparison of theory with experiment with respect to $A_2^{(E)}$ in Θ and also good solvents, for convenience.

Equation (8.119) predicts that $A_2^{(E)}M$ is linear in M^{-1}. Figure 8.22 shows plots of $A_2^{(E)}M$ against M^{-1} for a-PS in cyclohexane at 34.5°C (Θ) and in toluene at 15.0°C [46], a-PMMA in acetonitrile at 44.0°C (Θ) and in acetone at 25.0°C [47], and i-PMMA in acetone at 25.0°C [48]. Here, we note that the data for A_2 for the oligomers were obtained from light scattering measurements following the procedure described in Appendix 8.B, and that $A_2^{(E)} = A_2$ at the Θ temperature, as mentioned above, while the values of $A_2^{(E)}$ in the good solvents have been obtained from $A_2^{(E)} = A_2 - A_2^{(HW)}$ with values of $A_2^{(HW)}$ calculated from Eq. (8.104) with the values of the HW model parameters and λB (from $\langle S^2 \rangle$) and with the relation

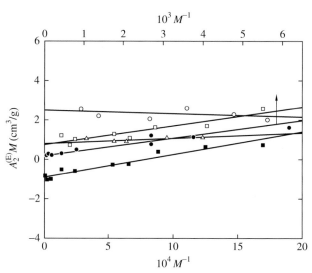

Fig. 8.22. Plots of $A_2^{(E)}M$ against M^{-1} for a-PS in cyclohexane at 34.5°C (Θ) ($\bullet$) and in toluene at 15.0°C ($\circ$) [46], a-PMMA in acetonitrile at 44.0°C (Θ) ($\blacksquare$) and in acetone at 25.0°C ($\square$) [47], and i-PMMA in acetone at 25.0°C ($\triangle$) [48]

$$L = M/M_{\mathrm{L}} \,. \tag{8.123}$$

The data points for each system are somewhat scattered but can be fitted by a straight line, and from its intercept and slope, $a_{2,1}$ and $a_{2,2}$ and hence $\beta_{2,1}$ and $\beta_{2,2}$ may be determined. The results so obtained for $\beta_{2,1}$ and $\beta_{2,2}$ taking the repeat unit as a single bead (with $M_0 = 104$ and 100 for PS and PMMA, respectively) are 16 and 260 Å^3 for a-PS in cyclohexane, 220 and 270 Å^3 for a-PS in toluene, -75 and 800 Å^3 for a-PMMA in acetonitrile, 62 and 910 Å^3 for a-PMMA in acetone, and 66 and 360 Å^3 for i-PMMA in acetone, respectively.

Figure 8.23 shows double-logarithmic plots of A_2 (in cm^3 mol/g^2) against M for a-PS in cyclohexane at $34.5°$C (Θ) and in toluene at $15.0°$C [46]. The solid curves represent the theoretical values calculated with the values of all the necessary parameters determined, and the dot-dashed curve represents those with $h = 1$ (for $\lambda L \lesssim 1$) for the latter system, for which the theoretical contributions of $A_2^{(\mathrm{HW})}$ and $A_2^{(\mathrm{E})}$ are shown by the dashed and dotted curves, respectively. It is seen that there is rather good agreement between theory and experiment, and that $A_2^{(\mathrm{E})}$ remains appreciable up to $M = 10^4$–10^5. Further, it is interesting to see that the theory predicts a maximum of A_2 in toluene at $15.0°$C at $M \simeq 150$, although this has not been confirmed experimentally. In this connection, we note that Sotobayashi and Ueberreiter [49] long ago found experimentally such behavior of A_2 for a-PS in naphthalene at $80.4°$C and obtained its negative value for the dimer. Figure 8.24 shows similar plots

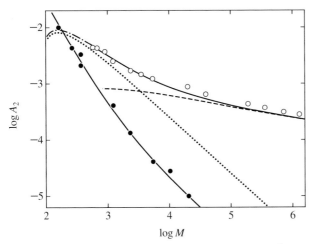

Fig. 8.23. Double-logarithmic plots of A_2 (in cm^3 mol/g^2) against M for a-PS in cyclohexane at $34.5°$C (Θ) ($\bullet$) and in toluene at $15.0°$C ($\circ$) [46]. The *solid and dot-dashed curves* represent the theoretical values of A_2, the *dashed and dotted curves* those of $A_2^{(\mathrm{HW})}$ and $A_2^{(\mathrm{E})}$ for the latter system, respectively

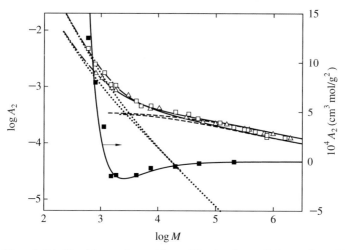

Fig. 8.24. Double- and semi-logarithmic plots of A_2 against M for a-PMMA in acetonitrile at $44.0°$C (Θ) ($\blacksquare$) and in acetone at $25.0°$C ($\square$) [47], and i-PMMA in acetone at $25.0°$C ($\triangle$) [48]. The curves have the same meaning as those in Fig. 8.23

for a-PMMA in acetonitrile at $44.0°$C (Θ) and in acetone at $25.0°$C [47] and i-PMMA in acetone at $25.0°$C [48]. The curves have the same meaning as those in Fig. 8.23. It is interesting to see that $A_{2,\Theta}$ for a-PMMA exhibits a minimum. We also note that Springer and co-workers [50] obtained data for A_2 for a-PMMA in acetone at $25.0°$C which exhibit its maximum at $M \simeq 380$.

Next we examine the behavior of the interpenetration function Ψ. Before doing this, we must first note that Fujita and co-workers [2, 14, 51, 52] were the first to find that for flexible polymers in good solvents Ψ increases from its asymptotic value for large M as M is decreased, and that Huber and Stockmayer [43] pointed out that this may be regarded as arising from chain stiffness. Of course, it should be considered at the present time that the increase in this *apparent* Ψ defined by Eq. (8.104) (with the whole A_2) with decreasing M for small M arises from the effects of chain ends as well as chain stiffness. Since Ψ is now defined for $A_2^{(\mathrm{HW})}$, its experimental values must be calculated from

$$\Psi = \frac{A_2^{(\mathrm{HW})} M^2}{4\pi^{3/2} N_{\mathrm{A}} \langle S^2 \rangle^{3/2}} \tag{8.124}$$

with *experimental* values of $A_2^{(\mathrm{HW})}$ obtained from $A_2^{(\mathrm{HW})} = A_2 - A_2^{(\mathrm{E})}$ with observed values of A_2 and values of $A_2^{(\mathrm{E})}$ calculated for $M \lesssim 10^5$ from Eq. (8.119) with values of $\beta_{2,1}$ and $\beta_{2,2}$ determined. (Note that $A_2^{(\mathrm{HW})} \simeq A_2$ for $M \gtrsim 10^5$.)

Values of Ψ so determined [14, 47, 53] are plotted against α_S^3 in Figs. 8.25 and 8.26 for a-PS and a-PMMA, respectively, where various types of circles

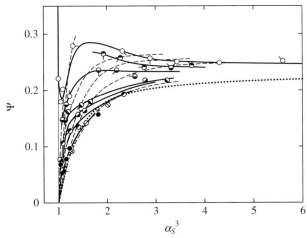

Fig. 8.25. Plots of Ψ against α_S^3 for a-PS; $\circ$, in toluene at 15.0°C; ◓, in n-butyl chloride at 15.0°C; ◒, in 4-*tert*-butyltoluene at 50.0°C; ◔, in cyclohexane (CH) at 55.0°C; ◕, in CH at 50.0°C; ◑, in CH at 45.0°C; ●, in CH at 40.0°C [14,47,53]; and ⊘, in CH at 42.0°C; ⊘, in CH at 38.0°C; ⊗, in CH at 36.0°C [14]. Various directions of pips indicate different values of M. The *solid and dashed curves* connect smoothly the data points at constant B and M, respectively. The *dotted curve* represents the TP theory values

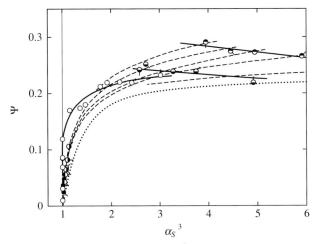

Fig. 8.26. Plots of Ψ against α_S^3 for a-PMMA [47]; $\circ$, in acetone at 25.0°C; ◓, in chloroform at 25.0°C; ◒, in nitroethane at 30.0°C; ◖, in acetonitrile (AN) at 55.0°C; ◗, in AN at 50.0°C; ●, in AN at 47.0°C (see legend for Fig. 8.25)

indicate different solvent conditions (different excluded-volume strength B), and various directions of pips attached to them indicate different values of M.

The solid and dashed curves connect smoothly the data points at constant B and M, respectively. There is semiquantitative agreement between theory and experiment, as seen from a comparison of Figs. 8.25 and 8.26 with Figs. 8.19 and 8.20, respectively. In particular, it is interesting to see that as α_S^3 (or M) is decreased in their respective good solvents of almost the same λB, toluene ($\lambda B = 0.26$) and acetone ($\lambda B = 0.22$), for a-PS Ψ increases steeply (at $\alpha_S = 1$) after passing through a maximum and then a minimum at $\alpha_S^3 \simeq 1$, while for a-PMMA it decreases monotonically and then drops suddenly after reaching a finite value at $\alpha_S = 1$ (except for the oligomers with very small M). Intermediate results have been obtained for i-PMMA [48], but we omit them. All these results indicate that Ψ as a function of α_S^3 depends strongly on chain stiffness and local chain conformation through λB and $\langle S^2 \rangle$.

8.4 Third Virial Coefficient

8.4.1 Perturbation Theory for the Random-Flight Chain

For convenience, we begin by considering the random-flight chain. Recently Norisuye and Nakamura [54] have developed the perturbation theory of the third virial coefficient A_3 for this chain in terms of the ternary-cluster integral (three-segment interaction) β_3 [1] as well as the binary-cluster integral β_2. Consider the chain composed of $n + 1$ identical beads as before. In the superposition approximation [1], A_3 may then be expanded for $n \gg 1$ as

$$A_3 = \frac{N_A^2 n^3}{3M^3}(\beta_3 - I_1\beta_2\beta_3 - I_2\beta_3^2 + J_1\beta_2^2 + J_2\beta_2^2\beta_3$$
$$+ J_3\beta_2\beta_3^2 + J_4\beta_3^3 + \cdots). \tag{8.125}$$

In Eq. (8.125), the leading term involving a single β_3 arises from a *single* contact among the i_1th bead of chain 1, the i_2th bead of chain 2, and i_3th bead of chain 3 [55,56], and J_1 had already been evaluated as [1,57,58]

$$J_1 = n^{-3}\sum P_0(\mathbf{0}_{k_2k_3})_{i_1i_2,j_1j_3} = \lambda_1\left(\frac{3}{2\pi a^2}\right)^{3/2}n^{3/2} \tag{8.126}$$

with $\lambda_1 = 1.664$, where $P_0(\mathbf{0}_{k_2k_3})_{i_1i_2,j_1j_3}$ is the (unperturbed) conditional probability density that when two initial contacts between the i_1th and i_2th beads, and between the j_1th and j_3th beads exist, there also exists an additional contact between the k_2th and k_3th beads [1]. [Note that Norisuye and Nakamura's notation is inappropriate; their $P(\mathbf{0}_{i_1j_2k_3})$ is not the probability density.]

Similarly, I_1, I_2, J_2, J_3, and J_4 (for $n \gg 1$) may be straightforwardly evaluated to be

$$I_1 = 3n^{-3} \sum P_0(\mathbf{0}_{j_1 j_2})_{i_1 i_2, i_1 i_3} = 6C_1 \left(\frac{3}{2\pi a^2}\right)^{3/2} n^{1/2},$$

$$I_2 = 6n^{-3} \sum_{j_1 < k_1} P_0(\mathbf{0}_{j_1 k_1}, \mathbf{0}_{k_1 j_2})_{i_1 i_2, i_1 i_3} = 24C_1 \left(\frac{3}{2\pi a^2}\right)^3 n^{1/2},$$

$$J_2 = 6n^{-3} \sum_{j_1 < k_1} P_0(\mathbf{0}_{j_1 k_1}, \mathbf{0}_{k_1 j_2})_{i_1 i_3, i_2 j_3} = 12\lambda_1 \left(\frac{3}{2\pi a^2}\right)^3 n^{3/2},$$

$$
\begin{aligned}
J_3 = 3n^{-3} \Bigg[&\sum_{\substack{i_1 < j_1 \\ k_1 < l_1}} P_0(\mathbf{0}_{i_1 j_1}, \mathbf{0}_{k_1 l_1}, \mathbf{0}_{l_1 j_2})_{i_1 i_3, i_2 j_3} \\
&+ \sum_{\substack{i_3 < j_3 \\ j_1 < k_1}} P_0(\mathbf{0}_{i_3 j_3}, \mathbf{0}_{j_1 k_1}, \mathbf{0}_{k_1 j_2})_{i_2 i_3, i_1 k_3} \\
&+ \sum_{\substack{i_2 < j_2 \\ j_1 < k_1}} P_0(\mathbf{0}_{i_2 j_2}, \mathbf{0}_{j_1 k_1}, \mathbf{0}_{k_1 k_2})_{i_2 i_3, i_1 j_3} \Bigg]
\end{aligned}
\tag{8.127}
$$

$$= 48\lambda_1 \left(\frac{3}{2\pi a^2}\right)^{9/2} n^{3/2},$$

$$
\begin{aligned}
J_4 = 2n^{-3} \Bigg[3 &\sum_{\substack{i_1 < j_1 \\ k_1 < l_1 \\ j_2 < k_2}} P_0(\mathbf{0}_{i_1 j_1}, \mathbf{0}_{k_1 l_1}, \mathbf{0}_{j_2 k_2}, \mathbf{0}_{k_2 j_3})_{i_1 i_2, k_1 i_3} \\
&+ \sum_{\substack{i_1 < j_1 \\ i_3 < j_3 \\ j_2 < k_2}} P_0(\mathbf{0}_{i_1 j_1}, \mathbf{0}_{i_3 j_3}, \mathbf{0}_{j_2 k_2}, \mathbf{0}_{k_2 k_3})_{i_1 i_2, k_1 i_3} \Bigg]
\end{aligned}
$$

$$= 64\lambda_1 \left(\frac{3}{2\pi a^2}\right)^6 n^{3/2},$$

where the sums have been replaced by integrals, which have been evaluated in the same manner as before [59], and C_1 $(= 2.865)$ is the coefficient of the double-contact term in Eq. (8.84). We note that the intramolecular excluded-volume effect does not affect I_1, I_2, J_2, J_3, and J_4.

From Eq. (8.125) with Eqs. (8.126) and (8.127), we may write A_3 in the form

$$A_3 = \frac{N_A^2 n^3}{3M^3} \left[\beta_3 H_1(z) + n\beta^2 H_2(z)\right] \tag{8.128}$$

with

$$H_1(z) = 1 - 6C_1 z + \mathcal{O}(z^2), \tag{8.129}$$

$$H_2(z) = \lambda_1 z + \mathcal{O}(z^2), \tag{8.130}$$

$$\beta = \beta_2 + 4\left(\frac{3}{2\pi a^2}\right)^{3/2} \beta_3, \tag{8.131}$$

where z is defined by Eq. (8.2) with $\langle R^2 \rangle_0 = na^2$ but with the *effective binary-cluster integral* β defined by Eq. (8.131).

The present evaluation applies to α_R (and perhaps also α_S) and A_2 [54, 56, 59, 60]. The results for α_R and A_2 are given by Eq. (8.52) and Eq. (8.82) with Eq. (8.84), respectively, with the redefined β and z (and with $\tilde{z} = z$ and $Q = C_1$). The Θ temperature is then redefined as the temperature at which A_2 for large M and the effective β vanish.

8.4.2 Effects of Chain Stiffness and Three-Segment Interactions

We consider the HW chain composed of $n+1$ beads, where the two end beads are assumed to be different from the $n - 1$ identical intermediate ones and also from each other in species as in Sect. 8.3.3. Throughout this subsection, all lengths are measured in units of λ^{-1} as before unless otherwise noted. Corresponding to Eq. (8.118) for A_2, A_3 in general may then be written, from Eq. (8.128), in the form

$$A_3 = A_{3,(2)}^{(\mathrm{HW})} + \Delta A_3^{(\mathrm{HW})} + A_{3,(3)}^{(1)} \tag{8.132}$$

with

$$A_{3,(2)}^{(\mathrm{HW})} = (N_{\mathrm{A}}^2 c_\infty^3 L^4 B^2 / 3M^3) H_2(z) \,, \tag{8.133}$$

$$\Delta A_3^{(\mathrm{HW})} = A_3^0 [H_1(z) - 1] \,, \tag{8.134}$$

$$A_{3,(3)}^{(1)} = A_3^0 + A_3^{(\mathrm{E})} \,, \tag{8.135}$$

where z is defined by Eq. (8.5), A_3^0 is given by

$$A_3^0 = \frac{N_{\mathrm{A}}^2 n^3 \beta_3}{3M^3} = \frac{N_{\mathrm{A}}^2 c_\infty^3 L^3 B_3}{3M^3} \tag{8.136}$$

with

$$B_3 = \frac{\beta_3}{a^3 c_\infty^3} \,, \tag{8.137}$$

and $A_3^{(\mathrm{E})}$ represents possible effects of chain ends. Thus A_3 at the Θ temperature, at which $B = z = 0$, is given by

$$A_{3,\Theta} = A_{3,(3)}^{(1)} \,. \tag{8.138}$$

Now it is convenient to introduce a factor g defined by

$$g \equiv A_3 / A_2^2 M$$
$$= g_2 + \Delta g_2 + g_3 \tag{8.139}$$

with

$$g_2 = 4H_2 / 3h^2 = U(L)z + \cdots \,, \tag{8.140}$$

$$\Delta g_2 = \frac{4B_3}{3LB^2h^2}\left[H_1(z) - 1\right], \tag{8.141}$$

$$g_3 = \frac{4B_3}{3LB^2h^2}\left(1 + \frac{A_3^{(E)}}{A_3^0}\right), \tag{8.142}$$

where h is given by Eq. (8.84), and the coefficient $U(L)$ approaches its coil-limiting value $4\lambda_1/3 = 2.219$ as L is increased. It is seen that Δg_2 and g_3 decrease as L (or M) is increased; the contribution of Δg_2 is smaller than that of g_3.

The coefficient $U(L)$ may be evaluated on the basis of the KP chain by a method similar to that in the case of the coefficient $Q(L)$ in A_2. This has been done by Norisuye and co-workers [61] as follows,

$$U(L) = \frac{32}{3}L^{-9/2}\int_0^L\int_0^L\int_0^L (L - s_1)(L - s_2)(L - s_3)$$
$$\times F(s_1, s_2, s_3)ds_1ds_2ds_3 \tag{8.143}$$

with

$$F(s_1, s_2, s_3) = (3/2\pi)^{-3/2}\int G(\mathbf{R}_1; s_1)G(\mathbf{R}_2; s_2)$$
$$\times G(\mathbf{R}_2 - \mathbf{R}_1; s_3)d\mathbf{R}_1 d\mathbf{R}_2 . \tag{8.144}$$

For simplicity, we give only the result for $L \gg 1$ and $d = 0.3$, that is,

$$U(L) = 2.219\left(1 + \frac{3.143}{L} - \frac{5.953}{L^{3/2}} + \cdots\right). \tag{8.145}$$

It is seen that the first-order deviation of $U(L)$ from its coil-limiting value is of order L^{-1}, so that the effect of chain stiffness on A_3 is less significant than that on A_2.

As in the case of the h function, we assume that g_2 is a function only of a certain excluded-volume parameter. In this case it is the parameter $\check{z}$ defined by

$$\check{z} = \left[\frac{U(L)}{2.219}\right]\bar{z}, \tag{8.146}$$

where $\bar{z}$ is given by Eq. (8.106). The functional form of $g_2(\check{z})$ to be combined with Eq. (8.57) may then be determined in such a way that in the coil limit values of $g_2(\bar{z})$ are as close as possible to those of Stockmayer and Casassa [62]. The result so obtained reads [61]

$$g_2(\check{z}) = 2.219\,\check{z}\,(1 + 18\,\check{z} + 12.6\,\check{z}^2)^{-0.5}. \tag{8.147}$$

Finally, we make a comparison of theory with experiment with respect to the factor g. Figure 8.27 shows plots of g against α_S^3 for a-PS for $M > 10^4$ in benzene at 25.0°C [63, 64], where the data for A_3 were obtained from light scattering measurements with the use of the Bawn plot [65, 66]. In this

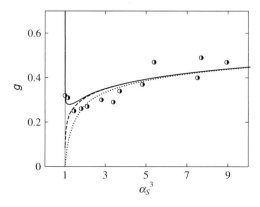

Fig. 8.27. Plots of g against α_S^3 for a-PS in benzene at 25.0°C [63,64]. The *solid curve* represents the theoretical values of $g = g_2 + g_3$ with $A_3^{(E)} = 0$, the *dashed curve* those of g_2, and the *dotted curve* the TP theory values of g_2 with $\check{z} = \bar{z}$ (see the text)

range of M (strictly for $M \gtrsim 10^5$), $A_3^{(E)}$ for a-PS may be neglected, as seen below, and Δg_2 may also be suppressed, as mentioned above. In the figure the solid curve represents the theoretical values calculated from Eq. (8.139) with Eqs. (8.142) and (8.147) (with $\Delta g_2 = A_3^{(E)} = 0$) with the values of the HW model parameters along with $B = 0.33$ (given in Table 8.3) and $B_3 = 0.031$. We note that the value of B_3 has been determined from the second of Eqs. (8.136) with the experimental value 4×10^{-4} cm^6 mol/g^3 of A_3^0 in cyclohexane at 34.5°C (Θ) [67], assuming that the solvent dependence of β_3 is small. The dashed curve represents the theoretical values of g_2 from Eq. (8.147), and the dotted curve represents its TP theory values with $\check{z} = \bar{z}$. Clearly the nonvanishing of g near the Θ temperature (at $\alpha_S \simeq 1$) arises from A_3^0, the effect of chain stiffness being of little significance there. In this connection, we note that earlier experimental studies gave $A_3 = 0$ at the Θ temperature, but it was later shown for several flexible polymers that this is not the case [66–70].

8.4.3 Effects of Chain Ends

In this subsection we evaluate the effects of chain ends on A_3, that is, the term $A_3^{(E)}$. As in the case of A_2, we take into account the effects only on *single*-contact terms [71]. From the general formulation of A_3 [1, 55], $A_{3,(3)}^{(1)}$ for the present model may be written in the form

$$A_{3,(3)}^{(1)} = (N_A^2/3M^3) \sum \beta_{3,klm} , \qquad (8.148)$$

corresponding to Eq. (8.112). In this case there are ten kinds of ternary clusters. We define *excess* ternary-cluster integrals β_{klm} by

$$\beta_{3,klm} = \beta_3 + \beta_{klm} \qquad (8.149)$$

with

$$\beta_3 \equiv \beta_{3,000} \,. \tag{8.150}$$

Further, we define *effective* excess ternary-cluster integrals $\beta_{3,1}$, $\beta_{3,2}$, and $\beta_{3,3}$ associated with the chain end beads by the equations

$$
\begin{aligned}
2\beta_{3,1} &= \beta_{001} + \beta_{002} \,, \\
4\beta_{3,2} &= \beta_{011} + 2\beta_{012} + \beta_{022} \,, \\
8\beta_{3,3} &= \beta_{111} + 3\beta_{112} + 3\beta_{122} + \beta_{222} \,.
\end{aligned}
\tag{8.151}
$$

Then Eq. (8.148) reduces to Eq. (8.135) with $A_3^{(E)}$ given by

$$A_3^{(E)} = a_{3,1} M^{-1} + a_{3,2} M^{-2} + a_{3,3} M^{-3} \,, \tag{8.152}$$

where

$$
\begin{aligned}
a_{3,1} &= 2 N_A^2 \beta_{3,1}/M_0^2 \,, \\
a_{3,2} &= 4 N_A^2 \Delta\beta_{3,2}/M_0 \,, \\
a_{3,3} &= \frac{8}{3} N_A^2 \Delta\beta_{3,3}
\end{aligned}
\tag{8.153}
$$

with

$$
\begin{aligned}
\Delta\beta_{3,2} &= \beta_{3,2} - 2\beta_{3,1} \,, \\
\Delta\beta_{3,3} &= \beta_{3,3} - 3\beta_{3,2} + 3\beta_{3,1} \,.
\end{aligned}
\tag{8.154}
$$

We make a comparison of theory with experiment with respect to $A_{3,\Theta} = A_{3,(3)}^{(1)}$. Figure 8.28 shows plots of $A_{3,\Theta}$ against $\log M$ for a-PS in cyclohexane at 34.5°C [53, 67] and a-PMMA in acetonitrile at 44.0°C [71], where the data were obtained from light scattering measurements with the use of the Bawn plot along with the procedure in Appendix 8.B for the oligomers. It is seen that $A_{3,\Theta}$ becomes A_3^0 independent of M for $M \gtrsim 10^4$ for a-PS and for $M \gtrsim 10^5$ for a-PMMA. The values of A_3^0 thus determined are 4.7×10^{-4}

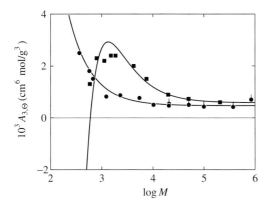

Fig. 8.28. Plots of $A_{3,\Theta}$ against $\log M$ for a-PS in cyclohexane at 34.5°C ($\bullet$, $\bullet$) [53,67] and a-PMMA in acetonitrile at 44.0°C ($\blacksquare$) [71]. The *solid curves* represent the respective best-fit theoretical values (see the text)

and 5.8×10^{-4} cm^6 mol/g^3, which give β_3 (per repeat unit) $= 4.4 \times 10^{-45}$ and 4.8×10^{-45} cm^6, for a-PS and a-PMMA, respectively. The solid curves represent the respective best-fit theoretical values calculated from Eq. (8.135) with Eq. (8.152) and with these values of A_3^0. The values of $\beta_{3,1}$, $\beta_{3,2}$, and $\beta_{3,3}$ (per repeat unit) thus obtained are 1.2×10^4, 2.4×10^4, and 3.6×10^4 Å^6 for a-PS, and 9.1×10^4, -1.6×10^5, and -2.4×10^5 Å^6 for a-PMMA, respectively. It is interesting to see that both theoretically and experimentally $A_{3,\Theta}$ exhibits a maximum for a-PMMA.

8.5 Some Remarks

8.5.1 Near the Θ Temperature

The α_S given by Eq. (8.57) has a singularity at $\tilde{z} = -0.1446$, and it cannot be applied to the range of negative $\tilde{z}$ far below the Θ temperature. Similarly, the function h given by Eq. (8.110) cannot be used for $z < 0$ since it has a singularity at $z = 0$. In the following discussion of α_S and A_2 near the Θ temperature, therefore, we tentatively adopt the perturbation theory. As is well known, for α_S it reads [1]

$$\alpha_S^2 = 1 + 1.276\,\tilde{z} - 2.082\,\tilde{z}^2 + \cdots . \tag{8.155}$$

As for h, if we simply assume that the expansion factor for each of the two chains in contact is also given by a function only of $\tilde{z}$, then the corresponding expansion of h may be given, from the conventional TP perturbation theory [1, 72], by [73]

$$h = 1 - 2.865\,\tilde{z} + 8.851\,\tilde{z}^2 + 5.077\,\tilde{z}\tilde{z} - \cdots . \tag{8.156}$$

Now, the parameter B (excluded-volume strength) may be rather accurately determined from α_S in non-Θ or good solvents, as done in Sect. 8.1.4. Near the Θ temperature, however, this determination becomes ambiguous since α_S is close to unity; it should then be determined from the single-contact term of $A_2^{(HW)}$ as follows [74]. In the oligomer region where the relation $h = 1$ holds, $A_2^{(HW)}$ is independent of M, so that we have, from Eq. (8.118) with Eq. (8.119),

$$(A_{2,i} - A_{2,j})/(M_i^{-1} - M_j^{-1}) = a_{2,1} + a_{2,2}(M_i^{-1} + M_j^{-1}), \tag{8.157}$$

where $A_{2,i}$ and $A_{2,j}$ are the second virial coefficients for different molecular weights M_i and M_j, respectively. Equation (8.157) indicates that $a_{2,1}$ and $a_{2,2}$ may be determined from the intercept and slope of the plot of the quantity on its left-hand side against $M_i^{-1} + M_j^{-1}$, respectively. Figure 8.29 shows plots of A_2 against M^{-1} for a-PS in such an oligomer region in cyclohexane at 15.0, 30.0, 34.5, and 50.0°C [74]. From the plots, we can determine $A_2^{(HW)}$

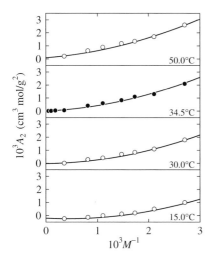

Fig. 8.29. Plots of A_2 against M^{-1} for a-PS in cyclohexane at the indicated values of T for the determination of B from the intercept [74] (see the text)

with $h = 1$ at each temperature so that the curve of A_2 as a function of M^{-1} calculated from Eq. (8.118) with Eq. (8.119) with these values of $A_2^{(\mathrm{HW})}$ (with $h = 1$), $a_{2,1}$, and $a_{2,2}$ gives a best fit to the data points. The solid curves in the figure represent the values so calculated. The intercept of each curve is then equal to $A_2^{(\mathrm{HW})}$ (with $h = 1$), that is, the prefactor (single-contact term) $(N_A c_\infty^{3/2} L^2 B / 2M^2)$, from which we can determine B at the corresponding temperature. The results thus obtained for β (in Å^3) per repeat unit for a-PS in cyclohexane [74] and also for a-PMMA in acetonitrile [75] are given by

$$\beta = 65\tau \qquad \text{for } \tau \geq 0$$
$$= 65\tau - 610\tau^2 \quad \text{for } \tau < 0 \quad \text{(a-PS)}, \tag{8.158}$$

$$\beta = 35\tau \qquad \text{(a-PMMA)} \tag{8.159}$$

with

$$\tau = 1 - \Theta/T. \tag{8.160}$$

We can then calculate z from Eq. (8.5) with Eq. (8.6) ($a = M_0/M_L$) and Eqs. (8.158)–(8.160) for a-PS and a-PMMA in the respective Θ solvents, and also calculate $A_2^{(\mathrm{E})}$ (with the above values of $a_{2,1}$ and $a_{2,2}$) to obtain *experimental* values of $A_2^{(\mathrm{HW})}$ for all values of M.

Values of $A_2^{(\mathrm{HW})} M^{1/2}$ ($A_2^{(\mathrm{HW})}$ in cm^3 mol/g^2) so obtained are plotted against the above-calculated z in Fig. 8.30 for a-PS (for various values of M) in cyclohexane near the Θ temperature [14, 74, 76]. The dashed straight line represents the theoretical values calculated from [73]

$$A_2^{(\mathrm{HW})} M^{1/2} = 0.323zh \quad \text{(a-PS)} \tag{8.161}$$

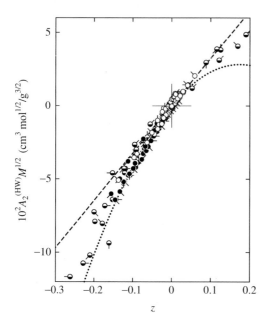

Fig. 8.30. Plots of $A_2^{(\mathrm{HW})} M^{1/2}$ against z for a-PS in cyclohexane near the Θ temperature ($\ominus$,$\bullet$,$\circ$) [14,74,76]. Various directions of pips indicate different values of M. The *dashed straight line* and the *dotted curve* represent the values with $h = 1$ and the first-order TP perturbation theory values, respectively (see the text)

with $h = 1$, and the dotted curve represents the values calculated from Eq. (8.161) with the first-order TP perturbation theory of h given by Eq. (8.156) with $\tilde{z} = \tilde{\tilde{z}} = z$ (that is, $h = 1 - 2.865z$). It is seen that all the data points nearly form a single-composite curve, indicating that the TP scheme is valid for $A_2^{(\mathrm{HW})}$ below Θ, the effect of chain stiffness on A_2 being of little significance there. The single-composite curve, although not explicitly shown, is located between the dashed and dotted curves and rather close to the latter. From the M independence of A_2 as a function of $|\tau|$ for a-PS (except for small $M < 5 \times 10^3$) below Θ, Fujita and co-workers [2,76] claimed that the TP theory of A_2 breaks down below Θ. However, it is now evident that their deduction is in error; this arises mainly from their assumption of the first line of Eqs. (8.158) for β below Θ in the theoretical calculation of A_2. The above independence for a-PS (for $M > 5 \times 10^3$) is due to a cancelation of the M dependence of $A_2^{(\mathrm{HW})}$ by that of $A_2^{(\mathrm{E})}$. A similar analysis has been made also for a-PMMA [75], for which the TP scheme is still valid for $A_2^{(\mathrm{HW})}$ below Θ and for which A_2 depends appreciably on M even there in contrast to the case of a-PS.

Next we examine the behavior of α_S below Θ in relation to the problem of the so-called coil-to-globule transition [77]. Before doing this, one remark should be made. It is now known that there are two types of the transition, that is, the gradual and sharp transitions, and that the former is observed in the stable state of the test solution, while the latter may be due to the metastable state, as claimed by Chu's group [78–80]. Thus the present analy-

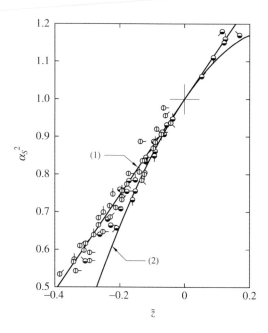

Fig. 8.31. Plots of α_S^2 against $\tilde{z}$ for a-PS in cyclohexane near the Θ temperature ($\bullet$,$\oplus$) [14,78,80]. Various directions of pips indicate different values of M. The *straight line* (1) and the *curve* (2) represent the first- and second-order perturbation theory values, respectively

sis is confined to the former case. Figure 8.31 shows plots of α_S^2 against $\tilde{z}$ for a-PS (for $M > 10^6$) in cyclohexane near the Θ temperature [14,78,80], where $\tilde{z}$ may be equated to z (because of large M) and the latter has been calculated above. The straight line (1) and the curve (2) represent the first- and second-order (TP) perturbation theory values calculated from Eq. (8.155), respectively. All the data points form a single-composite curve within experimental error, indicating that the TP scheme is valid for α_S (for large M) even below Θ, as was expected. This conclusion must be correct for the *stable* solution of *flexible* polymers (except for biological macromolecules with specific intramolecular interactions). The single-composite curve, although not explicitly shown, is located between the lines (1) and (2) and rather close to the former.

Historically, the phenomenon called the coil-to-globule transition was first suggested by Stockmayer [57] in 1960, and then Ptitsyn and co-workers [81] in 1968 were the first to treat it theoretically by taking account of β_3 as well as β_2 in the smoothed-density model [1]. In fact, Orofino and Flory [82] in 1957 had already presented such a smoothed-density (or mean-field) theory with consideration of both cluster (segment) interactions. Subsequently, following Ptitsyn, many theoreticians [77,83,84] have pursued this line to treat the coil-to-globule transition. The corresponding theory of A_2 was also developed by Orofino and Flory [82] and by Tanaka [85]. However, all these treatments lead to a non-TP theory, which cannot explain the well-established experimental results [66,67,73], in contrast to the above TP (or QTP) theory (Sect. 8.4.1)

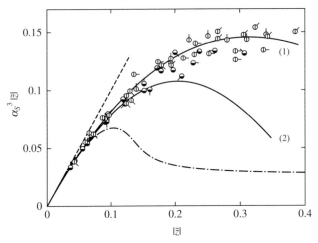

Fig. 8.32. Plots of $\alpha_S^3\,|\tilde{z}|$ against $|\tilde{z}|$ with the same data as those in Fig. 8.31. The *dashed straight line* indicates the initial slope of unity. The *solid curves* (1) and (2) represent the first- and second-order perturbation theory values, respectively, and the *dot-dashed curve* the values calculated from Eq. (8.162) with $z = \tilde{z}$ and $y = 0.07$

developed on the basis of the random-flight (or HW) chain taking account of the chain connectivity and also β_3 through the effective β.

In order to clarify the point further, values of $\alpha_S^3|\tilde{z}|$ are plotted against $|\tilde{z}|$ in Fig. 8.32 with the same data as those in Fig. 8.31 (except for those above Θ). The dashed straight line indicates the initial slope of unity, and the solid curves (1) and (2) correspond to those in Fig. 8.31. The dot-dashed curve represents the values calculated from the mean-field theory equation [2],

$$\alpha_S^5 - \alpha_S^3 - y\,\alpha_S^{-3} = 2.60\,z \tag{8.162}$$

with $z = \tilde{z}$ and $y = 0.07$, where y is a parameter proportional to β_3. Necessarily, the data points are located near the solid curve (1), corresponding to the results in Fig. 8.31. It is important to note that the so-called globule state (plateau region), $\alpha_S^3|z| = \text{const.}$ (for $\alpha_S \ll 1$), as predicted by Eq. (8.162) can never be observed for such stable solutions.

Finally, brief mention is made of A_3 near the Θ temperature. Equation (8.128) for A_3 predicts that for $\beta_3 > 0$, $\beta_3 H_1$ is positive for $T < \Theta$ and decreases with increasing T if β_3 is independent of T, while $n\beta^2 H_2$ increases with increasing T, indicating that A_3 as a function of T exhibits a positive minimum near the Θ temperature. This can be explicitly shown by assuming, for instance, Casassa–Markovitz-type equations [1] for H_1 and H_2 [54] and is in fact consistent with the experimental finding [67, 86].

8.5.2 More on Three-Segment Interactions

It can be shown that if the formulation for the random-flight chain with binary- and ternary-cluster interactions [59] is extended to the HW chain, Eq. (8.131) for the effective β may be replaced by

$$\beta = \beta_2 + C(\lambda a)^2(\beta_3/a^3) \tag{8.163}$$

(with lengths unreduced by λ^{-1}), where C are constants (dependent on M) at most of order unity but different in α_S and A_2 [73]. In the random-flight-chain limit of $\lambda a = 1$, Eq. (8.163) with $C = 4(3/2\pi)^{3/2}$ reduces to Eq. (8.131). On the other hand, in the stiff-chain limit of $\lambda a = 0$, the β_3 term may be completely neglected, so that $\beta = \beta_2$. This result is reasonable since the probability density for multiple contact between beads associated with β_3 is much smaller than that for the corresponding single contact associated with β_2. For flexible chains, the β_3 term, although not completely negligible, is rather small compared to the β_2 term, and may be neglected in the oligomer region (for small λL) because of chain stiffness, so that C and hence β must depend somewhat on L (or M). This may affect the M dependence of $A_2^{(E)}$.

Another factor may possibly affect the M dependence of $A_2^{(E)}$. Corresponding to Eq. (8.163), the expression for A_2 [56,67] derived for the random-flight chain with consideration of β_3 with the first-order deviation from the coil limit ($n \to \infty$) may be replaced by

$$A_2 = \frac{N_A n^2}{2M^2}\left[\beta - C'(\lambda a)^2\left(\frac{\beta_3}{a^3}\right)(\lambda L)^{-1/2} + \cdots\right], \tag{8.164}$$

where C' is a constant (dependent on M). This equation predicts that as M is decreased for $\beta_3 > 0$ (as in the case of a-PS), $A_{2,\Theta}$ (with $\beta = 0$) first decreases from zero and finally becomes zero in the oligomer region because of chain stiffness. However, higher-order corrections are required for small λL, and therefore Eq. (8.164) is incomplete. The mean-field theory equation for A_2 is also inadequate for the reason mentioned above, although it predicts that $A_{2,\Theta}$ increases with decreasing M for $\beta_3 > 0$.

Appendix 8.A Mean-Square Electric Dipole Moment

Experimentally, Marchal and Benoit [87] first showed that there is no excluded-volume effect on the mean-square electric dipole moment $\langle\mu^2\rangle$ for the chain having type-B (perpendicular) dipoles like polyoxyethyleneglycol and diethoxy polyethyleneglycol. On the theoretical side, Nagai and Ishikawa [88] and subsequently Doi [89] supported this conclusion on the basis of the Gaussian chain, that is, $\alpha_\mu = 1$ if $\langle \mathbf{R} \cdot \boldsymbol{\mu}\rangle_0 = 0$ with $\mathbf{R}$ the end-to-end vector distance and $\boldsymbol{\mu}$ the instantaneous (total) electric dipole moment vector. However, Mattice and Carpenter [90] have reported a Monte Carlo result in contradiction to the above conclusion on the basis of the RIS model; that is, α_μ is not equal to unity for the type-B chain of finite length, and moreover, it does not become unity even in the limit of $L \to \infty$. Mansfield [91] has then clarified that their result is due to the non-Gaussian nature of the chain, although not completely molecular-theoretically.

Thus, in this appendix we evaluate α_μ^2 (only its first-order perturbation coefficient) on the basis of the HW chain [92]. All lengths are measured in units of λ^{-1} as usual, and the same notation as that in Sect. 5.4.1 is used. By the use of Eq. (5.153), $\langle \mu^2 \rangle$ may be written in the form

$$\langle \mu^2 \rangle = \int_0^L \int_0^L \langle \tilde{\mathbf{m}}(t_1) \cdot \tilde{\mathbf{m}}(t_2) \rangle dt_1 dt_2 \,. \tag{8.A.1}$$

Corresponding to Eq. (8.4) with Eqs. (8.9) and (8.10) for α_R^2, the first-order perturbation theory of α_μ^2 may then be given by

$$\alpha_\mu^2 = 1 + K_\mu(L) z + \cdots \tag{8.A.2}$$

with

$$K_\mu(L) = \frac{F_\mu(L)}{L^{1/2} \langle \mu^2 \rangle_0} \,, \tag{8.A.3}$$

where

$$F_\mu(L) = \left(\frac{2\pi c_\infty}{3} \right)^{3/2} \int_0^L ds_1 \int_{s_1}^L ds_2 \Big\{ G(\mathbf{0}; s) \langle \mu^2 \rangle_0$$
$$- \int_0^L dt_1 \int_0^L dt_2 \int \left[\tilde{\mathbf{m}}(t_1) \cdot \tilde{\mathbf{m}}(t_2) \right] P_0(\Omega_1, \Omega_2, \mathbf{0}_{s_1 s_2}; L) d\Omega_1 d\Omega_2 \Big\} \tag{8.A.4}$$

with $\Omega_i = \Omega(t_i)$ $(i = 1, 2)$.

The asymptotic solution in the limit of $L \to \infty$ is then found analytically to be

$$\langle \mu^2 \rangle_0 = \left[\frac{4m^2 + (\kappa_0 m_\eta + \tau_0 m_\zeta)^2}{4 + \kappa_0^2 + \tau_0^2} \right] L + \mathcal{O}(L^0) \,, \tag{8.A.5}$$

$$K_\mu(L) = \frac{4}{3} \left[\frac{4 + \tau_0^2}{4m^2 + (\kappa_0 m_\eta + \tau_0 m_\zeta)^2} \right] \left(\frac{\kappa_0 \tau_0 m_\eta}{4 + \tau_0^2} + m_\zeta \right)^2 + \mathcal{O}(L^{-1/2}) \,. \tag{8.A.6}$$

For the HW chain having type-A (parallel) dipoles ($m_\xi = m_\eta = 0$), in the coil limit $K_\mu(L)$ is equal to 4/3, and therefore $\alpha_\mu = \alpha_R$, as seen from Eq. (8.A.6). For the HW chain having type-B dipoles ($m_\zeta = 0$), Eq. (8.A.6) reduces to

$$K_\mu(L) = \frac{4\kappa_0^2 \tau_0^2 m_\eta^2}{3(4m^2 + \kappa_0^2 m_\eta^2)(4 + \tau_0^2)} + \mathcal{O}(L^{-1/2}) \qquad (B). \tag{8.A.7}$$

It is seen from Eq. (8.A.7) that $K_\mu \neq 0$ if $m_\eta \neq 0$ and $\kappa_0 \tau_0 \neq 0$, so that α_μ then becomes infinitely large in the limit of $L \to \infty$. Such dependence of α_μ on L has not been pointed out by Mattice and Carpenter [90] and by Mansfield [91]. Further, this does not conflict with the above-mentioned result [88,89] for the Gaussian chain since the HW chain does not necessarily satisfy the condition $\langle \mathbf{R} \cdot \boldsymbol{\mu} \rangle_0 = 0$ even in the case of perpendicular dipoles [92]. It must also be noted that α_μ may possibly become a constant different from unity because of the term of order $L^{-1/2}$ in $K_\mu(L)$ if $\kappa_0 \tau_0 \neq 0$ for the type-B chain. This corresponds to the case pointed out by Mattice and Carpenter and by Mansfield.

Appendix 8.B Determination of the Virial Coefficients for Oligomers

For an accurate experimental determination of the (osmotic) second and third virial coefficients A_2 and A_3 for oligomers, light scattering measurements are preferable. Then, however, measurements must be carried out generally for optically anisotropic and rather concentrated solutions, and necessarily several problems are encountered. In this appendix we resolve them and present a procedure suitable for the present purpose [93].

We consider a binary solution which in general is optically anisotropic and not necessarily dilute. Let R^*_{Uv} be the reduced intensity of unpolarized scattered light for vertically polarized incident light, and let R^*_θ be the Rayleigh ratio, where the asterisk indicates the scattering from anisotropic scatterers, it being dropped for the isotropic scattering. The (isotropic) Rayleigh ratio $R_{\theta=0}$ at vanishing scattering angle θ, which is the first desired quantity, is obtained from

$$R_{\theta=0} = \left(1 - \tfrac{7}{6}\rho_\mathrm{u}\right) R^*_{\mathrm{Uv}}, \tag{8.B.1}$$

or

$$R_{\theta=0} = (1 + \rho_\mathrm{u})^{-1} \left(1 - \tfrac{7}{6}\rho_\mathrm{u}\right) R^*_{\theta=\pi/2} \tag{8.B.2}$$

with the observed R^*_{Uv} or $R^*_{\theta=\pi/2}$, where ρ_u is the depolarization ratio as defined as the ratio $I_{\mathrm{Hu}}/I_{\mathrm{Vu}}$ of the horizontal to vertical component of the scattered intensity at $\theta = \pi/2$ for unpolarized incident light. ρ_u may be determined from [94]

$$(1 + \cos^2 \theta) \frac{R^*_\theta}{R^*_{\theta=\pi/2}} = 1 + \left(\frac{1 - \rho_\mathrm{u}}{1 + \rho_\mathrm{u}}\right) \cos^2 \theta \tag{8.B.3}$$

with the observed R^*_θ. Note that these equations can readily be derived from the basic equations for I_{fi} in Sect. 5.3.2.

Now, according to the fluctuation theory [1,95,96], $R_{\theta=0}$ may be written in the form

$$R_{\theta=0} = R_\mathrm{d} + \Delta R_{\theta=0} \tag{8.B.4}$$

with R_d and $\Delta R_{\theta=0}$ being the density scattering (the Einstein–Smoluchowski term) and the composition scattering, respectively, and given by

$$R_d = \frac{2\pi^2 \tilde{n}^2 k_B T}{\lambda_0^4 \kappa_T} \left(\frac{\partial \tilde{n}}{\partial p} \right)_{T,m}^2 , \qquad (8.B.5)$$

$$\Delta R_{\theta=0} = -\frac{2\pi^2 \tilde{n}^2 k_B T V_0 c}{\lambda_0^4} \left(\frac{\partial \tilde{n}}{\partial c} \right)_{T,p}^2 \Big/ \left(\frac{\partial \mu_0}{\partial c} \right)_{T,p} , \qquad (8.B.6)$$

where λ_0 is the wavelength of the incident light in vacuum, κ_T the isothermal compressibility of the solution, $\tilde{n}$ the refractive index of the solution, p the pressure, m the ratio of the solute to solvent mass, V_0 the partial molecular volume of the solvent, c the mass concentration of the solution, and μ_0 the chemical potential of the solvent. We note that the molecular-theoretical basis of the term R_d has been given correctly by Fixman [97], and that the multiple scattering theory developed by Bullough [98] is in error [93].

We first rewrite Eq. (8.B.6). Under the osmotic condition, the chemical potential $\mu_0^0(T,p)$ of the pure solvent is equated to $\mu_0(T, p + \Pi, c)$ with Π the osmotic pressure, so that we have

$$\mu_0^0(T,p) = \mu_0(T,p,c) + \left(\frac{\partial \mu_0}{\partial p} \right)_{T,c} \Pi$$
$$+ \frac{1}{2} \left(\frac{\partial^2 \mu_0}{\partial p^2} \right)_{T,c} \Pi^2 + \frac{1}{3} \left(\frac{\partial^3 \mu_0}{\partial p^3} \right)_{T,c} \Pi^3 + \cdots . \qquad (8.B.7)$$

Differentiation of both sides of Eq. (8.B.7) with respect to c at constant T and p leads to

$$\left(\frac{\partial \mu_0}{\partial c} \right)_{T,p} = -V_0 \left(\frac{\partial \Pi}{\partial c} \right)_{T,p} - \Pi \left(\frac{\partial V_0}{\partial c} \right)_{T,p}$$
$$- \frac{1}{2} \Pi^2 \left(\frac{\partial^2 V_0}{\partial p \partial c} \right)_T - \Pi \left(\frac{\partial V_0}{\partial p} \right)_{T,c} \left(\frac{\partial \Pi}{\partial c} \right)_{T,p} \qquad (8.B.8)$$
$$- \frac{1}{3} \Pi^3 \left(\frac{\partial^3 V_0}{\partial p^2 \partial c} \right)_T - \Pi^2 \left(\frac{\partial^2 V_0}{\partial p^2} \right)_{T,c} \left(\frac{\partial \Pi}{\partial c} \right)_{T,p} + \cdots ,$$

where we have used the relation $(\partial \mu_0 / \partial p)_{T,c} = V_0$. In general, Π and V_0 may be expanded in powers of c as follows,

$$\frac{\Pi}{RT} = \frac{1}{M} c + A_2 c^2 + A_3 c^3 + \cdots , \qquad (8.B.9)$$

$$V_0 = V_0^0 \left[1 - \frac{1}{2} \left(\frac{\partial v_1}{\partial c} \right)_{T,p,0} c^2 + \cdots \right] , \qquad (8.B.10)$$

where R is the gas constant, V_0^0 the molecular volume of the pure solvent, v_1 the partial specific volume of the solute, and the subscript 0 on the derivative indicates its value at $c = 0$.

Substitution of Eq. (8.B.8) with Eqs. (8.B.9) and (8.B.10) into Eq. (8.B.6) leads to

$$\frac{Kc}{\Delta R_{\theta=0}} = \frac{1}{M} + 2A_2'c + 3A_3'c^2 + \cdots \tag{8.B.11}$$

with

$$K = \frac{2\pi^2 \tilde{n}^2}{N_A \lambda_0^4} \left(\frac{\partial \tilde{n}}{\partial c}\right)_{T,p}^2, \tag{8.B.12}$$

$$A_2 = A_2' + \frac{RT\kappa_{T,0}}{2M^2}, \tag{8.B.13}$$

$$A_3 = A_3' + \frac{1}{3M}\left(\frac{\partial v_1}{\partial c}\right)_{T,p,0} + \frac{RT\kappa_{T,0}A_2}{M} + \frac{RT}{2M^2}\left(\frac{\partial \kappa_T}{\partial c}\right)_{T,0}$$
$$+ \frac{(RT)^2}{3M^3}\left[\kappa_{T,0}^2 - \left(\frac{\partial \kappa_T}{\partial p}\right)_{T,0}\right], \tag{8.B.14}$$

where $\kappa_{T,0}$ is the isothermal compressibility of the pure solvent. Thus the desired virial coefficients A_2 and A_3 may be obtained from Eqs. (8.B.13) and (8.B.14) with the observed light-scattering virial coefficients A_2' and A_3', which are different from the former except for large M. We note that Eq. (8.B.13) is equivalent to a relation derived by Casassa and Eisenberg [99].

Next we consider the problem of determining R_d at *finite* concentrations, although indirectly. For this purpose, we adopt the Lorentz–Lorenz relation between $\tilde{n}$ and the solution density ρ_w [100],

$$\frac{\tilde{n}^2 - 1}{\tilde{n}^2 + 1} = \text{const.}\ \rho_w, \tag{8.B.15}$$

where we assume that the proportionality constant is independent of p. Equation (8.B.15) has been shown to be the best of such relations [93]. Differentiation of both sides of Eq. (8.B.15) with respect to p leads to

$$\kappa_T^{-1}\left(\frac{\partial \tilde{n}}{\partial p}\right)_{T,m} = \frac{(\tilde{n}^2 - 1)(\tilde{n}^2 + 2)}{6\tilde{n}}. \tag{8.B.16}$$

Substituting Eq. (8.B.16) into Eq. (8.B.5), we obtain

$$R_d = \frac{\kappa_T(\tilde{n}^2 - 1)^2(\tilde{n}^2 + 2)^2}{\kappa_{T,0}(\tilde{n}_0^2 - 1)^2(\tilde{n}_0^2 + 2)^2}R_{d,0}, \tag{8.B.17}$$

where $\tilde{n}_0$ and $R_{d,0}$ are the values of $\tilde{n}$ and R_d for the pure solvent, respectively.

Thus we may calculate R_d from Eq. (8.B.17) with the observed $R_{d,0}$, and then determine $\Delta R_{\theta=0}$ from Eq. (8.B.4) with this R_d and the observed $R_{\theta=0}$. Finally, we may determine M, A_2, and A_3 from Eq. (8.B.11) with Eqs. (8.B.12)–(8.B.14) by the use of the Berry square-root plot [101] or the Zimm plot [102] and also the Bawn plot [65, 66]. For the evaluation of the optical constant K given by Eq. (8.B.12), note that we must use values of $\tilde{n}$

and $(\partial \tilde{n}/\partial c)_{T,p}$ at finite concentrations c. For example, the results obtained for toluene (solute) in cyclohexane (solvent) at $25.0°C$ are $M = 93 \pm 4$ and $A_2 = 1.5 \times 10^{-3}$ cm^3 mol/g^2 (with $RT\kappa_{T,0}/2M^2 = 1.65 \times 10^{-4}$ cm^3 mol/g^2) [93]. (Note that the true M of toluene is 92.)

References

1. H. Yamakawa: *Modern Theory of Polymer Solutions* (Harper & Row, New York, 1971).
2. H. Fujita: *Polymer Solutions* (Elsevier, Amsterdam, 1990).
3. H. Yamakawa and W. H. Stockmayer: J. Chem. Phys. **57**, 2843 (1972).
4. H. Yamakawa and J. Shimada: J. Chem. Phys. **83**, 2607 (1985).
5. J. Shimada and H. Yamakawa: J. Chem. Phys. **85**, 591 (1986).
6. Z. Y. Chen and J. Noolandi: J. Chem. Phys. **96**, 1540 (1992).
7. F. Abe, Y. Einaga, T. Yoshizaki, and H. Yamakawa: Macromolecules **26**, 1884 (1993).
8. G. Weill and J. des Cloizeaux: J. Phys. (Paris) **40**, 99 (1979).
9. C. Domb and A. J. Barrett: Polymer **17**, 179 (1976).
10. P. J. Flory: *Statistical Mechanics of Chain Molecules* (Interscience, New York, 1969).
11. F. Abe, K. Horita, Y. Einaga, and H. Yamakawa: Macromolecules **27**, 725 (1994).
12. M. Kamijo, F. Abe, Y. Einaga, and H. Yamakawa: Macromolecules **28**, 1095 (1995).
13. K. Horita, F. Abe, Y. Einaga, and H. Yamakawa: Macromolecules **26**, 5067 (1993).
14. Y. Miyaki: Ph. D. Thesis (Osaka Univ., Osaka, Japan, 1981).
15. T. Arai, F. Abe, T. Yoshizaki, Y. Einaga, and H. Yamakawa: Macromolecules **28**, 3609 (1995).
16. F. Abe, Y. Einaga, and H. Yamakawa: Macromolecules **26**, 1891 (1993).
17. K. Horita, N. Sawatari, T. Yoshizaki, Y. Einaga, and H. Yamakawa: Macromolecules **28**, 4455 (1995).
18. T. Norisuye and H. Fujita: Polym. J. **14**, 143 (1982).
19. T. Norisuye, A. Tsuboi, and A. Teramoto: Polym. J. **28**, 357 (1996).
20. A. J. Barrett: Macromolecules **17**, 1566 (1984).
21. A. J. Barrett: Macromolecules **17**, 1561 (1984).
22. J. Shimada and H. Yamakawa: J. Polym. Sci., Polym. Phys. Ed. **16**, 1927 (1978).
23. W. H. Stockmayer and A. C. Albrecht: J. Polym. Sci. **32**, 215 (1958).
24. M. Fixman: J. Chem. Phys. **42**, 3831 (1965).
25. C. W. Pyun and M. Fixman: J. Chem. Phys. **42**, 3838 (1965).
26. H. Yamakawa and G. Tanaka: J. Chem. Phys. **55**, 3188 (1971).
27. H. Yamakawa and T. Yoshizaki: J. Chem. Phys. **91**, 7900 (1989).
28. H. Yamakawa and T. Yoshizaki: Macromolecules **28**, 3604 (1995).
29. T. Yoshizaki and H. Yamakawa: J. Chem. Phys. **105**, 5618 (1996).
30. B. H. Zimm: Macromolecules **24**, 592 (1980).
31. T. Yamada, H. Koyama, T. Yoshizaki, Y. Einaga, and H. Yamakawa: Macromolecules **26**, 2566 (1993).
32. B. K. Varma, Y. Fujita, M. Takahashi, and T. Nose: J. Polym. Sci., Polym. Phys. Ed. **22**, 1781 (1984).

33. L. J. Fetters, N. Hadjichristidis, J. S. Lindner, J. W. Mays, and W. W. Wilson: Macromolecules **24**, 3127 (1991).
34. Y. Tsunashima, M. Hirata, N. Nemoto, and M. Kurata: Macromolecules **21**, 1107 (1988).
35. T. Arai, N. Sawatari, T. Yoshizaki, Y. Einaga, and H. Yamakawa: Macromolecules **29**, 2309 (1996).
36. M. Osa, F. Abe, T. Yoshizaki, Y. Einaga, and H. Yamakawa: Macromolecules **29**, 2302 (1996).
37. T. Arai, F. Abe, T. Yoshizaki, Y. Einaga, and H. Yamakawa: Macromolecules **28**, 5485 (1995).
38. P. Vidakovic and F. Rondelez: Macromolecules **16**, 253 (1983).
39. H. Yamakawa: Macromolecules **25**, 1912 (1992).
40. A. J. Barrett: Macromolecules **18**, 196 (1985).
41. B. G. Nickel: Macromolecules **24**, 1358 (1991).
42. Z. Y. Chen and J. Noolandi: Macromolecules **25**, 4978 (1992).
43. K. Huber and W. H. Stockmayer: Macromolecules **20**, 1400 (1987).
44. T. Konishi, T. Yoshizaki, T. Saito, Y. Einaga, and H. Yamakawa: Macromolecules **23**, 290 (1990).
45. Y. Tamai, T. Konishi, Y. Einaga, M. Fujii, and H. Yamakawa: Macromolecules **23**, 4067 (1990).
46. Y. Einaga, F. Abe, and H. Yamakawa: Macromolecules **26**, 6243 (1993).
47. F. Abe, Y. Einaga, and H. Yamakawa: Macromolecules **27**, 3262 (1994).
48. M. Kamijo, F. Abe, Y. Einaga, and H. Yamakawa: Macromolecules **28**, 4159 (1995).
49. H. Sotobayashi and K. Ueberreiter: Z. Elektrochem. **67**, 178 (1963).
50. J. Springer, K. Ueberreiter, and E. Moeller: Z. Elektrochem. **69**, 494 (1965).
51. Y. Miyaki, Y. Einaga, and H. Fujita: Macromolecules **11**, 1180 (1978).
52. T. Hirosye, Y. Einaga, and H. Fujita: Polym. J. **11**, 819 (1979).
53. H. Yamakawa, F. Abe, and Y. Einaga: Macromolecules **26**, 1898 (1993).
54. T. Norisuye and Y. Nakamura: Macromolecules **27**, 2054 (1994).
55. B. H. Zimm: J. Chem. Phys. **14**, 164 (1946).
56. B. J. Cherayil, J. F. Douglas, and K. F. Freed: J. Chem. Phys. **83**, 5293 (1985).
57. W. H. Stockmayer: Makromol. Chem. **35**, 54 (1960).
58. H. Yamakawa: J. Chem. Phys. **42**, 1764 (1965).
59. H. Yamakawa: J. Chem. Phys. **45**, 2606 (1966).
60. T. Norisuye and Y. Nakamura: Polymer **34**, 1440 (1993).
61. T. Norisuye, Y. Nakamura, and K. Akasaka: Macromolecules **26**, 3791 (1993).
62. W. H. Stockmayer and E. F. Casassa: J. Chem. Phys. **20**, 1560 (1952).
63. T. Sato, T. Norisuye, and H. Fujita: J. Polym. Sci., Part B: Polym. Phys. **25**, 1 (1987).
64. Y. Nakamura, T. Norisuye, and A. Teramoto: J. Polym. Sci., Part B: Polym. Phys. **29**, 153 (1991).
65. C. E. H. Bawn, R. F. J. Freeman, and A. R. Kamalidin: Trans. Faraday Soc. **46**, 862 (1950).
66. T. Norisuye and H. Fujita: ChemTracts–Macromol. Chem. **2**, 293 (1991).
67. Y. Nakamura, T. Norisuye, and A. Teramoto: Macromolecules **24**, 4904 (1991).
68. H. Vink: Eur. Polym. J. **10**, 149 (1974).
69. B. L. Hager, G. C. Berry, and H. -H. Tsai: J. Polym. Sci., Part B: Polym. Phys. **25**, 387 (1987).
70. S. -J. Chen and G. C. Berry: Polymer **31**, 793 (1990).
71. H. Yamakawa, F. Abe, and Y. Einaga: Macromolecules **27**, 3272 (1994).
72. G. Tanaka and K. Šolc: Macromolecules **15**, 791 (1982).
73. H. Yamakawa: Macromolecules **26**, 5061 (1993).

74. H. Yamakawa, F. Abe, and Y. Einaga: Macromolecules **27**, 5704 (1994).

75. F. Abe, Y. Einaga, and H. Yamakawa: Macromolecules **28**, 694 (1995).

76. Z. Tong, S. Ohashi, Y. Einaga, and H. Fujita: Polym. J. **15**, 835 (1983).

77. C. Williams, F. Brochard, and H. L. Frisch: Ann. Rev. Phys. Chem. **32**, 433 (1981).

78. I. H. Park, Q. -W. Wang, and B. Chu: Macromolecules **20**, 1965 (1987).

79. B. Chu, I. H. Park, Q. -W. Wang, and C. Wu: Macromolecules **20**, 2833 (1987).

80. I. H. Park, L. J. Fetters, and B. Chu: Macromolecules **21**, 1178 (1988).

81. O. B. Ptitsyn, A. K. Kron, and Y. Y. Eizner: J. Polym. Sci., Part C **16**, 3509 (1968).

82. T. A. Orofino and P. J. Flory: J. Chem. Phys. **26**, 1067 (1957).

83. P. -G. de Gennes: J. Phys. Lett. **36**, L55 (1975).

84. I. C. Sanchez: Macromolecules **12**, 980 (1979).

85. F. Tanaka: J. Chem. Phys. **82**, 4707 (1985).

86. K. Akasaka, Y. Nakamura, T. Norisuye, and A. Teramoto: Polym. J. **26**, 363 (1994).

87. J. Marchal and H. Benoit: J. Chim. Phys. **52**, 818 (1955); J. Polym. Sci. **23**, 223 (1957).

88. K. Nagai and T. Ishikawa: Polym. J. **2**, 416 (1971).

89. M. Doi: Polym. J. **3**, 252 (1972).

90. W. L. Mattice and D. K. Carpenter: Macromolecules **17**, 625 (1984).

91. M. L. Mansfield: Macromolecules **19**, 1427 (1986).

92. T. Yoshizaki and H. Yamakawa: J. Chem .Phys. **98**, 4207 (1993).

93. Y. Einaga, F. Abe, and H. Yamakawa: J. Phys. Chem. **96**, 3948 (1992).

94. D. N. Rubingh and H. Yu: Macromolecules **9**, 681 (1976).

95. J. G. Kirkwood and R. J. Goldberg: J. Chem. Phys. **18**, 54 (1950).

96. W. H. Stockmayer: J. Chem. Phys. **18**, 58 (1950).

97. M. Fixman: J. Chem. Phys. **23**, 2074 (1955).

98. R. K. Bullough: Phil. Trans. R. Soc. **A254**, 397 (1962); Proc. R. Soc. **A275**, 271 (1963).

99. E. F. Casassa and H. Eisenberg: Adv. Protein Chem. **19**, 287 (1964).

100. H. A. Lorentz: Wiedem. Ann. **9**, 641 (1880); L. Lorenz: Wiedem. Ann. **11**, 70 (1881).

101. G. C. Berry: J. Chem. Phys. **44**, 4550 (1966).

102. B. H. Zimm: J. Chem. Phys. **16**, 1093 (1948).

9 Chain Dynamics

This chapter presents the foundation of the dynamics of unperturbed polymer chains in dilute solution on the basis of dynamic HW chain models within the framework of linear response theory. It is evident that the original (continuous) HW chain is not valid as a dynamic model; the discreteness must be, to some extent, recovered to introduce motional units into the chain. Thus diffusion equations for a time-dependent distribution function for the (constrained) chain are derived so as to be suitable for the treatments of its local and also global (to quasi-global) motions. The eigenvalue problems and time-correlation functions associated with the diffusion operators are then formulated by introducing several unavoidable approximations. Their applications to various dynamical properties are made in the next chapter. It is pertinent and instructive to begin by giving a general consideration of some aspects of polymer dynamics, followed by a brief description of the dynamics of conventional constrained bond chains.

9.1 General Consideration of Polymer Dynamics

The development of polymer dynamics is usually made in the classical diffusion limit, that is, on the Smoluchowski level, considering the time evolution of the distribution function only in coordinate space (of the phase space) [1–5]. The slow global motions of a single polymer chain in dilute solution may be well described by a simple, highly coarse-grained model. Among such models, the Rouse–Zimm spring-bead model [2–4, 6, 7] has retained a valid place for many years. It yields the same number of fundamental eigenvalues (relaxation rates) as that of beads in the chain. However, many more eigenvalues, or in general continuous spectra, are required to describe all kinds of chain motions, global to local. This must be a reflection of the chemical structure of the real chain. Its vibrational degrees of freedom are then classically treated, that is, constrained so that its bond lengths and bond angles are fixed at constant values. The adoption of such conventional bond chains leads to the development of the dynamics of constrained systems (in the diffusion limit), as initiated by Kirkwood [1]. As is well known, however, its final solutions are very difficult to obtain. Indeed, the spring-bead model was presented as a tractable replacement of the Kirkwood chain to avoid its difficulty.

Nevertheless, the Kirkwood approach must be pursued for the present purpose. The formal and standard procedure of imposing (holonomic) constraints on bond lengths and bond angles was essentially established by himself [1] and others [8, 9]. Subsequently, it was reformulated by Fixman and Kovac [10] in a form more convenient for the actual theoretical evaluation of individual dynamical properties. However, the evaluation still requires a preaveraging approximation to a constraining matrix involved in the diffusion operator, which leads to the unphysical result that the eigenvalues associated with the local motions become negative [11]. This may be regarded as arising from the fact that the approximation destroys, to some extent, the constraints imposed. On the other hand, it gives the well-known correct result for the chain without the constraints, that is, the spring-bead (or Gaussian) chain; or in other words, it has no serious effect on the evaluation of the eigenvalues associated with the global motions. This suggests that it is necessary to find an alternative way of introducing constraints which can describe the local motions even with the preaveraging approximation. However, this is impossible as far as the conventional bond chain is adopted, since there is only one way for it.

Now the HW chain can mimic the equilibrium conformational and steady-state transport behavior of individual real chains, both flexible and stiff, on the bond length or somewhat longer scales, as shown both theoretically and experimentally in the preceding chapters. Thus it fulfills the above requirement for the description of the local motions. However, the chain dynamics cannot be developed on the basis of the continuous HW chain model as it stands. In other words, it is not valid as a dynamic model unless the discreteness is, to some extent, recovered to introduce motional units into the chain. This can be done as follows. The two successive skeletal bonds in the real bond chain may form a rigid body, and therefore it may be regarded as composed of such rigid body elements, instead of bonds, joined successively. Indeed, the continuous HW chain may be obtained as a continuous limit of a discrete chain composed of rigid subbodies, or a coarse-grained discrete bond chain with coupled rotations, under certain conditions, as shown in Appendix 4.B. Thus we may construct a discrete chain of rigid subbodies and bonds of fixed length such that its equilibrium distribution obeys HW statistics. This is the *dynamic* HW model [12, 13] we adopt in the present and next chapters.

This model has various advantages. It facilitates the actual evaluation of dynamical properties for a given individual real chain, flexible or stiff. In fact, we can have $3N$ and $5N$ (or $6N$) eigenvalues for vector and tensor correlations, respectively, even in a crude approximation, where N is the number of subbodies in the chain, these being the motional units, each with three rotational degrees of freedom [14]. More important is the fact that the model enables us to introduce the constraints in it in two possible ways which are suitable for the treatments of the global and local motions, respectively [15], although necessarily the latter way leads to the negative global-mode eigenvalues [16].

Before proceeding to develop the dynamics of the dynamic HW chain, in the next section we give a brief description of the general formulation of the dynamics of conventional constrained bond chains along with some further remarks, for convenience. This may serve to make it easy to understand the later developments for the dynamic HW model.

9.2 Conventional Bond Chains

9.2.1 General Formulation – The Fixman–Kovac Chain

Consider a conventional bond chain composed of N beads and $N-1$ bonds, and let $\mathbf{q} = (q^1, q^2, \cdots, q^{3N})$ be its generalized coordinates. The subscripts s and h are used to indicate the unconstrained (soft) and constrained (hard) subspaces of $\mathbf{q}$, respectively, so that $\mathbf{q}_{\mathrm{s}} = (q^1, \cdots, q^m)$ and $\mathbf{q}_{\mathrm{h}} = (q^{m+1}, \cdots, q^{3N})$ denote the soft and hard coordinates, respectively. In the derivation of the diffusion equation satisfied by the time t-dependent distribution function $\Psi(\mathbf{q}_{\mathrm{s}}; t)$ there have been considered so far three types of constrained bond chains, which are referred to as types 1, 2, and 1′. For the type-1 chain, also called the Kramers chain [17], the constraints are imposed on the Lagrangian level so that the hard velocities $\dot{\mathbf{q}}_{\mathrm{h}}$ vanish [18, 19]. [It is in general different from a chain with vanishing hard conjugate momenta $\mathbf{p}_{\mathrm{h}} = (p_{m+1}, \cdots, p_{3N})$, which is unphysical since $\mathbf{p}$ is the covariant velocity vector.] For the type-2 chain, which is just the chain mentioned in the last section, the constraints are imposed on the Smoluchowski level so that the hard drift velocities $\mathbf{u}_{\mathrm{h}} = \langle \dot{\mathbf{q}}_{\mathrm{h}} \rangle_q$ vanish [8–10], where $\langle \ \rangle_q$ denotes an average over $\mathbf{p}$ and the solvent phase variables. A starting equation for the type-1 and -2 chains is the Liouville equation, while that for the type-1′ chain [20] is the Langevin equation without the inertia term but with constraints.

Now the RIS model in the equilibrium conformational study belongs to the type-2 chain, and the diffusion equations of this type have been standard in polymer dynamics. In this subsection we therefore consider the type-2 chain in some detail [21]. In the diffusion limit, the Liouville equation is reduced to the continuity equation for the distribution function $\Psi(\mathbf{q}; t)$ in the full $\mathbf{q}$ space [22–24],

$$\frac{\partial \Psi}{\partial t} = g^{-1/2} \nabla g^{1/2} \cdot \mathbf{J}, \tag{9.1}$$

where g is the metric determinant in this space, $\nabla = \partial/\partial \mathbf{q}$ is the gradient operator, and $\mathbf{J} = (J^1, \cdots, J^{3N}) = \Psi \mathbf{u}$ is the (contravariant) flux vector. Note that this Ψ is normalized as

$$\int \Psi g^{1/2} d\mathbf{q} = 1 \,. \tag{9.2}$$

In the field-free case, $\mathbf{u}$ or $\mathbf{J}$ may be determined from the force balance equation [2–4]

$$\boldsymbol{\zeta} \cdot \mathbf{u} = -\nabla(k_B T \ln \Psi + U) + \mathbf{P} \,, \tag{9.3}$$

or

$$\mathbf{J} = (k_B T)^{-1} \mathbf{D} \cdot (-k_B T \nabla \Psi - \Psi \nabla U + \Psi \mathbf{P}) \tag{9.4}$$

with

$$\mathbf{D} = k_B T \boldsymbol{\zeta}^{-1} \,, \tag{9.5}$$

where $U = U_s(\mathbf{q}_s)$ is the soft potential energy (not to be confused with the potential energy per unit contour length), $\mathbf{P} = -\nabla U_h$ is the constraining force vector, and $\boldsymbol{\zeta}$ and $\mathbf{D}$ are the friction and diffusion tensors, respectively. Note that in the $3N$-dimensional Cartesian space $\mathbf{D} = k_B T(\zeta^{-1}\mathbf{I} + \mathbf{T})$, where ζ is the translational friction constant of the bead, $\mathbf{I}$ is the unit tensor, and $\mathbf{T}$ is the Oseen hydrodynamic interaction tensor.

Following the Ikeda–Erpenbeck–Kirkwood procedure [8,9], the soft components of $\mathbf{J}$ may then be obtained from Eq. (9.3) by projection of $\boldsymbol{\zeta} \cdot \mathbf{u}$ onto the s subspace with $\mathbf{u}_h = \mathbf{0}$ and $\mathbf{P}_s = \mathbf{0}$,

$$\mathbf{J}_s = -(\mathbf{D}_{ss} - \mathbf{D}_{sh} \cdot \mathbf{D}_{hh}^{-1} \cdot \mathbf{D}_{hs}) \cdot \left[\nabla_s \Psi + (k_B T)^{-1} \Psi \nabla_s U\right] \tag{9.6}$$

with

$$\mathbf{J}_h = \mathbf{0} \,. \tag{9.7}$$

More conveniently, Eq. (9.6) may be obtained from Eq. (9.4) by projection of $\mathbf{J}$ onto the s and h subspaces and elimination of $\mathbf{P}_h$, following Fixman and Kovac [10]. (Note that $\mathbf{P}$ may be suppressed from the outset in the former route but not in the latter.)

For the type-2 chain with the constraints $\mathbf{q}_h = \mathbf{q}_h^0$, $\Psi(\mathbf{q}; t)$ may be written in the form

$$\Psi(\mathbf{q}; t) = \delta(\mathbf{q}_h - \mathbf{q}_h^0)\bar{\Psi}(\mathbf{q}_s; t) \,, \tag{9.8}$$

where δ is a Dirac delta function. Then the continuity Eq. (9.1) with Eq. (9.7) reduces to

$$\frac{\partial \bar{\Psi}}{\partial t} = g^{-1/2} \nabla_s g^{1/2} \cdot \mathbf{J}_s \tag{9.9}$$

with $\mathbf{J}_s$ being given by Eq. (9.6) with $\Psi = \bar{\Psi}$. This is the diffusion equation for the Fixman–Kovac (type-2) chain. The submatrix $\mathbf{D}_{hh}$ of $\mathbf{D}$ is the constraining matrix, and the prototype diffusion equation without the constraining term $\mathbf{D}_{sh} \cdot \mathbf{D}_{hh}^{-1} \cdot \mathbf{D}_{hs}$ in $\mathbf{J}_s$ is just the diffusion equation for the spring-bead (or Gaussian) chain. Thus the preaveraging of $\mathbf{D}_{hh}$ leads to the breakdown of the constraints, so that the diffusion Eq. (9.9) can then describe correctly the global motions but not the local ones (with the negative local-mode eigenvalues). We note that the constraints $\mathbf{q}_h = \mathbf{q}_h^0$ may be considered the so-called "flexible" constraints [25], although with infinitely large force constants.

9.2.2 Some Further Remarks

First, some remarks should be made on the other types of chains. The diffusion equation for the type-1 chain was derived by Bird and co-workers [18,19], although only in the free-draining case. The result is equivalent to that for the type-2 chain (with $\mathbf{T} = \mathbf{0}$) except for the metric determinant. In general, the metric determinant for the type-1 chain depends on the bead masses since the constraints are imposed on the Lagrangian level. In the case of identical beads, however, it becomes the metric determinant g_s in the s subspace. On the other hand, the diffusion equation for the type-1' chain, which was derived by Fixman [20] in his Brownian dynamics simulation study, does not involve the bead masses because of the suppression of the inertia term, and is equivalent to that for the type-2 chain with g_s in place of g. The (original) Kirkwood chain [1] is also of the type 1', although the constraining term $\mathbf{D}_{sh} \cdot \mathbf{D}_{hh}^{-1} \cdot \mathbf{D}_{hs}$ was erroneously dropped in his original expression for $\mathbf{J}_s$ [3,8]. In the free-draining case with identical beads, the type-1 and -1' chains are identical. The diffusion equations for them may be converted to that for the type-2 chain by addition of the metric potential U' given by

$$U' = k_B T \ln(g'_s)^{1/2} \tag{9.10}$$

to U, where g'_s is that part of g_s which depends on the internal soft coordinates [20]. The implication is that the simulation of the type-1' chain with this potential is equivalent to that of type 2. The constraints on the type-1 and -1' chains are the so-called "rigid" constraints [25].

Next, it is believed that the type-2 chain is the best, as mentioned above. Indeed, also in the Brownian dynamics simulation (based on the Langevin equation), Helfand and co-workers [26, 27] adopted chains with flexible constraints, and Weiner and co-workers [28, 29] used type-1' chains with the metric potential U' (and with the inertia term). Further, the evaluation of g_s required for the type-1 or -1' chain is a difficult problem for large N [20,30]. Although for the type-2 chain there is, of course, a difficulty in inversion of some matrices, it is greatly diminished by choosing soft coordinates expressed in an external coordinate system as in the case of the dynamic HW model (see the following sections). However, it is pertinent to note that these two types seem almost equivalent to each other for long enough, ordinary flexible chains [20].

Finally, brief mention must be made of the effects of chain stiffness. Clearly it arises from the structural constraints on bond lengths and angles along with the internal potential, as discussed in the preceding chapters. However, there have been several attempts [31–37] to approach the problem of stiff chain dynamics without imposition of constraints, some of which have already been shortly discussed in Appendix 3.C. In this book, we do not, of course, pursue this line.

9.3 Dynamic Helical Wormlike Chains

Consider a chain composed of N identical subbodies (beads), not necessarily spherical, joined successively with bonds of fixed length a, where their centers are located nearly on the contour of the continuous HW chain of length L. Suppose that each subbody has (mean) translational and rotatory friction constants ζ_t $(= \zeta)$ and ζ_r in a solvent of viscosity coefficient η_0. This is the dynamic HW model [12, 13]. Note that the bond length a is not equal to $L/N \equiv \Delta s$, which is equal to the spacing a introduced in Chap. 8. The relation between them is explicitly given below.

Now we introduce N localized Cartesian coordinate systems ($\mathbf{e}_{\xi_p}$, $\mathbf{e}_{\eta_p}$, $\mathbf{e}_{\zeta_p}$) $(p = 1, \cdots, N)$, the pth one being affixed to the pth subbody with the origin at its center and with $\mathbf{e}_{\zeta_p}$ in the direction of the pth bond vector $\mathbf{a}_p$ (from p to $p+1$). Let $\Omega_p = (\theta_p, \phi_p, \psi_p)$ $(p = 1, \cdots, N)$ be the Euler angles defining the orientation of the pth localized coordinate system with respect to an external coordinate system. Apart from its location, the configuration of the chain may be specified by $3N$ soft coordinates $\{\Omega_N\} = (\Omega_1, \cdots, \Omega_N)$.

The total potential energy $U_0(\{\Omega_N\})$ of the unperturbed chain without excluded volume may then be expressed as a sum of pair potentials $u(\Omega_p, \Omega_{p+1})$,

$$U_0(\{\Omega_N\}) = \sum_{p=1}^{N-1} u(\Omega_p, \Omega_{p+1}) \tag{9.11}$$

with

$$u(\Omega_p, \Omega_{p+1}) = -k_{\mathrm{B}}T \ln G(\Omega_{p+1} \,|\, \Omega_p; \Delta s)\,, \tag{9.12}$$

where G is the (equilibrium) Green function given by Eq. (4.106) (with Poisson's ratio $\sigma = 0$). Thus the equilibrium distribution function $\Psi_{\mathrm{eq}}(\{\Omega_N\})$ of $\{\Omega_N\}$ is given by

$$\Psi_{\mathrm{eq}}(\{\Omega_N\}) = \frac{e^{-U_0/k_{\mathrm{B}}T}}{\displaystyle\int e^{-U_0/k_{\mathrm{B}}T} d\{\Omega_N\}}$$

$$= (8\pi^2)^{-1} \prod_{p=1}^{N-1} G(\Omega_{p+1} \,|\, \Omega_p; \Delta s)\,. \tag{9.13}$$

In what follows, $\langle \ \rangle_{\mathrm{eq}}$ denotes an equilibrium average evaluated with Ψ_{eq}.

The dynamic HW chain is equivalent to a system of N coupled symmetric tops with constraints such that the rotation axis (ζ_p) of each one (p) points to the center of its successor $(p+1)$ with the fixed distance a between them, as depicted in Fig. 9.1.

The relation between a and Δs may be obtained by equating the mean-square end-to-end distance $\langle R^2(N)\rangle_{\mathrm{eq}}$ of the dynamic HW chain to that, $\langle R^2(L)\rangle_{\mathrm{eq}}$, of the corresponding continuous HW chain in the limit of $N \to \infty$. The result (in units of λ^{-1}) reads

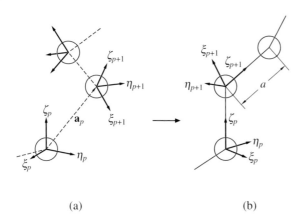

Fig. 9.1a,b. Construction of the (constrained) dynamic HW chain (b) from rigid subbodies (symmetric tops) without constraints (a)

(a)   (b)

$$a = (c_\infty \Delta s)^{1/2} \left\{ 1 + \frac{2\tau_0{}^2}{\nu^2 (e^{2\Delta s} - 1)} \right.$$

$$\left. + \frac{2\kappa_0{}^2}{\nu^2} \left[\frac{e^{2\Delta s} \cos(\nu \Delta s) - 1}{e^{4\Delta s} - 2e^{2\Delta s} \cos(\nu \Delta s) + 1} \right] \right\}^{-1/2}, \quad (9.14)$$

where c_∞ and ν are given by Eqs. (4.75) and (4.76), respectively. Thus the bond length a can be uniquely determined as a function of κ_0, τ_0, and Δs. As already mentioned, for flexible chains one subbody as a motional unit may be regarded as corresponding to two successive skeletal bonds of the real chain, that is, the repeat unit, so that Δs is chosen to be equal to M_0/M_L [see Eqs. (8.60)].

9.4 Diffusion Equations

We derive two representations of the diffusion equation for the dynamic HW chain having $3(N + 1)$ degrees of freedom, that is, three Cartesian coordinates $\mathbf{R}_c$ specifying its location and the N sets of Euler angles $\{\Omega_N\}$, by introducing the constraints in two ways. They are suitable for the treatments of the local and global motions, respectively. In each case, the derivation may be conveniently made in two steps, starting from the formulation in full Cartesian coordinate space. The first step is common to both cases. In what follows, all lengths are measured in units of λ^{-1} and $k_B T$ is chosen to be unity.

9.4.1 Space of Bond and Infinitesimal Rotation Vectors

We first consider the chain without constraints such that each of the N subbodies has six, translational and rotational, degrees of freedom, and add the $(N + 1)$th *imaginary* subbody having only three translational degrees of freedom (with ζ_t but with $\zeta_r = 0$), so that the magnitude of $\mathbf{a}_p$ $(p = 1, \cdots, N)$

is not always equal to a, nor does its direction always coincide with the ζ_p axis, as depicted in Fig. 9.1a [12]. The addition of the $(N+1)$th subbody of this nature to the chain serves to remove certain annoying asymmetry in the diffusion equation, its effect on the final result being small for large N. Let $\mathbf{r}_p = (r_{px}, r_{py}, r_{pz})$ be the vector position of the center of the pth subbody $(p = 1, \cdots, N+1)$ in the external Cartesian coordinate system $(\mathbf{e}_x, \mathbf{e}_y, \mathbf{e}_z)$, and let $d\boldsymbol{\chi}_p = (d\chi_{p\xi}, d\chi_{p\eta}, d\chi_{p\zeta})$ be its infinitesimal rotation vector in the pth localized coordinate system $(p = 1, \cdots, N)$ having the orientation Ω_p with respect to the former. The metric form in this $(6N+3)$-dimensional full Cartesian space $(d\{\mathbf{r}_{N+1}\}, d\{\boldsymbol{\chi}_N\})$ is

$$(dl)^2 = \sum_{p=1}^{N+1} (d\mathbf{r}_p)^2 + \sum_{p=1}^{N} (d\boldsymbol{\chi}_p)^2 . \tag{9.15}$$

The time-dependent distribution function $\Psi(\{\mathbf{r}_{N+1}\}, \{\Omega_N\}; t)$ for the chain satisfies the continuity equation in this space,

$$\frac{\partial \Psi}{\partial t} = -\sum_{p=1}^{N+1} \nabla_p^r \cdot \mathbf{J}_p^r - \sum_{p=1}^{N} \nabla_p^\chi \cdot \mathbf{J}_p^\chi , \tag{9.16}$$

where ∇_p^r and $\nabla_p^\chi = (\partial/\partial\chi_{p\xi}, \partial/\partial\chi_{p\eta}, \partial/\partial\chi_{p\zeta})$ are the gradient operators with respect to $\mathbf{r}_p$ and $d\boldsymbol{\chi}_p$, respectively, and $\mathbf{J}_p^r$ and $\mathbf{J}_p^\chi$ are the fluxes associated with them, respectively. Note that the fluxes $\mathbf{J}_p^\chi$ do not appear for conventional bond chains. If $\mathbf{V}_p$ and $\mathbf{W}_p$ are the translational and angular velocities of the pth subbody in the external coordinate system, respectively, $\mathbf{J}_p^r$ and $\mathbf{J}_p^\chi$ may be expressed as

$$\mathbf{J}_p^r = \Psi \mathbf{V}_p \qquad (p = 1, \cdots, N+1) , \tag{9.17}$$

$$\mathbf{J}_p^\chi = \Psi \mathbf{A}_p \cdot \mathbf{W}_p \qquad (p = 1, \cdots, N) , \tag{9.18}$$

where $\mathbf{A}_p = \mathbf{A}_p(\Omega_p)$ is the transformation matrix identical with the $\mathbf{Q}$ given by Eq. (4.96) with $(\theta_p, \phi_p, \psi_p)$ in place of $(\tilde{\theta}, \tilde{\phi}, \tilde{\psi})$.

If $\mathbf{V}_p^0$ is the unperturbed solvent velocity at $\mathbf{r}_p$, $\mathbf{V}_p$ and $\mathbf{W}_p$ may be written in the form

$$\mathbf{V}_p = \mathbf{V}_p^0 + \zeta_t^{-1} \mathbf{F}_p + \sum_{\substack{q=1 \\ \neq p}}^{N+1} \mathbf{T}_{pq} \cdot \mathbf{F}_q \qquad (p = 1, \cdots, N+1) , \tag{9.19}$$

$$\mathbf{W}_p = \mathbf{W}_p^0 + \zeta_r^{-1} \mathbf{T}_p \qquad (p = 1, \cdots, N) \tag{9.20}$$

with

$$\mathbf{W}_p^0 = \frac{1}{2} \nabla_p^r \times \mathbf{V}_p^0 , \tag{9.21}$$

where $\mathbf{F}_p$ and $\mathbf{T}_p$ are the frictional force and torque, respectively, exerted by the pth subbody on the solvent, and $\mathbf{T}_{pq} = \mathbf{T}(\mathbf{R}_{pq})$ with $\mathbf{R}_{pq} = \mathbf{r}_q - \mathbf{r}_p$ is the

Oseen hydrodynamic interaction (HI) tensor given by Eq. (6.4). We note that Eqs. (9.19) and (9.20) take into account correctly the HI between subbodies to terms of $\mathcal{O}(R_{pq}^{-1})$. In what follows, we use the preaveraged Oseen tensor,

$$\langle \mathbf{T}_{pq} \rangle = (6\pi\eta_0)^{-1} \langle R_{pq}^{-1} \rangle \mathbf{I}, \tag{9.22}$$

where $\mathbf{I}$ is the 3×3 unit tensor, $\langle\ \rangle$ denotes an average taken with Ψ, and $\langle \mathbf{T}_{pq} \rangle$ may be replaced by $\langle \mathbf{T}_{pq} \rangle_{\mathrm{eq}}$ in the regime of linear response. The effect of fluctuating HI (on the translational motion) is discussed in Appendix 9.A. With force balance equations like Eq. (9.3) for $\mathbf{F}_p$ and $\mathbf{A}_p \cdot \mathbf{T}_p$, Eq. (9.4) with $\mathbf{V}_p^0$ and an external potential $U_e(\{\mathbf{r}_{N+1}\}, \{\Omega_N\})$ may then be replaced by

$$\mathbf{J}_p^r = \sum_{q=1}^{N+1} D_{pq}\left(-\nabla_q^r \Psi - \Psi\nabla_q^r U + \Psi\mathbf{P}_q^r\right) + \Psi\mathbf{V}_p^0, \tag{9.23}$$

$$\mathbf{J}_p^\chi = \zeta_r^{-1}\left(-\nabla_p^\chi\Psi - \Psi\nabla_p^\chi U + \Psi\mathbf{P}_p^\chi\right) + \Psi\mathbf{A}_p \cdot \mathbf{W}_p^0, \tag{9.24}$$

where

$$D_{pq} = \delta_{pq}\zeta_t^{-1} + (1 - \delta_{pq})(6\pi\eta_0)^{-1}\langle R_{pq}^{-1} \rangle_{\mathrm{eq}}, \tag{9.25}$$

$$U = U_0 + U_e, \tag{9.26}$$

and $\mathbf{P}_p^r$ and $\mathbf{P}_p^\chi$ are the constraining forces on the pth subbody associated with $\mathbf{r}_p$ and $d\chi_p$, respectively. Equation (9.16) with Eqs. (9.23)–(9.26) gives the diffusion equation in $(d\{\mathbf{r}_{N+1}\}, d\{\chi_N\})$ space.

Now we transform $\{\mathbf{r}_{N+1}\}$ to bond coordinates. Since $d\{\mathbf{r}_{N+1}\}$ is separable from $d\{\chi_N\}$ in the above diffusion equation, we may consider only the former part. We put [38]

$$\mathbf{R}_c = \sum_{p=1}^{N+1} w_p\mathbf{r}_p, \tag{9.27}$$

$$\mathbf{a}_p = \mathbf{r}_{p+1} - \mathbf{r}_p \qquad (p = 1, \cdots, N), \tag{9.28}$$

where w_p are constants independent of the coordinates and satisfy

$$\sum_{p=1}^{N+1} w_p = 1. \tag{9.29}$$

We then have the transformation

$$\nabla_p^r = w_p\nabla_c + (1 - \delta_{p1})\nabla_{p-1}^a - (1 - \delta_{p(N+1)})\nabla_p^a$$
$$(p = 1, \cdots, N + 1), \tag{9.30}$$

where ∇_c and ∇_p^a are the gradient operators with respect to $\mathbf{R}_c$ and $\mathbf{a}_p$, respectively. The velocities $\mathbf{V}_p$ may be transformed to those, $\mathbf{V}_c$ and $\mathbf{v}_p$ ($p = 1, \cdots, N$), in $(\mathbf{R}_c, \{\mathbf{a}_N\})$ space of bond coordinates by the same contravariant law as Eqs. (9.27) and (9.28) for $\mathbf{r}_p$, and the frictional forces $\mathbf{F}_p$ to $\mathbf{F}_c$ and $\mathbf{f}_p$ in this space by the same covariant law as Eq. (9.30) for ∇_p^r. The constraining forces $\mathbf{P}_p^r$ may also be transformed to $\mathbf{p}_p^a$ (with $\mathbf{P}_c = \mathbf{0}$) by the same covariant law. (Note that there is not a constraining force associated with $\mathbf{R}_c$.)

If w_p is chosen to give

$$\sum_{q=1}^{N+1} w_p\left(-D_{qp} + D_{q(p+1)}\right) = 0 \quad (p = 1, \cdots, N),\tag{9.31}$$

then the desired diffusion equation for $\Psi(\mathbf{R}_c, \{\mathbf{a}_N\}, \{\Omega_N\}; t)$ in $(\mathbf{R}_c, \{\mathbf{a}_N\}, d\{\chi_N\})$ space, in which the metric determinant g is also unity, is obtained, from Eq. (9.16), as

$$\frac{\partial \Psi}{\partial t} = -\nabla_c \cdot \mathbf{J}_c - \sum_{p=1}^{N}(\nabla_p^a \cdot \mathbf{J}_p^a + \nabla_p^\chi \cdot \mathbf{J}_p^\chi),\tag{9.32}$$

where

$$\mathbf{J}_c = -D_c(\nabla_c\Psi + \Psi\nabla_c U_e) + \mathbf{V}_c^0\Psi,\tag{9.33}$$

$$\mathbf{J}_p^a = -\sum_{q=1}^{N} B_{pq}(\nabla_q^a\Psi + \Psi\nabla_q^a U - \Psi\mathbf{p}_q^a) + \mathbf{v}_p^0\Psi\tag{9.34}$$

with

$$D_c = \sum_{p,q=1}^{N+1} w_p w_q D_{pq},\tag{9.35}$$

$$B_{pq} = 2D_{pq} - D_{p(q+1)} - D_{(p+1)q}.\tag{9.36}$$

We note that if w_p satisfies Eq. (9.31), $\mathbf{R}_c$ is the Zimm center of resistance (in the scheme of preaveraged HI) [7], and that if $w_p = (N+1)^{-1}$, $\mathbf{R}_c$ is the molecular center of mass (see also Appendix 9.A).

9.4.2 Space of Euler Angles – Local Motions

In this subsection we derive, from Eq. (9.32), the final representation of the diffusion equation that is suitable for the description of the local motions [12]. We express the pth bond vector $\mathbf{a}_p$ as $\tilde{\Theta}_p = (\tilde{a}_p, \tilde{\theta}_p, \tilde{\phi}_p)$ in spherical polar coordinates in the pth localized Cartesian coordinate system, as depicted in Fig. 9.2. We transform the Cartesian coordinates $(\mathbf{R}_c, \{\mathbf{a}_N\}, d\{\chi_N\})$ to the curvilinear coordinates $(\mathbf{R}_c, \{\tilde{\Theta}_N\}, \{\Omega_N\})$ with

$$\begin{pmatrix} d\mathbf{a}_p \\ d\chi_p \end{pmatrix} = \mathbf{U}_P \cdot \begin{pmatrix} d\tilde{\Theta}_p \\ d\Omega_p \end{pmatrix},\tag{9.37}$$

where $\mathbf{U}_p$ is the transformation matrix but its explicit form is omitted. The metric determinant g in this space is given by

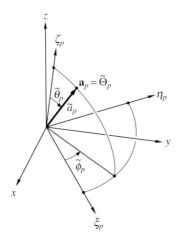

Fig. 9.2. The pth bond vector $\mathbf{a}_p$ expressed as $\tilde{\Theta}_p = (\tilde{a}_p, \tilde{\theta}_p, \tilde{\phi}_p)$ in spherical polar coordinates in the pth localized Cartesian coordinate system

$$g = \prod_{p=1}^{N} g_p \,, \tag{9.38}$$

where

$$g_p = |\mathbf{U}_p^{\,T} \cdot \mathbf{U}_p| = \tilde{a}_p^{\,4} \sin^2 \tilde{\theta}_p \sin^2 \theta_p \tag{9.39}$$

with the superscript T indicating the transpose.

The diffusion Eq. (9.32) may then be transformed to that in $(\mathbf{R}_c, \{\tilde{\Theta}_N\}, \{\Omega_N\})$ space,

$$\frac{\partial \Psi}{\partial t} = -\nabla_c \cdot \mathbf{J}_c - \sum_{p=1}^{N} g_p^{-1/2} (\nabla_p^{\tilde{\Theta}} g_p^{1/2} \cdot \mathbf{J}_P^{\tilde{\Theta}} + \nabla_p^{\Omega} g_p^{1/2} \cdot \mathbf{J}_p^{\Omega}) \,, \tag{9.40}$$

where $\nabla_p^{\tilde{\Theta}} = (\partial/\partial \tilde{a}_p, \partial/\partial \tilde{\theta}_p, \partial/\partial \tilde{\phi}_p)$ and $\nabla_p^{\Omega} = (\partial/\partial \theta_p, \partial/\partial \phi_p, \partial/\partial \psi_p)$ are the gradient operators with respect to $\tilde{\Theta}_p$ and Ω_p, respectively, and $\mathbf{J}_p^{\tilde{\Theta}}$ and $\mathbf{J}_p^{\Omega}$ are the fluxes associated with them, respectively. The fluxes $\mathbf{J}_p^a$ and $\mathbf{J}_p^{\chi}$ may be transformed to $\mathbf{J}_p^{\tilde{\Theta}}$ and $\mathbf{J}_p^{\Omega}$ by the contravariant law, and the gradient operators ∇_p^a and ∇_p^{χ} to $\nabla_p^{\tilde{\Theta}}$ and ∇_p^{Ω} by the covariant law,

$$\begin{pmatrix} \mathbf{J}_p^{\tilde{\Theta}} \\ \mathbf{J}_p^{\Omega} \end{pmatrix} = \mathbf{U}_p^{-1} \cdot \begin{pmatrix} \mathbf{J}_p^a \\ \mathbf{J}_p^{\chi} \end{pmatrix} \,, \tag{9.41}$$

$$\begin{pmatrix} \nabla_p^a \\ \nabla_p^{\chi} \end{pmatrix} = \mathbf{U}_p^{-1T} \cdot \begin{pmatrix} \nabla_p^{\tilde{\Theta}} \\ \nabla_p^{\Omega} \end{pmatrix} \,. \tag{9.42}$$

The constraining forces $\mathbf{p}_p^a$ and $\mathbf{P}_p^{\chi}$ involved in $\mathbf{J}_p^a$ and $\mathbf{J}_p^{\chi}$ may be transformed to $\mathbf{p}_p^{\tilde{\Theta}}$ and $\mathbf{P}_p^{\Omega}$ by the same covariant law. We impose the constraints $\tilde{\Theta}_p =$

$(a, 0, \tilde{\phi}_p)$ $(p = 1, \cdots, N)$, considering the constraining forces $\mathbf{p}_p^{\tilde{\Theta}}$ $(p = 1, \cdots, N)$ to make the fluxes $\mathbf{J}_p^{\tilde{\Theta}}$ vanish,

$$\mathbf{J}_p^{\tilde{\Theta}} = \mathbf{0} \qquad (p = 1, \cdots, N). \tag{9.43}$$

Then the solution for $\mathbf{p}_p^{\tilde{\Theta}}$ (with $\mathbf{P}_p^{\Omega} = \mathbf{0}$) is found from Eqs. (9.41) and (9.43), and the fluxes $\mathbf{J}_p^{\Omega}$ are obtained from Eq. (9.41) with Eq. (9.43) and the result for $\mathbf{p}_p^{\tilde{\Theta}}$.

Now, setting $\tilde{\Theta}_p = (a, 0, \tilde{\phi}_p)$, we write the distribution function $\Psi(\mathbf{R}_c, \{\tilde{\Theta}_N\}, \{\Omega_N\}; t)$ in the form like Eq. (9.8),

$$\Psi = \Psi_0(\{\tilde{\Theta}_N\})\bar{\Psi}(\mathbf{R}_c, \{\Omega_N\}; t) \tag{9.44}$$

with

$$\Psi_0 = \prod_{p=1}^{N} (2\pi\tilde{a}_p^{\;2} \sin\tilde{\theta}_p)^{-1}\delta(\tilde{a}_p - a)\delta(\tilde{\theta}_p). \tag{9.45}$$

The average of any configuration-dependent quantity α may then be calculated from

$$\langle\alpha\rangle = \int \alpha\Psi g^{1/2}d\mathbf{R}_c \prod_{p=1}^{N} d\tilde{a}_p d\tilde{\theta}_p d\tilde{\phi}_p d\theta_p d\phi_p d\psi_p$$

$$= \int \alpha\bar{\Psi}d\mathbf{R}_c d\{\Omega_N\}, \tag{9.46}$$

where we have used Eq. (9.38) with Eq. (9.39), and note that $\tilde{\Theta}_p = (a, 0, \tilde{\phi}_p)$ in $\bar{\Psi}$. It is also clear that Ψ_0 may be removed from the diffusion equation at the final stage as in Eq. (9.9). In what follows, we therefore denote $\bar{\Psi}$ by Ψ.

Thus, from Eq. (9.40) with Eq. (9.43) and the result for $\mathbf{J}_p^{\Omega}$, we obtain the desired diffusion equation for $\bar{\Psi} \equiv \Psi(\mathbf{R}_c, \{\Omega_N\}; t)$ in $(\mathbf{R}_c, \{\Omega_N\})$ space,

$$\frac{\partial\Psi}{\partial t} = D_c\nabla_c^2\Psi + \sum_{p,q=1}^{N} \mathbf{L}_p \cdot \{\mathbf{M}_{pq} \cdot [\zeta_r^{-1}(\mathbf{L}_q\Psi + \Psi\mathbf{L}_q U)$$

$$- \mathbf{A}_q \cdot \mathbf{W}_q^0\Psi] - \mathbf{N}_{pq} \cdot \mathbf{v}_q^0\Psi\} + \nabla_c \cdot (D_c\nabla_c U_e - \mathbf{V}_c^0)\Psi, \tag{9.47}$$

where $\mathbf{L}_p = (L_{p\xi}, L_{p\eta}, L_{p\zeta}) \; (= \nabla_p^\chi)$ is the angular momentum operator given by Eqs. (4.35) with $(\theta_p, \phi_p, \psi_p)$ in place of (θ, ϕ, ψ), and

$$\mathbf{M}_{pq} = \delta_{pq}\mathbf{I} - \mathbf{E}_p^T \cdot (\mathbf{C}^{-1})_{pq} \cdot \mathbf{E}_q, \tag{9.48}$$

$$\mathbf{N}_{pq} = \mathbf{E}_p^T \cdot (\mathbf{C}^{-1})_{pq}. \tag{9.49}$$

In Eqs. (9.48) and (9.49), $(\mathbf{C}^{-1})_{pq}$ is the pq element (3×3 matrix) of the inverse of the $3N \times 3N$ matrix $\mathbf{C}$ whose pq element is the 3×3 matrix $\mathbf{C}_{pq}$,

$$\mathbf{C}_{pq} = \zeta_r B_{pq}\mathbf{I} + \delta_{pq}\mathbf{E}_p \cdot \mathbf{E}_p^T, \tag{9.50}$$

and $\mathbf{E}_p$ is the 3×3 matrix,

$$\mathbf{E}_p = a(-\mathbf{e}_{\eta_p}, \mathbf{e}_{\xi_p}, \mathbf{0})$$

$$= a \begin{pmatrix} c_{\theta_p} c_{\phi_p} s_{\psi_p} + s_{\phi_p} c_{\psi_p} & c_{\theta_p} c_{\phi_p} c_{\psi_p} - s_{\phi_p} s_{\psi_p} & 0 \\ c_{\theta_p} s_{\phi_p} s_{\psi_p} - c_{\phi_p} c_{\psi_p} & c_{\theta_p} s_{\phi_p} c_{\psi_p} + c_{\phi_p} s_{\psi_p} & 0 \\ -s_{\theta_p} s_{\psi_p} & -s_{\theta_p} c_{\psi_p} & 0 \end{pmatrix} \qquad (9.51)$$

with $s_{\theta_p} = \sin\theta_p$, $c_{\theta_p} = \cos\theta_p$, and so on.

Clearly the above $\mathbf{C}^{-1}$ (or $\mathbf{C}$) is the constraining matrix. If we suppress the second term on the right-hand side of Eq. (9.48) for $\mathbf{M}_{pq}$, then Eq. (9.47) gives the prototype diffusion equation for the unconstrained system, that is, the system of N coupled rigid subbodies without the constraints, apart from the translational mode of the chain associated with its center of resistance.

Finally, we introduce the self-adjoint formulation of the diffusion equation. We factor Ψ into the equilibrium distribution function Ψ_{eq} given by the first line of Eqs. (9.13) and Φ,

$$\Psi = \Psi_{\mathrm{eq}} \Phi. \qquad (9.52)$$

In the field-free case ($U_{\mathrm{e}} = 0$ and $\mathbf{V}_p^0 = \mathbf{0}$), Eq. (9.47) reduces to

$$\left(\frac{\partial}{\partial t} - D_{\mathrm{c}} \nabla_{\mathrm{c}}^2 + \mathcal{L} \right) \Phi = 0, \qquad (9.53)$$

where $\mathcal{L}$ is the diffusion operator defined by

$$\mathcal{L} = -\zeta_{\mathrm{r}}^{-1} \Psi_{\mathrm{eq}}^{-1} \sum_{p,q=1}^{N} \mathbf{L}_p \Psi_{\mathrm{eq}} \cdot \mathbf{M}_{pq} \cdot \mathbf{L}_q. \qquad (9.54)$$

If the scalar product $\langle \alpha, \beta \rangle$ of any two functions α and β of $\{\Omega_N\}$ is defined with the weighting function Ψ_{eq} by

$$\langle \alpha, \beta \rangle = \int \Psi_{\mathrm{eq}} \alpha^* \beta d\{\Omega_N\} = \langle \alpha^* \beta \rangle_{\mathrm{eq}} \qquad (9.55)$$

with the asterisk indicating the complex conjugate, then the operator $\mathcal{L}$ becomes self-adjoint,

$$\langle \alpha, \mathcal{L}\beta \rangle = \langle \mathcal{L}\alpha, \beta \rangle$$

$$= \sum_{p,q=1}^{N} \left\langle (\mathbf{L}_p \alpha^*) \cdot \zeta_{\mathrm{r}}^{-1} \mathbf{M}_{pq} \cdot (\mathbf{L}_q \beta) \right\rangle_{\mathrm{eq}}. \qquad (9.56)$$

9.4.3 Space of Euler Angles – Global Motions

In this subsection we derive, from Eq. (9.32), the final representation of the diffusion equation that is suitable for the description of the global motions [15]. For convenience, consider the field-free case from the start. We express

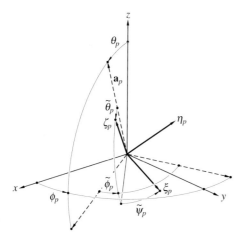

Fig. 9.3. The pth localized Cartesian coordinate system (*heavy lines*) and the pth intermediate Cartesian coordinate system (*dashed lines*) associated with the pth bond vector $\mathbf{a}_p$ expresses as $\Theta_p = (a_p, \theta_p, \phi_p)$ in spherical polar coordinates in the external coordinate system (see the text)

the pth bond vector $\mathbf{a}_p$ as $\Theta_p = (a_p, \theta_p, \phi_p)$ in spherical polar coordinates in the external coordinate system, and then introduce an intermediate Cartesian coordinate system whose orientation with respect to the former is determined by the Euler angles $(\theta_p, \phi_p, 0)$, in order to define the orientation of the pth localized coordinate system by two sets of the Euler angles $(\theta_p, \phi_p, 0)$ and $\tilde{\Omega}_p = (\tilde{\theta}_p, \tilde{\phi}_p, \tilde{\psi}_p)$ instead of the single Ω_p, the latter determining the orientation of the pth localized system with respect to the intermediate one, as depicted in Fig. 9.3. We transform the Cartesian coordinates $(\mathbf{R}_c, \{\mathbf{a}_N\}, d\{\chi_N\})$ to the curvilinear coordinates $(\mathbf{R}_c, \{\Theta_N\}, \{\tilde{\Omega}_N\})$. The metric determinant g in this space is given by Eq. (9.38) with

$$g_p = a_p{}^4 \sin^2 \theta_p \sin^2 \tilde{\theta}_p \,. \tag{9.57}$$

The diffusion Eq. (9.32) may then be transformed to that in $(\mathbf{R}_c, \{\Theta_N\}, \{\tilde{\Omega}_N\})$ space, corresponding to Eq. (9.40) with $\mathbf{J}_p^{\Theta} = (J_p^a, J_p^\theta, J_p^\phi)$ and $\mathbf{J}_p^{\tilde{\Omega}} = (J_p^{\tilde{\theta}}, J_p^{\tilde{\phi}}, J_p^{\tilde{\psi}})$. We impose the constraints $\Theta_p = (a, \theta_p, \phi_p)$ and $\tilde{\Omega}_p = (0, 0, \tilde{\psi}_p)$ $(p = 1, \cdots, N)$, by setting

$$J_p^a = J_p^{\tilde{\theta}} = J_p^{\tilde{\phi}} = 0 \qquad (p = 1, \cdots, N) \,. \tag{9.58}$$

We write the distribution function $\Psi(\mathbf{R}_c, \{\Theta_N\}, \{\tilde{\Omega}_N\})$ in the form

$$\Psi = \Psi_0(\{a_N\}, \{\tilde{\Theta}_N\}, \{\tilde{\phi}_N\}) \bar{\Psi}(\mathbf{R}_c, \{\Omega_N\}; t) \tag{9.59}$$

with

$$\Psi_0 = \prod_{p=1}^N (a_p{}^2 \sin \tilde{\theta}_p)^{-1} \delta(a_p - a) \delta(\tilde{\theta}_p) \delta(\tilde{\phi}_p) \,, \tag{9.60}$$

so that the average $\langle \alpha \rangle$ may be calculated from the second line of Eqs. (9.46).

Thus we obtain the desired diffusion equation for $\bar{\Psi} \equiv \Psi(\mathbf{R}_c, \{\Omega_N\}; t)$ in $(\mathbf{R}_c, \{\Omega_N\})$ space,

$$\frac{\partial \Psi}{\partial t} = D_c \nabla_c^2 \Psi + \sum_{p,q=1}^{N} (\sin \theta_p)^{-1} (\nabla_p^\Omega)^T \sin \theta_p \cdot \hat{\mathbf{M}}_{pq} \cdot (\nabla_q^\Omega \Psi + \Psi \nabla_q^\Omega U_0), \quad (9.61)$$

where

$$\hat{\mathbf{M}}_{pq} = a^{-2} B_{pq} \hat{\mathbf{U}}_p \cdot \hat{\mathbf{U}}_q^T + \delta_{pq} \zeta_r^{-1} \begin{pmatrix} 0 & 0 & 0 \\ 0 & 0 & 0 \\ 0 & 0 & 1 \end{pmatrix}$$

$$-a^{-2} \zeta_r \sum_{r,s=1}^{N} \hat{\mathbf{U}}_p \cdot B_{pr} (\mathbf{C}^{-1})_{rs} B_{sq} \cdot \hat{\mathbf{U}}_q^T \quad (9.62)$$

with

$$\hat{\mathbf{U}}_p = \begin{pmatrix} c_{\theta_p} c_{\phi_p} & c_{\theta_p} s_{\phi_p} & -s_{\theta_p} \\ -s_{\theta_p}^{-1} s_{\phi_p} & s_{\theta_p}^{-1} c_{\phi_p} & 0 \\ s_{\theta_p}^{-1} c_{\theta_p} s_{\phi_p} & -s_{\theta_p}^{-1} c_{\theta_p} c_{\phi_p} & 0 \end{pmatrix} \quad (9.63)$$

and with $\mathbf{C}$ being the same as that in Eq. (9.48).

In Eq. (9.62), $\mathbf{C}$ still has the meaning of the constraining matrix. The prototype of the diffusion Eq. (9.61) without the constraining term, that is, the third term on the right-hand side of Eq. (9.62), still involves the term proportional to B_{pq}, that is, the first term on the right-hand side of Eq. (9.62), so that it can give explicitly the Rouse–Zimm eigenvalues in the ground (global) state. Recall that the matrix B_{pq}, or its minor modification in the Zimm version [7], always appears in the dynamics of conventional bond chains [6,7,10]. We note that the second term on the right-hand side of Eq. (9.62) represents the (excited) rotational motion of each subbody about its ζ axis and is characteristic of the dynamic HW model. However, it is important to note that the diffusion Eqs. (9.47) and (9.61) are completely equivalent to each other (even in the non-field-free case), although the two representations are different [15].

In this case the diffusion operator $\mathcal{L}$ in Eq. (9.53) is given by

$$\mathcal{L} = -\Psi_{eq}^{-1} \sum_{p,q=1}^{N} (\sin \theta_p)^{-1} (\nabla_p^\Omega)^T \Psi_{eq} \sin \theta_p \cdot \hat{\mathbf{M}}_{pq} \cdot \nabla_q^\Omega \quad (9.64)$$

with

$$\langle \alpha, \mathcal{L}\beta \rangle = \langle \mathcal{L}\alpha, \beta \rangle$$

$$= \sum_{p,q=1}^{N} \left\langle [(\nabla_p^\Omega)^T \alpha^*] \cdot \hat{\mathbf{M}}_{pq} \cdot (\nabla_q^\Omega \beta) \right\rangle_{eq}. \quad (9.65)$$

In the second line of Eqs. (9.65), we have the relations

$$\hat{\mathbf{U}}_p^T \cdot \nabla_p^\Omega = a^{-1} \mathbf{E}_p \cdot \mathbf{L}_p$$
$$= -\mathbf{e}_{\eta_p} L_{p\xi} + \mathbf{e}_{\xi_p} L_{p\eta} . \tag{9.66}$$

From the second line of Eqs. (9.66) and the definition of the angular momentum operator given by Eq. (4.32), it is seen that the operator $\hat{\mathbf{U}}_p^T \cdot \nabla_p^\Omega$ changes infinitesimally the direction $\mathbf{e}_{\zeta_p}$ ($\equiv \mathbf{u}_p$) of the bond vector $\mathbf{a}_p$, so that it is just the gradient operator with respect to $\mathbf{u}_p$, that is,

$$\hat{\mathbf{U}}_p^T \cdot \nabla_p^\Omega = \nabla_p^u . \tag{9.67}$$

The gradient operator ∇_p^u is referred to as the *bond vector operator*, for convenience.

9.4.4 Approximation to the Constraining Matrix

In order to find solutions of the above diffusion equations, we must preaverage the constraining matrix $\mathbf{C}$ by replacing $\mathbf{C}_{pq}$ by $\langle \mathbf{C}_{pq} \rangle_{\text{eq}}$ [12,15]. We have, from Eqs. (9.51),

$$\langle \mathbf{E}_p \cdot \mathbf{E}_p^T \rangle_{\text{eq}} = \frac{2}{3} a^2 \mathbf{I} , \tag{9.68}$$

so that in the preaveraging approximation $\mathbf{C}_{pq}$ is given by

$$\mathbf{C}_{pq} = C_{pq} \mathbf{I} \tag{9.69}$$

with

$$C_{pq} = \frac{2}{3} \delta_{pq} a^2 + \zeta_{\text{r}} B_{pq} . \tag{9.70}$$

For instance, Eq. (9.48) then becomes

$$\mathbf{M}_{pq} = \delta_{pq} \mathbf{I} - (C^{-1})_{pq} \mathbf{E}_p^T \cdot \mathbf{E}_q , \tag{9.71}$$

where $(C^{-1})_{pq}$ is the pq element of the inverse of the $N \times N$ matrix C whose pq element is C_{pq}.

As mentioned in the last subsection, the physical contents of the two representations of the diffusion equation derived are exactly the same at the stage before making the preaveraging approximation in the constraining matrix $\mathbf{C}$ common to them. After the introduction of this approximation, however, they are no longer equivalent to each other but their physical contents become completely different from each other, as seen from the difference between the respective prototype diffusion equations, that is, the diffusion equation for the assembly of rigid subbodies in Eq. (9.47) and the one for the spring-bead-like model in Eq. (9.61). The constraints on the direction of $\mathbf{a}_p$ (or $\mathbf{e}_{\zeta_P}$) and its magnitude a_p may be, to some extent, destroyed by the preaveraging approximation, so that the directions of $\mathbf{a}_p$ and $\mathbf{e}_{\zeta_p}$ may not completely coincide with each other. Considering the fact that the coordinates θ_p and ϕ_p (in Ω_p) originally represent the direction of $\mathbf{e}_{\zeta_p}$ in the former case and that

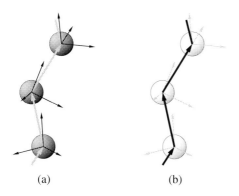

(a)   (b)

Fig. 9.4a,b. Illustrative comparison between: **a** the d-HW chain and;
b the c-HW chain. In each chain, the
heavily drawn part is being traced
(see the text)

of $\mathbf{a}_p$ in the latter, it may be mentioned that we are tracing the HW chain by
attaching probes to $\mathbf{e}_{\zeta_p}$ and $\mathbf{a}_p$ (with incomplete constraints), respectively.
The difference in the situation is illustratively shown in Fig. 9.4. Thus the
diffusion Eqs. (9.47) and (9.61) with the preaveraged $\mathbf{C}$ are suitable for the
description of the local and global (to quasi-global) motions, respectively.
The dynamic HW chains corresponding to these two equations, or Figs. 9.4a
and b, are referred to as the *discrete* HW (d-HW) and *coarse-grained* HW
(c-HW) chains, respectively, for convenience.

9.4.5 Formal Solutions

All dynamical properties in the regime of linear response may be expressed
by a standard method in terms of relevant time-correlation functions, and
therefore of the formal solutions of eigenvalue problems for the matrix representation of the diffusion operator $\mathcal{L}$ [12].

Let $G(\{\Omega_N\}, t \mid \{\Omega'_N\}, t')$ be the Green function of the linearized diffusion
Eq. (9.47) or (9.61) without the D_c term in the field-free case. This G represents the conditional probability density that the chain is found at $\{\Omega_N\}$ in
the configuration space at time t when it was at $\{\Omega'_N\}$ at time t' ($\leq t$), and
it satisfies the differential equation

$$\left(\frac{\partial}{\partial t} + \Psi_{\text{eq}} \mathcal{L} \Psi_{\text{eq}}^{-1}\right) G(\{\Omega_N\}, t \mid \{\Omega'_N\}, t') = \delta(t - t')\delta(\{\Omega_N\} - \{\Omega'_N\})$$

$$(9.72)$$

with $G = 0$ for $t < t'$, and

$$\delta(\{\Omega_N\} - \{\Omega'_N\}) = \prod_{p=1}^{N} \delta(\Omega_p - \Omega'_p).$$

$$(9.73)$$

If we define a function $\bar{G}(\{\Omega_N\}, t \mid \{\Omega'_N\}, t')$ by

$$G(\{\Omega_N\}, t \mid \{\Omega'_N\}, t) = \Psi_{\text{eq}}(\{\Omega_N\}) \bar{G}(\{\Omega_N\}, t \mid \{\Omega'_N\}, t'), \quad (9.74)$$

Eq. (9.72) reduces to

$$\left(\frac{\partial}{\partial t} + \mathcal{L}\right) \bar{G}(\{\Omega_N\}, t \mid \{\Omega'_N\}, t') = \left[\Psi_{\text{eq}}(\{\Omega_N\})\right]^{-1} \delta(t - t')$$
$$\times \delta(\{\Omega_N\} - \{\Omega'_N\}) \quad (9.75)$$

with $\bar{G} = 0$ for $t < t'$.

The formal solution of Eq. (9.75) may be written in the form

$$\bar{G}(\{\Omega_N\}, t \mid \{\Omega'_N\}, t') = \exp\left[-\mathcal{L}(t - t')\right] \left[\Psi_{\text{eq}}(\{\Omega_N\})\right]^{-1}$$
$$\times \delta(\{\Omega_N\} - \{\Omega'_N\}). \quad (9.76)$$

If the scalar product is defined by the first of Eqs. (9.55), the operator $\mathcal{L}$ is self-adjoint and has a complete orthonormal set of eigenfunctions ψ_ν and eigenvalues λ_ν with the weight Ψ_{eq}, that is,

$$\mathcal{L}\psi_\nu = \lambda_\nu \psi_\nu, \quad (9.77)$$

$$\langle \psi_\nu^* \psi_{\nu'} \rangle_{\text{eq}} = \delta_{\nu\nu'}. \quad (9.78)$$

With this set, we have the closure relation

$$\left[\Psi_{\text{eq}}(\{\Omega_N\})\right]^{-1} \delta(\{\Omega_N\} - \{\Omega'_N\}) = \sum_\nu \psi_\nu(\{\Omega_N\}) \psi_\nu^*(\{\Omega'_N\}). \quad (9.79)$$

Substitution of Eq. (9.79) into Eq. (9.76) and use of Eq. (9.77) leads to

$$\bar{G}(\{\Omega_N\}, t \mid \{\Omega'_N\}, t') = \sum_\nu \exp\left[-\lambda_\nu(t - t')\right]$$
$$\times \psi_\nu(\{\Omega_N\}) \psi_\nu^*(\{\Omega'_N\}). \quad (9.80)$$

Now we express ψ_ν in terms of the Wigner $\mathcal{D}$ functions of Ω_p as

$$\psi_\nu = \sum_\mu Q_{\mu\nu} D_\mu, \quad (9.81)$$

where

$$D_\mu = \sum_{p=1}^{N} \mathcal{D}_{l_p}^{m_p j_p}(\Omega_p) \quad (9.82)$$

with $\mu = (l_1, \cdots, l_N, m_1, \cdots, m_N, j_1, \cdots, j_N)$. Equation (9.81) gives a transformation from $\{D_\mu\}$ to $\{\psi_\nu\}$. Note that it is not unitary since $\{D_\mu\}$ and $\{\psi_\nu\}$ are orthonormal sets with different weights. In matrix notation, Eq. (9.81) may be written as

$$\psi = Q^T D. \quad (9.83)$$

From Eqs. (9.79), (9.80), and (9.83), we have

$$Q^\dagger E Q = 1 \,, \tag{9.84}$$

$$Q^\dagger L Q = \Lambda \,, \tag{9.85}$$

where the dagger indicates the adjoint, 1 and Λ are diagonal matrices with diagonal elements 1 and λ_ν, respectively, and the elements of the self-adjoint matrices E and L are given by

$$E_{\mu\mu'} = \langle D_\mu{}^* D_{\mu'} \rangle_{\text{eq}} \,, \tag{9.86}$$

$$L_{\mu\mu'} = \langle D_\mu{}^* \mathcal{L} D_{\mu'} \rangle_{\text{eq}} \,. \tag{9.87}$$

Finally, we define time-correlation functions $C_{\mu\mu'}(t)$ of D_μ by

$$C_{\mu\mu'}(t) = \left\langle D_\mu{}^*\left(\{\Omega_N\}, 0\right) D_{\mu'}\left(\{\Omega_N\}, t\right) \right\rangle_{\text{eq}} \,. \tag{9.88}$$

With the Green function G given by Eq. (9.74) with Eq. (9.76), $C_{\mu\mu'}(t)$ may then be evaluated in the usual fashion to be

$$\begin{aligned}
C_{\mu\mu'}(t) &= \langle D_\mu{}^* e^{-\mathcal{L}t} D_{\mu'} \rangle_{\text{eq}} \\
&= \sum_\mu e^{-\lambda_\nu t} (Q^{-1*})_{\nu\mu} (Q^{-1})_{\nu\mu'} \,,
\end{aligned} \tag{9.89}$$

where in the second line we have used Eqs. (9.77), (9.78), and (9.83).

9.5 Eigenvalue Problems and Time-Correlation Functions

9.5.1 Standard Basis Set

The problem is to solve the infinite-dimensional eigenvalue problem given by Eqs. (9.84) and (9.85). It may be greatly decoupled, or reduced to an infinite number of eigenvalue problems of much smaller size, by further transforming the basis set $\{D_\mu\}$ to a *standard basis set* [39] which is formed by the eigenfunctions of the *total* angular momentum operator of the entire chain [14].

We first note that $\mathcal{D}_l^{mj}(\Omega_p)$ are the simultaneous eigenfunctions of the square $\mathbf{L}_p{}^2$, the z component L_{pz}, and the ζ_p component $L_{p\zeta}$ of the angular momentum operator $\mathbf{L}_p$ with the eigenvalues $-l(l+1)$, im, and ij, respectively, as given by Eqs. (4.C.16). Then we construct from the set $\{D_\mu\}$ a new set of those basis functions which are simultaneous eigenfunctions of the square $\mathbf{L}^2$ and the z component L_z of the total angular momentum operator $\mathbf{L} = \mathbf{L}_1 + \mathbf{L}_2 + \cdots + \mathbf{L}_N$, which are linear combinations of products of $\mathcal{D}_{l_p}^{m_p j_p}(\Omega_p)$. For convenience, those basis functions which involve n $\mathcal{D}$ functions of $\Omega_{p_1}, \Omega_{p_2}, \cdots$, and Ω_{p_n} irrespective of the set $(p_1, \cdots, p_n) \equiv [p_n]$

$(p_1 < p_2 < \cdots < p_n)$ are referred to as the *n-body excitation basis functions*. For $n = 1$ and 2, they are given by

$$D_{L,[p]}^{M,j}(\Omega_p) = (8\pi^2)^{-(N-1)/2}\mathcal{D}_L^{Mj}(\Omega_p) \qquad (n = 1), \tag{9.90}$$

$$D_{L,(l_1 l_2)[p_1 p_2]}^{M,(j_1 j_2)}(\Omega_{p_1}, \Omega_{p_2}) = (8\pi^2)^{-(N-2)/2} \sum_{m_1, m_2} (l_1\, m_1\, l_2\, m_2 \,|\, l_1\, l_2\, L\, M)$$
$$\times \mathcal{D}_{l_1}^{m_1 j_1}(\Omega_{p_1})\mathcal{D}_{l_2}^{m_2 j_2}(\Omega_{p_2}) \qquad (n = 2), \tag{9.91}$$

where L and M are the total angular momentum and magnetic quantum numbers, respectively, and $(\cdots|\cdots)$ is the vector-coupling (VC) coefficient defined by Eq. (4.C.25). We note that the eigenvalues of $\mathbf{L}^2$ and L_z are $-L(L+1)$ and iM, respectively, and that $|l_1 - l_2| \leq L \leq l_1 + l_2$ and $M = m_1 + m_2 = -L, -L+1, \cdots, L$ for $n = 2$. The higher excitation basis functions may also be constructed by an application of the theory for the coupling of angular momentum vectors [40, 41]. We simply denote these functions by $D_{L,\gamma}^M$ with $\gamma = j[p]$ for $n = 1$, $\gamma = (l_1 l_2)(j_1 j_2)[p_1 p_2]$ for $n = 2$, and so on.

From the orthonormality of the $\mathcal{D}$ functions given by Eq. (4.C.8) and the unitarity of the VC coefficients given by Eq. (4.C.27), $D_{L,\gamma}^M$ are seen to have the orthonormality,

$$\int D_{L,\gamma}^{M*} D_{L',\gamma'}^{M'} d\{\Omega_N\} = \delta_{LL'}\delta_{MM'}\delta_{\gamma\gamma'}. \tag{9.92}$$

Thus, from the fact that $D_{L,\gamma}^M$ are the simultaneous eigenfunctions of $\mathbf{L}^2$ and L_z and satisfy the orthonormality of Eq. (9.92), it is seen that the set $\{D_{L,\gamma}^M\}$ is just a standard basis set in full Hilbert space [39], and therefore the desired set. We note that the transformation from $\{D_\mu\}$ to $\{D_{L,\gamma}^M\}$ is unitary and in fact orthogonal since the VC coefficient is real.

Now the scalar Ψ_{eq} and the scalar operator $\mathcal{L}$ are rotationally invariant and commute with the components of the total angular momentum operator $\mathbf{L}$. According to the theory of angular momentum [39], therefore, the standard representations E and L of the identity operator and the diffusion operator $\mathcal{L}$ with the weight Ψ_{eq} are diagonal in the quantum numbers L and M, and moreover, their diagonal elements are independent of M (a special case of the Wigner–Eckart theorem). This leads to $(2L+1)$-fold degeneracy with respect to M. The matrix elements $E_{\mu\mu'}$ and $L_{\mu\mu'}$ in the new basis set $\{D_{L,\gamma}^M\}$ may then be written in the form

$$E_{\mu\mu'} = \langle D_{L,\gamma}^{M*} D_{L',\gamma'}^{M'}\rangle_{eq} = \delta_{LL'}\delta_{MM'}E_{L,\gamma\gamma'}, \tag{9.93}$$

$$L_{\mu\mu'} = \langle D_{L,\gamma}^{M*} \mathcal{L} D_{L',\gamma'}^{M'}\rangle_{eq} = \delta_{LL'}\delta_{MM'}L_{L,\gamma\gamma'}. \tag{9.94}$$

Thus the elements $E_{L,\gamma\gamma'}$ and $L_{L,\gamma\gamma'}$ of the submatrices E_L and L_L may be evaluated simply from Eqs. (9.93) and (9.94) at $M = M' = 0$. Note also that E_L and L_L are self-adjoint, that is, $E_L = E_L^\dagger$ and $L_L = L_L^\dagger$.

We then introduce time-correlation functions $C_{\mu\mu'}(t)$ of the standard basis functions or a standard correlation matrix. It is just the standard representation of the operator $\exp(-\mathcal{L}t)$, and therefore diagonal in L and M, that is,

$$C_{\mu\mu'}(t) = \langle D_{L,\gamma}^{M*} e^{-\mathcal{L}t} D_{L',\gamma'}^{M'} \rangle_{\text{eq}} = \delta_{LL'}\delta_{MM'}C_{L,\gamma\gamma'}(t), \qquad (9.95)$$

where the submatrix elements $C_{L,\gamma\gamma'}(t)$ are independent of M and may be evaluated simply at $M = M' = 0$. Further, since $\mathcal{L}$ is a self-adjoint operator, the matrices C and C_L are seen to be self-adjoint, that is, $C_L = C_L^\dagger$, so that, in particular, $C_{L,\gamma\gamma}(t)$ are real.

Finally, we reformulate the eigenvalue problem in the standard representation. It is clear that because of Eqs. (9.93) and (9.94), the original full problem given by Eqs. (9.84) and (9.85) may be decoupled into those of smaller size for E_L and L_L, that is,

$$Q_L^\dagger E_L Q_L = 1_L, \qquad (9.96)$$

$$Q_L^\dagger L_L Q_L = \Lambda_L, \qquad (9.97)$$

where Q_L are diagonalizing matrices and are not unitary, and 1_L and Λ_L are diagonal matrices with diagonal elements 1 and λ_ν, respectively, the latter being submatrices of the original Λ. It is then easy to show that the correlation submatrices $C_L(t)$ are given by

$$C_L(t) = Q_L^{-1\dagger} \exp(-\Lambda_L t)Q_L^{-1}. \qquad (9.98)$$

The full standard representations E, L, and C (and also Q) are schematically shown in Fig. 9.5, where E_L, L_L, or C_L ($L = 0, 1, 2, \cdots$) appear in

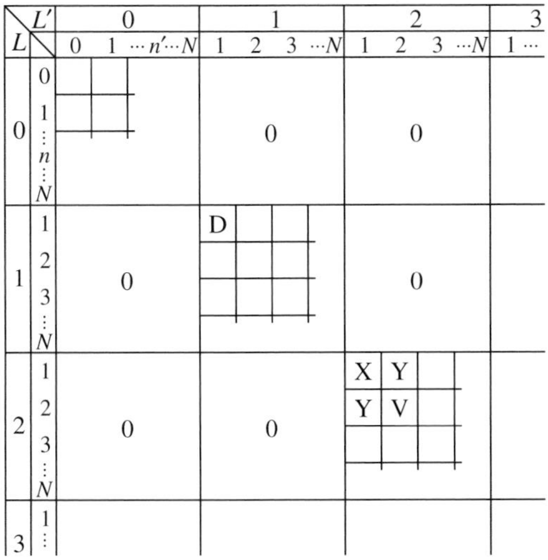

L \ L'	0			1				2				3	
	0	1	$\cdots n' \cdots N$	1	2	3	$\cdots N$	1	2	3	$\cdots N$	1	$\cdots$
0 1 : n : N					0				0				
1 2 3 : N		0		D					0				
2 2 3 : N		0			0			X Y Y V					
3 1 :													

Fig. 9.5. The full standard representations E, L, and C of the identity operator, the diffusion operator $\mathcal{L}$, and the operator $e^{-\mathcal{L}t}$, respectively (see the text)

the diagonal blocks (with $L = L'$), the submatrices in the off-diagonal blocks (with $L \neq L'$) are null matrices, and the M degeneracy has not been shown. In what follows, $L(n)$ denotes the n-body excitation for a given value of the quantum number L, or the corresponding subspace of the full Hilbert space. Note that $n = 0, 1, 2, \cdots, N$ for $L = 0$, and $n = 1, 2, \cdots, N$ for $L \neq 0$. In anticipation of results in the next chapter, we note that dielectric relaxation is associated with the subblock D in the figure, nuclear magnetic relaxation and fluorescence depolarization with the subblock X, dynamic light scattering with the subblock D or X, flow birefringence with the subblocks X and Y, and dynamic viscosity with the subblocks X, Y, and V.

9.5.2 Crude Subspace Approximation

Although we have reduced the size of the eigenvalue problem, that of the reduced one is still very large (infinite). We therefore introduce approximations to further reduce it [14]. In this subsection we first approximately decouple the space (strictly the subspace of the full Hilbert space) specified by the quantum number L into a subspace relevant to a given observable and its complementary space, for example, the subspace $1(1)$ and its complementary space $\{1(2), 1(3), \cdots, 1(N)\}$ in the case of dielectric relaxation. In this approximation, E_L, L_L, and Q_L become block-diagonal with the null off-diagonal blocks between these two subspaces, so that the problem may be solved only in the subblock D, X, or X+Y+V. Then the subspace $L(1)$ is $(2L + 1)N$-dimensional except for the M degeneracy (since $j = -L, -L+1, \cdots, L$ and $p = 1, \cdots, N$ in $\gamma = j[p]$), while the subspace $L(n)$ $(2 \leq n \leq L)$ is infinite dimensional. In this chapter we consider only the $L(1)$ problem. We note only that the subspace $\{2(1), 2(2)\}$ actually relevant to dynamic viscosity (and also flow birefringence) can be shown to be $6N$-dimensional.

In order to obtain the correlation matrix $C_{L(1)}(t)$ appearing in the subblock D $(L = 1)$ or X $(L = 2)$, we may solve the eigenvalue problem for the $(2L + 1)N \times (2L + 1)N$ submatrices $E_{L(1)}$ and $L_{L(1)}$ in the subspace $L(1)$, that is,

$$Q_{L(1)}^{\dagger} E_{L(1)} Q_{L(1)} = 1_{L(1)}, \qquad (9.99)$$

$$Q_{L(1)}^{\dagger} L_{L(1)} Q_{L(1)} = \Lambda_{L(1)}, \qquad (9.100)$$

$$C_{L(1)}(t) = Q_{L(1)}^{-1\dagger} \exp(-\Lambda_{L(1)} t) Q_{L(1)}^{-1}, \qquad (9.101)$$

instead of Eqs. (9.96)–(9.98), respectively. This is a *crude subspace approximation* [14]. Higher-order subspace approximations may be obtained if we solve the eigenvalue problem of somewhat larger size by augmenting the $L(1)$ subset with some basis functions suitably chosen from its complementary space. Note that at $t = 0$, the $C_{L(1)}(0)$ given by Eq. (9.101) is exactly correct even in the crude subspace approximation.

Now we show that the above subspace approximation (with or without augmentation) is equivalent to neglecting the memory term appearing in the projection of the full space dynamics onto the subspace $L(1)$ (with or without augmentation) by the projection operator method [42, 43]. Since the full Hilbert space has been decoupled with respect to L and M, we may consider the space specified by L from the start. Let $A(t)$ be some dynamical variable, and consider in general a subspace spanned by ν basis functions D_{L,γ_i}^M $(i = 1, \cdots, \nu)$. [Note that if $A(0)$ is confined in the space L, so also is $A(t)$.] We define the projection $\mathcal{P}A$ onto this subspace by

$$\mathcal{P}A = \sum_{i,j=1}^{\nu} D_{L,\gamma_i}^M (E_s^{-1})_{\gamma_i\gamma_j} \langle D_{L,\gamma_j}^{M*} A \rangle_{\text{eq}}, \qquad (9.102)$$

where the subscript s has been used to indicate the $\nu \times \nu$ submatrix in this subspace. If we take $A(t) = \exp(-\mathcal{L}t)D_{L,\gamma_k}^M$ $(k = 1, \cdots, \nu)$, then following the projection operator method [42, 43] we find the kinetic equation satisfied by the correlation submatrix $C_s(t)$,

$$\frac{\partial C_s(t)}{\partial t} = -L_s E_s^{-1} C_s(t) + \int_0^t K(t - t') C_s(t')dt' \qquad (9.103)$$

with $C_s(0) = E_s$, where the $\nu \times \nu$ memory kernel matrix $K = [K_{\gamma_i\gamma_j}(t)]$ is given by

$$K_{\gamma_i\gamma_j}(t) = \sum_{k=1}^{\nu} \langle D_{L,\gamma_i}^{M*} \mathcal{L} \exp[-(1 - \mathcal{P})\mathcal{L}t](1 - \mathcal{P})\mathcal{L}D_{L,\gamma_k}^M \rangle_{\text{eq}} (E_s^{-1})_{\gamma_k\gamma_j}.$$

$$(9.104)$$

Note that $\langle D_{L,\gamma}^M \rangle_{\text{eq}} = 0$ for the present case $(L \neq 0)$. If we neglect the memory term in Eq. (9.103), we obtain

$$\frac{\partial C_s(t)}{\partial t} = -L_s E_s^{-1} C_s(t) \qquad (9.105)$$

with $C_s(0) = E_s$. When $s = L(1)$, it is easy to show that the solution of Eq. (9.105) is identical with the $C_{L(1)}(t)$ approximated by Eq. (9.101). Thus we have shown the equivalence. Note that if we take the present full space L as the space s, we have $\mathcal{P} = 1$ and therefore $K = 0$, so that $C_L(t)$ exactly obeys Eq. (9.105) with E_L and L_L in place of E_s and L_s, respectively. In fact, this is consistent with Eq. (9.98).

Exact solution of Eq. (9.103) with the memory term is equivalent to finding the exact $C_s(t)$ by solving the full eigenvalue problem for E_L and L_L, and is also impossible. However, it is possible to take account of some interactions between the subspace and its complementary space by augmentation of the subspace with a small number of basis functions in the higher-order subspace approximations, as noted above, and this is equivalent to retaining partly the memory term after the projection onto the lowest subspace. This problem is treated in Sect. 9.5.4.

9.5.3 Block-Diagonal Approximation

In the last subsection we have shown that the problem is reduced to the $3N$- or $5N$-dimensional eigenvalue problem (for $L = 1$ or 2) in the $L(1)$ crude subspace approximation. For large N, therefore, we must introduce an additional approximation by a further transformation to another standard basis set [14]. The useful transformation is the one that approximately diagonalizes the matrix B defined by Eq. (9.36), and therefore also the matrix C defined by Eq. (9.70). For conventional bond chains, it is well known that B may be diagonalized in a good approximation with the orthogonal, symmetric matrix Q_{pk}^0 [11],

$$Q_{pk}^0 = \left(\frac{2}{N+1}\right)^{1/2} \sin\left(\frac{\pi pk}{N+1}\right) \quad (p, k = 1, \cdots, N), \quad (9.106)$$

which exactly diagonalizes the free-draining matrix B^0 equal to B with neglect of the second term on the right-hand side of Eq. (9.25) (that is, the Rouse matrix [6] except for the factor ζ_t^{-1}). For the dynamic HW model, we also adopt this approximation, that is,

$$(Q^0 B Q^0)_{kk'} = \delta_{kk'} \zeta_t^{-1} \lambda_k^B, \quad (9.107)$$

$$(Q^0 C Q^0)_{kk'} = \delta_{kk'} a^2 \lambda_k^C, \quad (9.108)$$

where

$$\lambda_k^C = \frac{2}{3} + \left(\frac{\zeta_r}{a^2 \zeta_t}\right) \lambda_k^B \quad (9.109)$$

with $\lambda_k^B = \zeta_t (Q^0 B Q^0)_{kk}$. Note that in the coil limit λ_k^B are just the Rouse–Zimm eigenvalues in the Hearst version [3,44]. In fact, it has been numerically shown that Eq. (9.107) is also a good approximation for the present model [16].

Now we transform the basis functions $D_{L,[p]}^{M,j}$ in the subspace $L(1)$ to new basis functions $F_{L,[k]}^{M,j}$ not only with Q^0 but also with the unnormalized $\mathcal{D}$ functions $\bar{\mathcal{D}}$ (as defined in Appendix 4.C) as follows,

$$F_{L,[k]}^{M,j}(\{\Omega_N\}) = \sum_{p=1}^{N} \sum_{j'=-L}^{L} Q_{pk}^0 \bar{\mathcal{D}}_L^{j'j}(\Omega_\alpha) D_{L,[p]}^{M,j'}(\Omega_p), \quad (9.110)$$

where L and M remain unchanged, and $\Omega_\alpha = (\alpha, -\frac{\pi}{2}, \frac{\pi}{2})$ with α being given by Eq. (4.101). It is seen that this new basis set is also a standard one in the subspace $\{1(1), 2(1), 3(1), \cdots\}$. It is referred to as the *standard Fourier basis set* (in this subspace), since Q^0 is just a Fourier sine transformation. Thus the standard Fourier representations of the identity and diffusion operators (with the weight) are also diagonal in L and M with the diagonal elements being independent of M, so that we may write them as

$$\langle F^{M,j*}_{L,[k]} F^{M',j'}_{L',[k']}\rangle_{\mathrm{eq}} = \delta_{LL'}\delta_{MM'}\bar{E}^{(j,j')}_{L,[k,k']}, \tag{9.111}$$

$$\langle F^{M,j*}_{L,[k]} \mathcal{L} F^{M',j'}_{L',[k']}\rangle_{\mathrm{eq}} = \delta_{LL'}\delta_{MM'}\bar{L}^{(j,j')}_{L,[k,k']}, \tag{9.112}$$

where we note that the elements $\bar{E}^{(j,j')}_{L,[k,k']}$ of the submatrix $\bar{E}_{L(1)}$ are the same for the two dynamic models, d- and c-HW, but the elements $\bar{L}^{(j,j')}_{L,[k,k']}$ of the submatrix $\bar{L}_{L(1)}$ are different.

The evaluation of these elements is straightforward, but we do not give the explicit expressions for them because of their length [14,45]. The results show that the matrix $\bar{E}_{L(1)}$ is diagonal in j, that for large N both $\bar{E}_{L(1)}$ and $\bar{L}_{L(1)}$ are approximately diagonal in k, and that in the case of the c-HW chain $\bar{L}_{L(1)}$ can be made *exactly* diagonal in k by further introducing the approximation that the *Fourier* bond vector operators are orthogonal to the Fourier basis functions, that is [45],

$$\left(\sum_{p=1}^{N} Q^0_{pk}\nabla^u_p\right)F^{M,j}_{L,[k']} = \delta_{kk'}\left(\sum_{p=1}^{N} Q^0_{pk}\nabla^u_p\right)F^{M,j}_{L,[k]} \qquad \text{(c-HW)}. \tag{9.113}$$

Thus the $(2L+1)N$-dimensional eigenvalue problem in the $L(1)$ crude subspace approximation given by Eqs. (9.99) and (9.100) may be reduced to N eigenvalue problems for the $(2L+1)\times(2L+1)$ matrices $\bar{E}_{L(1),[k]}$ and $\bar{L}_{L(1),[k]}$ $(k = 1, \cdots, N)$ whose jj' elements are $\delta_{jj'}\bar{E}^{(j,j)}_{L,[k,k]}$ and $\bar{L}^{(j,j')}_{L,[k,k]}$, respectively,

$$Q^{\dagger}_{L(1),[k]}\bar{E}_{L(1),[k]}Q_{L(1),[k]} = 1_{L(1),[k]}, \tag{9.114}$$

$$Q^{\dagger}_{L(1),[k]}\bar{L}_{L(1),[k]}Q_{L(1),[k]} = \Lambda_{L(1),[k]}, \tag{9.115}$$

where $1_{L(1),[k]}$ and $\Lambda_{L(1),[k]}$ are $(2L+1)\times(2L+1)$ diagonal matrices with diagonal elements 1 and $\lambda^j_{L,k}$ $(j = -L, -L+1, \cdots, L)$, respectively, and $Q_{L(1),[k]}$ is a diagonalizing matrix (not unitary). This is referred to as the *block-diagonal approximation*.

Since $\bar{E}_{L(1)}$ is diagonal in j, we may reduce the eigenvalue problem given by Eqs. (9.114) and (9.115) to that for a $(2L+1)\times(2L+1)$ self-adjoint matrix,

$$Q^{L\,\dagger}_{L(1),[k]}\left[(\bar{E}_{L(1),[k]})^{-1/2}\bar{L}_{L(1),[k]}(\bar{E}_{L(1),[k]})^{-1/2}\right]Q^L_{L(1),[k]} = \Lambda_{L(1),[k]}, \tag{9.116}$$

where $(\bar{E}_{L(1),[k]})^{-1/2}$ is the diagonal matrix with diagonal elements $(\bar{E}^{(j,j)}_{L,[k,k]})^{-1/2}$, and $Q^L_{L(1),[k]}$ is a unitary, diagonalizing matrix. Since the right-hand sides of Eqs. (9.115) and (9.116) are identical, the above two diagonalizing matrices are related to each other by

$$Q_{L(1),[k]} = (\bar{E}_{L(1),[k]})^{-1/2}Q^L_{L(1),[k]}. \tag{9.117}$$

The solutions of the three-dimensional ($L = 1$) and five-dimensional ($L = 2$) eigenvalue problems given by Eq. (9.116) can be analytically obtained, but we do not give the results [14, 45].

The correlation matrix $C_{L(1)}(t)$ in the crude subspace and block-diagonal approximations is obtained, from Eq. (9.101) with the elements $\bar{E}_{L,[k,k]}^{(j,j)}$ and the solution of Eq. (9.116), $\lambda_{L,k}^j$ and $Q_{L(1),[k]}^L$, as follows,

$$
C_{L,[p,p']}^{(j,j')}(t) = \sum_{k=1}^{N} \sum_{m,m',j''=-L}^{L} \bar{\mathcal{D}}_L^{jm}(\Omega_\alpha)\bar{\mathcal{D}}_L^{j'm'*}(\Omega_\alpha)Q_{pk}^0 Q_{p'k}^0
$$
$$
\times Q_{L,k}^{L,mj''} Q_{L,k}^{L,m'j''*}\left(\bar{E}_{L,[k,k]}^{(m,m)}\bar{E}_{L,[k,k]}^{(m',m')}\right)^{1/2} \exp(-\lambda_{L,k}^{j''}t)\,,
$$
$$(9.118)$$

where $Q_{L,k}^{L,jj'}$ is the jj' element of the unitary matrix $Q_{L(1),[k]}^L$. In contrast to the subspace approximation of Eq. (9.101) alone, the $C_{L(1)}(0)$ given by Eq. (9.118) is already approximate because of the block-diagonal approximation. For the KP chain ($\kappa_0 = 0$), both $\bar{E}_{L(1)}$ and $\bar{L}_{L(1)}$ become diagonal in j, so that we need not solve the eigenvalue problem given by Eqs. (9.114) and (9.115); that is, $Q_{L,k}^{L,jj'} = \delta_{jj'}$. Since we then also have $\bar{\mathcal{D}}_L^{jj'}(\Omega_\alpha) = \delta_{jj'}$, Eq. (9.118) reduces to

$$
C_{L,[p,p']}^{(j,j')}(t) = \delta_{jj'}\sum_{k=1}^{N} Q_{pk}^0 Q_{p'k}^0 \bar{E}_{L,[k,k]}^{(j,j)} \exp(-\lambda_{L,k}^j t) \qquad \text{(KP)} \quad (9.119)
$$

with

$$
\lambda_{L,k}^j = \lambda_{L,k}^{-j} = \left(\bar{E}_{L,[k,k]}^{(j,j)}\right)^{-1}\bar{L}_{L,[k,k]}^{(j,j)}\,. \tag{9.120}
$$

Finally, we must make some general remarks on the above $L = 1$ eigenvalues $\lambda_{1,k}^j$ ($j = -1, 0, 1$) or branches of the eigenvalue spectrum. Let the $j = 0$ branch be the lowest at small k and large N, and the eigenvalues $\lambda_{1,k}^0$ in this branch must be just the Rouse–Zimm dielectric relaxation rates. For both the d- and c-HW chains, they may be written in the form [14, 45]

$$
\lambda_{1,k}^0 = (2\zeta_r)^{-1}\left[f_k - (f_k^2 - g_k)^{1/2}\right]\,, \tag{9.121}
$$

but with

$$
g_0 < 0 \qquad\qquad \text{(d-HW)}\,, \tag{9.122}
$$
$$
g_k \propto \lambda_k^B \quad (k/N \ll 1) \qquad \text{(c-HW)}\,, \tag{9.123}
$$

so that $\lambda_{1,k}^0$ becomes negative at small wave numbers k for the d-HW chain, while we have $\lambda_{1,0}^0 = 0$ and $\lambda_{1,k}^0 > 0$ ($k \geq 1$) for the c-HW chain. This unphysical result for the d-HW chain arises from the preaveraging approximation in the constraining matrix (even without the block-diagonal approximation), indicating that it cannot describe correctly the global motions. In the next chapter, therefore, for the d-HW chain we use $\lambda_{1,k}^j - \lambda_{1,0}^0$ as the *corrected* $L = 1$ eigenvalues in all branches, for convenience. On the other hand, the above reasonable result at small wave numbers for the c-HW chain is due to the orthogonal approximation of Eq. (9.113). It can be shown that in the case of $L = 1$ (vector mode), Eq. (9.113) is exactly valid for the Gaussian (spring) bonds, so that it is indeed a good approximation at small k and large N for the c-HW chain [45]. In the case of $L = 2$ (tensor mode), however, it cannot be valid. Thus, in the next chapter, for the c-HW chain we consider only the $L = 1$ problems. We note that even with the orthogonal approximation, $\lambda_{1,0}^0$ for the d-HW chain cannot be made to vanish. As for the block-diagonal approximation, we further note that it has been numerically shown to be a good approximation also for the d-HW chain as far as the $L = 1$ eigenvalues for flexible chains with large N are concerned [16].

9.5.4 Higher-Order Subspace Approximation

In this subsection we briefly consider the correlation matrix $C_{L(1)}(t)$ in higher-order subspace approximations [46, 47], starting from Eq. (9.103) with $s = L(1)$, where we note that if the subspace $L(1)$ is ν_s-dimensional, then C_s, E_s, L_s, and K are $\nu_s \times \nu_s$ matrices.

Now we write the matrices E_L and L_L as

$$E_L = \begin{pmatrix} E_s & \epsilon E_i^\dagger \\ \epsilon E_i & E_c \end{pmatrix} , \tag{9.124}$$

and the like, where the subscript c indicates the subspace $\{L(2), \cdots, L(N)\}$ complementary to the relevant subspace $s = L(1)$, and the subscript i indicates the coupling (interaction) between the two subspaces. The parameter ϵ (small number) has been introduced to treat the interaction as perturbation and perform evaluation to terms of $\mathcal{O}(\epsilon^2)$, and it should be set equal to unity at the final stage. The subspace c is actually infinite dimensional, and in the higher-order subspace approximations we consider only its finite proper subspace, whose dimension is assumed to be ν_c. In what follows, we redefine the subspace c in Eq. (9.124) in such a way. Then E_c and L_c are $\nu_c \times \nu_c$ matrices, and E_i and L_i are $\nu_c \times \nu_s$ matrices. Let Q_s and Q_c be the matrices (not unitary) that simultaneously diagonalize E_s and L_s, and E_c and L_c, respectively. The memory kernel matrix $K(t)$ may then be evaluated to be

$$K(t) = \epsilon^2 Q_s^{-1\dagger} \Gamma_1^\dagger \exp(-\Lambda_c t) \Gamma_1 Q_s^\dagger \tag{9.125}$$

with

$$\Gamma_1 = Q_c^\dagger L_i Q_s - Q_c^\dagger E_i Q_s \Lambda_s , \tag{9.126}$$

where $\Lambda_s = \Lambda_{L(1)}$ is the diagonal matrix, whose diagonal elements are $\lambda_{L,k}^j$ in the block-diagonal approximation, Λ_c is a diagonal matrix with diagonal elements $\lambda_{c,\nu}$, and the elements of the $\nu_c \times \nu_s$ matrix Γ_1 are denoted by $\Gamma_{1,\nu j}$.

For flexible chains the solution thus obtained for $C_{L(1)}(t)$ in a good approximation has the following properties: (1) the amplitudes, or $Q_{L(1)}$, remain unchanged, (2) only the eigenvalues $\lambda_{L,k}^j$ are changed to $\bar{\lambda}_{L,k}^j$, and (3) the subspace c is only a small part of the subspace $L(2)$ with $\lambda_{c,\nu}$ being the $L(2)$ eigenvalues. Thus we may write $\lambda_{c,\nu}$ and $\Gamma_{1,\nu j}$ as

$$\lambda_{c,\nu} = \lambda_{L,(l_1 l_2)[k_1 k_2]}^{(j_1 j_2)}, \tag{9.127}$$

$$\Gamma_{1,\nu j} = \Gamma_{1,(L l_1 l_2 k_1 k_2 j_1 j_2)[Lkj]}. \tag{9.128}$$

The changed eigenvalues $\bar{\lambda}_{L,k}^j$ ($L = 1, 2$) may then be expressed as [47]

$$\bar{\lambda}_{L,k}^j = \lambda_{L,k}^j \left\{ 1 + \sum_{\nu} [(\lambda_{L,k}^j)^{-1} |\Gamma_{1,\nu j}|^2 + \Gamma_{1,\nu j}^* \Gamma_{2,\nu j}] \lambda_{c,\nu}^{-1} \right\}^{-1}, \tag{9.129}$$

where

$$\sum_{\nu} = \sum_{l_1=1}^{2} \sum_{l_2=l_1}^{l_1+L} \left\{ \delta_{l_1 l_2} \sum_{\substack{k_1,k_2=1 \\ k_1 < k_2}}^{N} \left[\delta_{k_1 k_2} \sum_{\substack{j_1,j_2=-l_1 \\ j_1 < j_2}}^{l_1} + (1 - \delta_{k_1 k_2}) \sum_{j_1,j_2=-l_1}^{l_1} \right] \right.$$
$$\left. + (1 - \delta_{l_1 l_2}) \sum_{k_1,k_2=1}^{N} \sum_{j_1=-l_1}^{l_1} \sum_{j_2=-l_2}^{l_2} \right\}, \tag{9.130}$$

$$\Gamma_{2,\nu j} = \frac{\Gamma_{1,\nu j}}{\lambda_{c,\nu} - \lambda_{L,k}^j} \quad \text{for } |\lambda_{c,\nu} - \lambda_{L,k}^j| \geq 0.2 \lambda_{L,k}^j$$

$$= 0 \quad \text{for } |\lambda_{c,\nu} - \lambda_{L,k}^j| < 0.2 \lambda_{L,k}^j. \tag{9.131}$$

It must however be noted that the higher-order subspace approximation is not very good for stiff chains, for which the eigenvalues are evaluated in the crude subspace approximation (see the next chapter).

Appendix 9.A Fluctuating Hydrodynamic Interaction

In this appendix we evaluate the effect of fluctuating hydrodynamic interaction (HI) on the translational diffusion coefficient D on the basis of the dynamic HW chain and also the Gaussian (spring-bead) chain with partially fluctuating (orientation-dependent) HI [48]. In Sect. 6.5.2 we have already given a brief survey of the theoretical investigations of the effects on D and also the intrinsic viscosity $[\eta]$.

For this purpose, we use a partially preaveraged form $\bar{\mathbf{T}}(\mathbf{r})$ of the Oseen HI tensor $\mathbf{T}(\mathbf{r})$ given by Eq. (6.4),

$$\overline{\mathbf{T}}(\mathbf{r}) = \frac{1}{8\pi\eta_0}\langle r^{-1}\rangle_{\text{eq}}(\mathbf{I} + \mathbf{e}_r\mathbf{e}_r)\,. \tag{9.A.1}$$

Note that this tensor has been averaged only over the magnitude r of the vector distance $\mathbf{r}$ retaining its anisotropic part $\mathbf{e}_r\mathbf{e}_r$, so that it can give correct results in the case of rigid bodies. It is then convenient to treat as perturbation the deviation of $\overline{\mathbf{T}}$ from the (isotropic) preaveraged Oseen tensor, which we denote by $T^{(0)}(\mathbf{r})\mathbf{I}$, so that we rewrite $\overline{\mathbf{T}}$ in the form

$$\overline{\mathbf{T}}(\mathbf{r}) = T^{(0)}(\mathbf{r})\mathbf{I} + \epsilon\mathbf{T}^{(1)}(\mathbf{r})\,, \tag{9.A.2}$$

where ϵ is a small perturbation parameter and $\mathbf{T}^{(1)}(\mathbf{r})$ is the fluctuating part given by

$$\mathbf{T}^{(1)}(\mathbf{r}) = \frac{1}{8\pi\eta_0}\langle r^{-1}\rangle_{\text{eq}}\left(\mathbf{e}_r\mathbf{e}_r - \tfrac{1}{3}\mathbf{I}\right)\,. \tag{9.A.3}$$

Evaluation is carried out to terms of $\mathcal{O}(\epsilon^2)$, and ϵ should be set equal to unity at the final stage of calculations.

We consider the field-free case. By the use of Eqs. (9.27), (9.29), and (9.31), the diffusion equation for the distribution function $\Psi(\mathbf{R}_c, \{\Omega_N\}; t)$ can then be derived, but we omit the result. Here, the point $\mathbf{R}_c$ has no special meaning in the present case of fluctuating HI. However, we still use it to specify the location of the chain, and refer to it as the Zimm center of resistance, for convenience.

Now recall that the (mean) translational diffusion coefficient of the center of mass has been shown to be dependent on time for the Gaussian chain [49–51] and also the trumbbell [52]. This is in general the case with the translation diffusion coefficient $D(t)$ of any point affixed to the polymer chain; it decreases with increasing t and becomes a constant $D(\infty)$ independent of the point after all internal motions have relaxed away. Indeed, $D(\infty)$ is measured in almost all experiments on D such as sedimentation and dynamic light scattering. In this connection, we note that although Fixman [53] has considered the effects of the constraints and therefore chain stiffness on the basis of the freely rotating chain, his analysis is essentially limited to $D(0)$. We may evaluate $D(t)$ (of $\mathbf{R}_c$) from a kinetic equation for the distribution function $\overline{\Psi}(\mathbf{R}_c; t)$ derived by the projection operator method [42,43], as done for the Gaussian chain [51] and the trumbbell [52], where $\overline{\Psi}(\mathbf{R}_c; t)$ is defined by

$$\overline{\Psi}(\mathbf{R}_c; t) = \int \Psi(\mathbf{R}_c, \{\Omega_N\}; t)d\{\Omega_N\}\,. \tag{9.A.4}$$

Thus, if we apply a projection operator defined by

$$\mathcal{P} = \Psi_{\text{eq}}\int d\{\Omega_N\}\,, \tag{9.A.5}$$

putting $\epsilon = 1$, and if we preaverage the constraining matrix $\mathbf{C}$, then we obtain, from Eq. (9.A.4), the length-coarse-grained kinetic equation (considering only terms of k_c^2 in Fourier space $\mathbf{k}_c$) satisfied by $\overline{\Psi}(\mathbf{R}_c; t)$,

$$\frac{\partial \overline{\Psi}(t)}{\partial t} = D_0 \nabla_c^2 \overline{\Psi}(t) - \frac{r_1^2 a^4}{12(N+1)} \sum_{k=1}^{N} S_k \int_0^t K_k(t-s) \nabla_c^2 \overline{\Psi}(s) ds \,,$$

$$(9.A.6)$$

where

$$D_0 = D_c^{(0)} - \frac{3r_1^2 r_2 a^2}{2\pi(N+1)\zeta_t \bar{a}^2} \sum_{k=1}^{N} S_k (\lambda_k^C)^{-1} \,,$$

$$(9.A.7)$$

$$S_k \mathbf{I} = \frac{2\pi(N+1)(\zeta_t \bar{a})^2}{3r_1^2 a^2} \sum_{p,q=1}^{N} Q_{pk}^0 Q_{qk}^0 \langle \mathbf{D}_{c,p}^{(1)} \cdot \mathbf{D}_{c,q}^{(1)} \rangle_{eq} \,,$$

$$(9.A.8)$$

$$K_k(t) = \sum_{j=-1}^{0} |R_{1,k}^{j0}|^2 \left(\frac{\lambda_{1,k}^j}{\lambda_k^B}\right)^2 \exp(-\lambda_{1,k}^j t)$$

$$(9.A.9)$$

with

$$\bar{a} = (c_\infty \Delta s)^{1/2} \,,$$

$$(9.A.10)$$

$$\lambda_k^C = \tfrac{2}{3} + r_2 \lambda_k^B \,,$$

$$(9.A.11)$$

$$R_{1,k}^{jj'} = (8\pi^2)^{N/2} \sum_{m=-1}^{1} Q_{1,k}^{L,mj*} (\bar{E}_{1,[k,k]}^{(m,m)})^{1/2} \bar{\mathcal{D}}_1^{j'm*}(\Omega_\alpha) \,,$$

$$(9.A.12)$$

and with $r_1 = \zeta_t/3\pi\eta_0 a$ and $r_2 = \zeta_r/a^2\zeta_t$. In Eq. (9.A.7), $D_c^{(0)}$ is identical with the D_c given by Eq. (9.35). In Eq. (9.A.8), Q_{pk}^0 is defined by Eq. (9.106) and $\mathbf{D}_{c,p}^{(1)}$ is defined by

$$\mathbf{D}_{c,p}^{(1)} = \sum_{q=1}^{N+1} w_q \left(-\mathbf{D}_{qp}^{(1)} + \mathbf{D}_{q(p+1)}^{(1)}\right)$$

$$(9.A.13)$$

with

$$\mathbf{D}_{pq}^{(1)} = (1 - \delta_{pq}) \mathbf{T}^{(1)}(\mathbf{R}_{pq}) \,.$$

$$(9.A.14)$$

In Eq. (9.A.9), λ_k^B is defined in Eq. (9.107), and the eigenvalues $\lambda_{1,k}^j$ and coefficients $|R_{1,k}^{j0}|^2$ arise from the correlation matrix $C_{L(1)}(t)$ given by Eq. (9.118). It is then important to see from Eq. (9.A.6) with Eq. (9.A.9) that if the fluctuation in HI is considered, the translational motion is coupled with the internal modes (all the Rouse vector modes) that are composed of the $j = 0$ and -1 branches of the dielectric eigenvalue spectrum [54] (see also Sect. 10.1).

When we consider the translational diffusion at the initial stage ($t = 0$), we may suppress the second term (the memory term) on the right-hand side of Eq. (9.A.6). On the other hand, when we consider the diffusion on the time scales sufficiently long compared to the "dielectric" relaxation times, we may apply the usual time-coarse-graining procedure. Thus we have, from Eq.(9.A.6),

$$\frac{\partial \overline{\Psi}(t)}{\partial t} = D(0)\nabla_c^2 \overline{\Psi}(t) \qquad (t \to 0) \tag{9.A.15}$$

$$= D(\infty)\nabla_c^2 \overline{\Psi}(t) \qquad (t \to \infty), \tag{9.A.16}$$

where $D(0) = D_0$ and $D(\infty)$ is given by

$$D(\infty) = D_0 - \frac{r_1^2 a^4}{12(N+1)} \sum_{k=1}^{N} S_k \int_0^\infty K_k(t)\,dt. \tag{9.A.17}$$

In what follows, we consider the case of large N. Then $D_c^{(0)}$ in Eq. (9.A.7) for D_0 is identical with the Zimm translational diffusion coefficient $D^{(Z)}$ (at $t = 0$ and ∞) [7],

$$D_c^{(0)} = D^{(Z)} = \frac{\Gamma(5/4)}{3\pi\Gamma(3/4)\eta_0 \langle S^2 \rangle^{1/2}}, \tag{9.A.18}$$

where Γ is the gamma function, and $\langle S^2 \rangle = \langle S^2 \rangle_{\text{eq}}$ is given by

$$\langle S^2 \rangle = \tfrac{1}{6} N \bar{a}^2. \tag{9.A.19}$$

Thus we can obtain, from Eqs. (9.A.7) and (9.A.17) with Eq. (9.A.18), for the translational diffusion coefficients $D(0)$ and $D(\infty)$

$$D(0) = D^{(Z)}(1 - \delta_0), \tag{9.A.20}$$

$$D(\infty) = D^{(Z)}(1 - \delta_0 - \delta_1), \tag{9.A.21}$$

where

$$\delta_0 = A r_1 r_2 a (c_\infty N \Delta s)^{-1/2} \sum_{k=1}^{N} S_k (\lambda_k^C)^{-1}, \tag{9.A.22}$$

$$\delta_1 = A r_1 a (c_\infty N \Delta s)^{-1/2} \sum_{k=1}^{N} S_k (\lambda_k^B)^{-1} \tag{9.A.23}$$

with

$$A = (3/8)^{1/2} \Gamma(3/4)/\pi\Gamma(5/4). \tag{9.A.24}$$

In Eq. (9.A.23), we have ignored the $j = -1$ branch of the eigenvalue spectrum, which is a minor contribution, with the amplitude of unity for the $j = 0$ branch, and put $\zeta_r \lambda_{1,k}^0 = 3r_2\lambda_k^B$ for all k to avoid the negative eigenvalues at small k [see also Eq. (9.121) with Eq. (9.123)]. Thus the translational motion may be correctly described, although we have derived the kinetic equation for the d-HW chain, for simplicity. We use the Gaussian approximation to evaluate the average in Eq. (9.A.8).

Now we have for the ratio ρ_∞ defined by Eqs. (6.130) and (6.132) with $D(\infty)$

$$\rho_\infty = \rho_\infty^{(Z)}(1 - \delta_0 - \delta_1), \tag{9.A.25}$$

where the subscript ∞ indicates the value for $N \to \infty$, and we note that $\rho_{\infty}^{(Z)} = 2\Gamma(5/4)/\Gamma(3/4) = 1.479$ is the Zimm value of ρ_{∞} for the center of resistance of the Gaussian chain at $t = 0$ and ∞ (with preaveraged HI). Note that the Kirkwood value $\rho_{\infty}^{(K)} = 8/3\sqrt{\pi} = 1.505$ is the value of ρ_{∞} for the center of mass of the Gaussian chain at $t = 0$ [1, 55]. Clearly the factor $1 - \delta_0 - \delta_1$ in Eq. (9.A.21) and (9.A.25) arises from the fluctuation in HI.

For flexible chains, the results of numerical calculations shows that

$$\delta_0 \simeq 0.02 \qquad (9.A.26)$$

independently of the HW model parameters, while δ_1 depends weakly on them, where we have put $r_1 = 1$. The values of ρ_{∞} thus calculated from Eq. (9.A.25) for several flexible polymers have already been given in Table 6.4. It is seen from the above analysis that the term δ_0 represents the decrease in D (from $D^{(Z)}$) at $t = 0$ and arises from the restriction of the chain motions by the constraints, while the term δ_1 represents the additional decrease at $t = \infty$ and arises from the coupling between the translational and internal motions, especially the long-wavelength internal motions, through the fluctuating part of the HI. Thus it may be considered that the preaveraging of the constraining matrix leads to an underestimate of δ_0 and therefore an overestimate of ρ_{∞} (see Table 6.4).

For stiff chains, which may be represented by the KP chain with $\kappa_0 = 0$ and $c_{\infty} = 1$, we simply consider the stiff-chain limit of $\lambda^{-1} \to \infty$ or $\Delta s \to 0$, so that we have $a = \Delta s$ from Eq. (9.14). It can then be shown that

$$\delta_0 = \delta_1 = 0 , \qquad (9.A.27)$$

$$\rho_{\infty} = \rho_{\infty}^{(Z)} = 1.479 \qquad (\lambda^{-1} \to \infty) . \qquad (9.A.28)$$

This value of ρ_{∞} is to be compared with the experimental values 1.50 for PHIC [56] and 1.48 for DNA [57, 58].

Finally, we briefly consider the case of the Gaussian (or spring-bead) chain composed of $N + 1$ identical beads connected with the effective bond length $\bar{a}$ given by Eq. (9.A.10). Following the same procedure as above, we can then obtain Eqs. (9.A.20), (9.A.21), and (9.A.25) but with

$$\delta_0 = 0 , \qquad (9.A.29)$$

$$\delta_1 = Ar_1 N^{-1/2} \sum_{k=1}^{N} S_k(\lambda_k^B)^{-1} \qquad \text{(Gaussian chain)} \qquad (9.A.30)$$

with A being given by Eq. (9.A.24) and with $r_1 = \zeta_t/3\pi\eta_0\bar{a}$. Thus δ_1 does not vanish even for the Gaussian chain (without constraints). Equation (9.A.23) for δ_1 (for the HW chain) becomes identical with Eq. (9.A.30) in the flexible-chain limit of $\Delta s \to \infty$, since we have, from Eq. (9.14), $a = \bar{a} = (c_{\infty}\Delta s)^{1/2}$ ($\Delta s \to \infty$). If we assume that the bead is the Stokes sphere of diameter d_b, that is, $\zeta_t = 3\pi\eta_0 d_b$, then we have $r_1 = d_b/\bar{a}$. It has been shown that as $d_b/\bar{a}$

is increased from 0.1 to 1.0, ρ_∞ decreases from 1.412 to 1.294; in particular, $\rho_\infty = 1.373$ for $d_b/\bar{a} = 0.3$ [48]. It is interesting to note that this value of ρ_∞ is somewhat larger than the corresponding Zimm Monte Carlo value 1.31 obtained for $d_b/\bar{a} = 0.27$ in the rigid-body ensemble approximation [59]. This difference may be regarded as arising from the initial decrease δ_0 in D due to the constraints in the latter case of the "rigid" Gaussian chain. In this connection, we note that Fixman [60] has shown that the rigid-body ensemble approximation gives the lower bound for ρ_∞ for a given model chain.

References

1. J. G. Kirkwood: Rec. Trav. Chim. **68**, 649 (1949); J. Polym. Sci. **12**, 1 (1954).
2. M. Fixman and W. H. Stockmayer: Ann. Rev. Phys. Chem. **21**, 407 (1970).
3. H. Yamakawa: *Modern Theory of Polymer Solutions* (Harper & Row, New York, 1971).
4. W. H. Stockmayer: In *Molecular Fluids—Fluides Moleculaires*, R. Balian and G. Weill, eds. (Gordon & Breach, New York, 1976), p. 107.
5. M. Doi and S. F. Edwards: *The Theory of Polymer Dynamics* (Clarendon Press, Oxford, 1986).
6. P. E. Rouse, Jr.: J. Chem. Phys. **21**, 1272 (1953).
7. B. H. Zimm: J. Chem. Phys. **24**, 269 (1956).
8. Y. Ikeda: Bull. Kobayashi Inst. Phys. Res. **6**, 44 (1956).
9. J. Erpenbeck and J. G. Kirkwood: J. Chem. Phys. **29**, 909 (1958); **38**, 1023 (1963).
10. M. Fixman and J. Kovac: J. Chem. Phys. **61**, 4939 (1974).
11. M. Fixman and G. T. Evans: J. Chem. Phys. **68**, 195 (1978).
12. H. Yamakawa and T. Yoshizaki: J. Chem. Phys. **75**, 1016 (1981).
13. H. Yamakawa: In *Molecular Conformation and Dynamics of Macromolecules in Condensed Systems*, M. Nagasawa, ed. (Elsevier, Amsterdam, 1988), p. 21.
14. H. Yamakawa, T. Yoshizaki, and J. Shimada: J. Chem. Phys. **78**, 560 (1983).
15. T. Yoshizaki and H. Yamakawa: J. Chem. Phys. **104**, 1120 (1996).
16. H. Yamakawa and T. Yoshizaki: J. Chem. Phys. **78**, 572 (1983).
17. H. A. Kramers: J. Chem. Phys. **14**, 415 (1946).
18. R. B. Bird, O. Hassager, R. C. Armstrong, and C. F. Curtiss: *Dynamics of Polymeric Liquids* (John Wiley, New York, 1977), Vol. 2.
19. C. F. Curtiss, R. B. Bird, and O. Hassager: Adv. Chem. Phys. **35**, 31 (1976).
20. M. Fixman: J. Chem. Phys. **69**, 1527 (1978).
21. H. Yamakawa: Ann. Rev. Phys. Chem. **35**, 23 (1984).
22. M. Doi and K. Okano: Polym. J. **5**, 216 (1973).
23. H. Yamakawa and G. Tanaka: J. Chem. Phys. **63**, 4967 (1975).
24. G. Wilemski: J. Stat. Phys. **14**, 153 (1976).
25. E. Helfand: J. Chem. Phys. **71**, 5000 (1979).
26. E. Helfand, Z. R. Wasserman, and T. A. Weber: Macromolecules **13**, 526 (1980).
27. E. Helfand, Z. R. Wasserman, T. A. Weber, J. Skolnick, and J. H. Runnels: J. Chem. Phys. **75**, 4441 (1981).
28. M. R. Pear and J. H. Weiner: J. Chem. Phys. **71**, 212 (1979).
29. D. Perchak and J. H. Weiner: Macromolecules **14**, 785 (1981).
30. M. Fixman: Proc. Natl. Acad. Sci. USA **71**, 3050 (1974).
31. R. A. Harris and J. E. Hearst: J. Chem. Phys. **44**, 2595 (1966).
32. N. Saito, K. Takahashi, and Y. Yunoki: J. Phys. Soc. Japan **22**, 219 (1967).

33. K. Soda: J. Phys. Soc. Japan **35**, 866 (1973); J. Chem. Phys. **95**, 9337 (1991).
34. M. Bixon and R. Zwanzig: J. Chem. Phys. **68**, 1896 (1978).
35. Yu. Ya. Gotlib and Yu. Ye. Svetlov: Polym. Sci. USSR **21**, 1682 (1979).
36. R. G. Winkler, P. Reineker, and L. Harnau: J. Chem. Phys. **101**, 8119 (1994).
37. L. Harnau, R. G. Winkler, and P. Reineker: J. Chem. Phys. **102**, 7750 (1995); **104**, 6355 (1996).
38. O. Hassager and R. B. Bird: J. Chem. Phys. **56**, 2498 (1972).
39. A. Messiah: *Quantum Mechanics* (North-Holland, Amsterdam, 1970), Vol. II.
40. A. R. Edmonds: *Angular Momentum in Quantum Mechanics* (Princeton Univ., Princeton, 1974).
41. A. P. Yutsis, I. B. Levinson, and V. V. Vanagas: *The Theory of Angular Momentum* (Israel Program for Scientific Translations, Jerusalem, 1962).
42. H. Mori: Prog. Theor. Phys. (Kyoto) **33**, 423 (1965).
43. R. Zwanzig: J. Chem. Phys. **60**, 2717 (1974); In *Molecular Fluids—Fluides Moleculaires*, R. Balian and G. Weill, eds. (Gordon & Breach, New York, 1976). p. 1.
44. J. E. Hearst: J. Chem. Phys. **37**, 2547 (1962).
45. T. Yoshizaki, M. Osa, and H. Yamakawa: J. Chem. Phys. **105**, 11268 (1996).
46. T. Yoshizaki and H. Yamakawa: J. Chem. Phys. **84**, 4684 (1986).
47. H. Yamakawa, T. Yoshizaki, and M. Fujii: J. Chem. Phys. **84**, 4693 (1986).
48. H. Yamakawa and T. Yoshizaki: J. Chem. Phys. **91**, 7900 (1989).
49. E. Dubois-Violette and P.-G. de Gennes: Physics **3**, 181 (1967).
50. M. Fixman: Macromolecules **14**, 1710 (1981).
51. A. Z. Akcasu: Macromolecules **15**, 1321 (1982).
52. K. Nagasaka and H. Yamakawa: J. Chem. Phys. **83**, 6480 (1985).
53. M. Fixman: Faraday Discuss. No. **83**, 199 (1987); J. Chem. Phys. **89**, 2442 (1988).
54. T. Yoshizaki and H. Yamakawa: J. Chem. Phys. **81**, 982 (1984).
55. J. G. Kirkwood and J. Riseman: J. Chem. Phys. **16**, 565 (1948).
56. H. Murakami, T. Norisuye, and H. Fujita: Macromolecules **13**, 345 (1980).
57. M. T. Record, Jr., C. P. Woodbury, and R. B. Inman: Biopolymers **14**, 393 (1975).
58. J. E. Godfrey and H. Eisenberg: Biophys. Chem. **5**, 301 (1976).
59. B. H. Zimm: Macromolecules **13**, 592 (1980).
60. M. Fixman: J. Chem. Phys. **78**, 1588 (1983).

10 Dynamical Properties

In this chapter various dynamical properties of unperturbed polymer chains in dilute solution in the regime of linear response are evaluated by the use of the time-correlation functions formulated in Chap. 9 on the basis of the dynamic HW chain. They include dielectric relaxation, nuclear magnetic relaxation, fluorescence depolarization, dynamic light scattering, and so on for both flexible and semiflexible polymers. Evaluation is carried out for the d-HW chain except for the first cumulant of the dynamic structure factor, which is evaluated for the c-HW chain. The eigenvalues are evaluated in the crude and also higher-order subspace approximations for the flexible d-HW chain but only in the crude approximation for the stiff d-HW chain and the c-HW chain. All dynamical properties considered in this chapter concern local chain motions except in the cases of the dielectric constant of semiflexible polymers and the first cumulant. A comparison of theory with experiment is made rather in detail along with a discussion of the approximations introduced in Chap. 9.

10.1 Dielectric Relaxation

10.1.1 Formulation

For flexible chains having parallel (type-A) dipoles, it is well known that their dielectric relaxation may be conveniently treated using the spring-bead model [1–4]. For chains having perpendicular (type-B) dipoles, the process is associated with the local chain motions, and therefore we should have recourse to the dynamic HW chain. We give the formulation generally applicable to both flexible and stiff chains having arbitrary dipoles [5].

Now let $\epsilon^* = \epsilon' - i\epsilon''$ be the excess complex dielectric constant as a function of angular frequency ω of the dilute solution over that of the solvent alone. If the effect of local fields is ignored, it may be expressed in terms of the dipole correlation function $M(t)$ (with t the time) as [6,7]

$$\frac{\epsilon^* - \epsilon_\infty}{\epsilon_0 - \epsilon_\infty} = 1 - i\omega \int_0^\infty e^{-i\omega t} \left[\frac{M(t)}{M(0)} \right] dt, \qquad (10.1)$$

where ϵ_0 and ϵ_∞ are the excess limiting low- and high-frequency dielectric constants, respectively. If $\boldsymbol{\mu}(t)$ is the instantaneous, field-free, dipole moment vector of the entire chain expressed in an external Cartesian coordinate system, $M(t)$ is given by

$$M(t) = \langle \boldsymbol{\mu}(0) \cdot \boldsymbol{\mu}(t) \rangle_{\mathrm{eq}} . \tag{10.2}$$

In what follows (in this chapter), we adopt the same dynamic HW chain and notation as those used in Chap. 9. Further, all lengths are measured in units of λ^{-1} and $k_B T$ is chosen to be unity unless otherwise noted.

Let $\mathbf{m}_p$ and $\tilde{\mathbf{m}}_p$ be the local electric dipole moment vectors attached to the pth subbody of the dynamic HW chain, expressed in the pth localized Cartesian coordinate system $(\mathbf{e}_{\xi_p}, \mathbf{e}_{\eta_p}, \mathbf{e}_{\zeta_p})$ affixed to it and the external one, respectively, and we have

$$\boldsymbol{\mu} = \sum_{p=1}^{N} \tilde{\mathbf{m}}_p . \tag{10.3}$$

We assume that their magnitudes are independent of p so that $|\mathbf{m}_p| = |\tilde{\mathbf{m}}_p| = m$. Further, suppose that the vector $\mathbf{m}_p$ is permitted to rotate about an axis, making a constant angle Δ with the axis, which has constant polar and azimuthal angles α and β (independent of p) in the pth localized coordinate system, as depicted in Fig. 10.1. Let $\gamma_p(t)$ be the (time-dependent) dihedral angle between the two planes containing the rotation axis and $\mathbf{e}_{\zeta_p}$, and the rotation axis and $\mathbf{m}_p$, respectively. The pth dipole moment vector expressed in a Cartesian coordinate system having the orientation defined by the Euler angles $(\alpha, \beta, \gamma_p)$ with respect to the pth localized coordinate system is independent of p. If we denote it by $\bar{\mathbf{m}}$, we have $\bar{\mathbf{m}} = (m \sin \Delta, 0, m \cos \Delta)$.

Since the scalar product $\tilde{\mathbf{m}}_p(0) \cdot \tilde{\mathbf{m}}_{p'}(t)$ may be expressed in terms of the spherical components $\tilde{m}_p^{(j)}$ $(j = 0, \pm 1)$ of $\tilde{\mathbf{m}}_p$ as in Eq. (5.156), $M(t)$ may be given, from Eq. (10.2) with Eq. (10.3), by

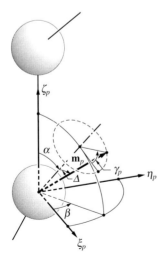

Fig. 10.1. Local dipole moment vector $\mathbf{m}_p$ in the pth localized Cartesian coordinate system

$$M(t) = \sum_{p,p'=1}^{N} \sum_{j=-1}^{1} \left\langle \tilde{m}_p^{(j)*}(0)\, \tilde{m}_{p'}^{(j)}(t) \right\rangle_{\text{eq}}. \tag{10.4}$$

By the use of Eq. (5.B.5), the component $\tilde{m}_p^{(j)}$ may be written as a sum of products of the Wigner functions $\mathcal{D}_1^{jk_1}(\Omega_p)$ and $\mathcal{D}_1^{k_1 k_2}(\alpha,\,\beta,\,\gamma_p)$ and the spherical components $\bar{m}^{(k_2)}$ of $\bar{\mathbf{m}}$ given by

$$\bar{m}^{(\pm 1)} = \mp \frac{1}{\sqrt{2}}\, m \sin \Delta\,,$$
$$\bar{m}^{(0)} = m \cos \Delta\,. \tag{10.5}$$

Now, if we assume that there are no correlations between the motion of each subbody (main-chain motion) and the rotational motion of the dipole moment vector about the rotation axis in it (side-chain motion) and also between the latter motions in different subbodies, $M(t)$ may be eventually expressed in terms of correlation functions $M^{jj'}(t)$, $M_s^{jj'}(t)$, and $C_{s1}^{jj'}(t)$ defined by

$$M^{jj'}(t) = (8\pi^2)^N \sum_{p,p'=1}^{N} C_{1,[p,p']}^{(j,j')}(t)\,, \tag{10.6}$$

$$M_s^{jj'}(t) = (8\pi^2)^N \sum_{p=1}^{N} C_{1,[p,p]}^{(j,j')}(t)\,, \tag{10.7}$$

$$C_{s1}^{jj'}(t) = \left\langle \exp\left[-ij\gamma_p(0)\right] \exp\left[ij'\gamma_p(t)\right] \right\rangle_{\text{eq}}. \tag{10.8}$$

In Eqs. (10.6) and (10.7), $C_{1,[p,p']}^{(j,j')}(t)$ are the 1(1) correlation functions given by Eq. (9.118) (for $\kappa_0 \neq 0$) and may be written in the form

$$C_{1,[p,p']}^{(j,j')}(t) = (8\pi^2)^{-N} \sum_{k=1}^{N} Q_{pk}^0 Q_{p'k}^0 \sum_{j''=-1}^{1} R_{1,k}^{j''j*} R_{1,k}^{j''j'} \exp(-\lambda_{1,k}^{j''} t)\,, \tag{10.9}$$

where Q_{pk}^0 is given by Eq. (9.106), and $R_{1,k}^{jj'}$ is given by

$$R_{1,k}^{jj'} = (8\pi^2)^{N/2} \sum_{m=-1}^{1} Q_{1,k}^{L,mj*} (\bar{E}_{1,[k,k]}^{(m,m)})^{1/2} \bar{\mathcal{D}}_1^{j'm*}(\Omega_\alpha)\,. \tag{10.10}$$

Taking the sums over p and p' in Eqs. (10.6) and (10.7), we then obtain

$$M^{jj'}(t) = \frac{2}{N+1} \sum_{k \text{ odd}} \cot^2\left[\frac{k\pi}{2(N+1)}\right] \sum_{j''=-1}^{1} R_{1,k}^{j''j*} R_{1,k}^{j''j'} \exp(-\lambda_{1,k}^{j''} t)\,,$$
$$\tag{10.11}$$

$$M_s^{jj'}(t) = \sum_{k=1}^{N} \sum_{j''=-1}^{1} R_{1,k}^{j''j*} R_{1,k}^{j''j'} \exp(-\lambda_{1,k}^{j''} t)\,. \tag{10.12}$$

For the KP chain ($\kappa_0 = 0$), $C^{(j,j')}_{1,[p,p']}(t)$ is given by Eq. (9.119), so that Eq. (10.10) reduces to

$$R^{jj'}_{1,k} = \delta_{jj'}(8\pi^2)^{N/2}(\bar{E}^{(j,j)}_{1,[k,k]})^{1/2} \qquad \text{(KP)}, \qquad (10.13)$$

and the eigenvalues $\lambda^j_{1,k}$ are given by Eqs. (9.120). We note that when the torsion dynamics of stiff chains is treated in later sections, the KP chain with nonvanishing Poisson's ratio σ is used. (Recall that in Appendix 9.A we have considered the KP stiff chain with $\sigma = 0$ for the treatment of the translational diffusion with fluctuating hydrodynamic interaction, to which only the Rouse vector modes make contribution.) All results for this KP chain may then be obtained from those with $\sigma = 0$ only if $L(L+1)$ is replaced by $L(L+1)+\sigma j^2$ in $\bar{E}^{(j,j)}_{L,[k,k]}$ and $\bar{L}^{(j,j)}_{L,[k,k]}$. (Note that we may set $\sigma = 0$ for $j = 0$.) Further, we note that the $j = \pm 1$ eigenvalues for it are degenerate, as seen from Eqs. (9.120).

As for $C^{jj'}_{s1}(t)$, which is associated with the rotational motion of the dipole moment vector about the rotation axis, it may be regarded as equivalent to that for a single-axis rotor on the above assumption. Whether its relaxation is due to stochastic diffusion among a very large number of equilibrium positions [8,9] or to random jumps between two or three equivalent equilibrium positions to either of the two adjacent ones [9,10], $C^{jj'}_{s1}(t)$ may then be written in the form

$$\begin{aligned} C^{jj'}_{s1}(t) &= \delta_{jj'} & \text{for } j = 0 \\ &= \delta_{jj'}e^{-t/\tau_{s1}} & \text{for } j = \pm 1 \end{aligned} \qquad (10.14)$$

with τ_{s1} the corresponding correlation time. Note that the jump rate is equal to $(n\tau_{s1})^{-1}$ for the n-state jump process ($n = 2, 3$).

Thus we obtain, from Eq. (10.1) with Eq. (10.4), the final result

$$\frac{\epsilon^* - \epsilon_\infty}{\epsilon_0 - \epsilon_\infty} = \frac{m^2}{M(0)}\left\{ \frac{2\cos^2\Delta}{N+1}\sum_{k \text{ odd}}\sum_{j=-1}^{1}\cot^2\left[\frac{k\pi}{2(N+1)}\right]\frac{A^j_{1,k}}{1+i\omega\tau^j_{1,k}} \right.$$

$$\left. +\frac{1}{2}\sin^2\Delta\sum_{k=1}^{N}\sum_{j=-1}^{1}\frac{A^j_{s1,k}}{1+i\omega\tau^j_{s1,k}} \right\} \qquad (10.15)$$

with

$$\begin{aligned} \tau^j_{1,k} &= (\lambda^j_{1,k})^{-1}, \\ \tau^j_{s1,k} &= (\lambda^j_{1,k} + \tau_{s1}^{-1})^{-1}, \end{aligned} \qquad (10.16)$$

where the coefficients $A^j_{1,k}$ and $A^j_{s1,k}$ are real, nonnegative, and dependent on α, β, and $R^{jj'}_{1,k}$ (independent of β for the KP chain), but we omit explicit expressions for them [5]. Note that $M(0) = \langle\mu^2\rangle_{\text{eq}}$, so that the final result is independent of m. When the dipole moment vectors are affixed rigidly to the subbodies ($\Delta = 0$), we note that if $\mathbf{m}_p$ is parallel to $\mathbf{e}_{\zeta_p}$ ($\alpha = 0$), then $A^j_{1,k} = 0$, and therefore the $j = 0$ and -1 branches of the eigenvalue spectrum make contribution to dielectric relaxation, and that if $\mathbf{m}_p$ is parallel to $\mathbf{e}_{\xi_p}$ ($\alpha = \pi/2$ and $\beta = 0$ or π), then only the $j = 1$ branch makes contribution. For the KP chain, when $\Delta = 0$, only the $j = 0$ or 1 branch makes contribution if $\mathbf{m}_p$ is parallel or perpendicular to $\mathbf{e}_{\zeta_p}$. Further, it is important to note that the eigenvalues at small k make main contribution (for both HW and KP) because of the factor $\cot^2\left[k\pi/2(N+1)\right]$ as far as the main-chain motion is concerned.

10.1.2 Eigenvalue Spectra and Mode Analysis

All numerical results for dielectric relaxation are obtained for the d-HW chain, since they are similar to those for the c-HW chain in the crude subspace approximation. It is then convenient to introduce instead of ζ_t and ζ_r the dimensionless parameters r_1 and r_2 defined by

$$r_1 = \zeta_t/3\pi\eta_0 a\,,$$
$$r_2 = \zeta_r/a^2\zeta_t\,. \tag{10.17}$$

As mentioned in Sect. 9.3, for flexible chains the repeat unit (whose molecular weight is M_0) is taken as the subbody, and then a may be calculated from Eq. (9.43) with $\Delta s = M_0/M_L$.

We first examine the behavior of the 1(1) eigenvalues $\lambda^j_{1,k}$ for flexible chains evaluated in the crude subspace approximation, taking as examples a-PS (with $f_r = 0.59$) and a-PMMA (with $f_r = 0.79$). The values of their HW model parameters are given in Table 5.1, and we have $a = 2.88$ and 2.78 Å for a-PS and a-PMMA, respectively. Figure 10.2 shows plots of the reduced eigenvalues $\tilde{\lambda}^j_{1,k} \equiv \zeta_r\lambda^j_{1,k}/k_BT$ (with $\lambda^j_{1,k}$ unreduced) against the reduced

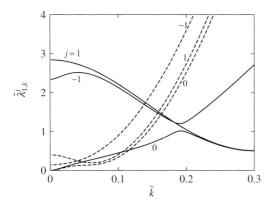

Fig. 10.2. Plots of $\tilde{\lambda}^j_{1,k}$ in the crude subspace approximation against $\tilde{k}$ for a-PS (*solid curves*) and a-PMMA (*dashed curves*), both with $r_1 = 1$, $r_2 = 10$, and $N = 999$

wave number $\tilde{k} \equiv k/(N+1)$ for a-PS (solid curves) and a-PMMA (dashed curves), both with $r_1 = 1$, $r_2 = 10$, and $N = 999$, where the correction for the negative eigenvalues has been made (see the last paragraph of Sect. 9.5.3). It is seen that an avoided crossing occurs between the $j = 0$ and -1 branches of the eigenvalue spectrum at $\tilde{k} \simeq 0.18$ and 0.03 for a-PS and a-PMMA, respectively. According to the results of mode analysis [5], the $j = 1$ branch is of purely local nature, while there is a mixing of global and local modes in each of the $j = 0$ and -1 branches. At small $\tilde{k}$, however, the $j = 0$ branch is mainly global, while the $j = -1$ branch is mainly local. Indeed, the $\tau_{1,k}^0$ given by the first of Eqs. (10.16) with small k may be regarded as the Rouse–Zimm dielectric relaxation times [4].

Next we consider stiff chains. All typical stiff chains such as DNA have helical structures and may be represented by the KP1 chain ($\kappa_0 = 0$ and $\tau_0 \neq 0$). Because of a structural symmetry about the helix axis (the ζ axis), their $\mathbf{m}_p$ may be regarded as parallel to it; that is, $\alpha = \Delta = 0$. Then only the $j = 0$ branch, which is purely global [5], is active for dielectric relaxation unless the side-chain motion exists, and moreover, it is independent of τ_0 and σ. However, it is important to note that even with the correction for the negative eigenvalues mentioned above, a few eigenvalues $\lambda_{1,k}^0$ at small k for $k \geq 1$ are still negative for small r_2 because of the preaveraging approximation in the constraining matrix (for the d-HW chain), so that we must assign a relatively (unreasonably) large value to r_2.

10.1.3 Comparison with Experiment

We make a comparison of theory with experiment with respect to the frequency dependences of the excess dielectric dispersion ϵ' and loss ϵ'', and a *dielectric correlation time* τ_D as defined as the reciprocal of the angular frequency ω_m corresponding to the maximum loss ω_m'' associated with the (net) main-chain motion. Their theoretical values are calculated from Eq. (10.15). The parameter r_2 and also Δ and τ_{s1} in the presence of side-chain motions are then determined to give good agreement between theory and experiment, assuming that $r_1 = 1$.

(a) Flexible Polymers

For flexible polymers, we analyze experimental data for polyoxyethylene (POE) [3, 11–13], atactic poly(p-chlorostyrene) (a-PPCS) [14, 15], atactic poly(methyl vinyl ketone) (a-PMVK) [16], s-PMMA [17], and i-PMMA [17]. For convenience, we assume the values of the HW model parameters determined from the RIS values for various equilibrium properties [18] except for s- and i-PMMAs, for which the values given in Table 5.1 are used, regarding the former as a-PMMA; that is, $\kappa_0 = 2.4$ and $\tau_0 = 0.5$ for POE, $\kappa_0 = 0.8$ and $\tau_0 = 2.3$ for a-PPCS (as s-PS), and $\kappa_0 = 0.1$ and $\tau_0 = 2.0$ for a-PMVK (as s-PMVK). We have $\alpha = 90°$ for all these polymers and $\beta = 180°$ except

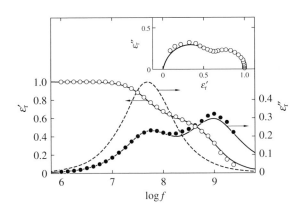

Fig. 10.3. Plots of ϵ_r' and ϵ_r'' against $\log f$ (f in Hz) for a-PMVK in dioxane at 20°C, the experimental data (*circles*) being due to Mashimo et al. [16]. The *solid curves* represent the theoretical values calculated from Eq. (10.15) in the crude subspace approximation with $r_1 = 1$, $r_2 = 7$, $N = 999$, $\Delta = 103°$, and $k_B T \tau_{s1}/\zeta_r = 0.5$, and the *dashed curve* represents the theoretical values of ϵ_r'' for $\Delta = 0°$. The inset shows the corresponding Cole–Cole plots

for i-PMMA, for which $\beta = 55°$. (We note that $A_{1,k}^j$ is independent of the sign of β for $\alpha = 90°$.) For these polymers except a-PMVK, no side-chain motions have been observed, so that we choose $\Delta = 0°$ except for it. Theoretical evaluation is carried out for $N = 999$ in all cases, all the experimental data above having been obtained for the molecular weight $M > 10^4$, in which range τ_D is independent of M.

Figure 10.3 shows plots of the reduced dispersion $\epsilon_r' \equiv (\epsilon' - \epsilon_\infty)/(\epsilon_0 - \epsilon_\infty)$ and reduced loss $\epsilon_r'' \equiv \epsilon''/(\epsilon_0 - \epsilon_\infty)$ against $\log f$ with $f = \omega/2\pi$ (in Hz) for a-PMVK in dioxane at 20°C, the experimental data (circles) being due to Mashimo et al. [16]. The solid curves represent the theoretical values calculated in the crude subspace approximation with $r_2 = 7$, $\Delta = 103°$, and $k_B T \tau_{s1}/\zeta_r = 0.5$, and the dashed curve represents the theoretical values of ϵ_r'' for $\Delta = 0°$ but with the other parameters remaining unchanged. The inset shows the corresponding Cole–Cole plots, the circles and curve representing the experimental and theoretical values, respectively. The loss peaks on the low- and high-frequency sides correspond to the main-chain and side-chain motions with the correlation times of 2.7 and 1.6 ns, respectively. However, the correlation time τ_D associated with the net main-chain motion is estimated to be 3.2 ns from the maximum of the dashed curve. There is seen to be rather good agreement between theory and experiment. In general, however, the observed loss curve is asymmetric about its peak (not of the Debye type), that is, somewhat broader on the high-frequency side for flexible chains without side-chain motions [19], indicating that there are several absorptions on that side. This cannot be easily explained by the HW theory even in the higher-order subspace approximation.

In Table 10.1 are given observed values of τ_D for the above five polymers and the values of r_2 obtained in the higher-order subspace approximation along with those in parentheses obtained in the crude approximation [5, 20].

Table 10.1. Values of τ_D, r_2, and d determined from dielectric relaxation for flexible polymers

Polymer	Solvent	Temp. (°C)	τ_D,obs (ns)	r_2	d (Å) From r	From chemical structures	Ref. (τ_D)
POE	Benzene	25	0.013	0.2 (0.3)[a]	2.5 (3.1)[a]	4.5	[3,11,12]
		20	0.019	0.3 (0.4)	3.1 (3.5)	...	[13]
a-PPCS	Benzene	25	4.7	42 (65)	16.1 (18.7)	12.0	[14]
		25.5	6.6	62 (95)	18.4 (21.3)	...	[15]
a-PMVK	Dioxane	20	3.1[b]	1.2 (7)	4.6 (8.8)	7.5	[16]
s-PMMA	Toluene	30	4.1	39 (74)	17.4 (21.7)	9.0	[17]
i-PMMA	Toluene	30	1.0	2.5 (6.8)	7.2 (10.5)	9.0	[17]

[a] The values in parentheses have been obtained in the crude subspace approximation
[b] Corresponding to the net main-chain motion

It is then interesting to estimate the size of the subbody from r_2. Clearly the product of r_1 and r_2 rather than their individual values (or ζ_r rather than ζ_t) plays an important role as far as the local motions are concerned [21]. From Eqs. (10.17), we have

$$r_1 r_2 \equiv r = \zeta_r / 3\pi\eta_0 a^3 . \tag{10.18}$$

It is reasonable here to regard the subbody as a spheroid having rotation axis of length a and diameter d. Then ζ_r must be the mean rotatory friction constant and is given by

$$\zeta_r = \frac{k_B T}{3} \left(\frac{2}{D_{r,1}} + \frac{1}{D_{r,3}} \right) , \tag{10.19}$$

where $D_{r,1}$ and $D_{r,3}$ are the rotatory diffusion coefficients $D_{r,1,(SD)}$ and $D_{r,3,(SD)}$ of the spheroid about the transverse axis and rotation axis and are given by Eqs. (6.A.47) and (6.A.48), respectively. Thus we may determine d from the values of a and r since r is a function of a/d.

In Table 10.1 are also given the values of d so determined along with those from the chemical structures [5,20]. It is seen that the values of d determined in the higher-order subspace approximation are smaller than those in the crude approximation, corresponding to the respective values of r (or r_2), and are closer to those determined from the chemical structures, as was expected, except for POE and a-PMVK.

(b) Semiflexible Polymers

For stiff chains ($\kappa_0 = 0$) with parallel dipoles ($\alpha = \Delta = 0$), τ_D reflects the global motions (end-over-end rotation, etc.) and its molecular weight M dependence becomes very important. In this subsection we consider primarily this problem.

As already mentioned, the eigenvalues are always positive (even for the d-HW chain) in the range of large r_2. In this range, τ_D is almost independent of r_2 [5]. Thus we adopt those values of τ_D as the theoretical ones. This independence is rather reasonable since the τ_D associated with the global motions should not depend on ζ_r (related to r_2) but on ζ_t. Then, since $r_1 = 1$, that is, $\zeta_t = 3\pi\eta_0 a$, the τ_D so evaluated may be regarded as the correlation time for the touched-bead model, each bead being a Stokes sphere of diameter a. We then replace the bead model by an equivalent cylinder model of diameter d by the use of the relation $d = 0.861a$ [22]. With values of $2k_BT\tau_D/\pi\eta_0 d^3$ so calculated (in the crude subspace approximation) as a function of $p = L/d$ for the KP cylinder of contour length L $(= N\Delta s \simeq Na)$ and diameter d, we may construct an interpolation formula for τ_D. The result reads [5]

$$\tau_D = \tau_{D,(R)} L^{-3}\left[L + \tfrac{1}{2}(e^{-2L} - 1)\right]^{3/2}$$
$$\times \left[1 + 0.539526 \ln(1 + L)\right] \quad (L \lesssim 30) \qquad (10.20)$$

with $\tau_{D,(R)} = 1/2D_{r,1,(R)}$, where $D_{r,1,(R)}$ is the rotatory diffusion coefficient of a spherocylinder and is given by Eq. (6.A.30) with Eq. (6.A.43) with $\epsilon = 1$, that is,

$$\tau_{D,(R)} = \frac{\pi\eta_0 L^3 F_r(p)}{6k_BT} \qquad (10.21)$$

with

$$F_r(p)^{-1} = \ln p + 2\ln 2 - \frac{11}{6} - \frac{8.25644}{\ln(1 + p)}$$
$$+13.0447\,p^{-1/4} - 62.6084\,p^{-1/2} + 174.0921\,p^{-3/4}$$
$$-218.8356\,p^{-1} + 140.2699\,p^{-5/4} - 33.2708\,p^{-3/2} . \quad (10.22)$$

Figure 10.4 shows double-logarithmic plots of $2k_BT\tau_D/\pi\eta_0 d^3$ against M for poly(γ-benzyl L-glutamate) (PBLG) in m-cresol at 25°C and PHIC in toluene at 25°C, the experimental data being due to Matsumoto et al. [23] and Takada et al. [24], respectively. The solid curves represent the best-fit theoretical values calculated from Eq. (10.20), and the dotted curves represent the values from Eq. (10.21) for the corresponding spherocylinders. The values of the model parameters thus determined are $M_L = 146$ Å^{-1} and $d = 28$ Å for PBLG and $\lambda^{-1} = 740$ Å, $M_L = 74.0$ Å^{-1}, and $d = 15$ Å for PHIC. The results for the latter are to be compared with those in Table 6.3. (For PBLG, λ^{-1} cannot be determined since the data are confined to the range of rigid rods.)

Finally, we make a comparison of theory with experiment with respect to the dispersion and loss curves, taking as an example the above PHIC sample with $M = 2.44 \times 10^5$. Figure 10.5 shows plots of ϵ'_r and ϵ''_r against $\log f$ (f in Hz) and the inset shows the corresponding Cole–Cole plots. The solid curves represent the theoretical values calculated (in the crude subspace approximation) with $a = 0.024$ and $N = 185$ (corresponding to the above

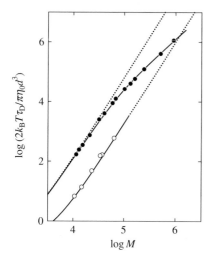

Fig. 10.4. Double-logarithmic plots of $2k_BT\tau_D/\pi\eta_0 d^3$ against M for PBLG in m-cresol at $25°C$ ($\circ$) and PHIC in toluene at $25°C$ ($\bullet$), the experimental data (*circles*) being due to Matsumoto et al. [23] and Takada et al. [24], respectively. The *solid curves* represent the best-fit theoretical values calculated from Eq. (10.20), and the *dotted curves* represent the values calculated from Eq. (10.21) for the corresponding spherocylinders (see the text)

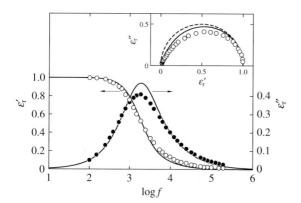

Fig. 10.5. Plots of ϵ'_r and ϵ''_r against $\log f$ (f in Hz) for the PHIC sample with $M = 2.44 \times 10^5$ in Fig. 10.4. The *solid curves* represent the theoretical values calculated from Eq. (10.15) in the crude subspace approximation with $a = 0.024$, $r_1 = 1$, $r_2 = 70$, and $N = 185$ (see the text). The *inset* shows the corresponding Cole–Cole plots, the *dashed curve* indicating the Debye curve

values of λ^{-1}, M_L, d, and M) and with $r_1 = 1$ and $r_2 = 70$, and the dashed curve in the inset indicates the Debye curve. The observed ϵ''_r is asymmetric and broader on the high-frequency side. The theory can better explain this fact than in the case of flexible polymers, but the agreement with experiment is not complete.

10.2 Nuclear Magnetic Relaxation

10.2.1 Formulation

In most nuclear magnetic relaxation experiments for polymers in dilute solution, ^{1}H, ^{13}C, ^{19}F, and ^{31}P are used as the probing nuclei. Then any relaxation mechanism other than the dipole interaction need not be considered for flexible chains, while the relaxation due to the anisotropic chemical shift cannot be ignored for some cases of stiff chains, especially for ^{31}P of DNA.

For the former, for simplicity, we consider only the heteronuclear dipolar interaction between two unlike spins I and S, with spin I observed and with spin S irradiated, where spin I represents ^{13}C, ^{19}F, and ^{31}P, and spin S represents ^{1}H, so that $I = S = 1/2$. If the internuclear distance r is constant (independent of t), the *spin-lattice relaxation time* T_1, the *spin-spin relaxation time* T_2, and the *nuclear Overhauser enhancement* NOE are given by [25–27]

$$T_1^{-1} = \frac{K^2}{20r^6}\left[J_0(\omega_S - \omega_I) + 3J_1(\omega_I) + 6J_2(\omega_S + \omega_I)\right], \quad (10.23)$$

$$T_2^{-1} = \frac{K^2}{40r^6}\left[4J_0(0) + J_0(\omega_S - \omega_I) + 3J_1(\omega_I) + 6J_1(\omega_S) + 6J_2(\omega_S + \omega_I)\right],$$
$$(10.24)$$

$$\text{NOE} = 1 + \frac{\gamma_S}{\gamma_I}\left[\frac{6J_2(\omega_S + \omega_I) - J_0(\omega_S - \omega_I)}{J_0(\omega_S - \omega_I) + 3J_1(\omega_I) + 6J_2(\omega_S + \omega_I)}\right] \quad (10.25)$$

with

$$K = \hbar\gamma_I\gamma_S, \quad (10.26)$$

where γ_I and γ_S are the gyromagnetic ratios of spins I and S, respectively, ω_I and ω_S are their Larmor angular frequencies, and $J_m(\omega)$ ($m = 0, 1, 2$) is the spectral density defined by

$$J_m(\omega) = 2\text{Re}\left[\int_0^\infty G_m(t)e^{-i\omega t}dt\right] \quad (10.27)$$

with Re indicating the real part and with $G_m(t)$ the autocorrelation function. In the case of the dipolar interaction, $G_m(t)$ is given by

$$G_m(t) = 8\pi^2\left\langle \mathcal{D}_2^{m0*}[\Omega(0)]\,\mathcal{D}_2^{m0}[\Omega(t)]\right\rangle_{\text{eq}}, \quad (10.28)$$

where $\Omega(t) = [\theta(t), \phi(t), 0]$ with $\theta(t)$ and $\phi(t)$ the polar and azimuthal angles, respectively, defining the instantaneous direction of the internuclear (spin–spin) vector in an external Cartesian coordinate system.

When there are two or more spins S that contribute to the relaxation of spin I, the above equations should be modified. If the internuclear distances of all spin pairs are constant and the same (or different), then T_1, T_2, and NOE are given by Eqs. (10.23)–(10.25), respectively, with the sum of spectral densities $J_m^{(i)}$ (or $r_i^{-6}J_m^{(i)}$) over spin pair i in place of J_m (or $r^{-6}J_m$). When there is the contribution of the anisotropic chemical shift to the relaxation of spin I, Eqs. (10.23)–(10.25) and (10.28) should be further modified, but we omit the explicit results [26, 28–31].

Now we derive an expression for $J_m(\omega)$ in the case of the dipolar interaction for the dynamic HW chain [31]. The electric dipole moment vector $\mathbf{m}_p$ in Fig. 10.1 may then be regarded as the internuclear (spin–spin) vector $I{\rightarrow}S$, so that its orientation in the localized coordinate system may be specified by the angles α, β, Δ, and $\gamma_p(t)$. In the following, we consider the case for which the only spin I on the pth subbody is observed. By a slight modification at the final stage, however, we can also obtain expressions for the case in which N or fewer identical spins I distributed uniformly or randomly on the N subbodies are observed. The orientation Ω in Eq. (10.28) may be represented by the successive rotations Ω_p, $(\alpha,\ \beta,\ \gamma_p)$, and $(\Delta,\ 0,\ 0)$ in this order, so that by the use of Eq. (4.C.13), $\mathcal{D}_2^{m0}(\Omega)$ may be written in terms of the $\mathcal{D}$ functions of these Euler angles. If we assume that the main-chain motion and the internal-rotational motion of the internuclear vector are independent of each other, $G_m(t)$ may then be expressed in terms of the correlation functions $C_{2,[p,p]}^{(j,j')}(t)$ and $C_{s2}^{jj'}(t)$, the latter being formally given by Eq. (10.8). In general, the 2(1) correlation function $C_{2,[p,p']}^{(j,j')}(t)$, which is given by Eq. (9.118), may be written in the form

$$C_{2,[p,p']}^{(j,j')}(t) = (8\pi^2)^{-N} \sum_{k=1}^{N} Q_{pk}^0 Q_{p'k}^0 \sum_{j''=-2}^{2} R_{2,k}^{j''j*} R_{2,k}^{j''j'} \exp(-\lambda_{2,k}^{j''}t),$$

$$(10.29)$$

where Q_{pk}^0 is given by Eq. (9.106), and $R_{2,k}^{jj'}$ are given by equations corresponding to Eqs. (10.10) ($\kappa_0 \neq 0$) and (10.13) ($\kappa_0 = 0$). Note that for the KP chain, the eigenvalues $\lambda_{2,k}^{j}$ and $\lambda_{2,k}^{-j}$ ($j = 1,\ 2$) are degenerate.

As for the correlation function $C_{s2}^{jj'}(t)$, which is associated with the rotational motion of the spin–spin vector about the rotation axis, we adopt only the random jump process. We then have

$$C_{s2}^{jj'}(t) = \delta_{jj'} \qquad \text{for } j = 0,\ \pm 2$$
$$= \delta_{jj'} e^{-t/\tau_{s2}} \qquad \text{for } j = \pm 1 \qquad \text{(two states)}, \qquad (10.30)$$
$$C_{s2}^{jj'}(t) = \delta_{jj'} \qquad \text{for } j = 0$$
$$= \delta_{jj'} e^{-t/\tau_{s2}} \qquad \text{for } j = \pm 1,\ \pm 2 \qquad \text{(three states)} \qquad (10.31)$$

with $(n\tau_{s2})^{-1}$ the jump rate for the n-state jump process ($n = 2,\ 3$) [9, 10, 32, 33].

Thus we obtain, from Eq. (10.26) with Eq. (10.27), for $J_m(\omega)$

$$J_m(\omega) = \frac{1}{2}(3\cos^2 \Delta - 1)^2 \sum_{k=1}^{N} (Q_{pk}^0)^2 \sum_{j=-2}^{2} \frac{A_{2,k}^j \tau_{2,k}^j}{1 + (\omega\tau_{2,k}^j)^2}$$

$$+\frac{3}{4}\sin^2 2\Delta \sum_{k=1}^{N}(Q^0_{pk})^2 \sum_{j=-2}^{2}\frac{A^j_{s21,k}\tau^j_{s21,k}}{1+(\omega\tau^j_{s21,k})^2}$$

$$+\frac{3}{4}\sin^4 \Delta \sum_{k=1}^{N}(Q^0_{pk})^2 \sum_{j=-2}^{2}\frac{A^j_{s22,k}\tau^j_{s22,k}}{1+(\omega\tau^j_{s22,k})^2} \qquad (10.32)$$

with

$$\begin{aligned}
\tau^j_{2,k} &= (\lambda^j_{2,k})^{-1}, \\
\tau^j_{s21,k} &= (\lambda^j_{2,k}+\tau^{-1}_{s2})^{-1}, \\
\tau^j_{s22,k} &= \tau^j_{2,k} \qquad \text{for two-state jumps} \\
&= \tau^j_{s21,k} \qquad \text{for three-state jumps},
\end{aligned} \qquad (10.33)$$

where $A^j_{2,k}$, $A^j_{s21,k}$, and $A^j_{s22,k}$ are dependent on α, β, and $R^{jj'}_{2,k}$ (independent of β for the KP chain), but we omit explicit expressions for them [31]. When $\Delta = 0$, we note that if the spin–spin vector is parallel to $\mathbf{e}_{\zeta_p}$ ($\alpha = 0$) or to $\mathbf{e}_{\xi_p}$ ($\alpha = \pi/2$ and $\beta = 0$ or π), then the $j = 0$, -1, and -2 branches of the eigenvalue spectrum make contribution. For the KP chain, when $\Delta = 0$, if the spin–spin vector is parallel to $\mathbf{e}_{\zeta_p}$, then only the $j = 0$ branch makes contribution as in the dielectric case; but if it is perpendicular to $\mathbf{e}_{\zeta_p}$, then the $j = 0$ and 2 branches make contribution in contrast to that case. Note that $G_m(\omega)$ and $J_m(\omega)$ are independent of m. Further, we note that for the case of N or fewer identical spins I under observation, that is, under conventional experimental conditions, the above $J_m(\omega)$ may be averaged over p, so that $(Q^0_{pk})^2$ may be replaced by N^{-1} in Eq. (10.32).

10.2.2 Eigenvalue Spectra and Amplitudes

All numerical results for nuclear magnetic relaxation are also obtained for the d-HW chain. In contrast to the dielectric case, the 2(1) eigenvalues $\lambda^j_{2,k}$ at all wave numbers k in general make contribution to magnetic relaxation, and therefore we examine the behavior of the amplitudes as well as the eigenvalues.

Figure 10.6 shows plots of the reduced eigenvalues $\tilde{\lambda}^j_{2,k} = \zeta_r\lambda^j_{2,k}/k_\mathrm{B}T$ in the crude subspace approximation against the reduced wave number $\tilde{k}$ for a-PS with $r_1 = 1$, $r_2 = 10$, and $N = 499$. It is seen that avoided crossings occur between the $j = 0$ and -1 branches at $\tilde{k} \simeq 0.2$, between the $j = -1$ and -2 branches at $\tilde{k} \simeq 0.4$, between the $j = 0$ and -1 branches at $\tilde{k} \simeq 0.5$, and between the $j = 1$ and 2 branches at $\tilde{k} \simeq 0.5$. According to the results of mode analysis [31], all the five branches for flexible chains are actually local, provided that $N \gtrsim 50$.

In the case of no internal rotations of the internuclear vector ($\Delta = 0$), the amplitudes are equal to $(Q^0_{pk})^2 A^j_{2,k}$, and therefore proportional to $A^j_{2,k}$

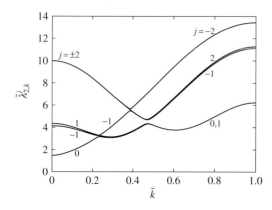

Fig. 10.6. Plots of $\tilde{\lambda}^j_{2,k}$ in the crude subspace approximation against $\tilde{k}$ for a-PS with $r_1 = 1$, $r_2 = 10$, and $N = 499$

when averaged over p, as seen from Eq. (10.32). Suppose that the methine carbons are observed for the above a-PS. The internuclear vector is then in the direction of C(methine)–H, and we have $\alpha = 90°$ and $\beta = 55°$. (We note that $A^j_{2,k}$ is independent of the sign of β for $\alpha = 90°$.) Figure 10.7 shows plots of $A^j_{2,k}$ against $\tilde{k}$ for a-PS in this case. It is seen that all branches except for $j = -1$ make main contribution to magnetic relaxation for $\tilde{k} \lesssim 0.5$, and the $j = 1$ branch for $\tilde{k} \gtrsim 0.5$. We note that the branches that make main contribution depend on the kind of polymer.

As for KP stiff chains (such as DNA), we note that the $j = 0$ branch is associated with the end-over-end rotation and bending (global), the $j = \pm 2$ degenerate branch with the torsional motions (local), and the $j = \pm 1$ degenerate branch with the coupled (bending and torsional) motions (mixed) [31]. Thus the $j = 0$ branch is independent of σ and the $j = \pm 2$ branch depends strongly on it. (All eigenvalues in the $j = 0$ branch are positive for large r_2 as in the dielectric case.)

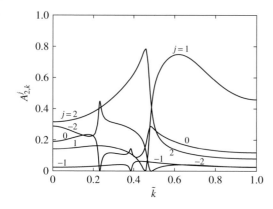

Fig. 10.7. Plots of $A^j_{2,k}$ against $\tilde{k}$ for the same a-PS as that in Fig. 10.6 with $\alpha = 90°$ and $\beta = 55°$

10.2.3 Spectral Densities

We examine the behavior of the spectral density $J_m(\omega)$ averaged over p only for flexible chains (with $\kappa_0 \neq 0$). Since it is independent of m, we suppress the subscript m in this subsection, for simplicity. Figure 10.8 shows double-logarithmic plots of the reduced spectral density $\tilde{J}(\omega) = k_B T J(\omega)/\zeta_r$ against the reduced angular frequency $\tilde{\omega} = \zeta_r \omega/k_B T$. The curve HW represents the values calculated from Eq. (10.32) with N^{-1} in place of $(Q_{pk}^0)^2$ for the above a-PS for the case of the methine carbons under observation. In general, $J(\omega)$ remains constant for small ω and is proportional to ω^{-2} for large ω. The vertical line segments attached to the curve indicate the values of $\tilde{\omega}$ equal to $\tilde{\lambda}_{2,k}^j$ corresponding to the maximum and minimum eigenvalues. (All eigenvalues are distributed between them.) The heavy part of the curve indicates the intermediate region, which corresponds to the range of those eigenvalues which make main contribution to $J(\omega)$.

Next we make a comparison of this $J(\omega)$ with those for the Jones–Stockmayer (JS) three-bond-motion model [34] and the Woessner (W) isotropic tumbling model [9] in the absence of internal rotations of the internuclear vectors. The JS spectral density $J^{JS}(\omega)$ may then be written in the form

$$J^{JS}(\omega) = 2 \sum_{k=1}^{s} \frac{G_k \tau_k}{1 + (\omega \tau_k)^2} \tag{10.34}$$

with

$$\tau_k^{-1} = 4w \sin^2\left[(2k-1)\pi/4s\right], \tag{10.35}$$

$$G_k = s^{-1} + 2s^{-1} \sum_{q=1}^{s} e^{-(\ln 9)q} \cos\left[(2k-1)q\pi/2s\right]. \tag{10.36}$$

Thus the basic JS model parameters are the number s of the correlation

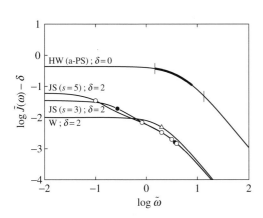

Fig. 10.8. Double-logarithmic plots of $\tilde{J}(\omega)$ against $\tilde{\omega}$. The curve HW represents the values calculated from Eq. (10.32) in the crude subspace approximation for the same a-PS as that in Fig. 10.7 for the case of the methine carbons under observation. The lower three curves represent the values of the JS and W models with $\zeta_r w/k_B T = 1$. The values of $\tilde{\omega}$ corresponding to the correlation times or their distribution range are also indicated (see the text)

times τ_k and the three-bond jump rate w. On the other hand, the spectral density for the W model (without internal rotations) is given by Eq. (10.34) with $s = 1$ and $G_k = 1$, that is, with a single correlation time. In Fig. 10.8 are included the results for the JS and W models calculated from Eq. (10.34) with $\zeta_r w / k_B T = 1.0$. The unfilled and filled circles and triangle represent the values of $\tilde{\omega}$ equal to $\tilde{\tau}_k^{-1} = \zeta_r / k_B T \tau_k$ for $s = 5$, 3, and 1, respectively, the eigenvalues being distributed nearly in the intermediate region. As s is increased, the distribution of correlation times and hence the intermediate region become wider, in particular, on the low-frequency side. (The results for the model of Monnerie and co-workers [35,36] are similar to those for the JS model.) It is then important to see that the intermediate region for the HW chain is narrow compared to that for the JS model, although wider than that for the W model. Thus it is anticipated that the HW model cannot explain T_1, T_2, NOE so consistently as the JS model (see the next subsection). We note that $J(\omega)$ for the flexible HW chain is almost independent of N for $N \gtrsim 50$.

10.2.4 Comparison with Experiment

(a) Flexible Polymers

We first make a comparison of theory with experiment with respect to T_1, T_2, and NOE for flexible polymers with the dipolar interaction. Their theoretical values are calculated from Eqs. (10.23)–(10.25) with Eq. (10.32) with N^{-1} in place of $(Q_{pk}^0)^2$. The parameter r_2 and also τ_{s2} in the presence of internal rotations are determined to give good agreement between theory and experiment for T_1. We analyze experimental data for POE [37], PIB [38], a-PS [39,40], a-PMVK [16], s-PMMA [41], a-PMMA [42], and i-PMMA [41,43]. The carbons under observation are the methine (CH), methylene (CH$_2$), methyl (CH$_3$), or C$_3$ (C$_5$) carbons, as indicated in Table 10.2. The values of the HW model parameters are the same as those in the dielectric case except for PIB and a-PMMA, for which the values given in Tables 6.3 (from [η]) and 5.1, respectively, are used. We have $\alpha = 90°$ for all these polymers, and $\beta = 55°$ in the absence of internal rotations except for s-, a-, and i-PMMAs, for which $\beta = 50°$. In the presence of internal rotations, we have $\beta = 125°$ and $\Delta = 70°$ with the three-state jumps for PIB, $\beta = 55°$ and $\Delta = 60°$ with the two-state jumps for a-PS, and $\beta = 125°$ and $\Delta = 70°$ with the three-state jumps for s-PMMA.

In Table 10.2 are given observed and calculated values of T_1, T_2, and NOE and the values of r_2 and $\tilde{\tau}_{s2} = k_B T \tau_{s2} / \zeta_r$ obtained in the higher-order subspace approximation along with those in parentheses obtained in the crude approximation [20,31]. For all these polymers, the values of r_2 determined in the former approximation are smaller than those in the latter, and there is better agreement between the observed and calculated values of NOE. As for

Table 10.2. Observed and calculated values of T_1, T_2, and NOE and estimates of the parameter r_2 for flexible polymers

Polymer	Solvent	Temp. (°C)	Nucleus C	Obsd. T_1 (ms)	T_2 (ms)	NOE	Calcd. T_1 (ms)	T_2 (ms)	NOE	r_2	$\tilde{\tau}_{s2}$	Ref. (Obs.)
POE	Benzene-d_6	30	CH$_2$	1560	...	...	1330 (1580)^a	1330 (1580)	2.99 (2.99)	0.8 (1.1)	...	[37]
PIB	CCl$_4$	40	CH$_2$	161	...	2.98	166 (161)	166 (161)	2.97 (2.97)	15 (38)	...	[38]
			CH$_3$	236	...	3.01	236 (238)	236 (238)	2.98 (2.98)	15 (38)	0.40 (0.12)	[38]
a-PS	Tetrachloro-ethylene	44	CH	77	...	...	77.3 (76.9)	75.3 (75.8)	2.78 (2.87)	24 (94)	...	[39]
			C$_3$,C$_5$	99	...	...	99.0 (98.6)	96.7 (97.4)	2.80 (2.89)	24 (94)	25 (0.23)	[39]
	Cyclohexane	40	CH	250^b	...	1.6	252 (250)	228 (234)	2.22 (2.44)	11 (37)	...	[40]
a-PMVK	Dioxane-d_6	26	CH	49	...	1.9	47.0 (47.7)	42.9 (44.2)	2.26 (2.41)	7.5 (45)	...	[16]
s-PMMA	Pyridine-d_6	38	CH$_2$	42	22	2.1	42.0 (42.0)	36.4 (40.6)	2.41 (2.71)	3.4 (36)	...	[41]
			CH$_3$	46	33	2.6	46.0 (45.8)	44.0 (45.2)	2.79 (2.89)	3.4 (36)	7.0 (0.68)	[41]
a-PMMA	Acetonitrile	44	CH$_2$	128^b	...	1.6	128 (128)	99.3 (116)	1.83 (2.17)	6.6 (47)	...	[42]
i-PMMA	Pyridine-d_6	38	CH$_2$	64	52	2.5	64.1 (63.8)	62.4 (62.9)	2.78 (2.88)	5.3 (23)	...	[41]
	Acetonitrile	35	CH$_2$	173^b	...	2.25	173 (175)	161 (167)	2.42 (2.59)	6.4 (24)	...	[43]

^{a} The values in parentheses have been obtained in the crude subspace approximation
^{b} $\omega_C/2\pi = 100$ MHz

Table 10.3. Values of d determined from nuclear magnetic relaxation for flexible polymers

	d (Å)	
Polymer	From r	From chemical structures
POE	4.8 (5.5)[a]	4.5
PIB	10.5 (14.5)	6.5
a-PS	11.8–15.5 (18.0–24.8)	11.0
a-PMVK	9.0 (16.8)	7.5
s-PMMA	7.3 (16.9)	9.0
a-PMMA	9.3 (18.8)	9.0
i-PMMA	9.6–10.2 (16.2–16.4)	9.0

[a] The values in parentheses have been obtained in the crude subspace approximation

T_2, the agreement cannot be markedly improved. This is due to the fact that the higher-order subspace approximation still fails to make the intermediate region of the spectral density $J_m(\omega)$ wider and give $J_m(0)$ large enough to explain T_2. Note that $J_m(0)$ is underestimated if the intermediate region is narrow. The values of the diameter d determined from r_2 (r) as in the dielectric case are given in Table 10.3. It is seen that the values determined in the higher-order subspace approximation are in good agreement with those from the chemical structures except for PIB.

The molecular weight dependences of T_1 and NOE for flexible polymers are discussed in connection with that of a correlation time τ_Γ for dynamic depolarized light scattering (optical anisotropy) in Sect. 10.4.

(b) Semiflexible Polymers

For semiflexible polymers, we only give the results [31] of analysis of experimental data for T_1 and T_2 obtained by Hogan and Jardetzky [44] for DNA (of 300 base pairs) in 1 mol/l NaCl at 23°C with ^{31}P under observation. Considering both the dipolar interaction and the anisotropic chemical shift and assuming $\lambda^{-1} = 1000$ Å and $a = l_{bp} = 3.4$ Å ($\kappa_0 = 0$ and $\tau_0 = 180$), we obtain $\sigma = 0$---0.3 and $r_2 = 120$ (in the crude subspace approximation). With these values, we obtain $d = 31.6$ Å and the values 2.0–2.9×10^{-19} erg cm for the torsional force constant β. The former value is somewhat too large compared to the values in Table 6.3, while the latter are to be compared with the values 2.4–3.0×10^{-19} erg cm from the J factor in Fig. 7.7. The result for d indicates the defect of the stiff d-HW (KP) chain. We note that a similar analysis was made by Allison et al. [30], who considered the end-over-end rotation and the torsional motion but not the bending and its coupling with the torsion corresponding to the $j = 0$ and 1 branches of the 2(1) eigenvalue spectrum of the HW chain.

10.3 Fluorescence Depolarization

10.3.1 Formulation

Suppose that a sample is excited by an infinitely short flash of plane-polarized light, incident along the x axis and polarized along the z axis of an external Cartesian coordinate system $(\mathbf{e}_x, \mathbf{e}_y, \mathbf{e}_z)$, and that the fluorescent light is observed from the direction of the y axis. The *emission anisotropy* $r(t)$ is defined by [45]

$$r(t) = \frac{I_z(t) - I_x(t)}{I_z(t) + 2I_x(t)}, \tag{10.37}$$

where I_z and I_x are the z and x components of the fluorescence emission intensity, respectively. The denominator $(I_z + 2I_x)$ is equal to the total intensity, which is proportional to $\exp(-t/\tau_f)$ with τ_f the fluorescence lifetime. The *average anisotropy* $\bar{r}$ observed in steady-state experiments is given by the ratio of the integrals of the numerator and denominator of Eq. (10.37) over t from 0 to ∞, so that we have

$$\bar{r} = \tau_f^{-1} \int_0^\infty r(t) e^{-t/\tau_f} dt. \tag{10.38}$$

Let $\mathbf{m}_a$ and $\mathbf{m}_e$ be the unit absorption and emission dipole moment vectors, respectively, of the fluorescent probe incorporated in the polymer chain. I_z and I_x may be expressed in terms of time-correlation functions as

$$I_z(t) = C e^{-t/\tau_f} \left\langle [\mathbf{e}_z \cdot \mathbf{m}_a(0)]^2 [\mathbf{e}_z \cdot \mathbf{m}_e(t)]^2 \right\rangle_{eq}, \tag{10.39}$$

$$I_x(t) = C e^{-t/\tau_f} \left\langle [\mathbf{e}_x \cdot \mathbf{m}_a(0)]^2 [\mathbf{e}_x \cdot \mathbf{m}_e(t)]^2 \right\rangle_{eq}, \tag{10.40}$$

where C is a constant independent of t. It can then be shown that $r(t)$ is given by [46]

$$r(t) = \frac{16\pi^2}{5} \left\langle \mathcal{D}_2^{00*}[\Omega_a(0)] \, \mathcal{D}_2^{00}[\Omega_e(t)] \right\rangle_{eq}$$
$$= 2 \left\langle P_2[\cos\theta_a(0)] \, P_2[\cos\theta_e(t)] \right\rangle_{eq}, \tag{10.41}$$

where $\Omega_a = (\theta_a, \phi_a, 0)$ $[\Omega_e = (\theta_e, \phi_e, 0)]$ with θ_a (θ_e) and ϕ_a (ϕ_e) the polar and azimuthal angles of $\mathbf{m}_a$ $(\mathbf{m}_e)$, respectively, in the external coordinate system, and $P_2(x)$ is the Legendre polynomial. Note that this result is equivalent to those of Wallach [32] and Szabo [47].

Now suppose that the probe is rigidly attached to the pth subbody of the dynamic HW chain, and assume that $\mathbf{m}_a$ and $\mathbf{m}_e$ are parallel to each other, for simplicity. Then $\mathbf{m}_e$ $(= \mathbf{m}_a)$ corresponds to the electric dipole moment vector and the internuclear vector (with $\Delta = 0$) in Fig. 10.1, and the $r(t)$ given by Eqs. (10.41) is equal to the magnetic autocorrelation function $G_0(t)$, as defined by Eq. (10.28), multiplied by the factor 2/5. Thus we readily find the results for the dynamic HW chain [46],

$$r(t) = \frac{2}{5} \sum_{k=1}^{N} (Q_{pk}^0)^2 \sum_{j=-2}^{2} A_{2,k}^j \exp(-\lambda_{2,k}^j t) , \tag{10.42}$$

$$\bar{r} = \frac{2}{5} \sum_{k=1}^{N} (Q_{pk}^0)^2 \sum_{j=-2}^{2} A_{2,k}^j \left(1 + \frac{\tau_f}{\tau_{2,k}^j} \right)^{-1} , \tag{10.43}$$

where $A_{2,k}^j$ is the same as that in Eq. (10.32), and $\tau_{2,k}^j$ is given by the first of Eqs. (10.33). For the case of some (a small number of) identical probes distributed randomly on the N subbodies, $(Q_{pk}^0)^2$ may be replaced by N^{-1} in Eqs. (10.42) and (10.43).

For DNA as an example of KP stiff chains, it is known that there is an initial, very rapid, but limited reorientation of the dipole arising from wobbling of a dye within its intercalation site. If we assume that the wobbling is the fluctuation in the angle β at constant α, then this effect may be taken into account by introducing the factor $\exp(-j^2 \beta_w^2/2)$ in front of $A_{2,k}^j$ in Eqs. (10.42) and (10.43), where β_w is a fluctuation width parameter.

10.3.2 Comparison with Experiment

(a) Flexible Polymers

The present theoretical values of $r(0) \equiv r_0$ is equal to $2/5$, but the observed values are very often smaller. Thus we regard r_0 as an adjustable parameter and calculate theoretical values of $r(t)$ and $\bar{r}$ from Eqs. (10.42) and (10.43), respectively, with r_0 in place of $2/5$.

We consider flexible polymers with perpendicular emission dipoles randomly distributed (with $\Delta = 0$). Figure 10.9 shows plots of $\log r(t)$ against t for a-PS with 9,10-diphenyl anthracene side groups ($\alpha = 90°$ and $\beta = 55°$) in ethylacetate–tripropionin mixtures with the indicated values of η_0 at 25°C [20, 46]. The shaded domains bounded by the dashed curves represent the experimental values obtained by Valeur and Monnerie [48]. The solid curves represent the theoretical values calculated in the higher-order subspace approximation for $r_1 = 1$, $r_2 = 80$, and $N = 199$ with $r_0 = 0.300$ and 0.270 for $\eta_0 = 0.075$ and 0.015 P, respectively, and the dot-dashed curves represent the theoretical values in the crude approximation for $r_2 = 300$ with $r_0 = 0.290$ and 0.260, respectively. The higher-order and crude subspace approximations give the values 23.4 and 36.7 Å for the diameter d, respectively.

As for $\bar{r}$, we analyze experimental data obtained by North and Soutar [49] for a-PMMA in toluene at 25°C, the dipoles having been incorporated by copolymerization with 9-vinyl anthracene ($\alpha = 90°$ and $\beta = 55°$). We then obtain $r_2 = 12$ (102), and thus $d = 11.5$ (24.1) Å in the higher-order (crude) subspace approximation from the observed $\bar{r} = 0.0530$, $r_0 = 0.277$, and $\tau_f = 5.5$ ns.

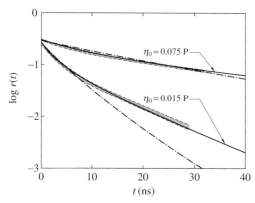

Fig. 10.9. Plots of $\log r(t)$ against t for a-PS with 9,10-diphenyl anthracene side groups ($\alpha = 90°$ and $\beta = 55°$) in ethylacetate–tripropionin mixtures with the indicated values of η_0 at 25°C. The *shaded domains* bounded by the *dashed curves* represent the experimental values obtained by Valeur and Monnerie [48]. The *solid curves* represent the theoretical values calculated in the higher-order subspace approximation for $r_1 = 1$, $r_2 = 80$, and $N = 199$ with $r_0 = 0.300$ and 0.270 for $\eta_0 = 0.075$ and 0.015 P, respectively, and the *dot-dashed curves* represent the theoretical values in the crude approximation for $r_2 = 300$ with $r_0 = 0.290$ and 0.260, respectively (see the text)

(b) Semiflexible Polymers

We consider as an example of semiflexible polymers DNA with the dyes (ethidium bromide) randomly distributed. Figure 10.10 shows plots of $r(t)$ against t for DNA (of about 10^4 base pairs) in 0.15 mol/l NaCl at 22°C (a) and in 0.01 mol/l NaCl at 23°C (b) [46]. The dashed curves represent the experimental values obtained by Millar et al. [50,51]. The solid curves represent the theoretical values calculated in the crude subspace approximation with $\kappa_0 = 0$, $\tau_0 = 200$, $\lambda^{-1} = 1100$ Å, $\sigma = -0.3$, $M_L = 195$ Å^{-1}, and $N = 999$ for case (a) and $\kappa_0 = 0$, $\tau_0 = 270$, $\lambda^{-1} = 1500$ Å, $\sigma = -0.3$, $M_L = 195$ Å^{-1}, and $N = 999$ for case (b) along with the indicated values of r_2, α, and β_w. With these values, we obtain $d = 31.6$–36.2 Å (from $r_2 = 120$–180 and $a = 3.4$ Å), and the values 3.2×10^{-19} and 4.4×10^{-19} erg cm for the torsional force constant β in cases (a) and (b), respectively. These values of d and β are somewhat larger than those from magnetic relaxation. It is seen from Fig. 10.10 that there is rather good agreement between theory and experiment, but that the theoretical $r(t)$ relaxes more rapidly than the experimental one for $t \gtrsim 40$ ns. This defect and also the above results for d arise from the fact that the d-HW (KP) chain cannot well describe the global, long-wavelength motions of semiflexible polymers in the cases of fluorescence depolarization as well as magnetic relaxation (in contrast to the dielectric case).

We note that the range of t in which the present $r(t)$ obeys the exponential $-t^{1/2}$ decay law is somewhat narrower than the prediction by Barkley and

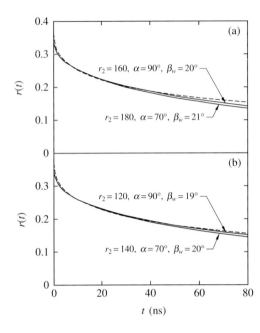

Fig. 10.10. Plots of $r(t)$ against t for DNA in 0.15 mol/l NaCl at 22°C (a) and in 0.01 mol/l NaCl at 23°C (b). The *dashed curves* represent the experimental values obtained by Millar et al. [50,51]. The *solid curves* represent the theoretical values calculated from Eq. (10.42) with the factor $\exp(-j\beta_w^2/2)$ in the crude subspace approximation with $\kappa_0 = 0$, $\sigma = -0.3$, $r_1 = 1$, and $N = 999$ for both cases and with $\tau_0 = 200$ and $\lambda^{-1} = 1100$ Å for case (a) and with $\tau_0 = 270$ and $\lambda^{-1} = 1500$ Å for case (b) along with the indicated values of r_2, α, and β_w

Zimm [52], who considered the bending and torsional motions on the basis of a continuous elastic model and gave an explanation of earlier data obtained by Wahl et al. [53].

10.4 Dynamic Depolarized Light Scattering

10.4.1 Formulation

We consider the scattering by a single dynamic HW chain in dilute solution [54]. Let $\boldsymbol{\alpha}_p$ and $\tilde{\boldsymbol{\alpha}}_p$ be the excess (local) polarizability tensors (over the mean polarizability of the solvent alone) of the pth subbody, expressed in the pth localized and external Cartesian coordinate systems, respectively. In dynamic depolarized light scattering measurements by the filter method [55], the ratio of the horizontal component I_{Hv} of the excess scattered intensity measured at the scattering angle θ to the intensity I_v^0 of monochromatic, vertically polarized incident light is determined as a function of the difference $\Delta\omega$ between the angular frequencies ω_f and ω_i of the scattered and incident light waves,

$$\Delta\omega = \omega_f - \omega_i . \tag{10.44}$$

If λ_0 is the wavelength of the light in vacuum and r is the distance from the center of the scattering volume to the detector, the ratio $I_{\mathrm{Hv}}(\Delta\omega)/I_v^0$ may be written in the form [55]

$$\frac{I_{\mathrm{Hv}}(\Delta\omega)}{I_{\mathrm{v}}^0} = \frac{16\pi^4 \bar{F}_{\mathrm{Hv}}(\Delta\omega)}{\lambda_0^4 r^2}, \tag{10.45}$$

corresponding to Eq. (5.70), with

$$\bar{F}_{\mathrm{Hv}}(\Delta\omega) = \frac{1}{2\pi} \sum_{p,p'=1}^{N} \int_{-\infty}^{\infty} \langle \alpha_{p,\mathrm{Hv}}(0)\alpha_{p',\mathrm{Hv}}(t)$$

$$\times \exp\{i\mathbf{k} \cdot [\mathbf{r}_{p'}(t) - \mathbf{r}_p(0)]\}\rangle_{\mathrm{eq}} e^{-i\Delta\omega t} dt, \tag{10.46}$$

where $\alpha_{p,\mathrm{Hv}}(t)$ is given by the discrete version of Eq. (5.73), $\mathbf{k}$ is the scattering vector whose magnitude k is given by Eq. (5.20), and $\mathbf{r}_p(t)$ is the vector position of the center of the pth subbody at time t.

It is then convenient to introduce a function $J_\Gamma(\Delta\omega)$ defined by

$$J_\Gamma(\Delta\omega) = 15\bar{F}_{\mathrm{Hv}}(\Delta\omega). \tag{10.47}$$

The total (reduced) intensity is obtained by integrating $\bar{F}_{\mathrm{Hv}}(\Delta\omega)$ over $\Delta\omega$, so that we have, from Eqs. (5.117) and (10.47), for the mean-square optical anisotropy $\langle \Gamma^2 \rangle$

$$\langle \Gamma^2 \rangle = \int_{-\infty}^{\infty} J_\Gamma(\Delta\omega; k = 0) d(\Delta\omega). \tag{10.48}$$

Now $\alpha_{p,\mathrm{Hv}}$ may be expressed in terms of the spherical components $\tilde{\alpha}_{p,2}^j$ ($j = 0, \pm 1, \pm 2$) of $\tilde{\boldsymbol{\alpha}}_p$, which are given by Eqs. (5.B.2) and are related to the spherical components α_2^j ($j = 0, \pm 1, \pm 2$) of $\boldsymbol{\alpha}_p$ (which are independent of p) by Eq. (5.B.6). Then $J_\Gamma(\Delta\omega)$ includes two kinds of contributions: the orientational fluctuations of the optically anisotropic subbodies, that is, the product $\mathcal{D}_2^{mj*}(\Omega_p, 0)\,\mathcal{D}_2^{m'j}(\Omega_{p'}, t)$ and the density fluctuation of the subbodies in the domain of a linear dimension of order $2\pi/k$, that is, the factor $\exp\{\ \}$ in Eq. (10.46). For such small k that $2\pi/k$ is much larger than the average chain dimension, the density fluctuation arises from the translational diffusion of the entire chain over the distance of $2\pi/k$, so that the relaxation of the product of the $\mathcal{D}$ functions may be considered to be much faster than that of the exponential factor, thereby leading to no correlation between the two kinds of relaxation. We may then assume that the density correlation function $\langle \exp\{\ \}\rangle_{\mathrm{eq}}$ does not relax at all during the orientational relaxation. It may therefore be replaced by its value at $t = 0$,

$$\langle \exp\{i\mathbf{k} \cdot [\mathbf{r}_{p'}(t) - \mathbf{r}_p(0)]\}\rangle_{\mathrm{eq}} \simeq I(\mathbf{k}; |p' - p|\Delta s), \tag{10.49}$$

where $I(\mathbf{k}; s)$ is the characteristic function for the (continuous) HW chain of contour length s. If $2\pi/k$ is much larger than the root-mean-square end-to-end distance, we may put $I(\mathbf{k}; |p' - p|\Delta s) \simeq 1$. In these approximations, Eq. (10.47) with Eq. (10.46) reduces to

$$J_\Gamma(\Delta\omega) = 6\pi(8\pi^2)^{N-1} \sum_{p,p'=1}^{N} \sum_{j,j'=-2}^{2} \alpha_2^{j*}\alpha_2^{j'} \int_{-\infty}^{\infty} C_{2,[p,p']}^{(j,j')}(t)e^{-i\Delta\omega t}dt ,$$

$$(10.50)$$

where $C_{2,[p,p']}^{(j,j')}(t)$ is given by Eq. (10.29).

Thus we obtain the final result

$$J_\Gamma(\Delta\omega) = \sum_{k \text{ odd}} \sum_{j=-2}^{2} \frac{A_k^j \tau_{2,k}^j}{1 + (\Delta\omega\tau_{2,k}^j)^2} , \qquad (10.51)$$

where $\tau_{2,k}^j$ is given by the first of Eqs. (10.33) and the amplitude A_k^j is given by

$$A_k^j = \frac{3}{2\pi(N+1)} \cot^2\left[\frac{k\pi}{2(N+1)}\right] \left| \sum_{j'=-2}^{2} \alpha_2^{j'} R_{2,k}^{jj'} \right|^2 . \qquad (10.52)$$

We note that the $\langle \Gamma^2 \rangle$ obtained from Eq. (10.48) with Eq. (10.50) or (10.51) is identical with the one given by Eq. (5.120) for $N \gg 1$ [54].

Finally, it is pertinent to make some remarks on the present results in relation to other theories. The spring-bead model (the coarse-grained bond chain) with the Kuhn–Grün expression [56] for the spring polarizability tensor is, in principle, inappropriate for the description of depolarized scattering since it cannot give the correct $\langle \Gamma^2 \rangle$ [54], although it can give the correct result for flow birefringence, as was derived by Zimm [1]. Ono and Okano [57] adopted this model to predict that the spectrum of the (forward) depolarized component is an equally weighted sum of Lorentzians each with a half-width at half-maximum (hwhm) inversely proportional to the Rouse–Zimm relaxation time [4], and thus their theory must be invalid for real chains. Recall that for the dynamic HW chain all branches of the 2(1) eigenvalue spectrum are local. According to the numerical results [54], J_Γ may be actually written in terms of a small number of eigenvalues $\lambda_{2,k}^j$ at small k which belong to two branches (for example, $j = 0$ and -1 for a-PS and $j = -1$ and -2 for a-PMMA), one (major) corresponding to the low-frequency modes and the other (minor) to the high-frequency modes. However, this does not necessarily correspond to the experimental results obtained by Bauer et al. [58] since their low-frequency modes are just the Rouse–Zimm modes. We note that the analysis of the low-frequency modes by Evans [59] on the basis of the Fixman–Kovac chain [60] also leads to the inclusion of the Rouse–Zimm modes.

10.4.2 Comparison with Experiment

We make a comparison of theory with experiment with respect to a *depolarized scattering correlation time* τ_Γ as defined as the reciprocal of the hwhm of

J_Γ. Figure 10.11 shows double-logarithmic plots of the ratio $\tau_\Gamma/\tau_\Gamma^0$ against the number of repeat units x for a-PS in cyclohexane at 34.5°C (Θ) [40], a-PMMA in acetonitrile at 44.0°C (Θ) [42], and i-PMMA in acetonitrile at 28.0°C (Θ) [43], where τ_Γ^0 is the τ_Γ of the monomer at the given temperature, that is, $\tau_\Gamma^0 = 0.00562$ ns for a-PS (cumene) and $\tau_\Gamma^0 = 0.00193$ and 0.00236 ns for a- and i-PMMAs (methyl isobutyrate), respectively. We note that in all cases, the observed $J_\Gamma(\Delta\omega)$ may be fitted by a single Lorentzian independently of x. The observed τ_Γ seems to level off in the limit of $x \to \infty$. In contrast to this, the experimental results obtained by Bauer et al. [58] and by Strehle et al. [61] for a-PS show that τ_Γ increases without limit with increasing x. The heavy solid, dashed, and dotted curves represent the theoretical values calculated from Eq. (10.51) in the higher-order subspace approximation (with the observed τ_Γ^0) for a-PS, a-PMMA, and i-PMMA ($a = 3.08$ Å), respectively, all with $r_2 = 8$ along with the polarizability tensors $\boldsymbol{\alpha}_0$ given by Eqs. (5.129)–(5.131). Here, the theoretical values for $x \lesssim 10$ have not been calculated because of the breakdown of the block-diagonal approximation. The theory also predicts that τ_Γ levels off. With the results for r_2, we obtain $d = 10.0,\ 10.0,$ and 11.0Å for these polymers, respectively.

For $x \lesssim 10$, we simply treat the chain (oligomer) as a rigid sphere having the hydrodynamic radius R_H. If we assume that the oligomer as a whole has a cylindrically symmetric polarizability tensor, for simplicity, then its

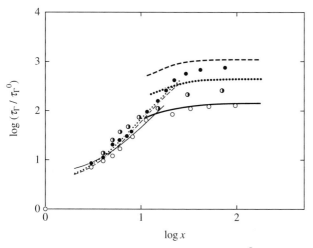

Fig. 10.11. Double-logarithmic plots of $\tau_\Gamma/\tau_\Gamma^0$ against x for a-PS in cyclohexane at 34.5°C (Θ) (○) [40], a-PMMA in acetonitrile at 44.0°C (Θ) (●) [42], and i-PMMA in acetonitrile at 28.0°C (Θ) (◑) [43], where τ_Γ^0 is the τ_Γ of the monomer. The *heavy solid, dashed, and dotted curves* represent the respective HW theoretical values calculated from Eq. (10.51) in the higher-order subspace approximation (with the observed τ_Γ^0), and the *light curves* represent the respective theoretical values calculated from Eq. (10.53) for the rigid sphere model (see the text)

relaxation time $\tau_{\Gamma,(S)}$ is equal to $(6D_r)^{-1}$ with D_r the rotatory diffusion coefficient and is given by

$$\tau_{\Gamma,(S)} = 4\pi\eta_0 R_H^3/3k_B T. \tag{10.53}$$

It is then reasonable to equate R_H to the apparent root-mean square radius of gyration $\langle S^2\rangle_s^{1/2}$ [40], which may be calculated from Eq. (5.A.8). In Fig. 10.11, the light curves represent the respective values calculated from Eq. (10.53) for the rigid sphere model. They are seen to reproduce satisfactorily the data points for $x \lesssim 10$.

We note that Hagerman and Zimm [62] evaluated τ_Γ for KP stiff chains by Monte Carlo methods, but that strictly their expression for the ratio of τ_Γ to its value $\tau_{\Gamma,(R)}$ for rods, which is identical with the ratio $\tau_D/\tau_{D,(R)}$ in their approach, is not correct since they used the Broesma equation for the rotatory diffusion coefficient of rods (see Appendix 6.A).

10.4.3 Correlation with Nuclear Magnetic Relaxation

In this subsection we show that there is strong correlation between nuclear magnetic relaxation and depolarized light scattering. For this purpose, an analysis is made of the dependences on x of T_1 and NOE in relation to τ_Γ along the same line as in the last subsection. Figure 10.12 shows plots of $n_{CH}T_1$ and NOE against $\log x$ for the intermediate methine carbon atoms for a-PS in cyclohexane at 40°C (with $n_{CH} = 1$) [40], where n_{CH} is the number of

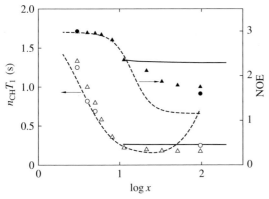

Fig. 10.12. Plots of $n_{CH}T_1$ and NOE against $\log x$ for the intermediate methine carbon atoms for a-PS in cyclohexane at 40°C (with $n_{CH} = 1$); ○, T_1; ●, NOE [40]. The *solid curves* represent the HW theoretical values calculated from Eqs. (10.23) and (10.25) with Eq. (10.32) in the higher-order subspace approximation, and the *dashed curves* represent the values for the rigid sphere model with Eq. (10.54), instead of with Eq. (10.32), with $\tau_{M,(S)}$ equal to $\tau_{\Gamma,(S)}$. The *unfilled and filled triangles* represent the values of T_1 and NOE, respectively, for the rigid sphere with $\tau_{M,(S)}$ equal to the scaled τ_Γ at 40°C (see the text)

C–H bonds associated with the carbon atom under observation. The unfilled and filled circles represent the experimental values of T_1 and NOE, respectively, and the solid curves represent the theoretical values calculated from Eqs. (10.23) and (10.25) with Eq. (10.32) in the higher-order subspace approximation with the same values of the model parameters as those in the last subsection (along with $\omega_C/2\pi = 100$ MHz, $\omega_H/2\pi = 400$ MHz, $r = 1.09$ Å, and $\eta_0 = 0.609$ cP). The theoretical values are again limited to the range of $x \gtrsim 10$. In this range, they are almost independent of x. The theoretical asymptotic value of T_1 in the limit of $x \to \infty$ is in good agreement with the experimental value, but that of NOE is appreciably larger than the experimental value (see also Table 10.2). We must note here that the x dependences of T_1 and NOE may depend on the frequency of the spectrometer used [40,63].

As in the case of J_Γ, we consider the rigid sphere model, to which a C$-$H internuclear vector is affixed. Its T_1 and NOE may then be calculated from Eqs. (10.23) and (10.25) with J_m given by [25]

$$J_m(\omega) = \frac{2\tau_{M,(S)}}{1 + (\omega\tau_{M,(S)})^2} , \tag{10.54}$$

where $\tau_{M,(S)}$ is identical with $\tau_{\Gamma,(S)}$ given by Eq. (10.53) with $R_H = \langle S^2 \rangle_s^{1/2}$. In Fig. 10.12, values of T_1 and NOE so calculated are represented by the respective dashed curves. The dashed curve for T_1 is in good agreement with the data points for the oligomers as in the case of τ_Γ. As mentioned in the last subsection, for the polymer–solvent systems in Fig. 10.11 J_Γ may be well represented in terms of a single relaxation time τ_Γ (a single Lorentzian) even for large x. It may therefore be expected that this is also the case with nuclear magnetic relaxation. Thus we calculate T_1 and NOE for the rigid sphere from Eqs. (10.23) and (10.25) with Eq. (10.54), where we equate $\tau_{M,(S)}$ to the (scaled) τ_Γ for a-PS in cyclohexane at 40°C. Values of T_1 and NOE so calculated for all a-PS fractions are represented by the unfilled and filled triangles, respectively, in Fig. 10.12. They agree well with the respective experimental values. This indicates that the two relaxation processes may give equivalent information about the local chain motions.

Figure 10.13 shows similar plots for the intermediate methylene carbon atoms for a-PMMA in acetonitrile at 44°C (with $n_{CH} = 2$) [42]. The solid curves represent the HW theoretical values (with $r = 1.09$ Å and $\eta_0 = 0.285$ cP) (see also Table 10.2). The unfilled and filled triangles represent the values for the rigid sphere with a single relaxation time but with $\tau_{M,(S)} = 0.6\,\tau_\Gamma$. The disagreement between $\tau_{M,(S)}$ and τ_Γ in this case arises from the fact that all the eigenvalues make contribution to T_1 and NOE in contrast to the case of J_Γ and there are differences in amplitudes between a-PS and a-PMMA, as mentioned in Sect. 10.2.2.

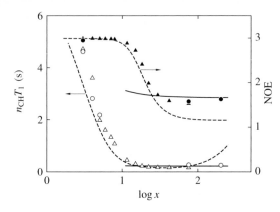

Fig. 10.13. Plots of $n_{CH}T_1$ and NOE against $\log x$ for the intermediate methylene carbon atoms for a-PMMA in acetonitrile at 44°C (with $n_{CH} = 2$); O, T_1; ●, NOE [42]. The *solid and dashed curves* have the same meaning as those in Fig. 10.12. The *unfilled and filled triangles* represent the values of T_1 and NOE, respectively, for the rigid sphere with $\tau_{M,(S)} = 0.6\,\tau_\Gamma$ (see the text)

10.5 First Cumulant of the Dynamic Structure Factor

10.5.1 Formulation

We first evaluate the *dynamic structure factor* $S(k,t)$ as a function of the magnitude k of the scattering vector **k** and time t on the basis of the dynamic HW chain [64]. Suppose that the N subbodies and also the $(N+1)$th imaginary one have identical isotropic polarizabilities. It may be written in the form [55]

$$S(k,t) = \frac{1}{(N+1)^2} \sum_{p,p'=1}^{N+1} \langle \exp\{i\mathbf{k} \cdot [\mathbf{r}_{p'}(t) - \mathbf{r}_p(0)]\} \rangle_{eq}. \qquad (10.55)$$

Note that the static structure factor $S(k,0)$ is identical with the scattering function $P(\mathbf{k};L)$ considered in Sect. 5.2. In order to carry out evaluation, we introduce the Gaussian approximation, that is, the approximation that the distribution of the quantity $\mathbf{r}_{p'}(t) - \mathbf{r}_p(0)$ is Gaussian. The equilibrium average in Eq. (10.55) may then be reduced to be

$$\langle \exp\{i\mathbf{k} \cdot [\mathbf{r}_{p'}(t) - \mathbf{r}_p(0)]\} \rangle_{eq} = \exp\left[-\frac{k^2}{6} \langle |\mathbf{r}_{p'}(t) - \mathbf{r}_p(0)|^2 \rangle_{eq}\right]. \qquad (10.56)$$

This approximation is not bad unless the chain is extremely stiff.

If we neglect the difference between the center of mass and the Zimm center of resistance $\mathbf{R}_c$, for simplicity, then $\mathbf{r}_p$ may be written in terms of $\mathbf{R}_c$ and the bond vectors $\mathbf{a}_p$ as

$$\mathbf{r}_p = \mathbf{R}_c + \sum_{q=1}^{N} u_{pq}\mathbf{a}_q, \qquad (10.57)$$

where

$$u_{pq} = \frac{q}{N+1} - h(q-p) \qquad (10.58)$$

with $h(x)$ the unit step function as before. Further, if we assume that the motion of $\mathbf{R}_c$ is independent of those of $\mathbf{a}_p$, the average $\langle |\mathbf{r}_{p'}(t) - \mathbf{r}_p(0)|^2 \rangle_{eq}$ in Eq. (10.56) may be written as

$$\langle |\mathbf{r}_{p'}(t) - \mathbf{r}_p(0)|^2 \rangle_{eq} = \langle |\mathbf{r}_{p'}(0) - \mathbf{r}_p(0)|^2 \rangle_{eq} + \langle |\mathbf{R}_c(t) - \mathbf{R}_c(0)|^2 \rangle_{eq}$$

$$+2 \sum_{q,q'=1}^{N} u_{pq} u_{p'q'} \left[\langle \mathbf{a}_{q'}(0) \cdot \mathbf{a}_q(0) \rangle_{eq} - \langle \mathbf{a}_{q'}(t) \cdot \mathbf{a}_q(0) \rangle_{eq} \right], \quad (10.59)$$

where the first average $\langle |\mathbf{r}_{p'}(0) - \mathbf{r}_p(0)|^2 \rangle_{eq}$ on the right-hand side may be equated to the mean-square end-to-end distance $\langle R^2(s) \rangle$ of the continuous HW chain of contour length $s = |p - p'| \Delta s$, and the second average may be given by

$$\langle |\mathbf{R}_c(t) - \mathbf{R}_c(0)|^2 \rangle_{eq} = 6Dt \quad (10.60)$$

with D being the translational diffusion coefficient of the center of mass in the approximation of Eq. (10.56).

By the use of the relations between the Cartesian components of $\mathbf{a}_p$ and the $\mathcal{D}$ functions $\mathcal{D}_1^{mj}(\Omega_p)$ [5], the time-correlation functions $\langle \mathbf{a}_{p'}(t) \cdot \mathbf{a}_p(0) \rangle_{eq}$ in Eq. (10.59) may be expressed in terms of the 1(1) correlation functions $C_{1,[p,p']}^{(j,j')}(t)$ as

$$\langle \mathbf{a}_{p'}(t) \cdot \mathbf{a}_p(0) \rangle_{eq} = (8\pi^2)^N a^2 C_{1,[p,p']}^{(0,0)}(t), \quad (10.61)$$

where $C_{1,[p,p']}^{(j,j')}(t)$ is given by Eq. (10.9). Substitution of Eq. (10.56) with Eqs. (10.59)–(10.61) into Eq. (10.55) leads to

$$S(k,t) = \frac{1}{(N+1)^2} \sum_{p,p'=1}^{N+1} \exp\left(-\frac{k^2}{6} \left\{ \langle R^2(|p-p'|\Delta s) \rangle + 6Dt \right. \right.$$

$$\left. \left. +2(8\pi^2)^N a^2 \sum_{q,q'=1}^{N} u_{pq} u_{p'q'} \left[C_{1,[q,q']}^{(0,0)}(0) - C_{1,[q,q']}^{(0,0)}(t) \right] \right\} \right).$$

$$(10.62)$$

Next we evaluate the *first cumulant*, that is, the initial decay rate of $S(k,t)$, which is defined by

$$\Omega(k) = -\left[\frac{d \ln S(k,t)}{dt} \right]_{t=0}. \quad (10.63)$$

From Eqs. (10.62) and (10.63), the dimensionless quantity $\eta_0 \Omega(k)/k_B T k^3$ as a function of the reduced magnitude $\bar{k}$ of the scattering vector may then be expressed as

$$\frac{\eta_0 \Omega(k)}{k_B T k^3} = \frac{1}{6\pi \bar{k}} [\rho + F(\bar{k})], \quad (10.64)$$

where ρ is defined by Eq. (6.130) with Eq. (6.132), and $\bar{k}$ and $F(\bar{k})$ are given by

$$\bar{k} = \langle S^2 \rangle^{1/2} k, \tag{10.65}$$

$$F(\bar{k}) = \frac{2\langle S^2 \rangle^{1/2}}{3 r_1 r_2 a S(\bar{k}, 0)} \sum_{K=1}^{N} A_K(\bar{k}) \left(|R_{1,K}^{00}|^2 \tilde{\lambda}_{1,K}^0 + |R_{1,K}^{(-1)0}|^2 \tilde{\lambda}_{1,K}^{(-1)} \right). \tag{10.66}$$

In Eq. (10.66), r_1 and r_2 are given by Eqs. (10.17), $R_{1,k}^{jj'}$ are given by Eqs. (10.10) ($\kappa_0 \neq 0$) and (10.13) ($\kappa_0 = 0$), $\tilde{\lambda}_{1,k}^j$ are the reduced 1(1) eigenvalues as before, and $S(\bar{k}, 0)$ and $A_K(\bar{k})$ are given by

$$S(\bar{k}, 0) = \frac{1}{N+1} + \frac{2}{(N+1)^2} \sum_{n=1}^{N} (N - n + 1) \exp\left[-\frac{\bar{k}^2}{6} \frac{\langle R^2(n\Delta s) \rangle}{\langle S^2(L) \rangle} \right], \tag{10.67}$$

$$
A_K(\bar{k}) = \frac{1}{4(N+1)^2} \operatorname{cosec}^2 \left[\frac{K\pi}{2(N+1)} \right]
$$
$$
\times \left\{ 1 + \frac{2}{N+1} \sum_{n=1}^{N} \left[(N - n + 1) \cos\left(\frac{nK\pi}{N+1} \right) \right. \right.
$$
$$
\left. \left. -\operatorname{cosec}\left(\frac{K\pi}{N+1} \right) \sin\left(\frac{nK\pi}{N+1} \right) \right] \exp\left[-\frac{\bar{k}^2}{6} \frac{\langle R^2(n\Delta s) \rangle}{\langle S^2(L) \rangle} \right] \right\}, \tag{10.68}
$$

where $\langle S^2(L) \rangle$ $(= \langle S^2 \rangle)$ is the mean-square radius of gyration of the continuous HW chain of total contour length L. Thus the two branches ($j = 0$ and -1) of the eigenvalue spectrum make contribution to $\Omega(k)$ for $\kappa_0 \neq 0$, and only the $j = 0$ branch does for $\kappa_0 = 0$, as seen from Eq. (10.66).

Now it is important to see from Eq. (10.64) that the so-called "universal" plot of $\eta_0 \Omega(k)/k_B T k^3$ against $\bar{k}$ depends on both ρ and $F(\bar{k})$, and therefore is not universal; it depends on the kind of polymer. According to the results of numerical calculation for the c-HW chain in the crude subspace approximation [64], the eigenvalues $\lambda_{1,k}^0$ in the $j = 0$ branch at small wave numbers k make main contribution to $\Omega(k)$. For KP stiff chains ($\kappa_0 = 0$), we may adopt the touched-bead model as in the case of the dielectric correlation time τ_D (for the d-HW chain) but with $r_1 = 1$ and $r_2 = 1/3$ for the Stokes bead. [Recall that because of the negative eigenvalues, in the cases of dielectric and magnetic relaxation and fluorescence depolarization for the stiff d-HW (KP) chain, an unreasonable value much larger than $1/3$ had to be assigned to r_2, or a value much smaller than d had to be assigned to a.] The numerical results also show that the universal plot for the KP chain depends appreciably on L; in particular, it exhibits no plateau region for small L, the dimensionless quantity above decreasing monotonically with increasing $\bar{k}$. We note that in the theoretical calculation, experimental values (if available) should be used for ρ in Eq. (10.64) since they are appreciably different from the Kirkwood or Zimm value and also dependent on N.

10.5.2 Comparison with Experiment

Figure 10.14 shows plots of $\eta_0\Omega(k)/k_BTk^3$ against $\bar{k}$ [64]. The circles represent the experimental values obtained by Tsunashima et al. [65] for a-PS samples with the molecular weights $M = 9.70 \times 10^6$ (unfilled circles), 5.53×10^6 (right-half-filled circles), and 2.42×10^6 (left-half-filled circles) in *trans*-decalin at 20.4°C, the triangles represent those (smoothed) obtained by Han and Akcasu [66] for a-PS with $M = 4.1 \times 10^4$–4.4×10^7 in cyclohexane at 35.0°C, and the squares represent those obtained by Soda and Wada [67] for DNA with $L = 2.24\ \mu\text{m}$ in 0.15 mol/l NaCl (with 0.015 mol/l trisodium citrate) at 25.0°C. The data points by Tsunashima et al. and by Han and Akcasu are somewhat different from each other, and those for DNA deviate appreciably upward from those for a-PS. This deviation arises from the large value 1.64 of ρ for DNA. The lower solid curve represents the theoretical values for a-PS calculated from Eq. (10.64) with Eqs. (10.66)–(10.68) (for the c-HW chain) with the values of its HW model parameters, $r_1 = 1$, $r_2 = 10$, $N + 1 = 10^4$,

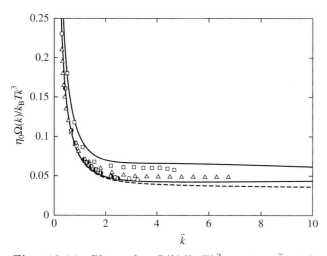

Fig. 10.14. Plots of $\eta_0\Omega(k)/k_BTk^3$ against $\bar{k}$ with experimental data by Tsunashima et al. [65] for a-PS with $M = 9.70 \times 10^6$ (○), 5.53×10^6 (◐), and 2.42×10^6 (◑) in *trans*-decalin at 20.4°C, those by Han and Akcasu [66] for a-PS with $M = 4.1 \times 10^4$–4.4×10^7 in cyclohexane at 35.0°C (△), and those by Soda and Wada [67] for DNA with $L = 2.24\ \mu\text{m}$ in 0.15 mol/l NaCl at 25.0°C (□). The *lower and upper solid curves* represent the theoretical values for a-PS and DNA, respectively, calculated from Eq. (10.64), and the *dashed curve* represents those for a-PMMA (see the text)

and the experimental value 1.27 of ρ [65] in *trans*-decalin at 20.4°C. We note that if we adopt the c-HW chain, such a reasonable value can be assigned to r_2 in the treatment of global modes even in the crude subspace approximation, that the theoretical $\Omega(k)$ is independent of N for such large N, and that this value of ρ agrees well with those by other workers in *trans*-decalin at 21.0°C [68] and in cyclohexane at 34.5°C [68, 69] (see also Table 6.4). The upper solid curve represents the theoretical values for DNA similarly calculated from Eq. (10.64) (for the touched-bead c-KP model) with $\lambda^{-1} = 1100$ Å, $d_b(= a)=$ 29 Å $(d = 0.861d_b = 25$ Å$)$, $N(= L/d_b)= 772$, and $\rho = 1.64$ (see Table 6.3). For comparison, the theoretical values for a-PMMA are represented by the dashed curve, which has been calculated with the values of its HW model parameters but with the same values of r_1, r_2, N, and ρ as those for a-PS.

It is seen that there is rather good agreement between theory and experiment. However, it may be fair to mention that this (apparent) agreement is mainly due to the use of the experimental values of ρ in the theoretical calculation from Eq. (10.64), since the preaveraged Oseen tensor has been used in the evaluation of $F(\bar{k})$. Now it is well known that for the spring-bead (Gaussian) chain in the nondraining limit, the plateau value $(1/6\pi)$ of $\eta_0\Omega(k)/k_BTk^3$ (in the k^3-region of Ω) with the preaveraged Oseen tensor [70, 71] is 15% smaller than that $(1/16)$ with the nonpreaveraged tensor [71, 72]. [Akcasu and co-workers [72–74] were the first to evaluate $\Omega(k)$ over the whole range of k on the basis of the Gaussian chain.] Considering this fact, the HW theoretical values would become somewhat larger if the nonpreaveraged Oseen tensor could be used. Then the resultant disagreement between theory and experiment might be attributed to the experimental difficulty in the determination of the true initial decay rate as pointed out by Stockmayer and co-workers [75, 76]. At any rate, important is the fact that the "universal" plot is not universal, as seen from Fig. 10.14; the differences in the plot among a-PS, a-PMMA, and DNA are due to those in ρ, chain stiffness, and local chain conformation. As for semiflexible chains, we note that Maeda and Fujime [77] and Harnau et al. [78] evaluated $\Omega(k)$ on the basis of the Harris–Hearst model [79] and its improved version [80, 81], respectively, and analyzed some experimental data. In these, however, the constraint on the contour length is relaxed (see Appendix 3.C).

Finally, we examine the behavior of $\Omega(k)$ from a different point of view. Figure 10.15 shows double-logarithmic plots of $\lambda^{-1}\eta_0\Omega(k)/k_BTk^2$ against $\lambda^{-1}k$ for a-PS with $\lambda^{-1} = 20.6$ Å [64]. The unfilled circles represent the experimental values by Tsunashima et al. for $M = 9.70 \times 10^6$ (in Fig. 10.14), the triangles represent those by Han and Akcasu for $M = 4.4 \times 10^7$ (in Fig. 10.14), and the filled circles represent those obtained by Nicholson et al. [82] (by the neutron spin-echo method) for a sample with $M = 5.5 \times 10^4$ in benzene-d_6 at 30.0°C. The solid curves represent the respective theoretical values, and the dot-dashed straight line has a slope of unity. [Note that the excluded-volume effect on $F(\bar{k})$ may be expected to be rather small for the

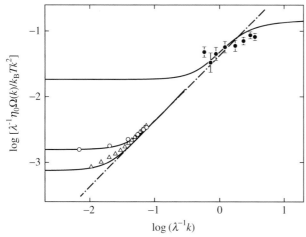

Fig. 10.15. Double-logarithmic plots of $\lambda^{-1}\eta_0\Omega(k)/k_BTk^2$ against $\lambda^{-1}k$ for a-PS with $\lambda^{-1} = 20.6$ Å with experimental data by Tsunashima et al. [65] for $M = 9.70\times 10^6$ in *trans*-decalin at 20.4°C ($\circ$), those by Han and Akcasu [66] for $M = 4.4\times 10^7$ in cyclohexane at 35.0°C ($\triangle$), and those by Nicholson et al. [82] for $M = 5.5\times 10^4$ in benzene-d_6 at 30.0°C ($\bullet$). The *solid curves* represent the respective theoretical values calculated from Eq. (10.64), and the *dot-dashed straight line* has a slope of unity (see the text)

last sample.] For small M (the last sample), it is interesting to see that the experimental data exhibit no k^3-region, being consistent with the theoretical prediction, and that the transition from the k^3- to k^2-region in the range of large k occurs at $\lambda^{-1}k \simeq 1$.

10.6 Some Remarks

10.6.1 Elementary Processes of Chain Motions

In the preceding sections of this chapter, it has been shown that the dynamic HW chain model may give a quantitative or semiquantitative explanation of experimental results for various dynamical properties of both flexible and semiflexible polymers in dilute solution. In the case of flexible chains, however, it is difficult to picture clearly the elementary processes of chain motions in contrast to the case of conventional bond chains [11, 83]. This is due to the coarse-graining made in the HW model (even in the d-HW chain). The situation may be manifested if the activation energy for local chain motions (conformational transitions) is considered. The simulation [84, 85] and experimental [11, 15, 63, 86] studies show that it is about 10 kJ mol^{-1}, nearly corresponding to the single *trans–gauche* barrier height. (This also indicates the nonexistence of the so-called crankshaft motion.) For comparison, if we consider the dielectric correlation time τ_D of, for example, a-PPCS on the basis of the dynamic HW model, the activation energy is estimated to be 3.1 kJ mol^{-1} from the Arrhenius plot of $\ln(\tau_D/\eta_0)$ against T^{-1}, assuming that λ^{-1} is proportional to T^{-1} [20]. The result is only comparable to the value 2.7 kJ mol^{-1} estimated from the Rouse–Zimm relaxation times ($\tau_D/\eta_0 \propto T^{-1}$).

10.6.2 Dynamic vs Static Chain Stiffness

In the preceding sections we have considered the dielectric correlation time τ_D and the depolarized scattering correlation time τ_Γ. We further introduce *magnetic* and *fluorescence correlation times* τ_M and τ_F defined by

$$\tau_M = \frac{1}{G_m(0)} \int_0^\infty G_m(t)dt = \frac{J_m(0)}{2G_m(0)}, \tag{10.69}$$

$$\tau_F = \frac{1}{r(0)} \int_0^\infty r(t)dt, \tag{10.70}$$

where $G_m(t)$ is the magnetic autocorrelation function, $J_m(\omega)$ is the spectral density, and $r(t)$ is the emission anisotropy as before. Note that τ_M is in general not an observable (except in the narrowing limit) in contrast to the other correlation times.

Now, for flexible chains, we consider the ratio of the correlation time τ_X (with $X = $ D, M, F, Γ) to that, τ_X^0, of the isolated single subbody as the spheroid having rotation axis of length a and diameter d, where we assume that the chain has perpendicular dipoles or (approximately cylindrically-symmetric) local polarizabilities. Clearly the ratio may be regarded as a measure of *dynamic* chain stiffness. We calculate it using the observed values of τ_D (in Table 10.1), theoretical values of τ_M computed from Eq. (10.69) in the higher-order subspace approximation (as in Table 10.2), observed values of τ_F [46], and the observed values of τ_Γ (in Sect. 10.4.2) along with theoretical values of τ_X^0 computed from [5, 31, 46]

$$\tau_D^0 = (D_{r,1} + D_{r,3})^{-1}, \tag{10.71}$$

$$\tau_M^0 = \tau_F^0 = \frac{1}{24}\left(\frac{1}{D_{r,1}} + \frac{9}{D_{r,1} + 2D_{r,3}}\right), \tag{10.72}$$

$$\tau_\Gamma^0 = \frac{1}{6D_{r,1}}, \tag{10.73}$$

where $D_{r,1}$ and $D_{r,3}$ are the same rotatory diffusion coefficients as those in Eq. (10.19), and the values of d from the chemical structures are used. We note that Eq. (10.73) can be derived from the depolarized component $I_{Hv}(t) \propto \exp(-6D_{r,1}t)$ [55].

Values of $\log(\tau_X/\tau_X^0)$ so calculated are plotted against the static stiffness parameter λ^{-1} in Fig. 10.16. The unfilled and filled circles, triangle, and squares represent the values for $X = $ D, M, F, and Γ, respectively. It is seen that the ratio τ_X/τ_X^0 is a monotonically increasing function of λ^{-1}, indicating that there is strong correlation between them.

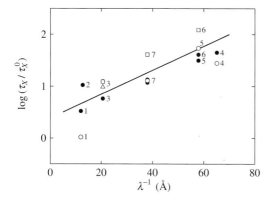

Fig. 10.16. Plots of $\log(\tau_X /\tau_X^0)$ against λ^{-1} for polymers with perpendicular dipoles or (approximately cylindrically-symmetric) local polarizabilities, where τ_X^0 is the τ_X of the isolated subbody, and $X = $ D ($\circ$), M ($\bullet$), F ($\triangle$), and Γ ($\square$). The polymers are identified by the numbers attached to the points: (1) POE, (2) PIB, (3) a-PS, (4) a-PMVK, (5) s-PMMA, (6) a-PMMA, and (7) i-PMMA

10.6.3 Dynamic Intrinsic Viscosity

In this final subsection we briefly discuss the dynamic intrinsic viscosity, that is, the real part $[\eta']$ of $[\eta] = [\eta'] - i[\eta'']$ [87]. The subspace $\{2(1), 2(2)\}$ actually relevant to viscosity becomes $6N$-dimensional if a new basis set, which is a hybrid of the one- and two-body excitation basis functions, is introduced. Then the eigenvalue problem may be reduced to N six-dimensional problems. Among the six branches of the eigenvalue spectrum $\lambda_{2(2),k}^J$ ($J = 1, \cdots, 6$), one global ($J = 1$) and two local ($J = 2, 3$) branches make contribution to $[\eta']$. Thus it may be written in the form

$$[\eta'] = [\eta]^{\text{glob}} + [\eta]^{\text{loc}} + [\eta]_\infty \tag{10.74}$$

with

$$[\eta]_\infty = [\eta]_C + [\eta]_E + \eta^* , \tag{10.75}$$

where $[\eta]^{\text{glob}}$ is the contribution from the $J = 1$ (Rouse–Zimm) branch, $[\eta]^{\text{loc}}$ is that from the $J = 2$ and 3 branches (at small wave numbers k), which arises from the interaction between the global and local modes, $[\eta]_C$ arises from the constraints and is independent of the angular frequency ω, and $[\eta]_E$ and η^* are the Einstein intrinsic viscosity and the specific interaction parameter in Eq. (6.134), respectively. Note that $[\eta]^{(\text{KR})}$ in Eq. (6.134) may now be written in the form

$$[\eta]^{(\text{KR})} = [\eta]_0^{\text{glob}} + [\eta]_0^{\text{loc}} + [\eta]_C , \tag{10.76}$$

where the subscript 0 indicates the value at $\omega = 0$.

Now the eigenvalues in the $J = 2$ and 3 branches are much larger than those in the $J = 1$ branch at small k, and therefore $[\eta]^{\text{loc}}$ still remains finite after $[\eta]^{\text{glob}}$ has relaxed away. The high-frequency plateau observed in viscoelastic experiments [88], the height of which we denote by $[\eta]^P$, may then be given by

$$[\eta]^P = [\eta]_0^{\text{loc}} + [\eta]_\infty . \tag{10.77}$$

In this connection, we note that Fixman and Evans [89] considered $[\eta]_0^{loc}$ to arise from the gap structure of the spectrum due to the interaction between the global and local modes, and that the effect of the constraints leading to $[\eta]_C$ was also considered by Doi et al. [90] and by Fixman and Evans [91]. As seen from Eq. (10.77) with Eq. (10.75), $[\eta]^P$ may possibly become negative for $\eta^* < 0$. Indeed, this has been observed experimentally in some cases [92]. In the case of $\eta^* = 0$, it has been shown that there is rather good agreement between theory and experiment for $[\eta]^P$ [87].

Finally, we note that the mechano-optic coefficient in oscillatory flow birefringence [93–95] may be expressed in terms of the eigenvalues in the above three branches of the viscoelastic spectrum and the five (local) branches of the magnetic one [96].

References

1. B. H. Zimm: J. Chem. Phys. **24**, 269 (1956).
2. W. H. Stockmayer and M. E. Baur: J. Am. Chem. Soc. **86**, 3485 (1964).
3. W. H. Stockmayer: Pure Appl. Chem. **15**, 539 (1967).
4. H. Yamakawa: *Modern Theory of Polymer Solutions* (Harper & Row, New York, 1971).
5. T. Yoshizaki and H. Yamakawa: J. Chem. Phys. **81**, 982 (1984).
6. G. Williams: Chem. Rev. **72**, 55 (1972).
7. J. E. Shore and R. Zwanzig: J. Chem. Phys. **63**, 5445 (1975).
8. P. Debye: *Polar Molecules* (Dover, New York, 1945).
9. D. E. Woessner: J. Chem. Phys. **36**, 1 (1962).
10. J. D. Hoffman and H. G. Pfeiffer: J. Chem. Phys. **22**, 132 (1954).
11. W. H. Stockmayer: Pure Appl. Chem. Suppl. Macromol. Chem. **8**, 379 (1973).
12. M. Davies, G. Williams, and G. D. Loveluck: Z. Elektrochem. **64**, 575 (1960).
13. S. Mashimo, S. Yagihara, and A. Chiba: Macromolecules **17**, 630 (1984).
14. W. H. Stockmayer and K. Matsuo: Macromolecules **5**, 766 (1972).
15. S. Mashimo: Macromolecules **9**, 91 (1976).
16. S. Mashimo, P. Winsor IV, R. H. Cole, K. Matsuo, and W. H. Stockmayer: Macromolecules **16**, 965 (1983).
17. Y. Iwasa, S. Mashimo, and A. Chiba: Polym. J. **8**, 401 (1976).
18. M. Fujii, K. Nagasaka, J. Shimada, and H. Yamakawa: Macromolecules **16**, 1613 (1983).
19. R. H. Cole, S. Mashimo, and P. Winsor IV: J. Phys. Chem. **84**, 786 (1980).
20. H. Yamakawa, T. Yoshizaki, and M. Fujii: J. Chem. Phys. **84**, 4693 (1986).
21. H. Yamakawa and T. Yoshizaki: J. Chem. Phys. **78**, 572 (1983).
22. H. Yamakawa: Macromolecules **16**, 1928 (1983).
23. T. Matsumoto, N. Nishioka, A. Teramoto, and H. Fujita: Macromolecules **7**, 824 (1974).
24. S. Takada, T. Itou, H. Chikiri, Y. Einaga, and A. Teramoto: Macromolecules **22**, 973 (1989).
25. I. Solomon: Phys. Rev. **99**, 559 (1955).
26. A. Abragam: *The Principles of Nuclear Magnetism* (Oxford Univ., London, 1961).
27. D. Doddrell, V. Glushko, and A. Allerhand: J. Chem. Phys. **56**, 3683 (1972).
28. W. T. Huntress, Jr.: J. Chem. Phys. **48**, 3524 (1968).

29. W. E. Hull and B. D. Sykes: J. Mol. Biol. **98**, 121 (1975).
30. S. A. Allison, J. H. Shibata, J. Wilcoxon, and J. M. Shurr: Biopolymers **21**, 729 (1982).
31. H. Yamakawa and M. Fujii: J. Chem. Phys. **81**, 997 (1984).
32. D. Wallach: J. Chem. Phys. **47**, 5258 (1967).
33. A. A. Jones: J. Polym. Sci., Polym. Phys. Ed. **15**, 863 (1977).
34. A. A. Jones and W. H. Stockmayer: J. Polym. Sci., Polym. Phys. Ed. **15**, 847 (1977).
35. B. Valeur, J. P. Jarry, F. Gény, and L. Monnerie: J. Polym. Sci., Polym. Phys. Ed. **13**, 667 (1975); **13**, 2251 (1975).
36. B. Valeur, L. Monnerie, and J. P. Jarry: J. Polym. Sci., Polym. Phys. Ed. **13**, 675 (1975).
37. F. Heatley and I. Walton: Polymer **17**, 1019 (1976).
38. Y. Inoue, A. Nishioka, and R. Chûjô: J. Polym. Sci., Polym. Phys. Ed. **11**, 2237 (1973).
39. A. Allerhand and R. K. Hailstone: J. Chem. Phys. **56**, 3718 (1972).
40. Y. Takaeda, T. Yoshizaki, and H. Yamakawa: Macromolecules **27**, 4248 (1994).
41. J. R. Lyerla, Jr., T. T. Horikawa, and D. E. Johnson: J. Am. Chem. Soc. **99**, 2463 (1977).
42. Y. Takaeda, T. Yoshizaki, and H. Yamakawa: Macromolecules **28**, 682 (1995).
43. Y. Naito, N. Sawatari, Y. Takaeda, T. Yoshizaki, and H. Yamakawa: Macromolecules **30**, 2751 (1997).
44. M. E. Hogan and O. Jardetzky: Proc. Natl. Acad. Sci. USA **76**, 6341 (1979).
45. A. Jablonski: Bull. Acad. Pol. Sci. Ser. Sci. Math. Astron. Phys. **8**, 259 (1960); Acta Phys. Pol. **28**, 717 (1965).
46. T. Yoshizaki, M. Fujii, and H. Yamakawa: J. Chem. Phys. **82**, 1003 (1985).
47. A. Szabo: J. Chem. Phys. **81**, 150 (1984).
48. B. Valeur and L. Monnerie: J. Polym. Sci., Polym. Phys. Ed. **14**, 11 (1976).
49. A. M. North and I. Soutar: J. Chem. Soc. Faraday Trans. 1 **68**, 1101 (1972).
50. D. P. Millar, R. J. Robbins, and A. H. Zewail: Proc. Natl. Acad. Sci. USA **77**, 5593 (1980).
51. D. P. Millar, R. J. Robbins, and A. H. Zewail: J. Chem. Phys. **74**, 4200 (1981); **76**, 2080 (1982).
52. M. D. Barkley and B. H. Zimm: J. Chem. Phys. **70**, 2991 (1979).
53. Ph. Wahl, J. Paoletti, and J.-B. LePecq: Proc. Natl. Acad. Sci. USA **65**, 417 (1970).
54. T. Yoshizaki and H. Yamakawa: J. Chem. Phys. **99**, 9145 (1993).
55. B. J. Berne and R. Pecora: *Dynamic Light Scattering* (Wiley, New York, 1976).
56. W. Kuhn and F. Grün: Kolloid Z. **101**, 248 (1942).
57. K. Ono and K. Okano: Japan J. Appl. Phys. **9**, 1356 (1970).
58. D. R. Bauer, J. I. Brauman, and R. Pecora: Macromolecules **8**, 443 (1975).
59. G. T. Evans: J. Chem. Phys. **71**, 2263 (1979).
60. M. Fixman and J. Kovac: J. Chem. Phys. **61**, 4939 (1974).
61. F. Strehle, Th. Dorfmüller, and D. Samios: Macromolecules **25**, 3569 (1992).
62. P. J. Hagerman and B. H. Zimm: Biopolymers **20**, 1481 (1981).
63. K. Matsuo, K. F. Kuhlmann, H. W.-H. Yang, F. Gény, W. H. Stockmayer, and A. A. Jones: J. Polym. Sci., Polym. Phys. Ed. **15**, 1347 (1977).
64. T. Yoshizaki, M. Osa, and H. Yamakawa: J. Chem. Phys. **106**, 2828 (1997).
65. Y. Tsunashima, N. Nemoto, and M. Kurata: Macromolecules **16**, 1184 (1983).
66. C. C. Han and A. Z. Akcasu: Macromolecules **14**, 1080 (1981).
67. K. Soda and A. Wada: Biophys. Chem. **20**, 185 (1984).
68. T. Konishi, T. Yoshizaki, and H. Yamakawa: Macromolecules **24**, 5614 (1991).
69. M. Schmidt and W. Burchard: Macromolecules **14**, 210 (1981).
70. E. Dubois-Viollete and P.-G. de Gennes: Physics **3**, 181 (1967).

71. W. Burchard, M. Schmidt, and W. H. Stockmayer: Macromolecules **13**, 580 (1980).
72. M. Benmouna and A. Z. Akcasu: Macromolecules **13**, 409 (1980).
73. A. Z. Akcasu and H. Gurol: J. Polym. Sci., Polym. Phys. Ed. **14**, 1 (1976).
74. A. Z. Akcasu, M. Benmouna, and C. C. Han: Polymer **21**, 866 (1980).
75. W. H. Stockmayer and B. Hammouda: Pure Appl. Chem. **56**, 1372 (1984).
76. M. Schmidt and W. H. Stockmayer: Macromolecules **17**, 509 (1984).
77. T. Maeda and S. Fujime: Macromolecules **17**, 2381 (1984).
78. L. Harnau, R. G. Winkler, and P. Reineker: J. Chem. Phys. **104**, 6355 (1996).
79. R. A. Harris and J. E. Hearst: J. Chem. Phys. **44**, 2595 (1966).
80. R. G. Winkler, P. Reineker, and L. Harnau: J. Chem. Phys. **101**, 8119 (1994).
81. L. Harnau, R. G. Winkler, and P. Reineker: J. Chem. Phys. **102**, 7750 (1995).
82. L. K. Nicholson, J. S. Higgins, and J. B. Hayter: Macromolecules **14**, 836 (1981).
83. E. Helfand: J. Chem. Phys. **54**, 4651 (1971).
84. E. Helfand, Z. R. Wasserman, and T. A. Weber: Macromolecules **13**, 526 (1980).
85. D. Perchak and J. H. Weiner: Macromolecules **14**, 785 (1981).
86. T.-P. Lias and H. Morawetz: Macromolecules **13**, 1228 (1980).
87. T. Yoshizaki and H. Yamakawa: J. Chem. Phys. **88**, 1313 (1988).
88. J. D. Ferry: *Viscoelastic Properties of Polymers*, 3rd ed. (Wiley, New York, 1980).
89. M. Fixman and G. T. Evans: J. Chem. Phys. **68**, 195 (1978).
90. M. Doi, H. Nakajima, and Y. Wada: Colloid Polym. Sci. **254**, 559 (1976).
91. M. Fixman and G. T. Evans: J. Chem. Phys. **64**, 3474 (1976).
92. V. F. Man: Ph. D. Thesis (Univ. of Wisconsin, Madison, 1984); P. A. Merchak: Ph. D. Thesis (Univ. of Wisconsin, Madison, 1987).
93. G. B. Thurston and J. L. Schrag: Trans. Soc. Rheol. **6**, 325 (1962).
94. T. P. Lodge, J. W. Miller, and J. L. Schrag: J. Polym. Sci., Polym. Phys. Ed. **20**, 1409 (1982).
95. T. P. Lodge and J. L. Schrag: Macromolecules **17**, 352 (1984).
96. K. Nagasaka, T. Yoshizaki, and H. Yamakawa: J. Chem. Phys. **90**, 5167 (1989).

Appendix I. Coefficients $A_{ij}^{(m)}$ in Eq. (3.72)

$$A_{00}^{(1)} = -\frac{1}{2}, \qquad A_{10}^{(1)} = 1, \qquad A_{11}^{(1)} = \frac{1}{2},$$

$$A_{00}^{(2)} = \frac{107}{54}, \qquad A_{10}^{(2)} = -\frac{26}{9}, \qquad A_{11}^{(2)} = -2,$$

$$A_{20}^{(2)} = \frac{5}{3}, \qquad A_{21}^{(2)} = -1, \qquad A_{22}^{(2)} = \frac{1}{54},$$

$$A_{00}^{(3)} = -\frac{6143}{324}, \qquad A_{10}^{(3)} = \frac{226}{9}, \qquad A_{11}^{(3)} = \frac{4743}{250},$$

$$A_{20}^{(3)} = -\frac{259}{18}, \qquad A_{21}^{(3)} = \frac{639}{50}, \qquad A_{22}^{(3)} = -\frac{1}{81},$$

$$A_{30}^{(3)} = \frac{35}{9}, \qquad A_{31}^{(3)} = \frac{21}{10}, \qquad A_{32}^{(3)} = -\frac{1}{54},$$

$$A_{33}^{(3)} = \frac{1}{4500},$$

$$A_{00}^{(4)} = \frac{123403}{375}, \qquad A_{10}^{(4)} = -\frac{281183}{675}, \qquad A_{11}^{(4)} = -\frac{18509371}{56250},$$

$$A_{20}^{(4)} = \frac{3554}{15}, \qquad A_{21}^{(4)} = -\frac{151042}{625}, \qquad A_{22}^{(4)} = -\frac{59}{3087},$$

$$A_{30}^{(4)} = -\frac{224}{3}, \qquad A_{31}^{(4)} = -\frac{7722}{125}, \qquad A_{32}^{(4)} = \frac{11}{1323},$$

$$A_{33}^{(4)} = -\frac{2}{28125}, \qquad A_{40}^{(4)} = \frac{35}{3}, \qquad A_{41}^{(4)} = -\frac{126}{25},$$

$$A_{42}^{(4)} = \frac{1}{63}, \qquad A_{43}^{(4)} = -\frac{1}{5625}, \qquad A_{44}^{(4)} = \frac{1}{771750},$$

$$A_{00}^{(5)} = -\frac{164016904}{18225}, \qquad A_{10}^{(5)} = \frac{67421951}{6075}, \qquad A_{11}^{(5)} = \frac{347266405529}{38587500},$$

$$A_{20}^{(5)} = -\frac{2545708}{405}, \qquad A_{21}^{(5)} = \frac{1902126692}{275625}, \qquad A_{22}^{(5)} = \frac{1255150}{12252303},$$

$$A_{30}^{(5)} = \frac{169301}{81}, \qquad A_{31}^{(5)} = \frac{18269383}{8750}, \qquad A_{32}^{(5)} = \frac{2545}{21609},$$

$$A_{33}^{(5)} = -\frac{31}{41006250}, \qquad A_{40}^{(5)} = -\frac{22715}{54}, \qquad A_{41}^{(5)} = \frac{35541}{125},$$

$$A_{42}^{(5)} = \frac{55}{9261}, \qquad A_{43}^{(5)} = \frac{61}{911250}, \qquad A_{44}^{(5)} = -\frac{1}{3781575},$$

$$A_{50}^{(5)} = \frac{385}{9}, \qquad A_{51}^{(5)} = \frac{693}{50}, \qquad A_{52}^{(5)} = -\frac{55}{3969},$$

$$A_{53}^{(5)} = \frac{11}{101250}, \qquad A_{54}^{(5)} = -\frac{1}{1080450}, \qquad A_{55}^{(5)} = \frac{1}{225042300}.$$

Appendix II. Coefficients $A_{ijk}^{(m)}$ in Eq. (4.81)

The coefficients $A_{ijk}^{(m)}$ in Eq. (4.81) may be written as functions of κ_0 and τ_0 in the form,

$$A_{ijk}^{(m)} = \sum_{n=0}^{m} a_{ijk}^{(m)n}(\nu) \frac{\kappa_0^{2n} \tau_0^{2(m-n)}}{\nu^{2m}}$$

with $\nu = (\kappa_0^2 + \tau_0^2)^{1/2}$. We note that the coefficients $a_{ijk}^{(m)n}$ have the symmetry property $a_{ij(-k)}^{(m)n} = a_{ijk}^{(m)n*}$ with the asterisk indicating the complex conjugate, and that

$$a_{ijk}^{(m)0} = A_{ij}^{(m)} \qquad \text{for } k = 0$$
$$= 0 \qquad \text{for } k \neq 0 \,,$$

where $A_{ij}^{(m)}$ are given in Appendix I. The coefficients $a_{ijk}^{(m)n}$ $(n \neq 0,\ k \geq 0)$ for $m = 1$ and 2 are given by

$$a_{000}^{(1)1} = -\frac{2(4 - \nu^2)}{(4 + \nu^2)^2}, \qquad a_{100}^{(1)1} = \frac{4}{4 + \nu^2},$$

$$a_{110}^{(1)1} = 0, \qquad a_{111}^{(1)1} = \frac{4 - \nu^2 - 4i\nu}{(4 + \nu^2)^2},$$

$$a_{000}^{(2)1} = \frac{8875008 - 118656\nu^2 - 100768\nu^4 - 8744\nu^6 - 154\nu^8}{27(4 + \nu^2)^3 (36 + \nu^2)^2},$$

$$a_{100}^{(2)1} = \frac{-29952 - 880\nu^2 + 20\nu^4}{9(4 + \nu^2)^2 (36 + \nu^2)}, \qquad a_{110}^{(2)1} = \frac{32 + 6\nu^2}{(16 + \nu^2)^2},$$

$$a_{111}^{(2)1} = \frac{-8448 - 208\nu^2 + 32\nu^4 + \nu^6}{4(4 + \nu^2)^3 (16 + \nu^2)} + i\frac{-1536 + 800\nu^2 + 76\nu^4 + \nu^6}{\nu(4 + \nu^2)^3 (16 + \nu^2)},$$

$$a_{200}^{(2)1} = \frac{40}{3(4 + \nu^2)}, \qquad a_{210}^{(2)1} = \frac{8}{16 + \nu^2},$$

$$a_{211}^{(2)1} = \frac{4 - \nu^2 - 4i\nu}{(4 + \nu^2)^2}, \qquad a_{220}^{(2)1} = -\frac{8}{27(16 + \nu^2)},$$

$$a_{221}^{(2)1} = \frac{36864 - 2368\nu^2 - 108\nu^4 - \nu^6 - 4i\nu(5376 + 152\nu^2 + \nu^4)}{4(16 + \nu^2)^2(36 + \nu^2)^2},$$

$$a_{222}^{(2)1} = 0,$$

$$a_{000}^{(2)2} = \frac{1109376 - 461088\nu^2 - 75272\nu^4 + 854\nu^6 + 130\nu^8}{27(4 + \nu^2)^4(9 + \nu^2)^2},$$

$$a_{100}^{(2)2} = \frac{-14976 + 2032\nu^2 + 244\nu^4}{9(4 + \nu^2)^3(9 + \nu^2)}, \qquad a_{110}^{(2)2} = 0,$$

$$a_{111}^{(2)2} = \frac{-65536 + 33024\nu^2 + 960\nu^4 + 16\nu^6 - 4\nu^8}{(4 + \nu^2)^4(16 + \nu^2)^2}$$
$$+ i\frac{-24576 + 68096\nu^2 - 5824\nu^4 - 448\nu^6 - 28\nu^8}{\nu(4 + \nu^2)^4(16 + \nu^4)^2},$$

$$a_{200}^{(2)2} = \frac{80}{3(4 + \nu^2)^2}, \qquad a_{210}^{(2)2} = 0,$$

$$a_{211}^{(2)2} = \frac{-4(32 - 60\nu^2 + 3\nu^4) + 8i\nu(36 - 11\nu^2)}{(4 + \nu^2)^3(16 + \nu^2)},$$

$$a_{220}^{(2)2} = \frac{32}{27(16 + \nu^2)^2}, \qquad a_{221}^{(2)2} = 0,$$

$$a_{222}^{(2)2} = \frac{144 - 73\nu^2 + \nu^4 - 14i\nu(12 - \nu^2)}{(9 + \nu^2)^2(16 + \nu^2)^2}.$$

Appendix III. Coefficients $E_{mn}(\kappa_0, \tau_0)$ and $D^{00,00}_{l_1 l_2 l_3, mn}(\kappa_0, \tau_0)$

$E_{11} = -\frac{2}{3}$,

$E_{12} = -\frac{1}{12}\kappa_0{}^2 + \frac{1}{3}$,

$E_{13} = \frac{1}{10}\kappa_0{}^2 - \frac{2}{15}$,

$E_{14} = \frac{1}{360}\kappa_0{}^4 + \frac{1}{360}\kappa_0{}^2\tau_0{}^2 - \frac{1}{15}\kappa_0{}^2 + \frac{2}{45}$,

$E_{15} = -\frac{1}{252}\kappa_0{}^4 - \frac{1}{252}\kappa_0{}^2\tau_0{}^2 + \frac{2}{63}\kappa_0{}^2 - \frac{4}{315}$,

$E_{22} = \frac{28}{45}$,

$E_{23} = \frac{8}{45}\kappa_0{}^2 - \frac{4}{5}$,

$E_{24} = \frac{1}{144}\kappa_0{}^4 - \frac{23}{60}\kappa_0{}^2 + \frac{73}{105}$,

$E_{25} = -\frac{11}{378}\kappa_0{}^4 - \frac{5}{756}\kappa_0{}^2\tau_0{}^2 + \frac{899}{1890}\kappa_0{}^2 - \frac{1396}{2835}$,

$E_{33} = -\frac{248}{315}$,

$E_{34} = -\frac{233}{630}\kappa_0{}^2 + \frac{988}{525}$,

$E_{35} = -\frac{11}{360}\kappa_0{}^4 + \frac{664}{525}\kappa_0{}^2 - \frac{13314}{4725}$,

$E_{44} = \frac{2032}{1575}$,

$E_{45} = \frac{1352}{1575}\kappa_0{}^2 - \frac{23264}{4725}$,

$E_{55} = -\frac{2336}{891}$,

$D^{00,00}_{202,01} = -\frac{8}{3}$,

$D^{00,00}_{202,02} = -\frac{11}{24}\kappa_0{}^2 + \frac{13}{3}$,

$D^{00,00}_{202,03} = \frac{8}{5}\kappa_0{}^2 - \frac{16}{3}$,

$D^{00,00}_{202,04} = \frac{47}{720}\kappa_0{}^4 + \frac{17}{720}\kappa_0{}^2\tau_0{}^2 - \frac{31}{10}\kappa_0{}^2 + \frac{242}{45}$,

$D^{00,00}_{202,05}(0, \tau_0) = -\frac{208}{45}$,

$D^{00,00}_{202,12} = -\frac{1}{12}\kappa_0{}^2 + \frac{127}{45}$,

$D^{00,00}_{202,13} = \frac{17}{18}\kappa_0{}^2 - \frac{736}{105}$,

$D^{00,00}_{202,14} = \frac{59}{1440}\kappa_0{}^4 + \frac{1}{360}\kappa_0{}^2\tau_0{}^2 - \frac{1301}{336}\kappa_0{}^2 + \frac{817}{63}$,

$D^{00,00}_{202,15}(0, \tau_0) = -\frac{55534}{2835}$,

$D^{00,00}_{202,23} = \frac{8}{45}\kappa_0{}^2 - \frac{1244}{315}$,

$D^{00,00}_{202,24} = \frac{1}{144}\kappa_0{}^4 - \frac{43}{21}\kappa_0{}^2 + \frac{4439}{315}$,

$D^{00,00}_{202,25}(0, \tau_0) = -\frac{519872}{14175}$,

$D^{00,00}_{202,34} = -\frac{233}{630}\kappa_0{}^2 + \frac{11092}{1575}$,

$D^{00,00}_{202,35}(0, \tau_0) = -\frac{1777438}{51975}$,

$D^{00,00}_{202,45}(0, \tau_0) = -\frac{2402912}{155925}$,

$D^{00,00}_{220,01} = -6$,

$D^{00,00}_{220,02} = -\frac{3}{2}\kappa_0{}^2 + 18$,

$D^{00,00}_{220,03} = 9\kappa_0{}^2 - 36$,

$D^{00,00}_{220,04} = \frac{1}{2}\kappa_0{}^4 + \frac{1}{8}\kappa_0{}^2\tau_0{}^2 - 27\kappa_0{}^2 + 54$,

$$D^{00,00}_{220,05}(0,\tau_0) = -\frac{324}{5}, \qquad\qquad D^{00,00}_{220,12} = -\frac{1}{12}\kappa_0{}^2 + \frac{19}{3},$$

$$D^{00,00}_{220,13} = \frac{13}{5}\kappa_0{}^2 - \frac{416}{15},$$

$$D^{00,00}_{220,14} = \frac{23}{180}\kappa_0{}^4 + \frac{1}{360}\kappa_0{}^2\tau_0{}^2 - \frac{559}{30}\kappa_0{}^2 + \frac{3548}{45},$$

$$D^{00,00}_{220,15}(0,\tau_0) = -\frac{1508}{9}, \qquad\qquad D^{00,00}_{220,23} = \frac{8}{45}\kappa_0{}^2 - \frac{44}{5},$$

$$D^{00,00}_{220,24} = \frac{1}{144}\kappa_0{}^4 - \frac{311}{60}\kappa_0{}^2 + \frac{5599}{105}, \qquad D^{00,00}_{220,25}(0,\tau_0) = -\frac{581626}{2835},$$

$$D^{00,00}_{220,34} = -\frac{233}{630}\kappa_0{}^2 + \frac{8128}{525}, \qquad\qquad D^{00,00}_{220,35}(0,\tau_0) = -\frac{195166}{1575},$$

$$D^{00,00}_{220,45}(0,\tau_0) = -\frac{157184}{4725},$$

$$D^{00,00}_{222,01} = -\frac{17}{3}, \qquad\qquad D^{00,00}_{222,02} = -\frac{29}{24}\kappa_0{}^2 + \frac{107}{6},$$

$$D^{00,00}_{222,03} = \frac{309}{40}\kappa_0{}^2 - \frac{124}{3},$$

$$D^{00,00}_{222,04} = \frac{17}{45}\kappa_0{}^4 + \frac{31}{360}\kappa_0{}^2\tau_0{}^2 - \frac{1643}{60}\kappa_0{}^2 + \frac{3536}{45},$$

$$D^{00,00}_{222,05}(0,\tau_0) = -\frac{40606}{315}, \qquad\qquad D^{00,00}_{222,12} = -\frac{1}{12}\kappa_0{}^2 + \frac{262}{45},$$

$$D^{00,00}_{222,13} = \frac{79}{36}\kappa_0{}^2 - \frac{2773}{105},$$

$$D^{00,00}_{222,14} = \frac{149}{1440}\kappa_0{}^4 + \frac{1}{360}\kappa_0{}^2\tau_0{}^2 - \frac{1815}{112}\kappa_0{}^2 + \frac{21475}{252},$$

$$D^{00,00}_{222,15}(0,\tau_0) = -\frac{250531}{1134}, \qquad\qquad D^{00,00}_{222,23} = \frac{8}{45}\kappa_0{}^2 - \frac{2504}{315},$$

$$D^{00,00}_{222,24} = \frac{1}{144}\kappa_0{}^4 - \frac{467}{105}\kappa_0{}^2 + \frac{15569}{315}, \qquad D^{00,00}_{222,25}(0,\tau_0) = -\frac{429653}{2025},$$

$$D^{00,00}_{222,34} = -\frac{233}{630}\kappa_0{}^2 + \frac{21802}{1575}, \qquad\qquad D^{00,00}_{222,35}(0,\tau_0) = -\frac{5861782}{51975},$$

$$D^{00,00}_{222,45}(0,\tau_0) = -\frac{4612592}{155925},$$

$$D^{00,00}_{224,01} = -4, \qquad\qquad D^{00,00}_{224,02} = -\frac{5}{12}\kappa_0{}^2 + \frac{146}{15},$$

$$D^{00,00}_{224,03} = \frac{73}{30}\kappa_0{}^2 - \frac{2038}{105},$$

$$D^{00,00}_{224,04} = \frac{63}{64}\kappa_0{}^4 - \frac{1}{120}\kappa_0{}^2\tau_0{}^2 - \frac{2441}{280}\kappa_0{}^2 + \frac{7481}{210},$$

$$D^{00,00}_{224,05}(0,\tau_0) = -\frac{59303}{945}, \qquad\qquad D^{00,00}_{224,12} = -\frac{1}{12}\kappa_0{}^2 + \frac{19}{5},$$

$$D^{00,00}_{224,13} = \frac{91}{90}\kappa_0{}^2 - \frac{4138}{315},$$

$$D^{00,00}_{224,14} = \frac{3}{80}\kappa_0{}^4 + \frac{1}{360}\kappa_0{}^2\tau_0{}^2 - \frac{14837}{2520}\kappa_0{}^2 + \frac{173717}{4725},$$

$$D^{00,00}_{224,15}(0,\tau_0) = -\frac{1602416}{17325}, \qquad\qquad D^{00,00}_{224,23} = \frac{8}{45}\kappa_0{}^2 - \frac{172}{35},$$

$$D^{00,00}_{224,24} = \frac{1}{144}\kappa_0{}^4 - \frac{445}{84}\kappa_0{}^2 + \frac{21949}{945}, \qquad D^{00,00}_{224,25}(0,\tau_0) = -\frac{181040}{2079},$$

$$D^{00,00}_{224,34} = -\frac{233}{630}\kappa_0{}^2 + \frac{1452}{175}, \qquad\qquad D^{00,00}_{224,35}(0,\tau_0) = -\frac{42442}{825},$$

$$D^{00,00}_{224,45}(0,\tau_0) = -\frac{303008}{17325}.$$

Appendix IV. Coefficients a_{ij}^k in Eq. (6.31)

i	j	a_{ij}^2	a_{ij}^3	a_{ij}^4	a_{ij}^5	a_{ij}^6	a_{ij}^7
0	0	-2.7049	$-7.5400(-1)$	6.1401	-6.6199	2.6941	$4.1447(-2)$
0	1	$1.5233(1)^a$	$9.4768(-1)$	-2.2437	$-5.9720(1)$	$9.3801(1)$	$-4.2218(1)$
0	2	$-9.3705(1)$	$2.0811(1)$	$-8.8606(1)$	$3.6688(2)$	$-3.3357(2)$	$1.0364(2)$
0	3	$3.4199(2)$	$-2.0445(2)$	$6.7654(2)$	$-1.0032(3)$	$-1.2857(2)$	$3.7135(2)$
0	4	$-6.1943(2)$	$5.4622(2)$	$-1.6914(3)$	$1.3957(3)$	$1.9124(3)$	$-1.6006(3)$
0	5	$5.4254(2)$	$-5.6789(2)$	$1.7137(3)$	$-1.0037(3)$	$-2.5173(3)$	$1.8642(3)$
0	6	$-1.8490(2)$	$2.0504(2)$	$-6.1311(2)$	$3.1010(2)$	$9.7025(2)$	$-6.9619(2)$
1	0	9.1142	3.4651	$-2.5624(1)$	$2.9550(1)$	$-1.2770(1)$	$3.8899(-1)$
1	1	$-5.3595(1)$	-6.9304	$2.1913(1)$	$2.0709(2)$	$-3.5688(2)$	$1.6667(2)$
1	2	$3.0376(2)$	$-3.0288(1)$	$1.3036(2)$	$-1.0721(3)$	$1.2051(3)$	$-4.4621(2)$
1	3	$-1.0880(3)$	$5.4423(2)$	$-1.5530(3)$	$2.2622(3)$	$8.0190(2)$	$-1.1734(3)$
1	4	$1.9786(3)$	$-1.6278(3)$	$4.4380(3)$	$-2.4375(3)$	$-7.5809(3)$	$5.4732(3)$
1	5	$-1.7484(3)$	$1.7824(3)$	$-4.8472(3)$	$1.6007(3)$	$9.4110(3)$	$-6.3560(3)$
1	6	$6.0051(2)$	$-6.6501(2)$	$1.8343(3)$	$-5.9341(2)$	$-3.4610(3)$	$2.3325(3)$
2	0	$-1.0953(1)$	-5.1542	$3.6013(1)$	$-4.3831(1)$	$1.9952(1)$	-1.1709
2	1	$5.4313(1)$	$1.3826(1)$	$-3.1847(1)$	$-2.7059(2)$	$4.9588(2)$	$-2.3698(2)$
2	2	$-2.2099(2)$	$-3.2259(1)$	$-4.7715(1)$	$1.2764(3)$	$-1.8389(3)$	$7.8458(2)$
2	3	$6.8539(2)$	$-3.6187(2)$	$1.3094(2)$	$-2.2413(3)$	$1.1757(2)$	$6.0981(2)$
2	4	$-1.1983(3)$	$1.4531(3)$	$-4.4486(3)$	$2.2324(3)$	$6.9490(3)$	$-5.0433(3)$
2	5	$1.0698(3)$	$-1.7667(3)$	$5.3715(3)$	$-1.9864(3)$	$-8.7553(3)$	$6.0542(3)$
2	6	$-3.8042(2)$	$6.9945(2)$	$-2.1919(3)$	$1.0478(3)$	$2.9927(3)$	$-2.1592(3)$
3	0	5.7440	3.4159	$-2.2711(1)$	$2.9061(1)$	$-1.3378(1)$	$8.6677(-1)$
3	1	$-1.9872(1)$	$-1.0809(1)$	$1.0667(1)$	$1.8760(2)$	$-3.4288(2)$	$1.6405(2)$
3	2	-3.7479	$5.6583(1)$	$7.4575(1)$	$-9.6990(2)$	$1.5440(3)$	$-6.9034(2)$
3	3	$1.5736(2)$	$3.5420(1)$	$-8.7510(2)$	$2.1083(3)$	$-1.9366(3)$	$5.8246(2)$
3	4	$-3.5702(2)$	$-5.2768(2)$	$2.9171(3)$	$-3.1763(3)$	$-7.4079(1)$	$9.8854(2)$
3	5	$3.0368(2)$	$7.8272(2)$	$-3.6467(3)$	$3.4184(3)$	$9.2878(2)$	$-1.5527(3)$
3	6	$-8.6201(1)$	$-3.4025(2)$	$1.5476(3)$	$-1.6139(3)$	$-8.5762(1)$	$4.9886(2)$
4	0	-1.4876	-1.1070	7.0707	-9.2818	4.0545	$-1.1502(-1)$
4	1	2.5094	3.7734	2.0359	$-7.4027(1)$	$1.2920(2)$	$-6.0976(1)$
4	2	$3.8402(1)$	$-2.4593(1)$	$-6.8657(1)$	$4.5264(2)$	$-7.0566(2)$	$3.1418(2)$
4	3	$-2.0277(2)$	$2.4499(1)$	$4.4154(2)$	$-1.2793(3)$	$1.5600(3)$	$-6.0554(2)$
4	4	$4.0109(2)$	$9.9426(1)$	$-1.2757(3)$	$2.3455(3)$	$-2.0101(3)$	$5.8374(2)$
4	5	$-3.4870(2)$	$-1.9649(2)$	$1.5435(3)$	$-2.5035(3)$	$1.7772(3)$	$-4.0787(2)$
4	6	$1.1115(2)$	$9.4812(1)$	$-6.5280(2)$	$1.0764(3)$	$-7.6449(2)$	$1.8051(2)$
5	0	$2.0156(-1)$	$1.8587(-1)$	-1.1416	1.4900	$-5.5695(-1)$	$-4.5479(-2)$
5	1	$1.0037(-1)$	$-6.4465(-1)$	-1.5497	$1.5971(1)$	$-2.6405(1)$	$1.2233(1)$
5	2	$-1.1430(1)$	4.6045	$2.2698(1)$	$-1.1165(2)$	$1.6633(2)$	$-7.2814(1)$
5	3	$5.4302(1)$	-6.9392	$-1.1903(2)$	$3.6679(2)$	$-4.6797(2)$	$1.8802(2)$
5	4	$-1.0524(2)$	$-1.1949(2)$	$3.1002(2)$	$-7.1974(2)$	$7.7277(2)$	$-2.7903(2)$
5	5	$9.1596(1)$	$3.0585(1)$	$-3.6016(2)$	$7.5604(2)$	$-7.3586(2)$	$2.4845(2)$
5	6	$2.9583(1)$	$-1.5919(1)$	$1.4990(2)$	$-3.1092(2)$	$2.9400(2)$	$-9.7735(1)$
6	0	$-1.3692(-2)$	$-1.1546(-2)$	$9.1506(-2)$	$-1.1498(-1)$	$3.0642(-2)$	$1.1544(-2)$
6	1	$-4.5337(-2)$	$5.2776(-2)$	$2.5063(-1)$	-1.7173	2.7092	-1.2318
6	2	1.3124	$-3.9410(-1)$	-3.0626	$1.3099(1)$	$-1.8776(1)$	8.0822
6	3	-5.9817	$6.4804(-1)$	$1.4823(1)$	$-4.6670(1)$	$5.9690(1)$	$-2.4081(1)$
6	4	$1.1478(1)$	$9.8534(-1)$	$-3.6396(1)$	$9.4233(1)$	$-1.0758(2)$	$4.0635(1)$
6	5	-9.9799	-2.7398	$4.1058(1)$	$-9.8016(1)$	$1.0452(2)$	$-3.7915(1)$

(Continued on next page)

(Continued)

6 6	3.2356	1.4718	$-1.6848(1)$	$3.9416(1)$	$-4.0848(1)$	$1.4602(1)$
7 0	$3.6668(-4)$	$5.1292(-4)$	$-2.8821(-3)$	$3.3886(-3)$	$-3.7515(-4)$	$-6.7520(-4)$
7 1	$2.5887(-3)$	$-1.6554(-3)$	$-1.2765(-2)$	$7.1257(-2)$	$-1.0837(-1)$	$4.8466(-2)$
7 2	$-5.3926(-2)$	$1.2641(-2)$	$1.4489(-1)$	$-5.7451(-1)$	$8.0042(-1)$	$-3.3976(-1)$
7 3	$2.4002(-1)$	$-2.0081(-2)$	$-6.7528(-1)$	2.1445	-2.7276	1.0976
7 4	$-4.5749(-1)$	$-4.0030(-2)$	1.6025	-4.3922	5.1268	-1.9646
7 5	$3.9713(-1)$	$1.0440(-1)$	-1.7729	4.5412	-5.0238	1.8683
7 6	$-1.2890(-1)$	$-5.6148(-2)$	$7.1994(-1)$	-1.8035	1.9439	$-7.1361(-1)$

[a] $a(n)$ means $a \times 10^n$

Appendix V. Coefficients a_{ij}^{kl} in Eq. (6.122)

j k l	a_{1j}^{kl}	a_{2j}^{kl}	a_{3j}^{kl}	a_{4j}^{kl}	a_{5j}^{kl}	a_{6j}^{kl}
0 0 0	$4.3740(-2)^{\mathrm{a}}$	$-5.7005(-3)$	1.5783	-6.1714	9.3510	-2.1546
0 0 1	$-2.6683(-2)$	$5.0153(-3)$	$-4.8764(-1)$	2.2984	-3.8827	$9.4255(-1)$
0 0 2	$5.4865(-3)$	$-9.9676(-4)$	$5.5772(-2)$	$-3.0618(-1)$	$5.7135(-1)$	$-1.5618(-1)$
0 0 3	$-3.5146(-4)$	$5.6241(-5)$	$-2.2400(-3)$	$1.4108(-2)$	$-2.8824(-2)$	$9.0350(-3)$
0 1 0	$-9.8759(-3)$	$9.3855(-3)$	$7.9863(-2)$	-1.6074	6.2428	-5.5149
0 1 1	$8.6995(-3)$	$-8.0343(-3)$	$1.1442(-1)$	$4.2438(-1)$	-2.8537	2.6941
0 1 2	$-1.9806(-3)$	$2.0749(-3)$	$-3.3841(-2)$	$-1.5011(-2)$	$4.1581(-1)$	$-4.2806(-1)$
0 1 3	$1.3292(-4)$	$-1.6323(-4)$	$2.3114(-3)$	$-1.3467(-3)$	$-2.0080(-2)$	$2.2502(-2)$
0 2 0	$8.2175(-3)$	$-1.3063(-2)$	$-2.7468(-1)$	$7.5428(-1)$	-1.4470	1.7712
0 2 1	$-8.0286(-3)$	$1.1449(-2)$	$1.9329(-1)$	$-6.2368(-1)$	$9.0207(-1)$	$-8.7677(-1)$
0 2 2	$2.1003(-3)$	$-2.8623(-3)$	$-3.8882(-2)$	$1.4285(-1)$	$-1.7885(-1)$	$1.4077(-1)$
0 2 3	$-1.6036(-4)$	$2.0273(-4)$	$2.3772(-3)$	$-9.6046(-3)$	$1.1084(-2)$	$-7.2736(-3)$
0 3 0	$-1.1933(-2)$	$8.4024(-3)$	$-2.0610(-1)$	1.9123	-2.7095	$9.5591(-1)$
0 3 1	$8.5897(-3)$	$-7.8143(-3)$	$1.2410(-1)$	-1.0929	1.5158	$-5.0654(-1)$
0 3 2	$-1.8595(-3)$	$2.0735(-3)$	$-2.3313(-2)$	$2.0051(-1)$	$-2.7595(-1)$	$8.8414(-2)$
0 3 3	$1.1845(-4)$	$-1.6266(-4)$	$1.4186(-3)$	$-1.1941(-2)$	$1.6443(-2)$	$-5.0939(-3)$
1 0 0	$3.4750(-1)$	$-8.3845(-2)$	1.1307	$3.5292(1)$	$-8.9519(1)$	$5.8430(1)$
1 0 1	$-2.3098(-1)$	$7.2274(-2)$	-2.9049	-5.7669	$3.3006(1)$	$-2.9224(1)$
1 0 2	$5.9275(-2)$	$-2.9808(-2)$	$7.2091(-1)$	$9.6138(-2)$	-5.1080	5.6318
1 0 3	$-4.7492(-3)$	$3.3823(-3)$	$-4.1161(-2)$	$-1.1742(-2)$	$3.1940(-1)$	$-3.7437(-1)$
1 1 0	$-9.9003(-2)$	$-1.1472(-2)$	-1.6532	$3.2134(1)$	$-1.0126(2)$	$7.7624(1)$
1 1 1	$1.1904(-1)$	$2.6621(-2)$	$1.2605(-1)$	$-1.4103(1)$	$5.2055(1)$	$-4.3340(1)$
1 1 2	$-3.0261(-2)$	$-1.6424(-2)$	$-2.7407(-2)$	2.9139	$-1.0186(1)$	8.7944
1 1 3	$1.3699(-3)$	$2.4692(-3)$	$2.1815(-2)$	$-2.6778(-1)$	$7.2290(-1)$	$-6.0322(-1)$
1 2 0	$-1.0729(-1)$	$3.9537(-2)$	$7.1921(-1)$	$1.1626(1)$	$-3.8148(1)$	$2.8939(1)$
1 2 1	$1.0803(-1)$	$-4.4421(-2)$	$-6.3187(-1)$	-4.7301	$2.0883(1)$	$-1.8974(1)$
1 2 2	$-3.1278(-2)$	$1.1258(-2)$	$9.4846(-2)$	$7.8525(-1)$	-4.0122	4.1390
1 2 3	$2.7356(-3)$	$-5.8429(-4)$	$7.6219(-3)$	$-7.8339(-2)$	$2.7683(-1)$	$-2.9243(-1)$
1 3 0	$-1.1283(-1)$	$6.4364(-2)$	7.5854	$-6.7640(1)$	$9.0024(1)$	$-1.2847(1)$
1 3 1	$5.3218(-2)$	$3.8384(-2)$	-4.9448	$4.2303(1)$	$-5.5395(1)$	7.5210
1 3 2	$-9.9555(-3)$	$-1.9128(-2)$	$9.7517(-1)$	-8.1131	$1.0367(1)$	-1.1041
1 3 3	$1.1307(-3)$	$2.0767(-3)$	$-5.9088(-2)$	$4.9207(-1)$	$-6.2191(-1)$	$4.7344(-2)$
2 0 0	$-8.7321(-2)$	$-8.2387(-1)$	5.2142	$-2.3419(2)$	$9.1804(2)$	$-8.5028(2)$
2 0 1	$-4.4202(-1)$	$8.7523(-1)$	6.7368	$2.2585(1)$	$-3.1183(2)$	$3.4289(2)$
2 0 2	$3.0158(-1)$	$-2.6845(-1)$	-2.7114	$1.4998(1)$	$1.9266(1)$	$-3.6278(1)$
2 0 3	$-4.0674(-2)$	$2.9185(-2)$	$3.6072(-1)$	-2.3310	1.2420	$6.5231(-1)$
2 1 0	1.6450	$-8.3105(-1)$	$1.4327(1)$	$-2.6724(2)$	$1.1038(3)$	$-1.0509(3)$
2 1 1	-1.6847	$8.2167(-1)$	$1.0634(1)$	7.1381	$-3.8962(2)$	$4.5837(2)$
2 1 2	$5.6846(-1)$	$-2.5421(-1)$	-6.7373	$3.1702(1)$	$1.8997(1)$	$-5.3356(1)$
2 1 3	$-6.2738(-2)$	$2.9075(-2)$	$9.5012(-1)$	-4.7219	2.7531	1.1859
2 2 0	-1.1464	$8.2756(-1)$	$1.1256(1)$	$-2.0060(2)$	$5.6208(2)$	$-3.6057(2)$
2 2 1	$9.4501(-1)$	$-7.3284(-1)$	3.2532	$7.0140(1)$	$-2.5178(2)$	$1.5880(2)$
2 2 2	$-2.4519(-1)$	$2.0333(-1)$	-3.8184	1.0338	$2.6357(1)$	$-1.6509(1)$
2 2 3	$1.9771(-2)$	$-1.6959(-2)$	$6.0246(-1)$	-1.4806	$1.5466(-1)$	$5.0002(-2)$
2 3 0	$7.5857(-1)$	$4.0032(-1)$	$-2.9017(1)$	-6.6804	$2.3024(2)$	$-1.9069(2)$
2 3 1	$-9.4206(-1)$	$-3.6059(-1)$	$2.0700(1)$	$-3.3473(1)$	$-2.1948(1)$	9.9298
2 3 2	$2.6331(-1)$	$1.1902(-1)$	-4.0263	9.1264	-9.9517	$1.3157(1)$
2 3 3	$-1.2637(-2)$	$-1.2501(-2)$	$2.6890(-1)$	$-6.4069(-1)$	1.1255	-1.4794

(Continued on next page)

(Continued)

3	0	0	6.5184(−1)	−6.2144(−1)	6.6916	−3.2445(1)	2.1835(2)	−2.4382(2)
3	0	1	−7.6704(−1)	6.2903(−1)	−4.0015	−7.0646	−4.9882(1)	6.9552(1)
3	0	2	2.8098(−1)	−2.1177(−1)	4.2592(−1)	8.7627	−8.4234	2.0351
3	0	3	−3.0253(−2)	2.3329(−2)	7.9989(−2)	−1.2561	1.7532	−1.0013
3	1	0	4.8598(−1)	−3.5124(−1)	−1.4982	−4.4677	1.8852(2)	−2.5407(2)
3	1	1	−4.9148(−1)	3.9521(−1)	7.8212	−5.7644(1)	−2.4845	7.4645(1)
3	1	2	2.0234(−1)	−1.4935(−1)	−3.8395	2.7669(1)	−2.7516(1)	4.3975
3	1	3	−2.8119(−2)	1.9441(−2)	5.5524(−1)	−3.2869	3.8792	−1.4953
3	2	0	−6.5991(−1)	3.7925(−1)	4.8386	−5.1154(1)	1.2690(2)	−5.8648(1)
3	2	1	5.8571(−1)	−3.5870(−1)	1.7015	1.0476(1)	−3.7466(1)	2.2645
3	2	2	−1.6411(−1)	9.9268(−2)	−1.9989	4.8801	−3.6094	7.6738
3	2	3	1.4356(−2)	−7.6591(−3)	3.4061(−1)	−1.0747	1.1634	−1.0411
3	3	0	7.2831(−2)	2.8616(−1)	−9.9441(−2)	−1.1154(2)	2.0480(2)	−4.5199(1)
3	3	1	−2.6777(−1)	−1.6349(−1)	4.5855(−1)	5.6364(1)	−7.5629(1)	−2.1635(1)
3	3	2	8.2542(−2)	3.2414(−2)	−3.8354(−2)	−9.9394	7.7662	1.1992(1)
3	3	3	−1.2987(−3)	−2.4754(−3)	1.4545(−2)	5.8013(−1)	−2.3605(−1)	−1.0952
4	0	0	−5.0646(−1)	−2.9893	1.5330(1)	−9.6644(2)	4.1540(3)	−4.1107(3)
4	0	1	−1.6542	3.1431	2.5105(1)	1.4837(2)	−1.6212(3)	1.8403(3)
4	0	2	1.2742	−9.6873(−1)	−1.0146(1)	4.5487(1)	1.5667(2)	−2.4129(2)
4	0	3	−1.6679(−1)	1.0564(−1)	1.3436	−8.2377	−3.2100(−1)	8.9303
4	1	0	4.7927	−3.1651	7.3469(1)	−1.2451(3)	4.9249(3)	−4.7293(3)
4	1	1	−5.0663	3.1295	3.0783(1)	1.4788(2)	−1.9232(3)	2.2140(3)
4	1	2	1.8082	−9.5242(−1)	−2.3998(1)	1.0003(2)	1.5901(2)	−3.0079(2)
4	1	3	−2.1016(−1)	1.0774(−1)	3.4843	−1.6826(1)	4.9561	1.1102(1)
4	2	0	−2.1684	3.1140	5.0778(1)	−8.1729(2)	2.3649(3)	−1.5871(3)
4	2	1	1.5108	−2.7877	8.5505	2.9690(2)	−1.0986(3)	7.4775(2)
4	2	2	−3.1496(−1)	7.7728(−1)	−1.4006(1)	−5.0408(−1)	1.2687(2)	−9.3232(1)
4	2	3	1.8133(−2)	−6.4934(−2)	2.2531	−5.4765	−9.3697(−1)	2.2453
4	3	0	2.9167	1.3875	−9.4950(1)	−3.0581(2)	1.5824(3)	−1.2009(3)
4	3	1	−4.0135	−1.1704	7.1417(1)	2.1381(1)	−4.7280(2)	3.0264(2)
4	3	2	1.2295	3.9572(−1)	−1.4168(1)	9.2382	3.1019(1)	2.4680
4	3	3	−7.7263(−2)	−4.3431(−2)	9.6412(−1)	−1.0662	4.5089(−1)	−2.8997

[a] $a(n)$ means $a \times 10^n$

Glossary of Abbreviations

a-	Atactic
c-	Coarse-grained (dynamic HW chain)
d-	Discrete (dynamic HW chain)
i-	Isotactic
s-	Syndiotactic
DNA	Deoxyribonucleic acid
GPC	Gel permeation chromatography
HI	Hydrodynamic interaction
HW	Helical wormlike (chain)
IAIV	Isoamyl isovalerate
KP	Kratky–Porod (wormlike chain)
KR	Kirkwood–Riseman (approximation, equation)
MEK	Methyl ethyl ketone, 2-Butanone
MTPS	*Modern Theory of Polymer Solutions*
NMR	Nuclear magnetic relaxation
NOE	Nuclear Overhauser enhancement
OB	Oseen–Burgers (procedure)
PBIC	Poly(n-butyl isocyanate)
PBLG	Poly(γ-benzyl L-glutamate)
PDMS	Poly(dimethylsiloxane)
PHIC	Poly(n-hexyl isocyanate)
PIB	Polyisobutylene
PIP	*cis*-Polyisoprene
PM	Polymethylene
PMMA	Poly(methyl methacrylate)
PMVK	Poly(methyl vinyl ketone)
POE	Polyoxyethylene

PPCS	Poly(p-chlorostyrene)
PS	Polystyrene
QTP	Quasi-two-parameter (scheme, theory)
RIS	Rotational isomeric state (model)
SANS	Small-angle neutron scattering
SAXS	Small-angle X-ray scattering
THF	Tetrahydrofuran
TP	Two-parameter (theory)
WKB	Wentzel–Kramers–Brillouin (approximation)
YSS	Yamakawa–Stockmayer–Shimada (scheme, theory)

Author Index

The *italic* number is the page on which the complete literature citation is listed

Subject Index

G.R. Strobl

The Physics of Polymers
Concepts for Understanding Their Structures and Behavior
2nd corr. ed. 1997. XII, 439 pages. 218 figures, 5 tables.
Softcover DM 58,-
ISBN 3-540-63203-4

Polymer Physics is one of the key lectures not only in polymer science
but also in materials science. Strobl presents in his textbook the elements
of polymer physics to the necessary extent in a very didactical way. His
main focus lays on the concepts of polymer physics, not on theoretical
aspects or mere physical methods. He has written the book in a personal
style evaluating the concepts he is dealing with. Every student of polymer
and materials science will be happy to have it on his shelf.

H. Pasch, B. Trathnigg

HPLC of Polymers
1997. Approx. 220 pages. 122 figures, 29 tables.
(Springer Laboratory)
Hardcover DM 128,-
ISBN 3-540-61689-6

Polymers are mainly characterized by molar mass, chemical composition, functionality and architecture. The determination of the complex
structure of polymers by chromatographic and spectroscopic methods
is one of the major concerns of polymer analysis and characterization.
This lab manual describes the experimental approach to the chromatographic analysis of polymers. Different chromatographic methods,
their theoretical background, equipment, experimental procedures and
applications are discussed. The book will enable polymer chemists,
physicists and material scientists as well as students of macromolecular
and analytical science to optimize chromatographic conditions for a
specific separation problem. Special emphasis is given to the description of applications for homo- and copolymers and polymer blends.

■ ■ ■ ■ ■ ■ ■ ■ ■ ■

Please order from
Springer-Verlag Berlin
Fax: + 49 / 30 / 8 27 87- 301
e-mail: orders@springer.de
or through your bookseller

Prices subject to change without notice.
In EU countries the local VAT is effective.

Springer

Springer-Verlag, P. O. Box 31 13 40, D-10643 Berlin, Germany

M. Kamachi, A. Nakamura (Eds.)

New Macromolecular Architecture and Functions

Proceedings of the OUMS' 95
Toyonaka, Osaka,
Japan, 2 - 5 June, 1995

1996. VIII, 207 pages.
136 figures, 28 tables.
Hardcover DM 298,-
ISBN 3-540-61473-7

This volume summarizes the main lectures presented at the Osaka University Macromolecular Symposium OUMS '95 on **New Macromolecular Architectures and Functions,** where the following three topics were discussed: (1) controlled polymerizations, (2) macromolecular organized systems, and (3) biomimetic polymers. Either of these topics in itself is a hot issue at present and frequently taken up as a main theme at a particular symposium. The present symposium invited leading scientists in these fields as guest speakers and is expected to attract interests of a significant range of readers.

J.F. Rabek

Photodegradation of Polymers

Physical Characteristics and Applications

1996. VII, 212 pages.
94 figures, 26 tables.
Hardcover DM 148,-
ISBN 3-540-60716-1

In this book on physical characteristics and practical aspects of polymer photodegradation Rabek emphasizes the experimental work on the subject. The most important feature of the book is the physical interpretation of polymer degradation, e.g. mechanism of UV/light absorption, formation of excited states, energy transfer mechanism, kinetics, dependence on physical properties of macromolecules and polymer matrices, formation of mechanical defects, and practics during environmental ageing. He also includes some aspects of polymer photodegradation in environmental and space condition.

Please order from
Springer-Verlag Berlin
Fax: + 49 / 30 / 8 27 87- 301
e-mail: orders@springer.de
or through your bookseller

Prices subject to change without notice.
In EU countries the local VAT is effective.

Springer

Springer-Verlag, P. O. Box 31 13 40, D-10643 Berlin, Germany

T
1 Month